Will

# AMERICAN CLOCKS

## Volume 1

Tran Duy Ly

Arlington Book Company, Inc.
1999

# AMERICAN CLOCKS
Volume 1

Tran Duy Ly

EDITING by **THOMAS J. SPITTLER**

HISTORY by **CHRIS H. BAILEY**

INTRODUCTION by **LARRY RAPPAPORT**

©1989, 1999 Tran Duy Ly
All Rights Reserved

Second Edition Published in 1999 by **Arlington Book Company, Inc.**
2706 Elsmore Street, Fairfax, Viriginia 22031-1409 USA

http://www.arlingtonbooks.com
email: **info@arlingtonbooks.com**

**ISBN** 0-930163-39-7

To Larry & Barbara Funk
Good friends whose help over the years
is so appreciated.

Tran Duy Ly

---

The Arlington Book Company proudly presents the most comprehensive publication available to Clock Collectors, "**American Clocks - Volume 1**" by **Tran Duy Ly**.

It reproduces illustrations from manufacturers' original catalogs, including descriptions, specifications, and cost from Ansonia to Welch, from 30 hour spring driven movements to pin wheel weight movements to self-winding clocks in a carefully arranged guide where clocks can be found easily and quickly.

You will find the same features that have made all our publications distinctive and easy-to-use as well as new aspects that will ensure our books will remain progressive and popular. Please notice the use of exact illustrations from the manufacturer's original catalogs including descriptions, specifications and original costs. This lends a sense of confidence that the clocks within are shown with their original hands, pendulums, glasses, finials and all other case trim.

This valuable information guide will assist you as you consider whether or not to buy a particular clock. The separate up-to-date price guide, compiled by a panel of national experts, will help you to get the best possible pricing advice quickly and simply.

**Arlington Book Company, Inc.**

# INTRODUCTION

The updated reprint in colorful hard cover is a beauty and a must in every clock collector's library. I am happy that it is being introduced in this bright hard cover edition because I used my soft cover version so often that the cover and pages are badly frayed.

As a major collector of American regulator clocks, I could not function without this valuable guide. This beautifully illustrated volume is arranged in logical, easy to follow, alphabetical order, and features nineteen clock companies including Ansonia, Chelsea, Gilbert, Howard, Ingraham, Ithaca, Kroeber, New Haven, Seth Thomas, Waltham, Waterbury, and Welch. The section, **Important Tips**, has been expanded from five to sixteen pages. This section is a valuable tool for all beginning collectors and serves the advanced collector well by keeping him up-to-date on changes in the clock marketplace. There is also a carefully prepared index that serves as an outstanding guide to find what you need in a hurry.

In addition, an historical essay on each company prepared by Chris Bailey, a leading authority on American clock history of the American Clock & Watch Museum at Bristol, Connecticut is provided as well as an on target price guide which was compiled by a national panel of 32 antique clock experts.

If you need to identify a specific clock for completeness and authenticity, this is your one stop shopping center. The large type, as well as the quality of illustrations, leaves nothing to chance. Over the years, Tran Duy Ly has become the master of clock identification and price guide books. This new edition of American Clocks is another example of Mr. Ly's expertise. This improved example is an important addition to the clock collector's library and will serve as a source of information for years to come.

The fact that an up-to-date price guide is available makes this **American Clocks Volume 1** even more indispensable to clock collectors. Our hat is off to Tran Duy Ly and the continuing oustanding job he does in contacting clock experts everyday for clock information and price trends. The clock community owes him a huge debt of gratitude.

**LARRY RAPPAPORT**
Cutchogue, New York

# ACKNOWLEDGEMENTS

I would like to take this opportunity to express my sincere appreciation to the following friends who provided me with valuable material, information and up-to-date prices without which this publication could not have been accomplished.

### Tran Duy Ly

**AMERICAN CLOCK AND WATCH MUSEUM**
Bristol, CT

**BAILEY**, Chris H.
FNAWCC*
Bristol, CT

**BERGER**, Steven & Marsha
FNAWCC
Fountain Hills, AZ

**BLACKWELL**, Dana, J.
FNAWCC*
Naugatuck, CT

**BJORNESTAD**, Arthur
FNAWCC*
Westlake, CA

**BOXENHORN**, Jerold
S. Bellmore, NY

**CASTRO**, Mariano
Springfield, VA

**CILLI**, Joan & Mario
Middletown, NJ

**CRUM**, Elmer
Hudson, FL

**CRUTSINGER**, Larry
Norfolk, VA

**DAVIS**, Lee
FNAWCC*
York, PA

**DEMETER**, Andrew
Topsfield, MA

**DYSON**, James
Virginia Beach, VA

**ELLER**, James
Greensboro, NC

**FRIED**, Henry B.
FNAWCC*
Flushing, NY

**FUNK**, Larry & Barbara
Northfield, IL

**GORRELL**, David
Annapolis, MD

**KAZEMEKAS Jr**, Ed.
Wolcott, CT

**KNIGHT**, Tracy, J.
Fairfax, VA

**KRAUSE**, James
Hungtington Beach, CA

**LEE**, J. David
Delanson, NY

**MARREN**, Tom
Kirkwood, MO

**MAZZELLA**, Tony
Springfield, VA

**Mc DERMOTT**, Philip A.
St. Louis, MO

**MITTELSTADT**, William
Sheboygan, WI

**MORGAN**, David
FNAWCC*
Whitehouse, NJ

**NAWCC., INC.**
Columbia, PA

**O'BRIEN**, Gene
St. Louis, MO

**PETRUCELLI**, Steve
Cranbury, NJ

**RAPPAPORT**, Larry
Cutchogue, NY

**READ**, R. B.
Vienna, VA

**RICHEY**, Earl
Alexandria, VA

**SADOWSKI**, Steve
Maspeth, NY

**SCANLON**, Loren
Modesto, CA

**SERNAK**, Joseph
Stirling, NJ

**SIMMONS**, J. W.
FNAWCC*
Eagle Rock, VA

**SPITTLER**, Tom & Sonya
FNAWCC
New Carlisle, OH

**STOFFERS**, Joyce
Bristol, CT

**SWETSKY**, Martin
FNAWCC
Brooklyn, NY

**SUMMAR**, Donald
Columbia, PA

**TANNER**, John
Upland, CA

**WEBBER**, Harvey
Hampton, NH

**WEBBER**, Robert S.
Hampton, NH

**WIEMER**, Roger
Springfiled, MO

**WONG**, Dennis
Falls River, MA

**WRIGHT**, Douglas
Werthersfield, CT

\* Star Fellow

# CONTENTS

|  | | |
|---|---|---|
| Page | 4 | Introduction |
| | 5 | Acknowledgements |
| | 6 | Contents |
| | 7 | Important Tips |
| | 313 | Index |
| | | |
| | 10 | Ansonia Clock Co. |
| | 74 | Boston Clock Co. |
| | 77 | Chelsea Clock Co. |
| | 85 | Gilbert Clock Co. |
| | 116 | Howard Watch & Clock Co. |
| | 140 | Ingraham & Co. |
| | 145 | Inthaca Calendar Clock Co. |
| | 149 | Kroeber Clock Co. |
| | 164 | New Haven Clock Co. |
| | 189 | New York Standard Watch Co. |
| | 190 | Parker & Whipple Manufacturing Co. |
| | 194 | Russell & Jones Clock Co. |
| | 198 | Seth Thomas Clock Co. |
| | 239 | Terry Clock Co. |
| | 242 | Tiffany Electric Manufacturing Co. |
| | 243 | Waltham Clock Co. |
| | 258 | Warren Telechron Co. |
| | 259 | Waterbury Clock Co. |
| | 303 | Welch Manufacturing Co. |
| | 321 | Important Tips continued |

# IMPORTANT TIPS

The finest American-made clocks are pictured in this book which provides a comprehensive guide to the American clocks that are most desired by collectors and have consistently appreciated in value over the years. Use of Catalog illustrations has permitted the display of thousands of different clocks, many more than would be possible if it were necessary to rely solely upon photographs of clocks in private collections. The use of catalog illustrations has also permitted the display of modifications and changes in the appearance of many models, showing clearly that the changes were made by the company and are not the result of modification or restoration occurring after the clock left the factory.

The clocks in this book are organized alphabetically by company, including: Ansonia, Boston, Chelsea, Gilbert, Howard, Ingraham, Ithaca, Kroeber, New Haven, New York Standard, Parker and Whipple, Russell and Jones, Seth Thomas, Terry, Tiffany Never Wind, Warren, Waltham, Waterbury and Welch.

**American Clocks Volume 1** is formatted with you the reader in mind.
- Clocks of the same type are grouped together.
- Clocks with similar appearance are grouped next to each other.
- Beside each clock is the year of the catalog from which the illustration was taken and this represents the date of production. However, some popular clocks may also have been produced a few years earlier or later. You will also sometimes note two dates for a clock. While the clock was produced in both years, one date represents one catalog which had the better illustration, while the other date represents a different catalog which had a more detailed description.
- Most clocks also have their original selling prices, descriptions, specifications and dimensions. Please note that sometimes there are two different sets of dimensions given with a single clock illustration, because the "same" clock was made in two sizes.
- All models are completely and professionally indexed.
- A separate price list has been compiled from the top expert horologists in the field and provides **professional estimates** for antique clocks in *original mint condition.* There will be periodic updates so that this book will be a lasting source of reference.

This book is a must for every antique clock collector's library. Whether you are a novice or experienced collector, a speculator or an investor, this book will be easy for you to use. It is designed to provide you with a great deal of the information you will need to make wise choices in your clock buying experiences. Today's clock marketplace demands knowledgeable and often times fast decision making. Guided by the information provided in this book, you will be better able to meet that challenge.

When the opportunity to purchase a clock arises, remember that clock prices are subjective. For example, if the clock is in original mint condition and brings you joy, don't spend too much time bargaining. Three years ago, a friend of mine, B. M. of Sheboygan, Wisconsin, bought an Ansonia Florida Group for $3,400.00. The clock was in very good condition and was one he had been trying to acquire for years. However, the catalog price for the clock at that time was only $2,200.00. Today the price for this clock is $3,500.00. Did he pay too much? I say a fervent "No!" The fact that this was a rare clock and completed a collection made that price realistic. So remember, when the chance to buy the clock you've been waiting for arises, don't let it slip by. A clock worth $1,000.00 to you may be worth $1,500.00 to the person behind you. My major rule has always been, "if you like it... buy it....," it's likely that you won't live long enough to find another like it. If you read and understand my **"Important Tips"** that follow, you will be confident that your buying decision is a good one.

From the individual who buys an antique clock for decoration, to the young person who wants one clock "like Grandma had," to the investor or experienced collector who belongs to the National Association of Watch and Clock Collectors (NAWCC) and searches for the one missing model with which to finish his collection; the demand for clocks is high. The competition for the limited supply of good antique clocks requires the buyer be prepared and as knowledgeable as possible. Preparation is the assurance that a purchase won't be a disappointment.

It is obvious that collecting clocks appeals to a vast cross section of our population. The fun of seeking and finding that special timepiece is now shared by more people than ever before. For example, there are now many more members of the NAWCC, all of whom are out there competing for what seems to be fewer and fewer good clocks and watches. This means that you might have to exhibit some patience and often times leniency when deciding to make a new purchase. You might now consider a repainted dial or a replaced finial on a **rare** clock. It's more difficult to find what you want and you may have to exercise prudent but open minded caution.

To help you in your search, it may be advisable to establish relationships with specific clock dealers whom you can trust, and can contact from time to time to make them aware of your clock needs. Keep in mind "Out of sight, out of mind..." is the old expression. Keeping in touch is your responsibility. In fact, it might be a good idea to contact these dealers prior to any local or regional clock or antique shows to arrange to have them bring clocks for your acceptance or refusal prior to showing them to everyone else.

As with any collection, it is advisable to consider staying within certain areas of specialty so that you know what to look for and where to find it and to whom you can sell it when necessary. An example would be the collector who specializes in American Weight Driven Regulators, Electric, Grandfather, Novelty or unusual French clocks.

There are those clocks, however, that are so collectable and always in demand that they should be purchased at every opportunity when reasonably priced and in good condition. Some current examples of these "opportunity" clocks are: Ansonia Bobbing Doll, Inkstand, Regulators No. 8, 9, and 18, and Swing Clocks, Gilbert Regulators No. 12 and 20, Seth Thomas Regulators No. 4, 5, 7, 10, and 10 Double Time, 12, 14, 15, 16, 19, 60, and 63; and some others such as the Gale Fancy Case Astronomical Gallery Calendar, the Howard Regulators No. 6, 36, and 60; the Ithaca No. 1 Sweep Second Calendar, and other rare, or early American and European clocks; also clocks that do something besides tell time, such as Musical, Annular, Conical, Industrial, Mystery, and rare Swinging clocks.

With the exception of "opportunity" clocks, it may not be advantageous to collect more than one specimen of a specific clock. If at a future date you liquidate your collection, selling multiples of the same clock may not bring the high price you would have received by selling just one. Remember the old saying, "cheaper by the dozen." The more examples you have of the same model, the more ordinary they might appear to be.

Learn to make quick decisions... the person next to you might walk off with your prize. The guidelines in this book will help you avoid loosing that prize through indecision. It should also help you avoid expensive mistakes and will continue to help you throughout your collecting career. At whatever level your interest in clocks is -- whether hobbyist, investor, or serious collector, these **"Important Tips"** should always be a part of your decision making process. And remember that clock prices fluctuate according to price trends, supply and demand and economic conditions. Clocks purchased during a peak period may not have comparable value later. Buy wisely, and always buy what you like. You may have to live with your purchase a long time.

I feel there are three main factors that determine the price of antique clocks.
- **Rarity**
- **Desirability**
- **Condition**

**Rarity**, the first factor affecting price is defined in terms of supply or availability. A "rare" clock is usually expensive, seldom found, available only in limited numbers, and chances are, will never lose its value.

**Desirability** is the second key factor affecting the antique clock market. With the right style and design, some clocks are just more beautifully decorative and just better suited for use in any home or office. These are not common ordinary clocks. Clocks in this category might be the Crystal Regulators, Hanging Regulators, Calendar and Longcase clocks. These clocks all have something special like size and special features which increase their appeal.

**Size.** It used to be the rule that Longcase Clocks or Grandfather Clocks and Hanging Regulator Clocks under eight feet were the most desirable and practical. However now with the many new homes built with cathedral ceilings, very large clocks are in demand. Often times, the bigger the better. There is still one category of Grandfather or Longcase Clock that remains somewhat less desirable and difficult to sell, that is one that is just over eight feet tall. It is not considered large enough to be truly impressive in a room with cathedral ceilings yet it is too large to fit in most modern homes. Some manufacturers produced large Regulator clocks in two styles: one to stand on the floor, and one to hang on a wall. Of the clocks designed with two styles, the Hanging Regulator style is considered to be the most desirable.

Large clocks are generally more desirable than standard size models. Likewise, miniatures are also more valuable than the common sized models. In the case of marble clocks the small sizes again, are the most desirable as they weigh less and are easier to display in the average home.

**Special Features.** Besides size, many clocks possess features that make them more desirable. Special features are those elements that make the timepiece stand out from the ordinary. Clocks under glass domes are very attractive and beveled glass on a Crystal Regulator or Carriage Clock is a definite plus. However note that a cracked dome is almost impossible to replace as domes are mostly non-standard in size. You may be able to have a replacement custom made at considerable cost. On-the-other-hand, beveled glass can be replaced without much difficulty at reasonable cost. Movement features such as long duration - more than eight days, music or unusual chimes and complicated escapements are all very desirable. Historical and horological significance, regional factors and brand name recognition also can make a clock more interesting more desirable and more expensive.

**Condition.** Third and perhaps the most important factor when considering a clock's value is condition. The condition of an antique clock can not be emphasized enough in determining a realistic market value. Whether you are buying, selling, trading, collecting, investing or speculating in antique clocks, the condition is the critical factor. Only clocks in *original mint condition* and in good running order will command top prices. A replaced or damaged dial, faded or incomplete label, missing parts, damaged case, dull finish, worn-off gold gilt, chipped beveled glass, tablets which are either cracked, flaked or repainted, or missing trim will greatly reduce the price of the clock. Obviously, originality implies the clock maintains its original parts - no replaced pieces, and no "marriages," i.e. movements changed from one clock to another.

Of these three determining factors, **rarity, desirability** and **condition**, it is surely condition that causes the most discussion and controversy. When does repair to a clock become an alteration and restoration become excessive? When does preservation go beyond "keeping from further decay?" By definition, to restore is to ,"put back into a former or original state; put back into service" while preservation is, "to keep from deteriorating - maintaining." The greater the amount of work done to a clock, the more controversy there is.

When analyzing the condition of an antique clock, there-by determining it's value, you must decide whether it's originality has been compromised and if so, to what degree it will affect the price you are willing

See **IMPORTANT TIPS**, page **321**

# ANSONIA CLOCK COMPANY
## By CHRIS H. BAILEY, FNAWCC*

Historically, Ansonia Clock Company did not have its roots in Ansonia, the Connecticut town after which it was named, but some 35 miles northeast in the great clockmaking town of Bristol. In 1841, Theodore Terry, nephew of Eli Terry the man who had started the manufacture of inexpensive clocks in the first decade of the 19th century, formed a partnership with one Franklin C. Andrews. The new firm of Terry & Andrews was to tool up and manufacture inexpensive brass clocks.

On August 13, 1841, Theodore Terry of Bristol and Franklin C. Andrews of New York purchased two parcels of land, one noted ". . . as being the same ground my sawmill stood & one other building that was burnt..." from Theodore's father, retired clockmaker Samuel Terry, for $3,000.00. Samuel Terry extended a mortgage for the entire property to Terry & Andrews which was payable ten years from date, only interest due annually. {1} A history of clockmaking published in the *Bristol Herald* in 1890 noted". . . In 1841, the factory built by Samuel Terry near the rolling mill dam, was burned while occupied by Ray & Carpenter making 'OG' cases and Terry & Andrews making movements. Mr. Terry built up the shop again and it was occupied by Theodore Terry and Franklin Andrews, under the company name of Terry & Andrews." {2}

During the remainder of the 1840's the Terry & Andrews business prospered and 30-hour and 8-day spring-driven models were added to the line of 30-hour weight-driven clocks. After the destruction of the Chauncey Jerome factory in 1845 and Jerome's subsequent removal of the remainder of his business to New Haven, Conn., Terry & Andrews became the largest clock manufacturers in Bristol. In June of 1850, they reported $50,000 invested in capital, were employing 58 hands and had produced about 25,000 clocks, valued at $75,000, in the previous 12 months. {3} One report on Bristol manufacturers in 1850 noted Terry & Andrews "are called as safe a firm as any in town, pay well, you can trust them all they ask for . . ." {4}

By 1850, Terry & Andrews was annually using 58 tons of brass, so it is not surprising that Theodore Terry was one of several Bristol clockmakers who were incorporators of the Bristol Brass & Clock Company in April, 1850 and subscribed for 266 shares, an investment of $6,650.00. However, about this time Anson G. Phelps, a wealthy industrialist from New York who operated large foundry operations at Birmingham (now Derby), CT, persuaded Terry & Andrews to leave Bristol and become allies with his foundry operations.

Terry & Andrews paid up their mortgage to Samuel Terry on April 13, 1850 {5} and on the following May 11th sold half interest in their firm to Anson G. Phelps for $7,200 {6}. Two months later Theodore Terry sold his entire interest in the Bristol Brass & Clock Company, probably labelling him "traitor" to its others investors.

On May 7, 1850, Anson G. Phelps, Theodore Terry and Franklin C. Andrews formed a joint stock corporation known as the "Ansonia Clock Company" for the manufacture and sale of clocks, movements and related wares. Some $100,000 capital was authorized with 4,000 shares sold at $25 each. Phelps held controlling interest with 1,334 shares while Terry and Andrews held 1,333 each. {7} Theodore Terry was chosen President and his son Hubbell P. Terry became secretary. {8} The new location, Ansonia, was a village in the town of Derby, CT which Anson Phelps had named after himself. The firm eventually utilized a two-story factory for the clock business which was in use by January of 1851, though Phelps did not sell the property to the company and leased them the water rights until April 12th of that year. {8} This land was noted as being on the northwest corner of a lot on which also stood a stone factory owned by Phelps which was then being used by the Jerome Manufacturing Company of New Haven. {9}

During the first year of business, there was a definite transition between the old firm of Terry & Andrews and the Ansonia Clock Company. A few clocks are known with a full printed label of Terry & Andrews, *Ansonia*, and many clocks are known with labels and/or movements marked Terry & Andrews, *Bristol*, yet with dials or otherwise marked Ansonia Clock Company, Ansonia.

Franklin Andrews sold out all but one of his shares in the business on December 20, 1851 {10} and even sold that single share to H. P. Terry on November 18, 1852. {11} He remained at Jersey City, NJ until his death in January of 1881 and operated a New York clock store for many years. {12}

During January of 1853, Anson Phelps sold 1,000 shares of his Ansonia Clock Company stock to his son-in-law, James B. Stokes who was trustee for the firm of Phelps, Dodge & Co. The day preceeding Phelps' death, he transferred his remaining 1,000 shares to Stokes. {13} Phelps died at his New York City home on November 30, 1853 at the age of 73. A rich and powerful man indeed, he bequeathed $371,000 to charitable institutions as well as $100,000 to his son and $5,000 each to his 24 grandchildren. {14}

Ansonia was one of three Connecticut clock manufacturing firms which exhibited at the New York World's Fair which opened on July 4, 1853. Ansonia primarily exhibited their cast iron cased clocks ornamented with paint and mother-of-pearl. The other two exhibitors were the Jerome Manufacturing Company of New Haven and the Litchfield Manufacturing Company of Litchfield, the latter specializing in papier mache clock cases.

With a transfer by James B. Stokes of his 1,000 shares of clock company stock to Phelps, Dodge & Company during January of 1854, the Phelps firm became the controlling stockholder with 2,000 shares while Theodore Terry held 1,999 and Hubbell P. Terry held one. {15} Ansonia's business apparently proceeded well until about November of 1854, when the factory was reduced to ashes at a loss estimated at $120,000. A meeting of Directors was held at Bridgeport, CT on November 15, 1854, at which time the following resolution was passed:

"That in consequence of the destruction of the Building by fire it has been agreed to sell the land & ruins to A. G. Phelps [Jr.], Wm. E. Dodge, Daniel James, James Stokes, Wm. E. Downs, Junr. & D. Willis James for the sum of Eight Thousand Dollars and that the President be authorized to make and execute a deed of the same." {16}

The above resolution and sale on November 16, 1854 {17} effectively ended the original Ansonia Clock Company and sold the balance of the firm to the Directors of Phelps, Dodge & Company. Theodore Terry was thereafter approached by P. T. Barnum, the great showman, and formed a partnership known as the Terry & Barnum Manufacturing Company to manufacture clocks at East Bridgeport, CT. However, Terry machinery and stock were for the most part destroyed and full scale manufactured was never achived. Terry left the firm when Barnum became financially entanged with the Jerome Manufacturing Company and went bankrupt in March of 1856. {18}

Theodore Terry also purchased interest in the Terryville Manufacturing Company, formed by his cousin Silas B. Terry in 1853. This union resulted in a few torsion escapement "candlestick" clocks with "Terryville Manufacturing Company" impressed on the mainspring cover, but original pressed metal dials marked "Ansonia Clock Company." Theodore became a major stockholder of the Terryville firm in March of 1854 and soon after became its President. Theodore and his son Hubbell P. Terry ran the operation for about five years after which S. B. Terry resigned as General Manager of the firm in the fall of 1854. After 1860 Theodore Terry went to Pennsylvania for a few years and became involved in the oil business but returned to Connecticut by 1864 and died June 18, 1881 at New Haven. {19}

For the 15 years following the 1854 fire, the history of clock manufacture at Ansonia is more difficult to follow. No official clock company was formed during these years though the parent firm of Phelps, Dodge & Company continued to manufacture clock movements and sell some cased clocks, but in relatively small numbers. Clocks from this period are rarely seen today, but are usually labelled "Ansonia Brass Company" (see pg. 442) often showing the factory which was destroyed by fire in 1854 on the labels. A few clocks were also labelled "Ansonia Brass & Battery Company" during this period. The Ansonia Brass & Battery Mill, another operation of Phelps, Dodge & Company, reported in June of 1860 the manufacture of 22,000 clock movements, but only 2,000 finished clocks during the previous 12 months. {20} They were primarily a movement supplier to the trade.

However, the clockmaking business became a major operation with the reorganization of the Ansonia Brass & Battery Company as the Ansonia Brass &

Copper Company on February 11, 1869. {20} Fourteen months later, 150 were employed and some 90,000 pound of brass had been utilized between June of 1869 and June 1870 to produce 83,503 clocks valued at $200,000. {21} Clocks during this period were labelled "Ansonia Brass & Copper Company," a name more commonly seen as the production of finished clocks was much greater during this period than the previous ones (1854-69). The earliest price list of this firm, now known, is dated January 1, 1873 and offers 45 models including 4 timepieces, 12 lever wall clocks, 9 one-day and 8 eight-day spring-driven mantel clocks, 5 weight-driven shelf clocks and 7 office clocks. Fourteen different movements were being produced. {22}

After eight years, another reorganization took place separating the brass mills and the clockmaking operation. On December 21, 1877, a joint stock corporation was formed at New York City adopting the original name, "Ansonia Clock Company". The incorporators were primarily the officers of Phelps, Dodge & Company, but one important exception was Henry J. Davies of Brooklyn, a man whose influence and leadership would be strongly felt in coming years. {23}

Little is known about Henry J. Davies in spite of his being involved in the clock business for some years in New York. Davies was New York plant manager of the clockmaking establishment of George A. Jones & Company, by November of 1870 and perhaps earlier. In 1873 he gained control of the Jones operation in New York and probably sold old stock of Jones clocks. He also manufactured and sold many walnut parlor clocks which he had originally designed for Jones, through the American Clock Company. The Ansonia Brass & Copper Company also offered some of his models and were major suppliers of movements to Davies.

Ansonia bought H. J. Davies' clock business at the time they reorganized and formed the "new" Ansonia Clock Company in December of 1877. Davies was made General Manager of the new Ansonia Clock Company which officially began business Jan. 1, 1878. Davies initially purchased 800 of the 4,000 shares of stock issued. In 1878 the new firm had two superintendents, Walter D. Davies and Edward Davies, who were likely sons of Henry. No doubt the construction of new factory facilities in Brooklyn, New York and the gradual moving of the business from Ansonia, Connecticut to Brooklyn, New York between 1877 and 1883 was largely due to Davies. Davies is also believed to be responsible for many of the designs of Ansonia Clocks between 1878 and 1884.

H. J. Davies was no longer included in a list of Ansonia stockholder dated Jan. 1, 1885, although W. D. Davies was owner of 100 shares. It appears Henry J. Davies died or had left the company in 1884. The last of H. J. Davies' 46 patents or trademarks granted since 1873 was on Jan. 15, 1884, so Davies may well have died in 1884.

In April of 1879, a large factory was commenced at Brooklyn, New York and its new machinery installed in the spring of 1880. {26} Brooklyn soon became the major manufacturing site, though operations at Ansonia, Conn. were not totally shut down until about 1883.

During the 12 months prior to June 1, 1880, the Ansonia operation in Connecticut utilized $200,000 worth of materials to produce an estimated $400,000 worth of clocks. They still had 100 men and 25 women working there. At Brooklyn they had 360 hands working, though they reported only $440,000 worth of clocks produced from $280,000 in raw materials, indicating they had not been in production the entire 12 months period. Both factories paid a total of $260,000 in wages that year and a skilled worker was earning $2.50 a day and ordinary laborers $1.25 a day. {24}

However, it was only a few months after this reporting on October 27, 1880, the Brooklyn factory was totally destroyed by fire. The *Hartford Times*, claimed the fire was caused by an explosion of leaking gas and noted the loss at $750,000 with only $250,000 insured. {25} *The New York* Times reported:

"The fire is supposed to have originated in the drying-room, where a large amount of wood was seasoning. About 2 o'clock in the morning the night watchman, while on the fourth floor, heard a dull heavy sound, like that made by an explosion. ... A strong wind was blowing, and before the engines arrived the flames had made considerable headway. ... Three additional engines arrived, but no efforts could save the building or its contents. ... When the roof gave way with a terrific crash, a great pillar of fire reared itself up and lighted the heavens for miles around. At 7 o'clock yesterday morning all that remained of the factory was the burned and blackened walls." {26}

Despite the setback, the firm immediately erected a new building and within a few years the firm's entire clockmaking operation was centered at Brooklyn. Annual statements during the 1880's show the firm remained financially strong. By 1886, they reported $600,000 worth of stock on hand, a quarter of a million dollars due them for sales and no debts. {27}

By January of 1883, Ansonia had sales offices in New York, Chicago and London. By 1886, they offered 228 clock models and by 1914, this number had grown to almost 450! They had become especially noted for their iron cased clocks, often with white metal figurines, clocks with imported china cases and crystal regulators. Non-jewelled watches were added to their line in 1894 and they produced an estimated 10 million of these by 1929. By 1914, the firm had agents in Australia, New Zealand, Japan, China, India and exported large quantities of clocks to these and eighteen other countries. {28}

During World War I, few trade materials seem to have been published, but the 1920 catalog shows their offerings had greatly changed in numbers and quality. The items which formerly had comprised most of their catalog such as black iron clocks, figurine clocks and china cased clocks had all been discontinued. Only 136 clocks and 9 watches were offered in 1920, down from a high of 450 in 1914. By 1927, the number had dropped to 47 clocks and 3 watches.

At a lecture presented in Swampscott, Mass. on April 13, 1968, Edward Ingraham, former President of the E. Ingraham Company, noted the following, which may account, in part, for Ansonia's rapid decline:

"After a clock meeting, possibly about 1910, in which members of the industry agreed to establish certain prices, our then Sales Manager, Elmer E. Stockton, caught one of their [Ansonia's] officials telephoning to customers evidently offering deals at 'old pricing.' [Stockton] evidently confirmed this because he went after all of Ansonia's principal accounts and ended up taking away all of that business from Ansonia, which having lost its volume business, went down and out." {29}

Though the above must be classed as here-say, it appears to have merit. An observation of the product line would suggest this event may occurred after 1910, more likely about 1915, since the decline seems great between 1915 and 1920. It is significant that their decline occurred prior to the Great Depression and, in fact, the sale of the company took place some months prior to the stock market crash.

By the mid-1920's, the company was definitely in trouble. In July of 1926, they sold their five-story Brooklyn warehouse. The end culminated with the sale of the firm to Soviet Russia's Amtorg Trading Corporation. This sale, along with that of Canton, Ohio's Dueber-Hampden Watch Company, was noted in the *New York Times* on September 10, 1929. {30} Some workmen from both Canton and Brooklyn went to Russia for up to 18 months to get the machinery in operation and train Russian workers. These operations became the nucleus of Moscow's factories No. 1 and No. 2, both established in 1930. They were in full operation in 1956 and are likely so today as Russia is still producing mechanical timekeepers. {31}

An interesting letter was written in 1947 to Mr. Edward Ingraham, President of the E. Ingraham Company, by Mr. E. Cantelo White, formerly President of Ansonia and then associated with the Tork Clock Company. Mr. White, who then still owned the Ansonia trademark stated:

"The sale of machinery to the Amtorg Corporation, which you refer, was negotiated by me in 1929. The clock orders which we felt obligated to fill were not completed until 1930. The machinery was boxed and shipped about the end of that year. ... We have about completed arrangements for a greatly increased production of Ansonia Clocks in 1948, including types suitable for export." {32}

Plans for the 1948 re-introduction of Ansonia clocks into the marketplace by the White family did not materialize, although Mr. White maintained an office under the firm's name for many years. However, in 1969 the Nofziger family of Lynnwood, Washington re-registered the Ansonia trademark and have since manufactured clocks as the Ansonia Clock Company, Inc.

# REFERENCES

1. Bristol Land Records (hereafter BLR), Vol. 21, p. 13 and mortgage Vol. 21, p. 14.
2. *The Bristol Herald*, July 10, 1890, p. 1.
3. Federal Industrial Census of Bristol, Hartford County, CT taken June 1, 1850
4. Report of Tracy Peck, published in *The Timepiece Journal*, Vol. 3, No. 2, pp. 30-31.
5. BLR, Vol. 25, p. 5.
6. BLR, Vol. 22, p. 367.
7. Ansonia Land Records (hereafter ALR), Vol. 32, p. 372.
8. ALR, Vol. 32, p. 406.
9. ALR, Vol. 32, p. 459; Vol. 36, p. 111
10. ALR, Vol. 32, p. 416.
11. ALR, Vol. 38, p. 3.
12. *The Bristol Press*, January 20, 1881.
13. ALR, Vol. 38, pp. 23, 70.
14. *Tercentenary Portrait and History of the Lower Naugatuck Valley*, by Leo T. Molloy. Ansonia, CT: Emerson Bros., Inc., 1935. p. 160.
15. ALR, Vol. 38, p. 83.
16. ALR, Vol. 38, p. 154.
17. ALR, Vol. 39, p. 391.
18. Bankruptcy Proceedings, Terry & Barnum Mfg. Co., 1856, house at Connecticut State Library, Hartford.
19. Bristol Press, June 23, 1881.
20. ALR, Bk. 43, p. 588
21. 1850 Federal Industrial Census of Conn. Original volumes at Conn. State Library, Hartford.
22. Price List of Ansonia Clock Company, January 1, 1873 in materials of Edward Ingraham Library, American Clock & Watch Museum, Inc., Bristol, CT.
23. ALR, Bk. 56, p. 76.
24. 1880 Federal Industrial Census of Conn.
25. *The Hartford Times*, Thursday, October 28, 1880.
26. *The New York Times*, October 28, 1880, p. 8, col. 4.
27. ALR, Bk. 56, pp. 553, 571; Bk. 61, pp. 175, 388, 575; Bk. 66, pp. 106, 187, 479; Bk. 70, pp. 160, 451; Bk. 70, p. 146.
28. Data compiled by author from trade catalogs in the Edward Ingraham Library, American Clock & Watch Museum.
29. "Reminiscences of Edward Ingraham, Clockmaker." Unpublished collection in Edward Ingraham Library, American Clock & Watch Museum.
30. *The New York Times*, September 10, 1929, p. 1, col. 2.
31. "The Russian Horological Industry" published in *The Timepiece Journal*, Vol. 3, No. 8 and Vol. 4, No. 1.
32. Letter from Mr. E. Cantelo White of the Ansonia Clock Company to Edward Ingraham, dated November 14, 1947.

# CHRONOLOGY

| DATE | EVENT |
|---|---|
| August 13, 1841 | Firm of Terry & Andrews formed in Bristol, Conn. by Theodore Terry and Franklin Andrews. |
| About April, 1850 | Brass manufacturer Anson G. Phelps of then Birmingham, now Derby, Conn. persuades Terry & Andrews to ally with his brass foundry. |
| May 7, 1850 | Anson G. Phelps, Theodore Terry and Franklin Andrews form Ansonia Clock Co. located at Ansonia, Conn., a village in the town of Derby, Conn. |
| May 11, 1850 | Anson G. Phelps buys half interest in Terry & Andrews for $7,200.00. |
| Fall, 1850 | Operations move from Bristol to Ansonia, Conn. |
| December 20, 1851 | Franklin Andrews sells out of Ansonia Clock Co. |
| January, 1853 | Anson Phelps sells his shares of Ansonia Clock Co. to his son-in-law James B. Stokes. |
| November 30, 1853 | Anson G. Phelps dies. |
| January, 1854 | James B. Stokes sells his shares in Ansonia Clock Co. to Phelps, Dodge & Co. |
| November, 1854 | Ansonia Clock Co. factory burns and "land & ruins" sold to six individuals who were the directors of Phelps, Dodge & Co. thus ending the original Ansonia Clock Co. |
| Period 1854 to 1869 | Phelps, Dodge & Co. produce many movements and a few complete clocks labeled "Ansonia Brass Company" and fewer labeled "Ansonia Brass & Battery Company." |
| February 11, 1869 | Ansonia Brass & Battery Co. reorganized as Ansonia Brass & Copper Company, still under Phelps, Dodge & Co. |
| December 21, 1877 | Joint stock company formed in New York City adopting original name Ansonia Clock Co. Incorporators were Phelps, Dodge & Co. and Henry J. Davis of Brooklyn, N.Y. |
| April 1879 | Large factory for Ansonia Clock Co. under construction in Brooklyn. |
| Spring 1880 | New machinery installed in new Ansonia Clock Co. factory in Brooklyn. |
| October 27, 1880 | New Brooklyn factory of Ansonia Clock Co. destroyed by fire shortly after going into production. |
| Early 1881 | Brooklyn factory rebuilt. Operations still continue at old factory in Ansonia, Conn. |
| 1883 | Operations of Ansonia Clock Co. now solely at Brooklyn. |
| January, 1883 | Ansonia Clock Co. has sales offices in New York, Chicago and London. |
| 1894 | Non-jewelled watches added to product line. |
| 1914 | Sales agents in Australia, New Zealand, Japan, China, India and 18 other countries. |
| World War I period | Ansonia Clock Co. goes into rapid decline. |
| July 1926 | Ansonia Clock Co. sells five story Brooklyn warehouse. |
| September 10, 1929 | New York Times reports sale of Ansonia Clock Co. to Soviet Union's (Russia's) Amtorg Trading Corporation. Note: This is prior to the stock market crash signalling the start of the Great Depression in October 1929. |
| 1930 | Machinery installed and clocks being produced in Moscow. |
| Post 1929 | Mr. E. Cantelo White, president of Ansonia Clock Co. and later associated with the Tork Clock Company, retained ownership of Ansonia trademark. |

# ANSONIA

The charm of these banjo clocks lies in the glasses which are made of high quality crystal clear glass. If either of the glasses are broken, chipped, cracked or replaced the price is **substantially** reduced.

The year entered beside each clock is the year of the catalog from which the illustration was taken. Some clocks may have been produced a few years earlier.

Eight-Day Movements. Genuine Mahogany Cases, Dull Hand Rubbed. Silver Plated Dials. Arabic Numerals. Beveled Chipped Glasses in Panels and Lower Doors. Eagles and Side Ornaments Bronze Finish.

DUAL CHIMETONE — 1

EIGHT-DAY TIME — 2

WESTMINSTER CHIME — 3

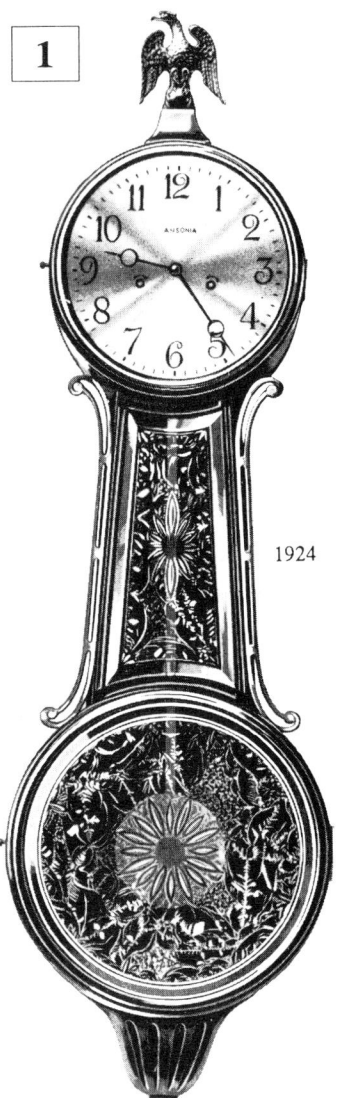
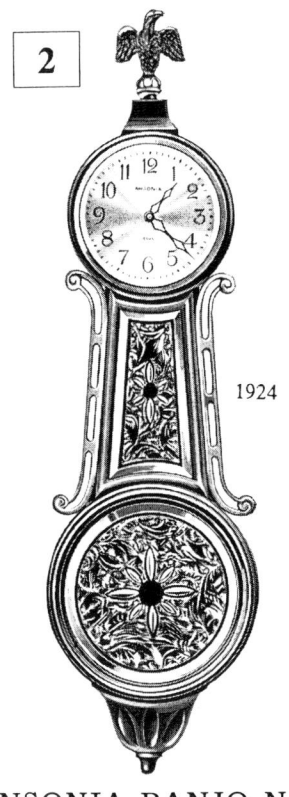
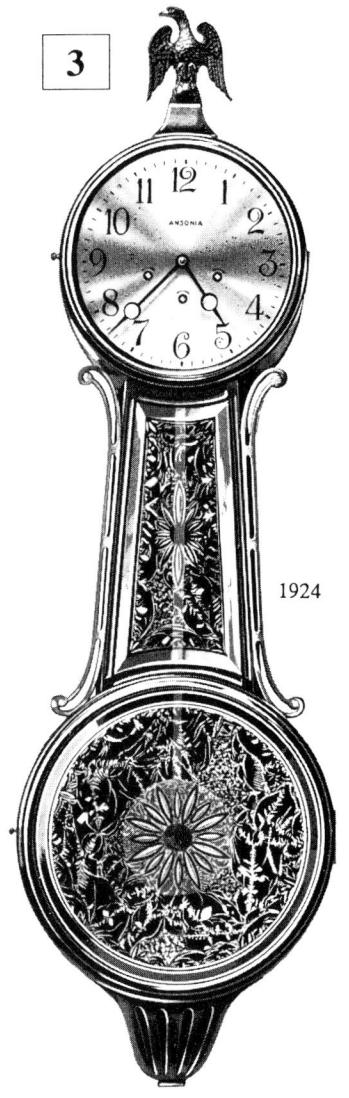

1924

### ANSONIA BANJO No. 1
Time Lever Movement
Height 18½ inches   Width 5¼ inches
Dial 3½ inches

### ANSONIA BANJO No. 2
Pendulum Movement
Height 40½ inches   Width 12 inches
Dial 8¼ inches
Dual Chimetone Strike

Hour and Half-Hour Dual Chimetone Strike is Recorded on Two Sweet-Toned Chime Rods.

### ANSONIA BANJO No. 3
Pendulum Movement
Height 40½ inches   Width 12 inches
Dial 8¼ inches

**Full Westminster Quarter Hour Chime on Melodious True-Toned Rods.**

# ANSONIA

17

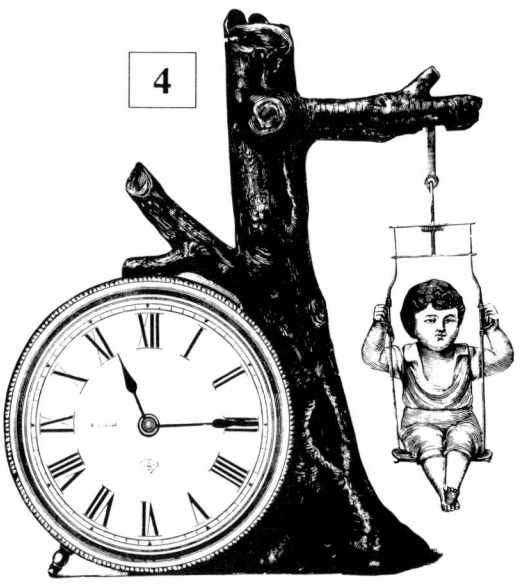

**SWING No. 2.**
ONE-DAY TIME.   1884
Automatic Swing.
Made in Fancy Brass and Bronze.
Dial, 4 inches.   Height, 8 inches.
List, each, $5.00.

1884

**SWING No. 1.**
ONE-DAY TIME.
Automatic Swing.
Made in Fancy Brass.
Dial, 4 inches.   Height, 11½ inches.
List, each, $5.00.

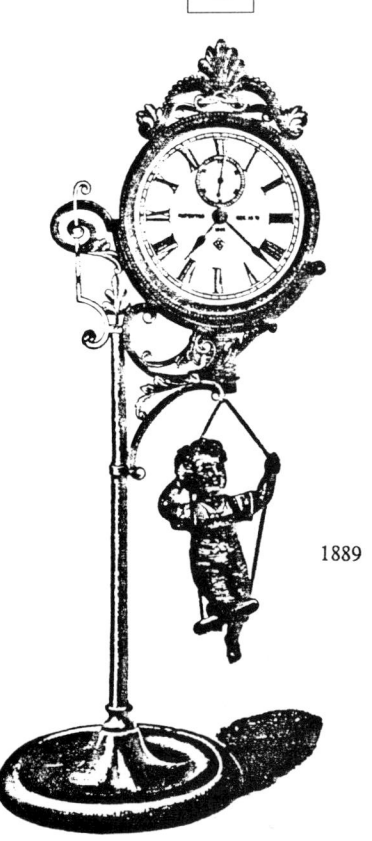

1889

**JUMPER No. 1.**
One Day, Time.
Dial, 4 inches.   Height, 15½ inches.
Automatic Motion.

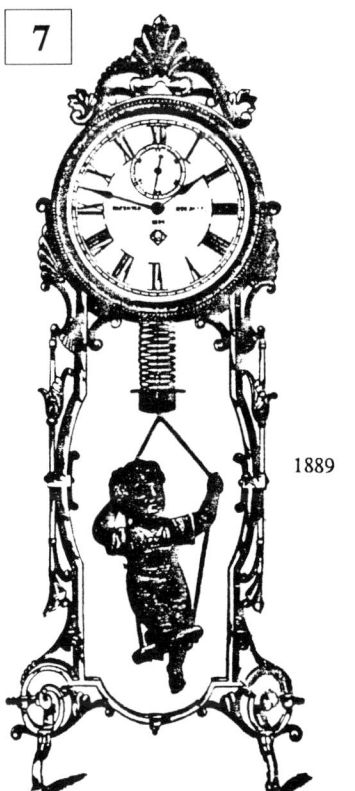

1889

**JUMPER No. 2.**
One Day, Time.
Dial, 4 inches.   Height, 14½ inches.
Automatic Motion.

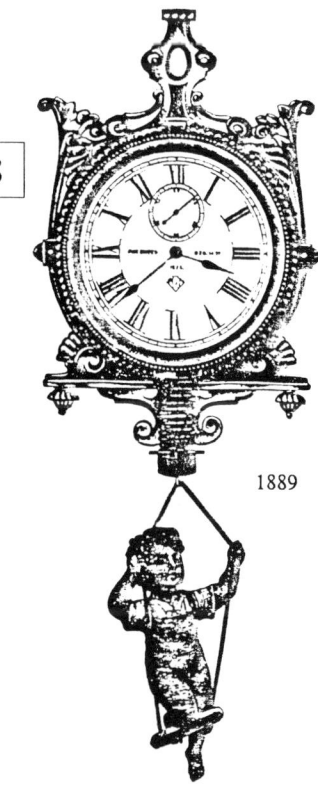

1889

**JUMPER No. 3.**
One Day, Time.
Dial, 4 inches.   Height, 15 inches.
Automatic Motion.

*The most important component of these "Bobbing Doll" clocks is the **bisque** doll. Most are either broken, chipped, repaired or reproduction which **substantially** reduces the price from one that is all original.*

*The year entered beside each clock is the year of the catalog from which the illustration was taken. The clock may have been produced in an earlier year.*

# ANSONIA

Most white metal and brass ornamentals and clock cases are gilded or silver plated. Great care must be taken when cleaning these surfaces. The thin layer of gilding or silver plating will be easily removed by cleaning solutions and strong rubbing. If you believe cleaning is necessary, hire the services of a professional and avoid the expense of re-gilding or re-plating.

**9**

1894

**CABINET D.**
Antique Oak, with Brass Trimmings.
Height, 23 inches.   Width, 12 inches.
Sash and Dial, 5¾ inches.
List, each, $18.00.

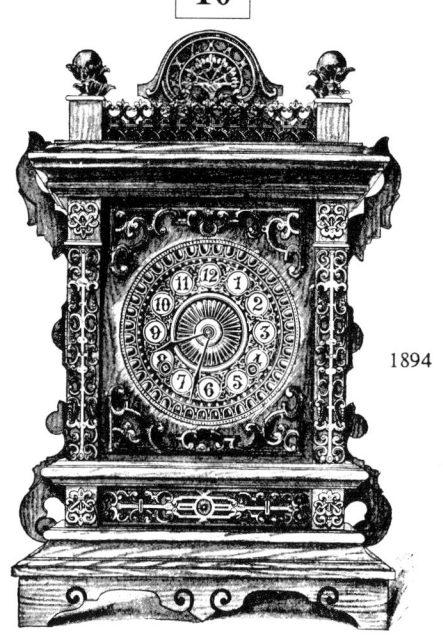

**10**

1894

**CABINET F.**
Antique Oak, with Brass Trimmings.
Height, 19 inches.   Width, 12 inches.
Sash and Dial, 5¾ inches.
List, each, $15.50.

**11**

1894

**CABINET ANTIQUE, No. 1.**
Polished Oak or Mahogany,
with
Antique Brass Trimmings.
EIGHT-DAY, HALF-HOUR OLD ENGLISH BELL STRIKE. BELL ON TOP.
Dial, 5 inches.   Height, 18¾ inches.
List, each, $35.00.

**12**

1894

**CABINET ANTIQUE.**
Polished Mahogany or Oak,
with
Antique Brass Trimmings.
EIGHT-DAY, HALF-HOUR GONG STRIKE.
Dial, 5 inches.   Height, 20 inches.
List, each, $35.00.

**13**

1894

**SENATOR.**
Antique Oak or Mahogany,
with
Antique Brass Trimmings.
EIGHT-DAY, HALF-HOUR GONG STRIKE.
Height, 22 inches.   Width, 19 inches.
Silver Dial, 6 inches.   List, each, $45.00.

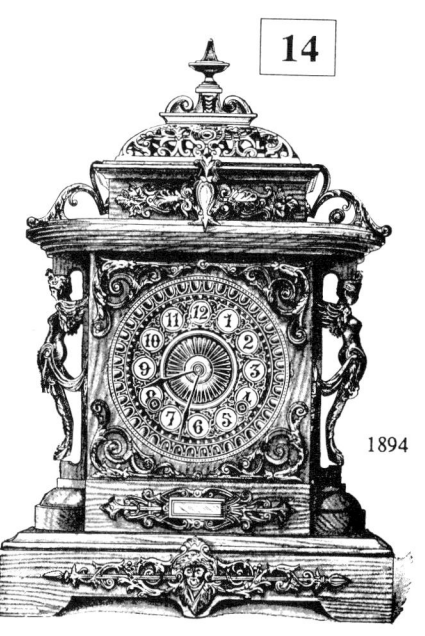

**14**

1894

**CABINET E.**
Antique Oak, with Brass Trimmings.
Height, 18 inches.   Width, 12 inches.
Sash and Dial, 5¾ inches.
List, each, $17.00.

# ANSONIA

*The year entered beside each clock is the year of the catalog from which the illustration was taken. Some clocks may have been produced a few years earlier.*

**15**

**BRILLIANT.**
Dial, - - - - - - - 2½ inches.

1880

**16**

1880

**ELLIPTICAL CARRIAGE.**
ONE DAY TIME, ALARM.
Dial, 2½ inches.   Height, 6 inches.

**17**

1901

**BONANZA**
Eight-Day Time.
Dial, 1½ inches.   Height, 5⅛ inches.
Porcelain Dial, Arabic or Roman, Beveled Glass.
Finished in Rich Gold, with Jeweled Settings ....................List, each, $12 75
Leather Carrying Case, $1 50 List additional.

*Most white metal and brass ornamentals and clock cases are gilded or silver plated. Great care must be taken when cleaning these surfaces. The thin layer of gilding or silver plating will be easily removed by cleaning solutions and strong rubbing. If you believe cleaning is necessary, hire the services of a professional and avoid the expense of re-gilding or re-plating.*

**18**

1901

**BON-TON**
Eight-Day Time.
Dial, 1½ inches.   Height, 5½ inches.
Porcelain Dial, Arabic or Roman, Beveled Glass.
Finished in Antique Brass or Silver,
                                    List, each, $8 05
Finished in Rich Gold........List, each, 10 50
Leather Carrying Case, $1 50 List additional.

**19**

1901

**COMET**
One-Day, Half-Hour Strike and Alarm.
Dial, 3 inches.   Height, 6½ inches.
Finished in Antique Brass or Silver, Beveled Glass.
List, each ...............................$6 50

**20**

1901

**SATELLITE**
Eight-Day Time.
Dial, 1½ inches.   Height, 5⅝ inches.
Porcelain Dial, Arabic or Roman, Beveled Glass Front, Sides and Back.
Finished in Rich Gold........List, each, $10 50

# ANSONIA

*Clocks under glass domes are very attractive — but, these glass domes are mostly nonstandard and almost impossible to find. They can, however, be custom-made at considerable cost. The beveled glass on Crystal Regulator or Carriage Clocks, if cracked or chipped, can be replaced without difficulty at reasonable cost.*

1883

**CRYSTAL PALACE, No. 1 EXTRA.**

Eight Day, Strike.
Dial, 6 inches.
Height .................. 18½ inches.

1880

PATENT APPLIED FOR
**HELMSMEN.**
With or without shade.
Dial, - - - - - - - - - - - - - - 6 inches.

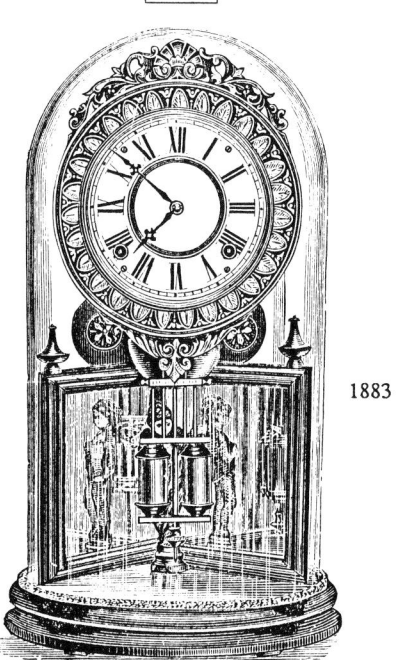

1883

**CRYSTAL PALACE, No. 2 EXTRA.**

Eight Day, Strike.
Dial, 6 inches.
Height ...................... 18 inches.

I have no knowledge whether the Helmsmen Crystal Palace "Wheel" moves automatically, like the Mechanical clock on page 206 of Tran Duy Ly's "Ansonia Clocks, History, Identification and Price Guide," but if it does the suggested price with original glass dome would be three times higher.

# ANSONIA

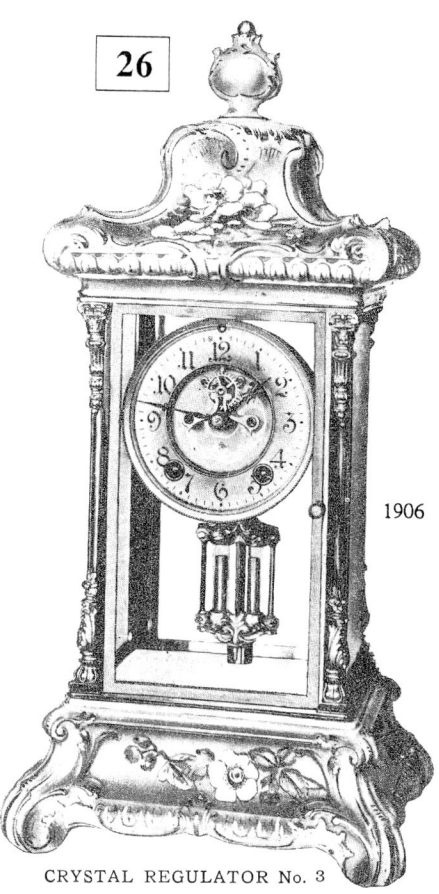

**24** — CRYSTAL REGULATOR No. 1
Height, 17½ inches. Width, 9 inches. 1906

**25** — CRYSTAL REGULATOR No. 2
Height, 17½ inches. Width, 9¼ inches. 1906

**26** — CRYSTAL REGULATOR No. 3
Height, 18¼ inches. Width, 9¼ inches. 1906

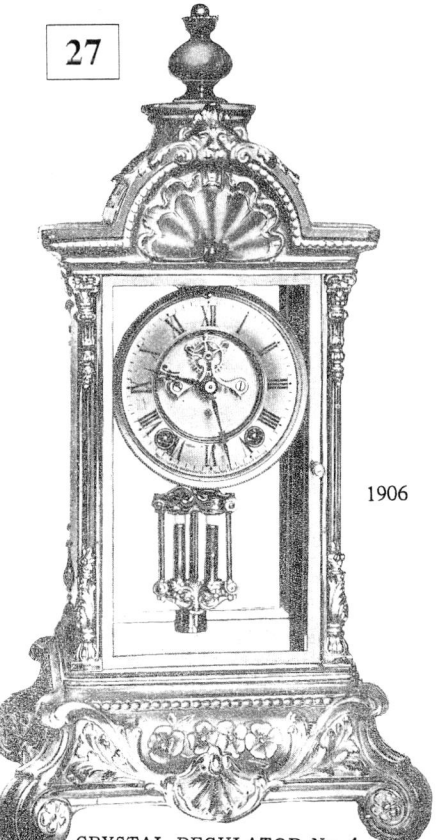
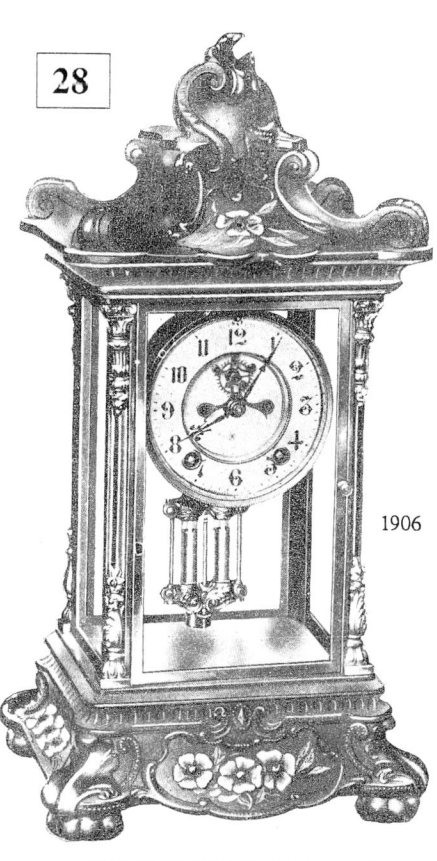
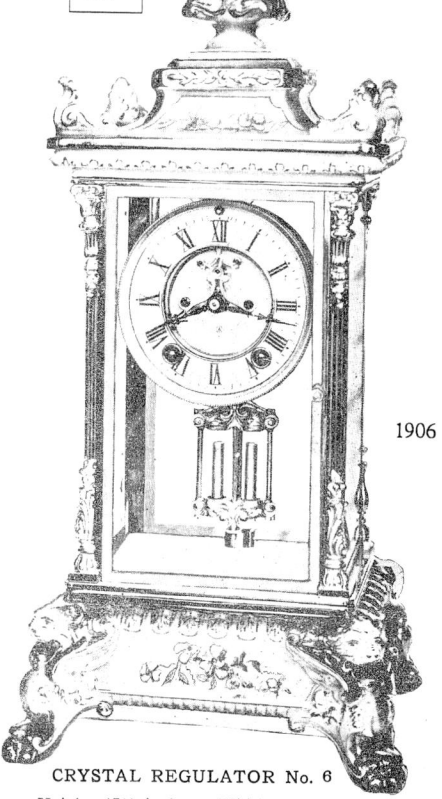

**27** — CRYSTAL REGULATOR No. 4
Height, 17¾ inches. Width, 9½ inches. 1906

**28** — CRYSTAL REGULATOR No. 5
Height, 17½ inches. Width, 8½ inches. 1906

**29** — CRYSTAL REGULATOR No. 6
Height, 17½ inches. Width, 9 inches. 1906

# ANSONIA

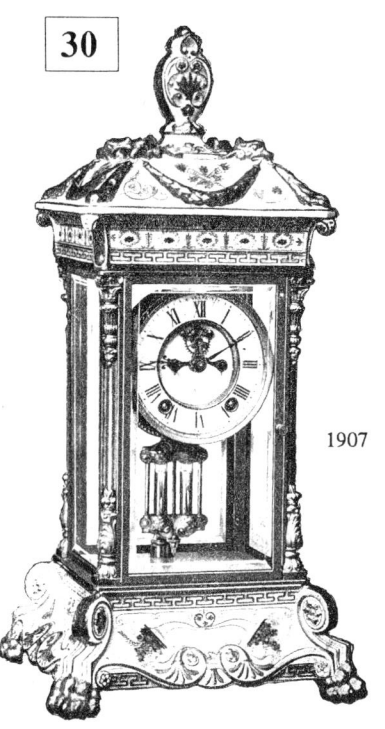

**CRYSTAL REGULATOR No. 7**
Height, 18½ inches.  Width, 9½ inches.
Polished Brass, Rich Gold Ornaments, Royal
Bonn Porcelain Top and Base.
Rich Colors.
List, each ................................$47.50
Empire Decoration.
List, each ................................ 58.00

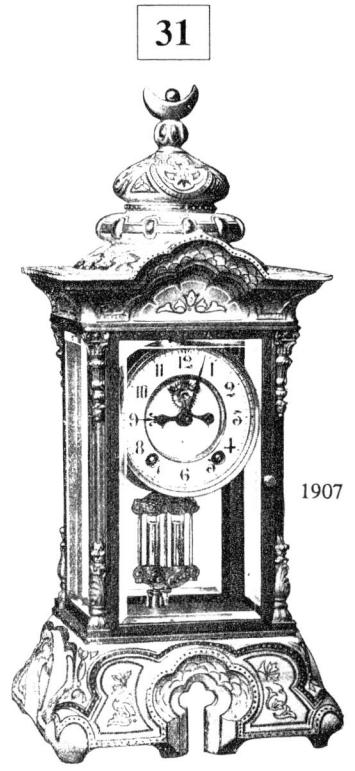

**CRYSTAL REGULATOR No. 8**
Height, 18½ inches.  Width, 9 inches.
Polished Brass, Rich Gold Ornaments, Royal
Bonn Porcelain Top and Base.
Rich Colors.
List, each ................................$47.50

**UTOPIA**
Height, 17 Inches.
Width, 8¾ Inches.
Dial, 4 Inches.
Polished Brass.  Rich Gold Ornaments.
Cut Glass Columns.  Convex Doors.
**List, Each, $95.00**

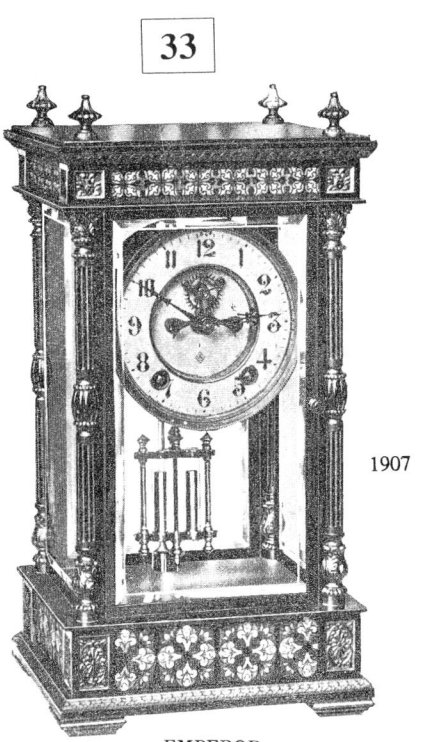

**EMPEROR**
Height, 13 inches.  Width, 7 inches.
Polished Brass, Rich Gold Ornaments,
Cloisonne inlay.
List, each ................................$63.00

**EARL**
Height, 12 Inches     Width, 7 Inches
Dial, 4 Inches
Finished in Polished Brass
Rich Gold Ornaments

**GONFALON**
Height, 14½ Inches     Width, 12 Inches

# ANSONIA

Most white metal and brass ornamentals and clock cases are gilded or silver plated. Great care must be taken when cleaning these surfaces. The thin layer of gilding or silver plating will be easily removed by cleaning solutions and strong rubbing. If you believe cleaning is necessary, hire the services of a professional and avoid the expense of re-gilding or re-plating.

1907

**FLORAL**
Height, 16½ inches.  Width, 8¼ inches.
Finished in Syrian Bronze......List, each, $41.00
Finished in Rich Gold..........List, each, 42.00
Jeweled Sash and Pendulum, $4.20 List Additional.

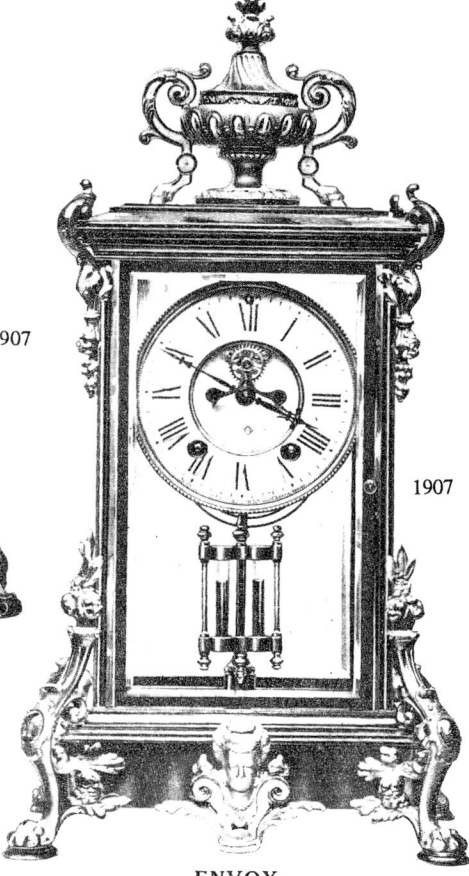

1907

**ENVOY**
Height, 19 inches.  Width, 9¾ inches.
Dial, 5 inches.
Real Bronze, Rich Gold Ornaments.
List, each ............................$65.00
Can be supplied with Solid Porcelain Dial without Visible Escapement.

1914

**JUPITER**
Height, 17 Inches.
Width, 8¾ Inches.
Dial, 4 Inches.
Polished Brass. Rich Gold Ornaments.
Cut Glass Columns. Convex Doors.
**List, Each, $70.00**

1901

**DINORAH**
Eight-Day, Half-Hour Gong Strike.
Height, 11½ inches.  Width, 8¾ inches.
Porcelain Visible Escapement Dial, 4½ inches, Arabic or Roman.
Mercurial Pendulum.
Beveled Plate Glass Front, Sides and Back.
Finished in Rich Gold.
List, each ............................$50 00
Jeweled Sash and Pendulum, $4 20 List additional.

**DINORAH**

1901

**MARTHA**
Eight-Day, Half-Hour Gong Strike.
Lever Movement.
Height, 10½ inches.  Width, 8¾ inches.
Porcelain Dial, 4½ inches, Arabic or Roman.
Beveled Plate Glass Front, Sides and Back.
Finished in Rich Gold.
List, each ............................$50 00
Jeweled Sash, $3 00 List additional.

# ANSONIA

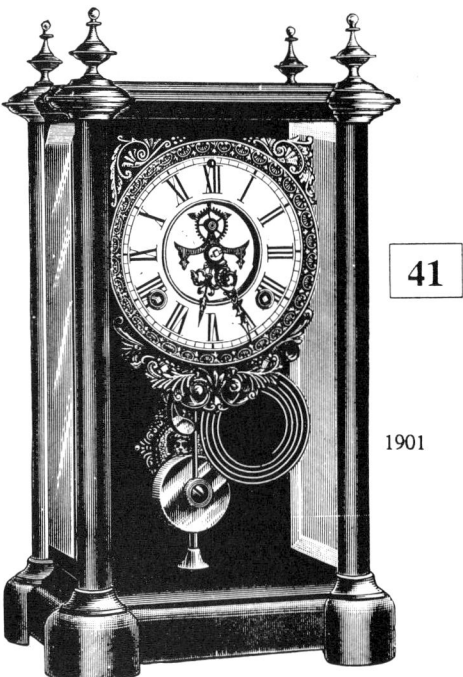

41 — 1901

### SYMBOL
Eight-Day, Half-Hour Gong Strike.
Height, 15¼ inches.  Width, 9 inches.
Dial, 5½ inches.
Porcelain Visible Escapement Dial,
Arabic or Roman.
Beveled Plate Glass Front and Sides.
Finished in Brass ............List, each, $28 50
Finished in Rich Gold ........List, each, 38 50

*Most white metal and brass ornamentals are gilded or silver plated. Great care must be taken when cleaning these surfaces. The thin layer of gilding or silver plating will be easily removed by cleaning solutions and strong rubbing. If you believe cleaning is necessary, hire the services of a professional and avoid the expense of re-gilding or re-plating.*

42 — 1901

### APEX
Eight-Day, Half-Hour Gong Strike.
Height, 18½ inches.  Width, 10½ inches.
Porcelain Visible Escapement Dial, 4½ inches,
Arabic or Roman.
Beveled Plate Glass Front, Sides and Back.
Finished in Rich Gold.
List, each, .............................$92 00
Jeweled Sash and Pendulum, $4 20 List
additional.

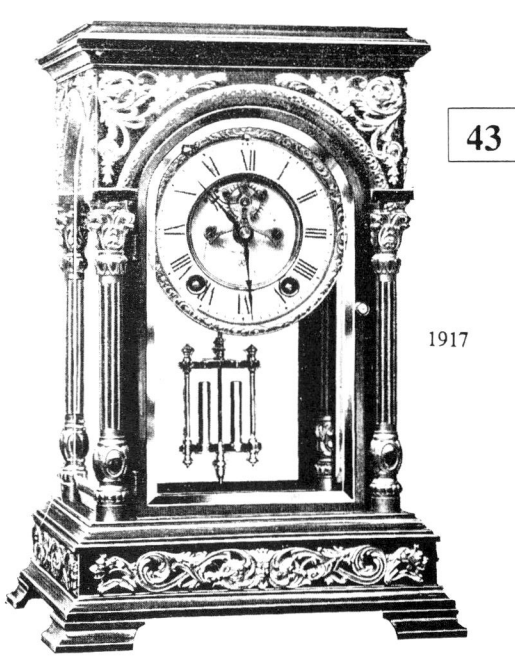

43 — 1917

### SIRIUS
Height, 13 Inches.
Width, 9 Inches.
Dial, 4 Inches.
Polished Brass.
Rich Gold Ornaments.

# ANSONIA

*For comprehensive listing of Ansonia Clocks, see Tran Duy Ly "Ansonia Clocks - A Guide to Identification & Prices."*

*The year entered beside each clock is the year of the catalog from which the illustration was taken. Some clocks may have been produced a few years earlier.*

### LUCIA
Height, 15 inches.   Width, 8¾ inches.
Porcelain Visible Escapement Dial, 4½ inches,
Arabic or Roman.
Mercurial Pendulum.
Finished in Rich Gold.
List, each ................................$56 00
Jeweled Sash and Pendulum, $4 20 List additional.

### REGAL
Eight-Day, Half-Hour Gong Strike.
Height, 18½ inches.   Width, 10½ inches.
Porcelain Visible Escapement Dial, 4½ inches,
Arabic or Roman.
Beveled Plate Glass Front, Sides and Back.
Finished in Rich Gold.
List, each, ................................$96 00

### NORMA
Lever Movement.
Height, 14 inches.   Width, 8¾ inches.
Porcelain Dial, 4½ inches,
Arabic or Roman.
Finished in Rich Gold.
List, each ................................$56 00
Jeweled Sash, $3 00 List additional.

# ANSONIA

*For comprehensive listing of Ansonia Clocks, see Tran Duy Ly "Ansonia Clocks - A Guide to Identification & Prices."*

**FISHER AND HUNTER.**
Height, 21½ inches.  Width, 19 inches.
Japanese Bronze.  List, each, $47.00.
Silver.  List, each, $51.70.

**FISHER**

Height, 22 inches.  Width, 15 inches.
Finished in Gilt or Japanese Bronze,
List, each, $28.00
Finished in Barbedienne Bronze,
List, each, 30.80
Finished in Syrian Bronze....List, each, 32.20

**HUNTER**

Height, 22½ inches.  Width, 15 inches.
Finished in Gilt or Japanese Bronze,
List, each, $28.00
Finished in Barbedienne Bronze,
List, each, 30.80
Finished in Syrian Bronze....List, each, 32.20

**HUNTER**

# ANSONIA

**MELODY AND MOTION.**
Height, 20½ inches.   Width, 25½ inches.
Japanese Bronze.   List, each, $57.00.
Silver.   List, each, $62.70.

**MOTION.**
Height, 23¾ inches.   Width, 19½ inches.
Japanese Bronze.   List, each, $43.00.
Silver.   List, each, $47.30.

**MELODY.**
Height, 23¾ inches.   Width, 19½ inches.
Japanese Bronze.   List, each, $43.00.
Silver.   List, each, $47.30.

# ANSONIA

*For comprehensive listing of Ansonia Clocks, see Tran Duy Ly "Ansonia Clocks - A Guide to Identification & Prices."*

*The year entered beside each clock is the year of the catalog from which the illustration was taken. Some clocks may have been produced a few years earlier.*

**DON CÆSAR AND DON JUAN.**
Height, 20½ inches.   Width, 25½ inches.
Japanese Bronze.   List, each, $47.00.
Silver.   List, each, $51.70.

**DON CÆSAR.**
Height, 22 inches.   Width, 19½ inches.
Japanese Bronze.   List, each, $38.00.
Silver.   List, each, $41.80.

**DON JUAN.**
Height, 22 inches.   Width, 19½ inches.
Japanese Bronze.   List, each, $38.00.
Silver.   List, each, $41.80.

# ANSONIA

*Most white metal and brass ornamentals and clock cases are gilded or silver plated. Great care must be taken when cleaning these surfaces. The thin layer of gilding or silver plating will be easily removed by cleaning solutions and strong rubbing. If you believe cleaning is necessary, hire the services of a professional and avoid the expense of re-gilding or re-plating.*

**PIZARRO AND CORTEZ.**
Height, 20½ inches.   Width, 25¾ inches.
Japanese Bronze.   List, each, $57.00.
Silver.   List, each, $62.70.

**CORTEZ.**
Height, 21¾ inches.   Width, 19½ inches.
Japanese Bronze.   List, each, $42.00.
Silver.   List, each, $46.20.

**PIZARRO.**
Height, 21¾ inches.   Width, 19½ inches.
Japanese Bronze.   List, each, $42.00.
Silver.   List, each, $46.20.

# ANSONIA

**59** FAUST.
Height, 22 inches. Width, 24 inches.
Japanese Bronze. List, each, $60.00.
Silver. List, each, $66.00.

**60** REUBENS.
Height, 16¼ inches. Width, 21 inches.
Japanese Bronze, List, $32.00. Silver, $35.20.

*Most white metal and brass ornamentals and clock cases are gilded or silver plated. Great care must be taken when cleaning these surfaces. The thin layer of gilding or silver plating will be easily removed by cleaning solutions and strong rubbing. If you believe cleaning is necessary, hire the services of a professional and avoid the expense of re-gilding or re-plating.*

**61** OPERA.
Height, 16¼ inches. Width, 21 inches.
Japanese Bronze, List, $35.00. Silver, $38.50.

**62** SUITOR.
Height, 25 inches. Width, 24 inches.
Japanese Bronze. List, each, $60.00.
Silver. List, each, $66.00.

# ANSONIA

*Most white metal and brass ornamentals and clock cases are gilded or silver plated. Great care must be taken when cleaning these surfaces. The thin layer of gilding or silver plating will be easily removed by cleaning solutions and strong rubbing. If you believe cleaning is necessary, hire the services of a professional and avoid the expense of re-gilding or re-plating.*

**63**

**COUNTESS.**
Height, 25 inches.    Width, 24 inches.
Japanese Bronze.  List, each, $60.00.
Silver.  List, each, $66.00.

**64**

**NEWTON.**
Height, 15 inches.    Width, 17½ inches.
Japanese Bronze, List, $30.00.  Silver, $33.00.

**65**

**MOZART.**
Height, 14½ inches.    Width, 18 inches.
Japanese Bronze, List, $30.00.  Silver, $33.00.

**66**

**WALTZ.**
Height, 22 inches.    Width, 24 inches.
Japanese Bronze.  List, each, $60.00.
Silver.  List, each, $66.00.

# ANSONIA

### 67

**EROS.**
Height, 15 inches.   Width, 17½ inches.
Japanese Bronze.   List, each, $30.00.
Silver.   List, each, $33.00.

*Most white metal and brass ornamentals are gilded or silver plated. Great care must be taken when cleaning these surfaces. The thin layer of gilding or silver plating will be easily removed by cleaning solutions and strong rubbing. If you believe cleaning is necessary, hire the services of a professional and avoid the expense of re-gilding or re-plating.*

### 68

**IDYL.**
Height, 22 inches.   Width, 24 inches.
Japanese Bronze.   List, each, $65.00.
Silver.   List, each, $71.50.

### 69

**AIDA.**
Height, 23 inches.   Width, 24 inches.
Japanese Bronze.   List, each, $80.00.
Silver.   List, each, $88.00.

### 70

**PHILOSOPHER.**
Height, 15 inches.   Width, 17½ inches.
Japanese Bronze.   List, each, $30.00.
Silver.   List, each, $33.00.

# ANSONIA

**FANTASY.**
Height, 15 inches.   Width, 17½ inches.
Japanese Bronze.   List, each, $30.00.
Silver.   List, each, $33.00.

**TERPSICHORE.**
Height, 25 inches.   Width, 24 inches.
Japanese Bronze.   List, each, $65.00.
Silver.   List, each, $71.50.

**VOCALISTS.**
Height, 22 inches.   Width, 24 inches.
Japanese Bronze.   List, each, $60.00.
Silver.   List, each, $66.00.

**MERCURY.**
Height, 15 inches.   Width, 17½ inches.
Japanese Bronze.   List, each, $30.00.
Silver.   List, each, $33.00.

# ANSONIA

**75**

MUSICIAN.
Height, 22 inches.    Width, 24 inches.
Japanese Bronze.  List, each, $60.00.
Silver.  List, each, $66.00.

**76**

CUPID.
Height, 15 inches.    Width, 17½ inches.
Japanese Bronze.  List, each, $30.00.
Silver.  List, each, $33.00.

**77**

SHAKESPEARE.
Height, 15 inches.    Width, 17½ inches.
Japanese Bronze.  List, each, $30.00.
Silver.  List, each, $33.00.

**78**

POET.
Height, 22 inches.    Width, 24 inches.
Japanese Bronze.  List, each, $60.00.
Silver.  List, each, $66.00.

# ANSONIA

**79**

SUMMER AND WINTER.
Height, 22 inches.   Width, 24 inches.
Japanese Bronze.   List, each, $70.00.
Silver.   List, each, $77.00.

**80**

VICTORY.
Height, 15 inches.   Width, 17½ inches.
Japanese Bronze.   List, each, $30.00.
Silver.   List, each, $33.00.

*Most white metal and brass ornamentals are gilded or silver plated. Great care must be taken when cleaning these surfaces. The thin layer of gilding or silver plating will be easily removed by cleaning solutions and strong rubbing. If you believe cleaning is necessary, hire the services of a professional and avoid the expense of re-gilding or re-plating.*

**81**

SUMMER.
Height, 20½ inches.   Width, 20 inches.
Japanese Bronze.   List, each, $50.00.
Silver.   List, each, $55.00.

**82**

WINTER.
Height, 20½ inches.   Width, 20 inches.
Japanese Bronze.   List, each, $50.00.
Silver.   List, each, $55.00.

# ANSONIA

*For comprehensive listing of Ansonia Clocks, see Tran Duy Ly "Ansonia Clocks - A Guide to Identification & Prices."*

BARD.
Height, 19 inches. Width, 20 inches.
Japanese Bronze. List, each, $50.00.
Silver. List, each, $55.00.

PROSPERITY.
Height, 19 inches. Width, 20 inches.
Japanese Bronze. List, each, $50.00.
Silver. List, each, $55.00.

TENOR.
Height, 22 inches. Width, 20 inches.
Japanese Bronze. List, each, $50.00.
Silver. List, each, $55.00.

SOLO.
Height, 21½ inches. Width, 20 inches.
Japanese Bronze. List, each, $50.00.
Silver. List, each, $55.00.

# ANSONIA

37

*For comprehensive listing of Ansonia Clocks, see Tran Duy Ly "Ansonia Clocks - A Guide to Identification & Prices."*

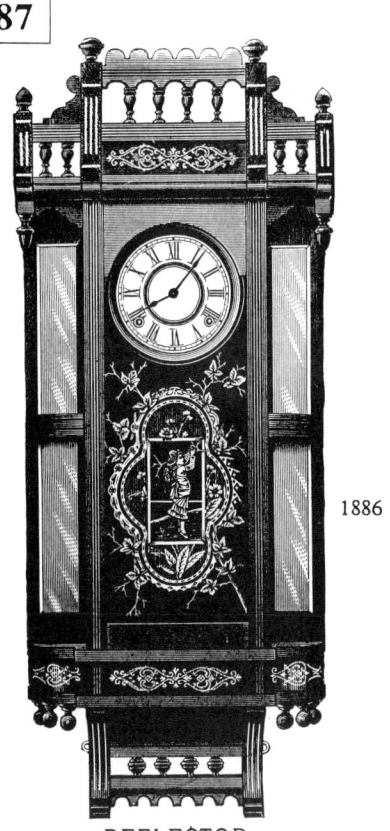

1886

**REFLECTOR.**
Ebony or Mahogany.
EIGHT DAY, GONG, STRIKE.
Dial, 6 inches. Height, 35 inches.

1883

**MAJOR.**
Eight Day, Spring Strike. Dial, 6 inches.
Height....................29 inches.

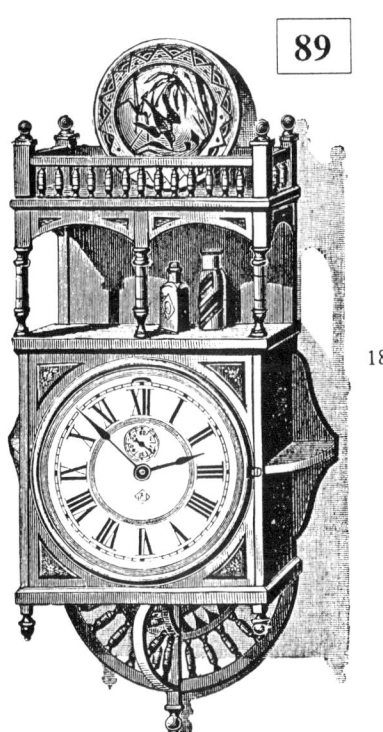

1883

**WHAT-NOT.**
Eight Day, Time.
Dial, 8 inches.
Height........................30 inches.

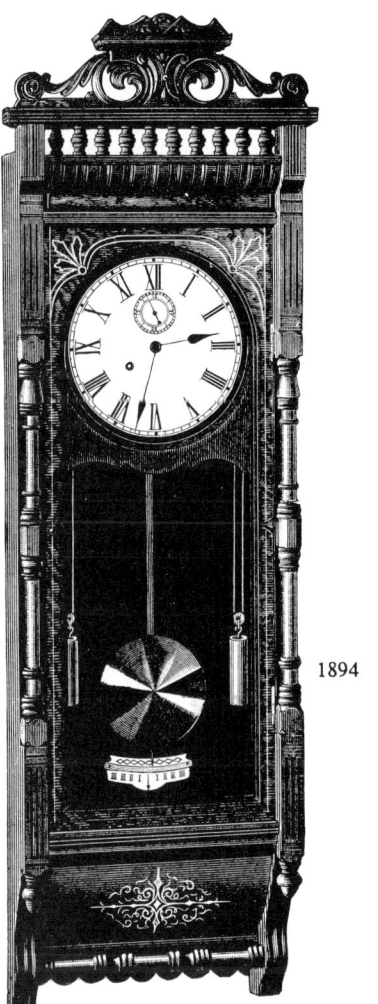

1894

**SANTA FE.**
Finished in Black Walnut, Oak or Mahogany.
Dial, 10 inches. Height, 52 inches.
Eight-Day Spring Gong Strike. List, each, $31.50.
Eight-Day Weight Time. List, each, $35.00.

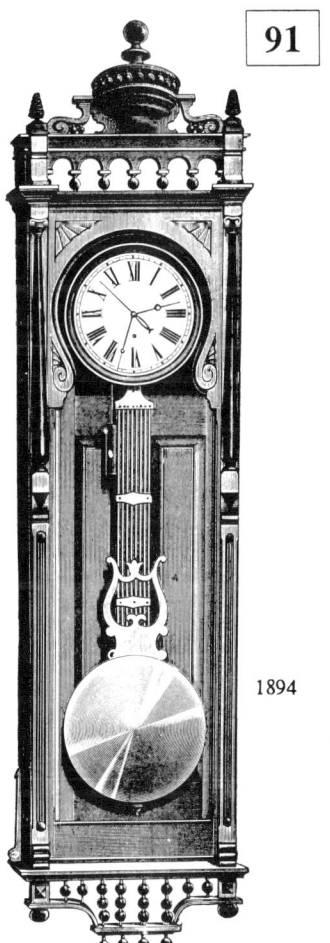

1894

**REGULATOR No. 18.**
Black Walnut, Oak or Cherry.
Dial, 12 inches. Height, 90 inches.
Eight-Day Weight Time. List, each, $104.00.

# ANSONIA

*Except for those fabulous clocks for which you wouldn't mind cutting a hole in the ceiling, Grandfather or Hanging Clocks under eight feet are the most desirable and practical. Some manufacturers produced certain of these clocks in two styles—one to stand on the floor, and the other to hang on a wall. Of the clocks designed with two styles, the hanging style is considered to be the most valuable.*

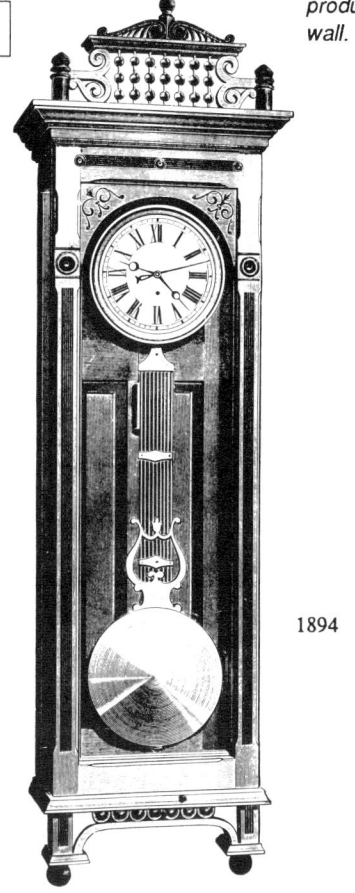

1894

**REGULATOR No. 16.**
Black Walnut, Oak or Cherry.
Dial, 12 inches.   Height, 84 inches.
Eight-Day Weight Time.   List, each, $89.00.

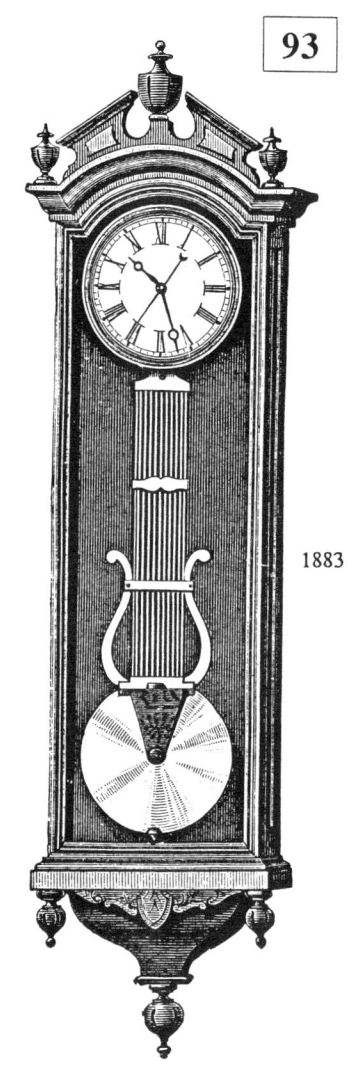

1883

**REGULATOR NO. 4.**

Eight Day, Weight Time.
Dial, 12 inches.
Height............ 84 inches.

**FOYER No. 2**
Height, 32½ inches.   Dial, 16 inches.
Eight-Day Time ............List, each, $30.00
Eight-Day, Gong Strike.......List, each, 31.40

1901

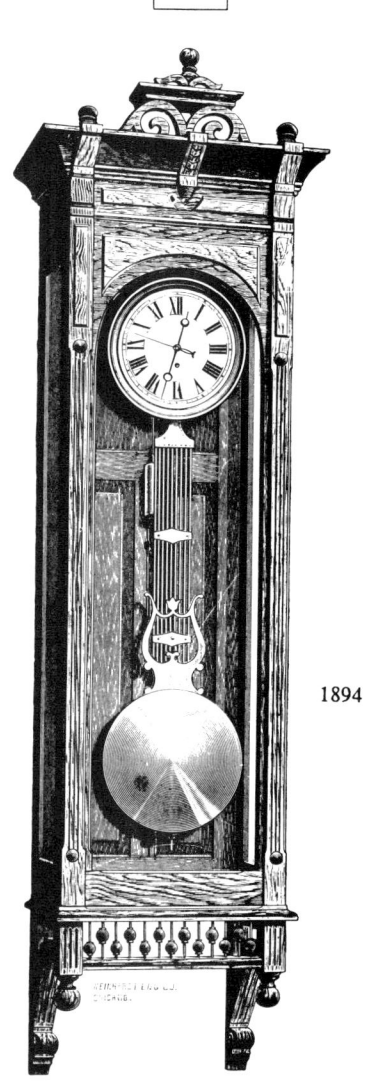

1894

**REGULATOR No. 17.**
Black Walnut, Oak or Cherry.
Dial, 12 inches.   Height, 92 inches.
Eight-Day Weight Time.   List, each, $94.00.

# ANSONIA

**96**

**FOYER No. 1**
Height, 39 inches. Dial, 18 inches.
Eight-Day Time ............List, each, $39.00
Eight-Day, Gong Strike......List, each, 41.50

*The year entered beside each clock is the year of the catalog from which the illustration was taken. Some clocks may have been produced a few years earlier.*

**98**

**99**

1901

FOYER No. 1

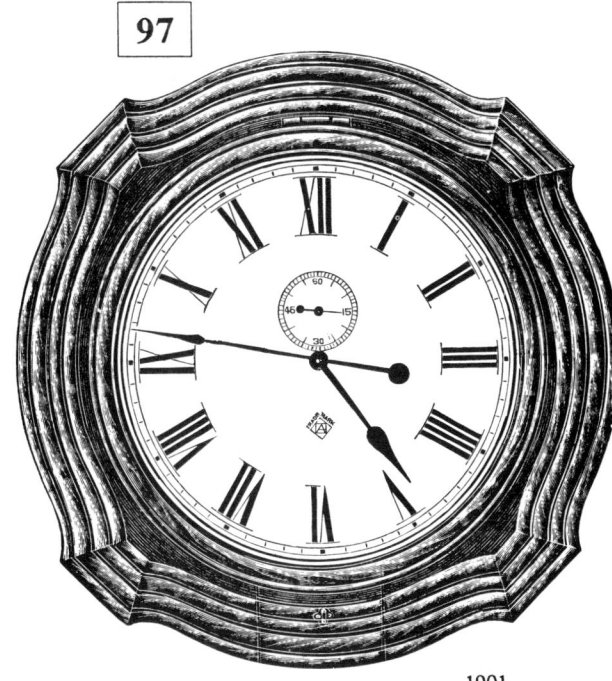

1901

**GALLERY**

Height, 28 inches. Dial, 18 inches.
Eight-Day Time ............List, each, $35.00
Eight-Day, Gong Strike......List, each, 37.50

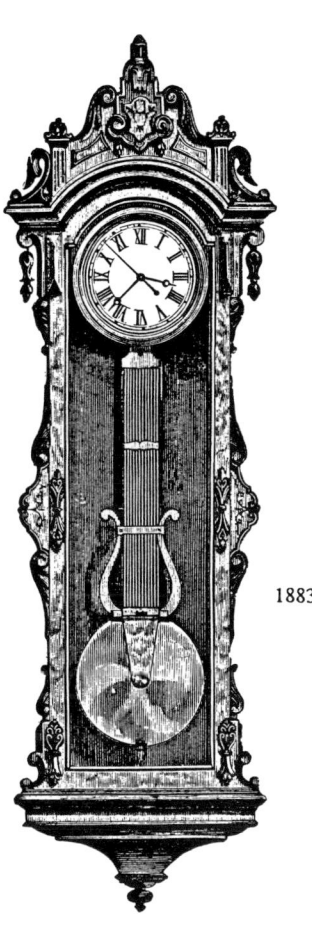

1883

**REGULATOR, No. 14.**

Eight Day, Weight Time.
Dial, 12 inches.
Height................84 inches.

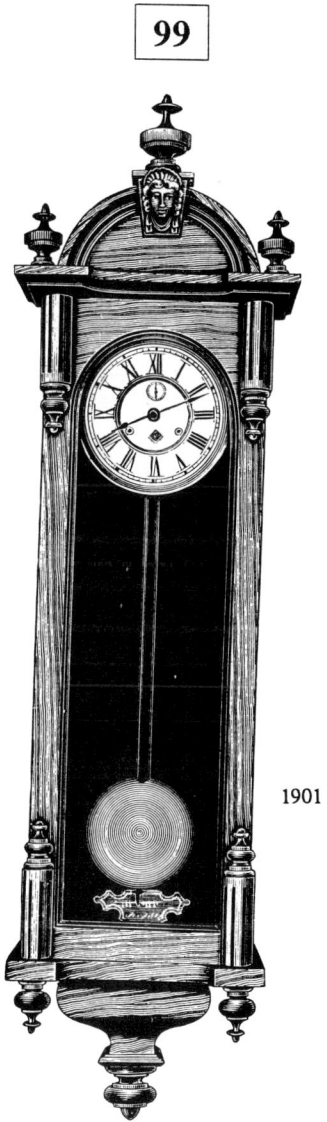

1901

**CAPITOL**

Finished in Black Walnut, Mahogany or Oak.
Dial, 8 inches.   Height, 54 inches.
Eight-Day Spring Time.
List, each ...........................$16 50
Eight-Day Spring, Half-Hour Gong Strike.
List, each ...........................$17 90
Eight-Day Weight Time.
List, each............................$20 00

# ANSONIA

*Most white metal and brass ornamentals and clock cases are gilded or silver plated. Great care must be taken when cleaning these surfaces. The thin layer of gilding or silver plating will be easily removed by cleaning solutions and strong rubbing. If you believe cleaning is necessary, hire the services of a professional and avoid the expense of re-gilding or re-plating.*

*For comprehensive listing of Ansonia Clocks, see Tran Duy Ly "Ansonia Clocks - A Guide to Identification & Prices."*

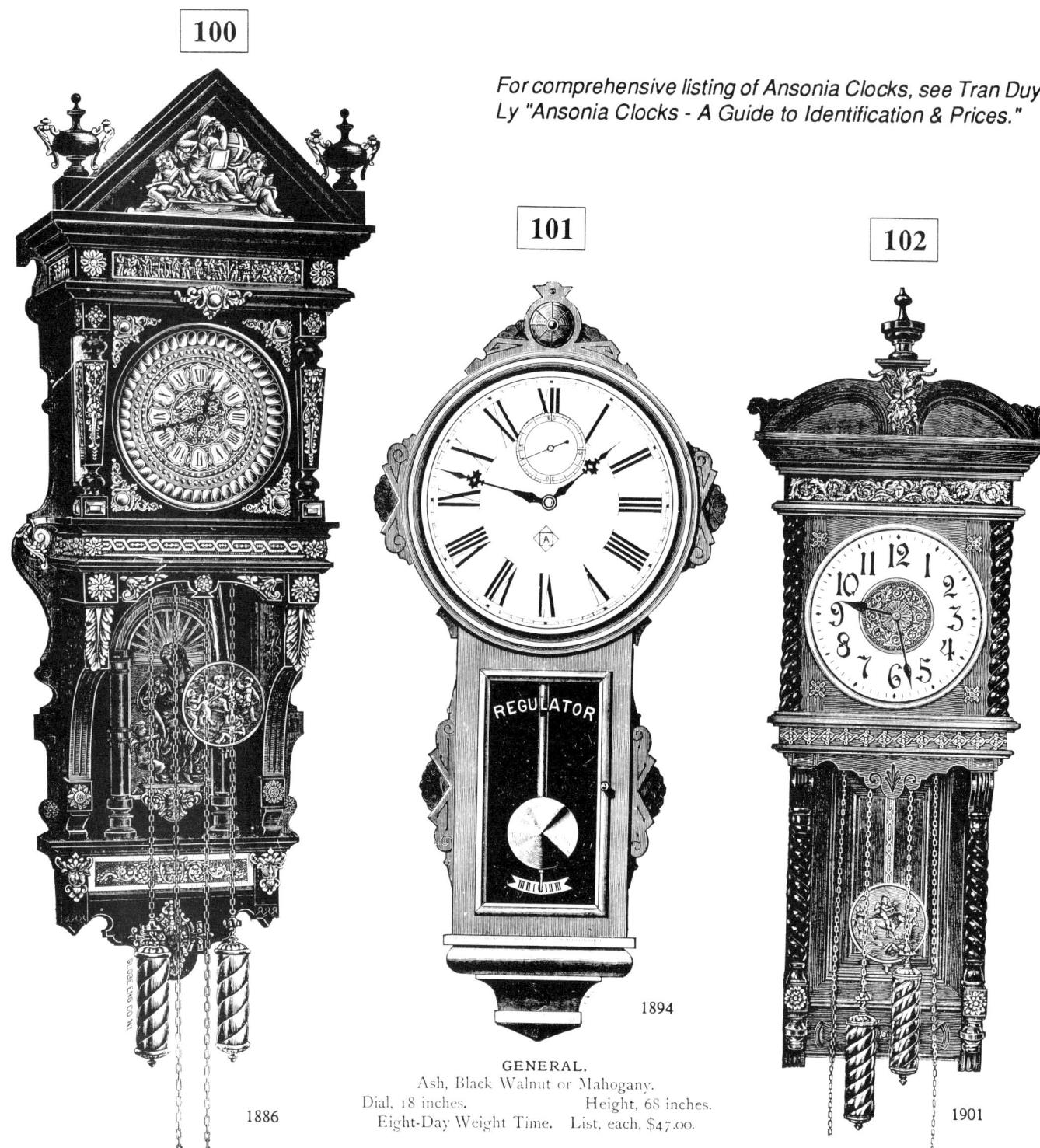

**100** — 1886

**ANTIQUE HANGING CLOCK.**
Finished in Oak or Mahogany, with Antique Brass Trimmings.
EIGHT DAY.
HALF-HOUR, CATHEDRAL GONG, WEIGHT, STRIKE.
Dial, 9½ inches. Height, 46½ inches.

**101** — 1894

**GENERAL.**
Ash, Black Walnut or Mahogany.
Dial, 18 inches. Height, 68 inches.
Eight-Day Weight Time. List, each, $47.00.

**102** — 1901

**NIOBE**
Oak.
Eight-Day, Half-Hour Gong Strike.
Silver Dial, 10 inches. Height, 45 inches.
Antique Brass Trimmings.
List, each .................. $37.50

# ANSONIA

*Except for those fabulous clocks for which you wouldn't mind cutting a hole in the ceiling, Grandfather or Hanging Clocks under eight feet are the most desirable and practical. Some manufacturers produced certain of these clocks in two styles: one to stand on the floor, and the other to hang on a wall. Of the clocks designed with two styles, the hanging style is considered to be the most valuable.*

*The year entered beside each clock is the year of the catalog from which the illustration was taken. Some clocks may have been produced a few years earlier.*

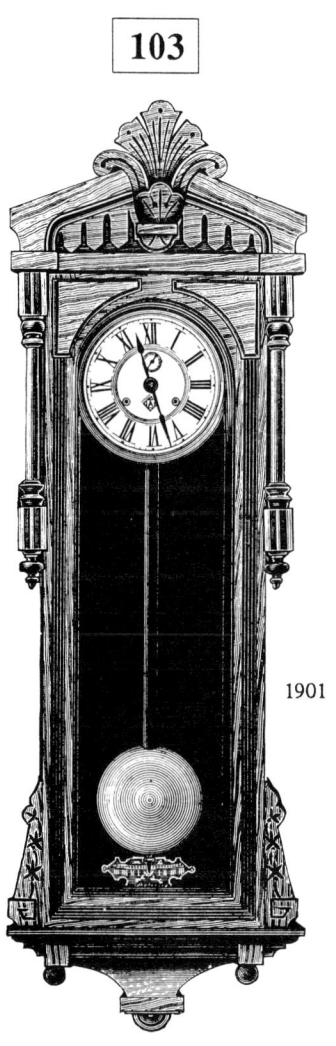

1901

1883

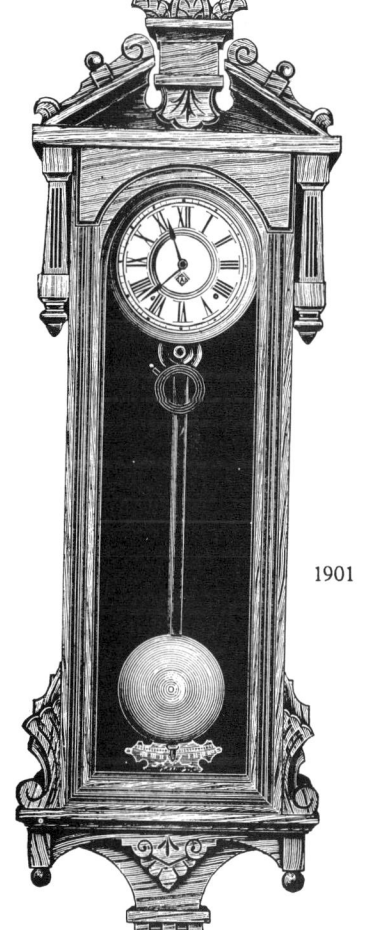

1901

### BAGDAD

Finished in Black Walnut, Mahogany or Oak.

Dial, 8 inches.　　Height, 50½ inches.

Eight-Day Spring Time.
List, each .................................$18 50
Eight-Day Spring, Half-Hour Gong Strike.
List, each .................................$19 90
Eight-Day Weight Time.
List, each .................................$22 00

### REGULATOR, No. 9.

Eight Day, Weight Time.

Dial, 16 inches.

Height................89 inches.

### PROMPT

Finished in Black Walnut, Mahogany or Oak.

Dial, 8 inches.　　Height, 50 inches.

Eight-Day Spring Time.
List, each .................................$17 50
Eight-Day Spring, Half-Hour Gong Strike.
List, each .................................$18 90
Eight-Day Weight Time.
List, each .................................$21 00

# ANSONIA

For comprehensive listing of Calendar Clocks, see Tran Duy Ly "Calendar Clocks - A Guide to Identification & Prices."

**106** — 1874

**Drop Extra Calendar.**
EIGHT DAY, STRIKE, CALENDAR.
12 Inch Dial.
Height, - - 26 Inches.

**107** — 1894

**HABANA.**
Dial, 6 inches.         Height, 26 inches.
Eight-Day Strike.   List, each, $7.50.
Eight-Day Gong.    List, each, $7.90.
Eight-Day Alarm.   List, each, $8.00.

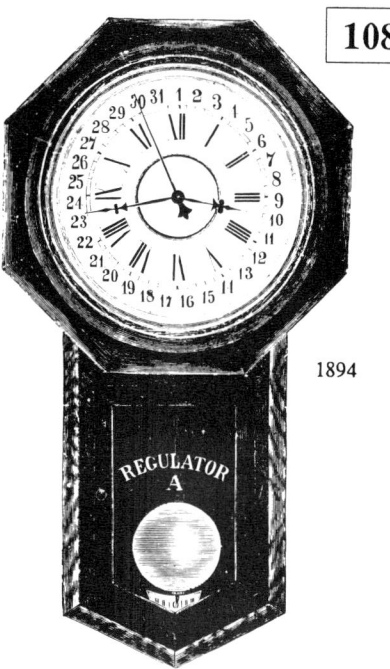

**108** — 1894

**REGULATOR A.**
CALENDAR.
Black Walnut or Ash.
Dial, 12 inches.    Height, 32 inches.
Eight-Day Time, Calendar.    List, each, $9.50.
Eight-Day Strike, Calendar.   List, each, $10.90.

**109** — 1874

**Drop Extra.**
EIGHT DAY, STRIKE.
Dial, 12 inches.   Height, 26 inches.

**110** — 1920

**24-INCH GALLERY**
Dark Oak or Mahogany
Height, 30½ Inches    Dial, 24 Inches
8-Day Hour and Half-Hour Gong Strike
When fitted with Gong Strike the Dial has no Second Hand

30-Day Time

8-Day Time

# ANSONIA

*For comprehensive listing of Ansonia Clocks, see Tran Duy Ly "Ansonia Clocks - A Guide to Identification & Prices."*

43

**111**

**112**

**113**

1914

1883

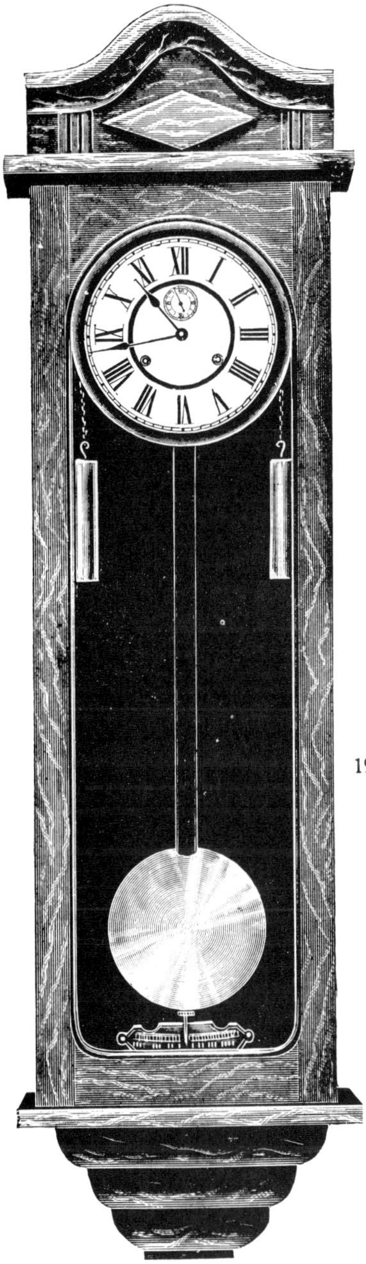

1914

### YORK

Mahogany or Oak.
Dial, 8 Inches.
Height, 46¼ Inches.
8-Day Spring Time.
**List, Each, $14.60**
8-Day Spring, Hour and Half-Hour Gong Strike.
**List, Each, $16.00**
8-Day Weight Time.
**List, Each, $18.10**

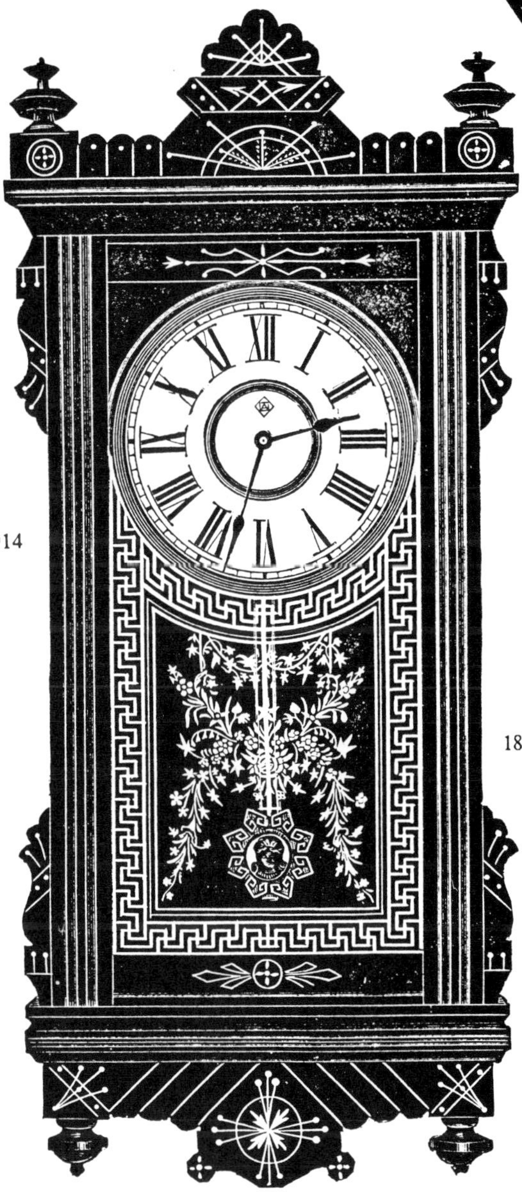

### DISPATCH.

Ebony Finish.
Eight Day, Spring Strike.
Dial, 8 inches.
Height.............................30 inches.

### COLONEL

Black Walnut, Mahogany or Oak.
Arabic or Roman Dial, 18 Inches.
Height, 61 Inches.
8-Day Weight Time.
**List, Each, $47.00**

*The year entered beside each clock is the year of the catalog from which the illustration was taken. Some clocks may have been produced a few years earlier.*

# ANSONIA

**114**

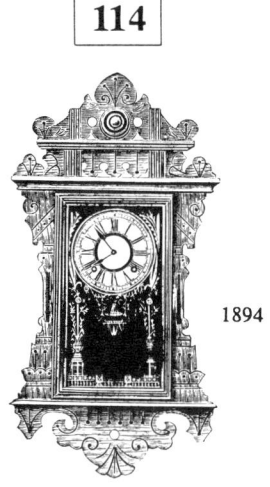
1894

**MEXICO.**
Dial, 6 inches. Height, 27¼ inches.
Eight-Day Strike. List, each, $6.50.
Eight-Day Gong. List, each, $6.90.
Eight-Day Alarm. List, each, $7.00.

**115**

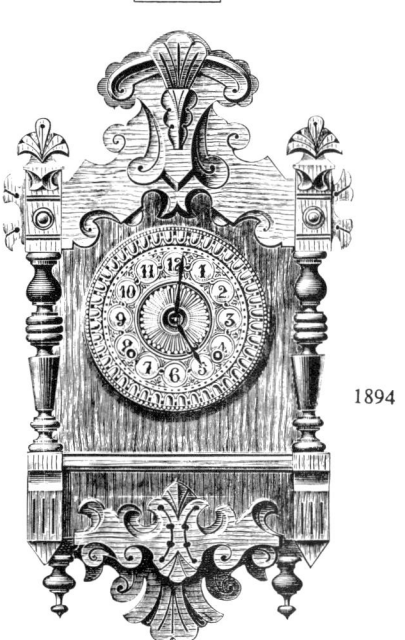
1894

**ULSTER.**
EIGHT-DAY, HALF-HOUR GONG STRIKE.
Oak.
Height, 19¼ inches. Width, 11 inches.
Sash and Dial, 5¾ inches.
List, each, $7.75.

**116**

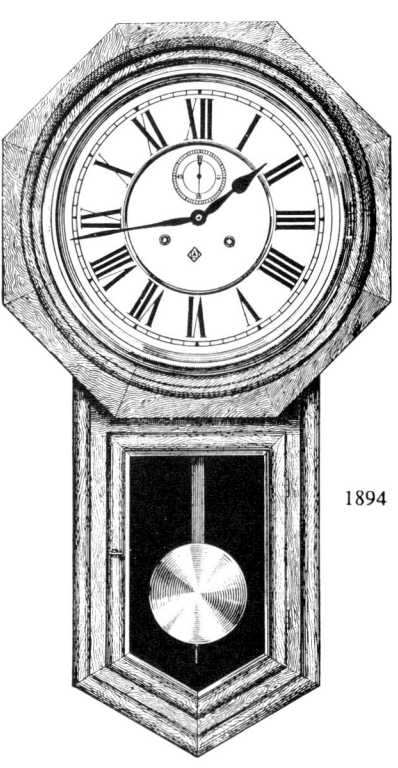
1894

**OFFICE REGULATOR**
Ash or Black Walnut.
Dial, 12 inches. Height, 32 inches.
Eight-Day Time.
List, each .................$10.40
Eight-Day, Hour or Half-Hour Gong Strike.
List, each ................. 11.80
Packed singly and strapped in packages of 3.

**117**

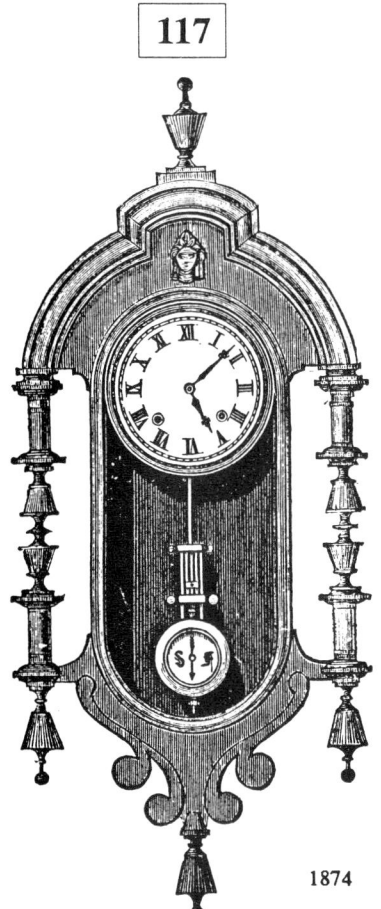
1874

**Roman Gothic.**
Eight Day Time.
" Strike.
Height, - - - 30 Inches.

**118**

1894

**UPTON.**
EIGHT-DAY, HALF-HOUR GONG STRIKE.
Oak.
Height, 20 inches. Width, 10¼ inches.
Sash and Dial, 5¾ inches.
List, each, $7.75.

**119**

1894

**TAMPICO.**
Dial, 6 inches. Height, 25 inches.
Eight-Day Strike. List, each, $6.50.
Eight-Day Gong. List, each, $6.90.
Eight-Day Alarm. List, each, $7.00.

# ANSONIA

**120**

*For comprehensive listing of Ansonia Clocks, see Tran Duy Ly "Ansonia Clocks - A Guide to Identification & Prices."*

**UTOPIA.**
EIGHT-DAY, HALF-HOUR GONG STRIKE.
Oak.
Height, 18¼ inches.   Width, 10½ inches.
Sash and Dial, 5¾ inches.
List, each, $7.75.

1894

UTOPIA.

**121**

1874

**HANGING MONOGRAM**
WALNUT.   Spring.
EIGHT DAY, TIME.   EIGHT DAY, STRIKE
Height, 29 inches.

**123**

**124**

1874

**New York.**
SPRING.
Eight Day, Spring Time.
Eight Day, Spring Strike.

Height,   -   26 Inches.

1914

**STANDARD REGULATOR**
Height 37 inches   Width 15¼ inches
Eight-Day Weight Time.   Has Maintaining Power.   Seventy-six Beats to the Minute.

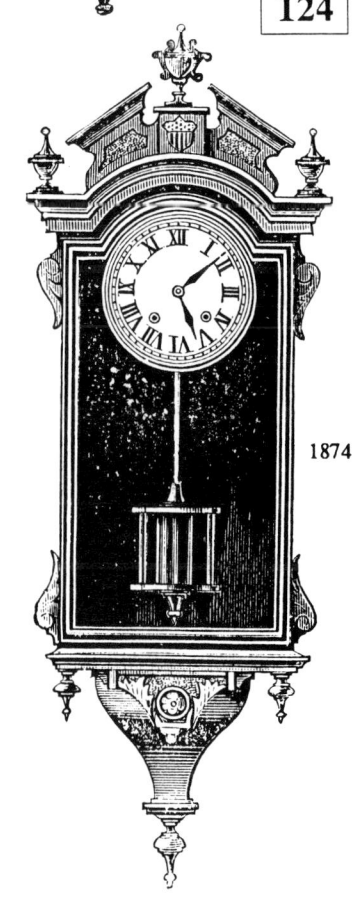

1874

**Ansonia.**
Eight Day Time.
"   Strike.

Height,   -   -   37 Inches.

45

# ANSONIA

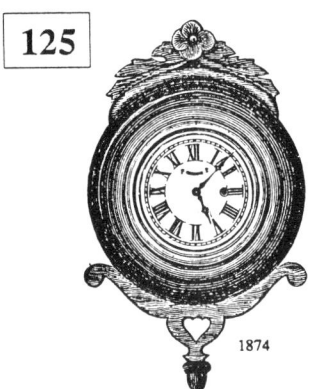

*The year entered beside each clock is the year of the catalog from which the illustration was taken. Some clocks may have been produced a few years earlier.*

**125** — 1874

## Bouquet Lever.
One Day Time.
Four Inch Dial

**126** — 1901

### 10-INCH KOBE, CAL.
Mosaic.
Dial, 10 inches. Height, 21½ inches.
Eight-Day Time.
List, each ............................... $7.20
Eight-Day, Hour or Half-Hour Strike.
List, each ............................... 8.30
Gong, 30 cents List Additional.

**127** — 1886

## MECCA.
Black Walnut or Mahogany.
EIGHT DAY, WEIGHT, TIME.
Porcelain Dial, 8 inches. Height, 58 inches

**128** — 1906

### QUEEN ISABELLA
Oak or Black Walnut.
Dial, 8 inches. Height, 38½ inches.
Eight-Day Time.
List, each .........................$11.00
Eight-Day, Half-Hour Gong Strike.
List, each ............................ 12.40

**129** — 1880

**BROOKLYN.**   Dial - - - - 12 inches.

# ANSONIA

**130**

**QUEEN MAB**
Oak or Black Walnut.
Dial, 8 inches. Height, 36¼ inches.
Eight-Day Time.
List, each .................................$11.00
Eight-Day, Half-Hour Gong Strike.
List, each ................................. 12.40

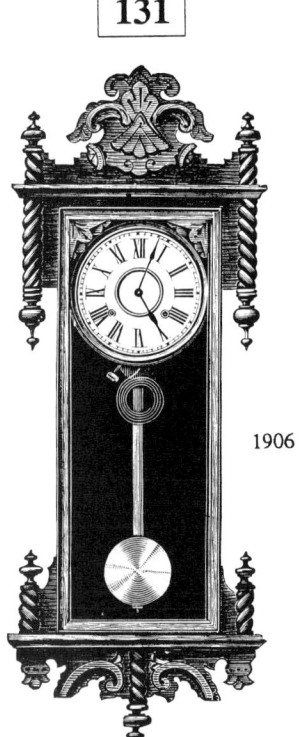

**131**

1906

**PARA**
Black Walnut.
Dial, 8 inches. Height, 39 inches.
Eight-Day Time.
List, each .................................$11.50
Eight-Day, Half-Hour Gong Strike.
List, each ................................. 12.90

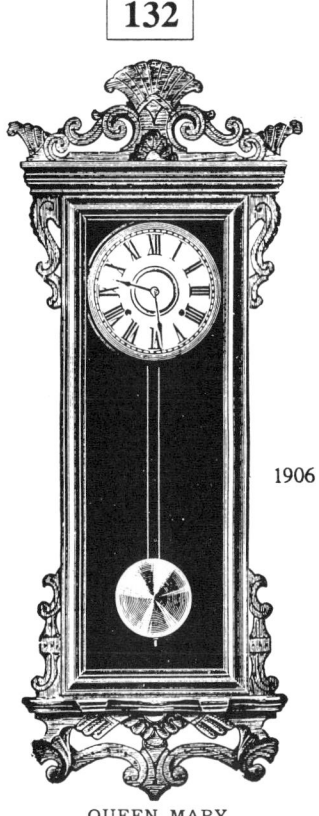

**132**

1906

**QUEEN MARY**
Oak or Black Walnut.
Dial, 8 inches. Height, 42 inches.
Eight-Day Time.
List, each .................................$10.50
Eight-Day, Half-Hour Gong Strike.
List, each ................................. 11.90

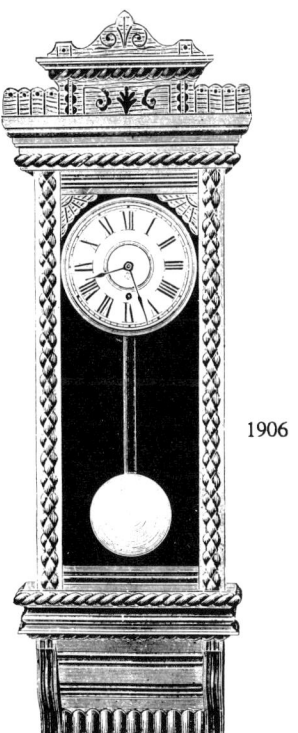

**133**

1906

**QUEEN JANE**
Oak or Black Walnut.
Dial, 8 inches. Height, 41 inches.
Eight-Day Time.
List, each .................................$12.00
Eight-Day, Half-Hour Gong Strike.
List, each ................................. 13.40

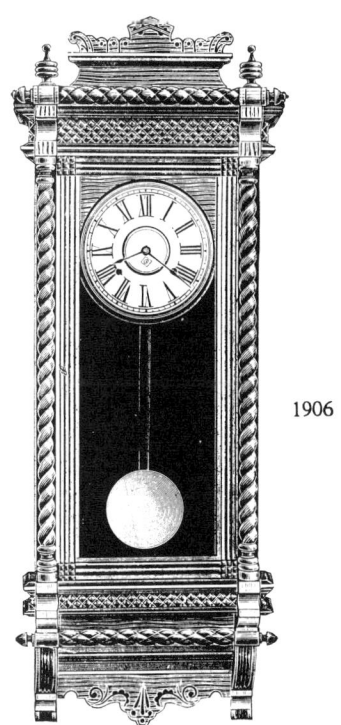

**134**

1906

**QUEEN CHARLOTTE**
Oak or Black Walnut.
Dial, 8 inches. Height, 40 inches.
Eight-Day Time.
List, each .................................$15.00
Eight-Day, Half-Hour Gong Strike.
List, each ................................. 16.40

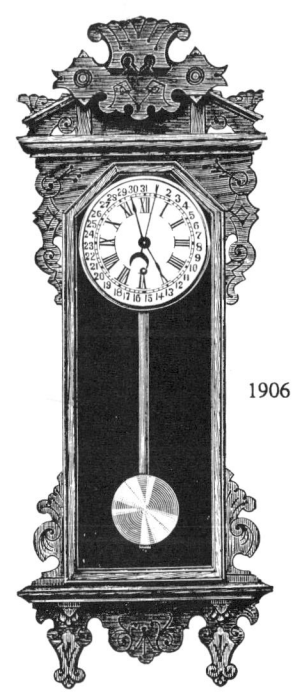

**135**

1906

**BAHIA**
Oak.
Dial, 8 inches. Height, 38½ inches.
Eight-Day Time.
List, each .................................$11.50
Eight-Day, Half-Hour Gong Strike.
List, each ................................. 12.90

# ANSONIA

For comprehensive listing of Ansonia Clocks, see Tran Duy Ly "Ansonia Clocks - A Guide to Identification & Prices."

**FORREST.**
Black Walnut.
Dial, 8 inches.    Height, 41 inches.
Eight-Day Spring Time.    List, each, $9.00.
Eight-Day Spring Gong Strike.    List, each, $11.00.

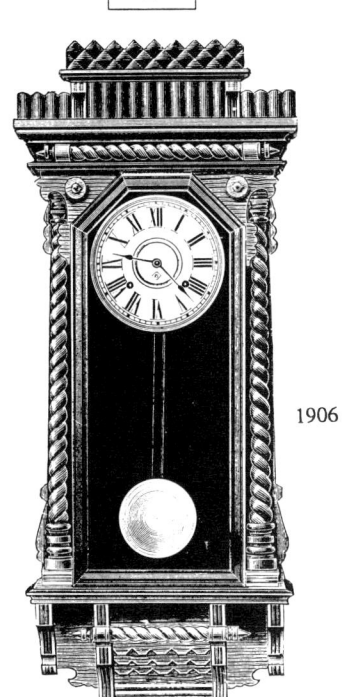

1906

**QUEEN ANNE**
Oak or Black Walnut.
Dial, 8 inches.   Height, 40¼ inches.
Eight-Day Time.
List, each ............................. $14.50
Eight-Day, Half-Hour Gong Strike.
List, each .............................  15.90

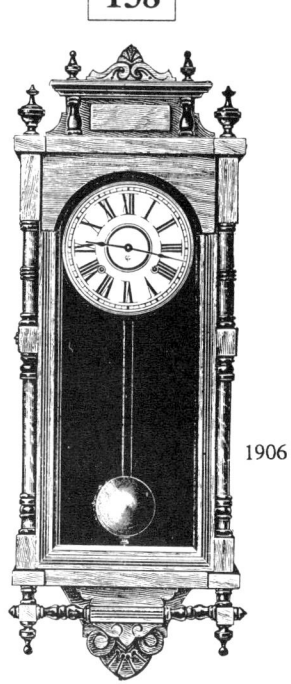

1906

**QUEEN ELIZABETH**
Oak, Black Walnut or Mahogany.
Dial, 8 inches.   Height, 37 inches.
Eight-Day Time.
List, each ............................. $11.00
Eight-Day, Half-Hour Gong Strike.
List, each .............................  12.40

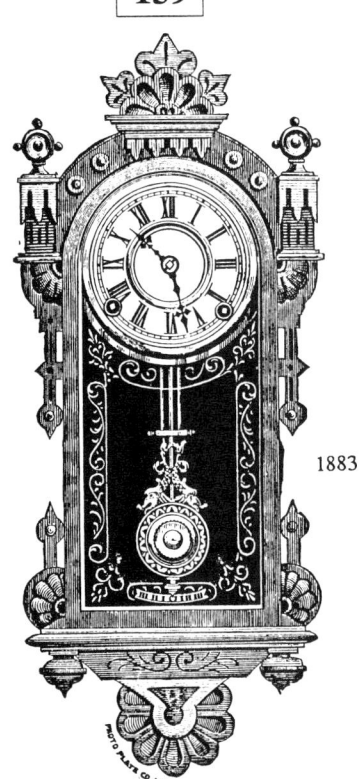

1883

**MISSISSIPPI.**
Eight Day, Spring Strike.
Dial, 6 inches.
Height.................... 28 inches.

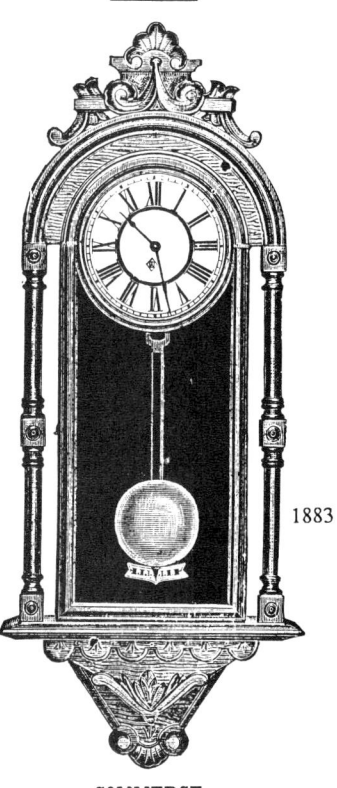

1883

**COMMERCE.**
Eight Day Spring Time, Eight Day Spring Strike.
Dial, 8 inches.
Height.................... 36 inches.

# ANSONIA

*Most white metal and brass ornamentals and clock cases are gilded or silver plated. Great care must be taken when cleaning these surfaces. The thin layer of gilding or silver plating will be easily removed by cleaning solutions and strong rubbing. If you believe cleaning is necessary, hire the services of a professional and avoid the expense of re-gilding or re-plating.*

141

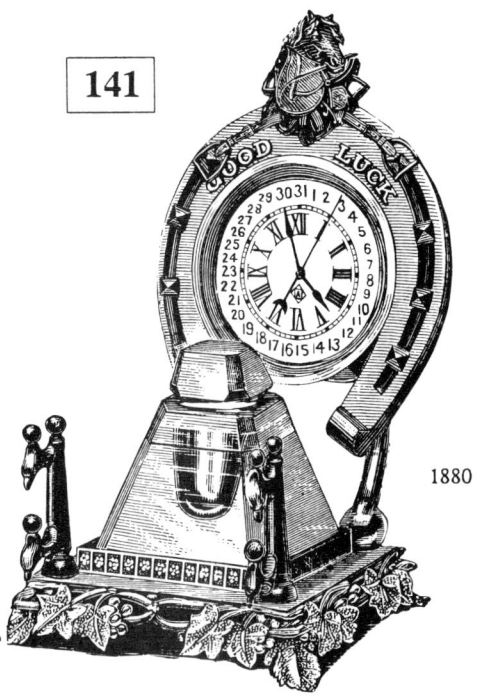

1880

142

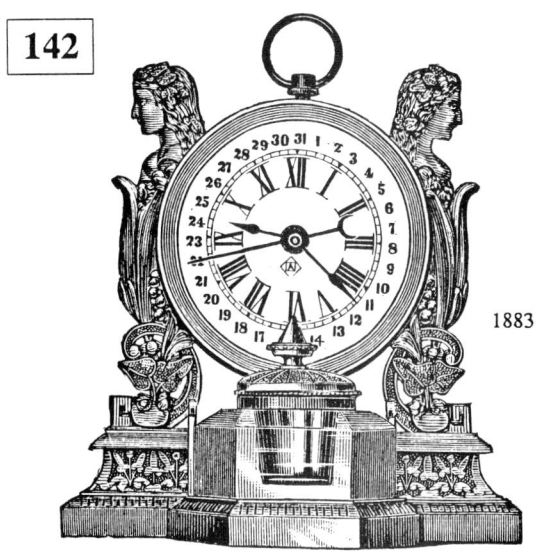

1883

### LILY INK.
Bronze and Nickel Finish.
One Day, Time, Calendar.
Dial, 4 inches.
Height..............................7 inches.

### GOOD LUCK INKSTAND.
Dial - - - - - - 3 inches.

143

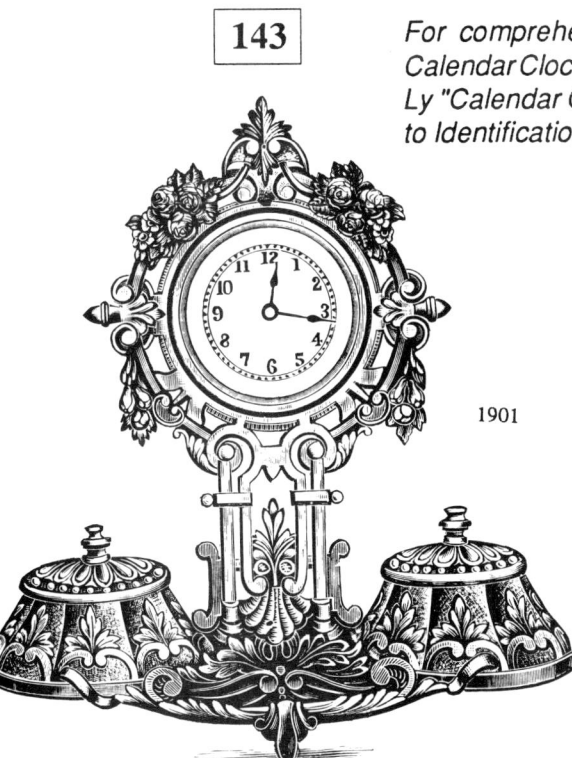

1901

*For comprehensive listing of Calendar Clocks, see Tran Duy Ly "Calendar Clocks - A Guide to Identification & Prices."*

144

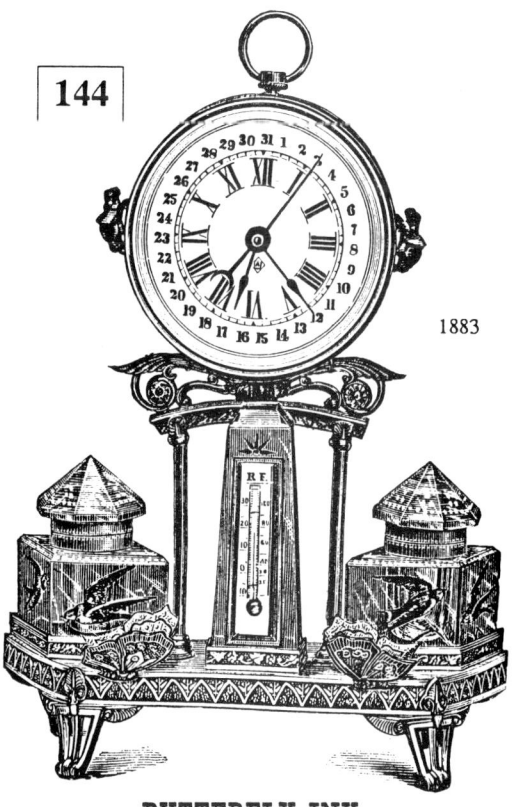

1883

### BEE INK
One-Day Time.
Dial, 2 inches.   Height, 8¼ inches.
Finished in Gilt.
Plain Dial ..................List, each, $5 25
Fancy Dial ................List, each,  5 50

### BUTTERFLY INK.
Bronze or Silver, with Nickel Finish, Colored Enameled
Bottles and Thermometer.
One Day, Time, Calendar.
Dial, 4 inches.
Height..............................10½ inches.

# ANSONIA

*For comprehensive listing of Ansonia Clocks, see Tran Duy Ly "Ansonia Clocks - A Guide to Identification & Prices."*

1906

**CAMPANIA**

Height, 15¾ inches.  Dial, 5 inches.  Width, 10½ inches.
Visible Escapement.
Real Bronze.

Finished in Brass .................................... List, each, $39.00
Finished in Rich Gold ................................ List, each, 55.00

1906

**GERMANIC**

Height, 14 inches.  Dial, 4¼ inches.  Width, 8 inches.
Real Bronze.

Finished in Brass .................................... List, each, $26.00
Finished in Syrian Bronze ............................ List, each, 28.00
Finished in Rich Gold ................................ List, each, 35.00
Finished in Rich Gold, with Jeweled Sash ............. List, each, 38.00

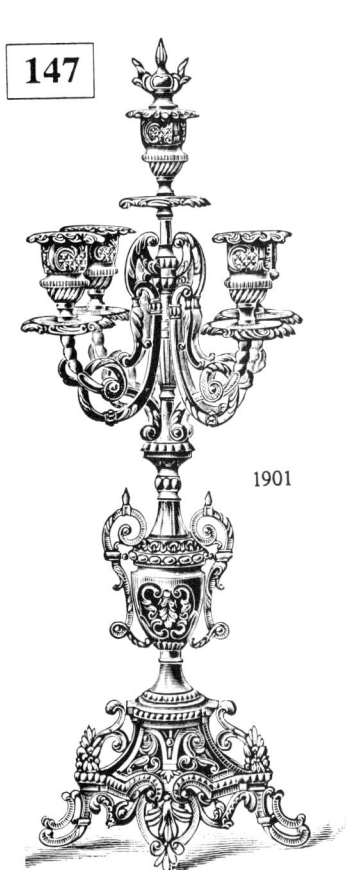

1901

**CANDELABRA No. 1163**

Height, 18¾ inches.

Finished in Brass ........ List, per pair, $27 50
Finished in Rich Gold .... List, per pair, 37 00

1901

**TAURIC**

Eight-Day, Half-Hour Gong Strike.
Height, 17½ inches.  Width, 10 inches.
Dial, 4¼ inches.
French or Rococo Sash, Beveled Glass, Porcelain
Dial, Arabic or Roman.

Finished in Brass ............ List, each, $26 00
 Set complete, $53 50
Finished in Rich Gold ........ List, each, $35 00
 Set complete, $72 00
Jeweled Setting, $3 00 List additional.

1901

**CANDELABRA No. 1163**

Height, 18¾ inches.

Finished in Brass ........ List, per pair, $27 50
Finished in Rich Gold .... List, per pair, 37 00

# ANSONIA

*Only clocks in ORIGINAL MINT condition and in good running order will command top prices. A replaced or damaged dial, missing parts, dull finish, worn-off gold gild, re-painted or chipped beveled glass will reduce prices accordingly.*

**149**

**150**

1906

**CANDELABRA No. 1158**

Height, 21½ inches.

Real Bronze.

Finished in Brass..........List, per pair, $60.00
Finished in Rich Gold.....List, per pair, 70.50

**TEUTONIC**

Eight-Day, Half-Hour Gong Strike.
Height, 17½ inches. Width, 12¼ inches.
French or Rococo Sash, Beveled Glass, Porcelain Dial, 5 inches, Arabic or Roman,
Visible Escapement.
Real Bronze.

Finished in Brass.

List Price of Clock.....................$34.50
List Price of Set Complete............. 94.50

Finished in Rich Gold.

List Price of Clock..................... 48.50
List Price of Set Complete.............119.00

**CANDELABRA No. 1158**

Height, 21½ inches.

Real Bronze.

Finished in Brass..........List, per pair, $60.00
Finished in Rich Gold.....List, per pair, 70.50

**151**

1894

**PEORIA.**
Silver or Barbedienne.
Height, 19½ inches. Width, 11½ inches.
List, each, $45.00.

*Most white metal and brass ornamentals and clock cases are gilded or silver plated. Great care must be taken when cleaning these surfaces. The thin layer of gilding or silver plating will be easily removed by cleaning solutions and strong rubbing. If you believe cleaning is necessary, hire the services of a professional and avoid the expense of re-gilding or re-plating.*

**152**

1894

**EMPORIA.**
Silver or Barbedienne.
Height, 18 inches. Width, 9 inches.
List, each, $38.00.

# ANSONIA

**CANDELABRA No. 1156**

Height, 22½ inches.
Finished in Rich Gold.
List, per pair ..........................$70 00

**MAJESTIC**

Eight-Day, Half-Hour Gong Strike.
Real Bronze.
Height, 19 inches.   Width, 10 inches.
Dial, 5 inches.
French or Rococo Sash, Beveled Glass, Porcelain
Visible Escapement Dial, Arabic or Roman.
Finished in Rich Gold.
List Price of Clock .....................$68 00
List Price of Set complete .............138 00

**LATONIA, WITH FIGURE 1091.**
Barbedienne or Silver.
Height, 19 inches.   Width, 19 inches.
Dial, 4½ inches.   List, each, $62.50.
List, without figure, $52.00.

**CANDELABRA No. 1156**

Height, 22½ inches.
Finished in Rich Gold.
List, per pair ..........................$70 00

**LATONIA, WITH FIGURE 1091.**

**ELITE**

Height 15½ inches.   Width, 9 inches.
Dial, 5 inches.
Finished in Barbedienne or Syrian Bronze,
List, each, $35.00
Finished in Rich Gold........List, each, 42.50

# ANSONIA

**157**

**CANDELABRA No. 1066**

Height, 22 inches.
Finished in Brass.
List, per pair .......................... $67.50

**158**

**REGENT**

Eight-Day, Half-Hour Gong Strike.
Height 22½ inches. Width, 9½ inches.
French or Rococo Sash, Beveled Glass, Porcelain Dial, 5 inches, Arabic or Roman,
Visible Escapement.
Finished in Brass.
List Price of Clock .................... $67.50
List Price of Set Complete ............ 135.00

**CANDELABRA No. 1066**

Height, 22 inches.
Finished in Brass.
List, per pair .......................... $67.50

**159**

**MIRANDA AND FIGURE No. 1008**
Height, 16½ inches. Width, 11¼ inches.
Dial, 5 inches.
List ................................... $24.30
List Price of Clock without Figure ...... 21.00

*Most white metal and brass ornamentals and clock cases are gilded or silver plated. Great care must be taken when cleaning these surfaces. The thin layer of gilding or silver plating will be easily removed by cleaning solutions and strong rubbing. If you believe cleaning is necessary, hire the services of a professional and avoid the expense of re-gilding or re-plating.*

**LYDIA**
Height, 19½ inches. Width, 11¾ inches.
Dial, 5½ inches.
Finished in Barbedienne or Syrian Bronze,
List, each, $40.00

**160**

**LYDIA.**

# ANSONIA

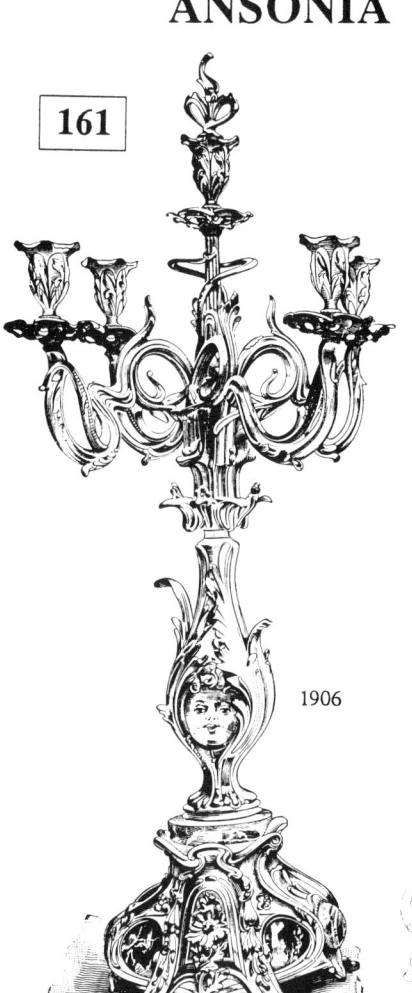

**CANDELABRA No. 1179**
Height, 28½ inches.
Brazilian Green Onyx Inlay.
Finished same as Clock.
List, per pair............................$64.00

**SUPERBA AND No. 1180**
Eight-Day, Half-Hour Gong Strike.
Height, 28 inches. Width, 15 inches.
French or Rococo Sash, Beveled Glass, Porcelain Dial, 4¾ inches, Arabic or Roman.
Brazilian Green Onyx Inlay.
Finished as follows: Art Nouveau, Barbedienne, Queen's Gray, Real Bronze, Syrian.
List Price of Clock.....................$66.00
List Price of Set Complete.............130.00

**RENAISSANCE.**
EIGHT-DAY, HALF-HOUR GONG STRIKE.
Brass, Silver, or Brass Antique.    List price, Set, $80.00.

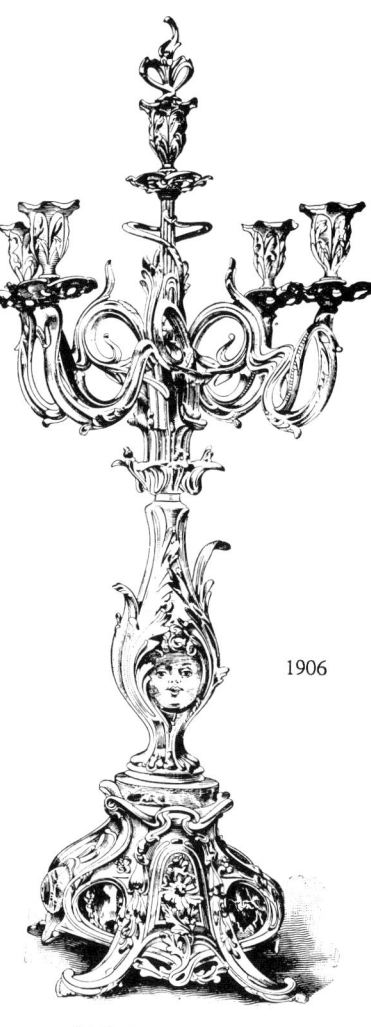

**CANDELABRA No. 1179**
Height, 28½ inches.
Brazilian Green Onyx Inlay.
Finished same as Clock.
List, per pair............................$64.00

**FRANCONIA.**
Silver or Barbedienne.
Height, 18 inches.    Width, 11 inches.
List, each, $42.00.

**FRANCONIA.**

# ANSONIA

*Only clocks in ORIGINAL MINT condition and in good running order will command top prices. A replaced or damaged dial, missing parts, dull finish, worn-off gold gild, re-painted or chipped beveled glass will reduce prices accordingly.*

**165**

1906

**HENRY IV**
Height, 18 inches.  Width, 12½ inches.
Dial, 5 inches.
Finished in Antique Brass or Gilt.
List, each ..................................$43.00

**167**

1901

**ROYAL**
Eight-Day, Half-Hour Gong Strike.
Height, 17¾ inches.    Width, 9½ inches.
Dial, 4¼ inches.
French or Rococo Sash, Beveled Glass, Porcelain
Dial, Arabic or Roman.
Finished in Rich Gold with Inlaid Decorated
Porcelain Panel.
List Price of Clock, $28.75.   Set complete, $57.50.
Jeweled Setting, $3.00 List additional.
With Hand Painted Figure Panel, $2.00 List additional.

**166**

1906

**SIBYL AND INDUSTRY**
Height, 19¼ inches.  Width, 10¾ inches.
Dial, 4¼ inches.
List, each ..................................$40.00

**168**

1906

**UNDINE AND FLORA**
Height, 20¼ inches.  Dial, 4¾ inches.  Width, 16½ inches.
Finished in Art Nouveau, Barbedienne, Queen's Gray, Real Bronze,
Syrian.
List, each ..................................$47.00

**169**

1906

**UNDINE AND CIRCE**
Height, 24 inches.  Dial, 4¾ inches.  Width, 16½ inches.
Finished in Art Nouveau, Barbedienne, Queen's Gray, Real Bronze,
Syrian.
List, each ..................................$52.00

# ANSONIA

*For comprehensive listing of Ansonia Clocks, see Tran Duy Ly "Ansonia Clocks - A Guide to Identification & Prices."*

**170**

**FLORIDA WITH FIGURE, No. 1078**

Eight-Day, Half-Hour Gong Strike.
Height, 31½ inches.   Dial, 5½ inches.   Width, 12¼ inches.
French or Rococo Sash, Beveled Glass, Porcelain Visible Escapement Dial, Arabic or Roman.
Finished in Barbedienne or Syrian Bronze........List, complete, $102.50

**171**

**FLORIDA AND GROUP, No. 1089.**
Barbedienne or Silver.
Height, 28 inches.   List, each, $100.00.

**172**

**FLORIDA WITH GROUP, No. 1090**

Eight-Day, Half-Hour Gong Strike.
Height, 35 inches.   Dial, 5½ inches.   Width, 12¼ inches.
French or Rococo Sash, Beveled Glass, Porcelain Visible Escapement Dial, Arabic or Roman.
Finished in Barbedienne or Syrian Bronze........List, complete, $95.00

# ANSONIA

*Most white metal and brass ornamentals and clock cases are gilded or silver plated. Great care must be taken when cleaning these surfaces. The thin layer of gilding or silver plating will be easily removed by cleaning solutions and strong rubbing. If you believe cleaning is necessary, hire the services of a professional and avoid the expense of re-gilding or re-plating.*

*Only clocks in ORIGINAL MINT condition and in good running order will command top prices. A replaced or damaged dial, missing parts, dull finish, worn-off gold gild, re-painted or chipped beveled glass will reduce prices accordingly.*

**ETRURIA AND No. 1184**
Eight-Day, Half-Hour Gong Strike.
Height, 28½ inches. Width, 16½ inches.
French or Rococo Sash, Beveled Glass, Porcelain Dial, 4¾ inches, Arabic or Roman.
Brazilian Green Onyx Inlay.
Finished as follows: Art Nouveau, Barbedienne, Queen's Gray, Real Bronze, Syrian.
List Price of Clock.....................$57.50
List Price of Set Complete.............120.00

**SIBYL AND WINTER**
Eight-Day, Half-Hour Gong Strike.
Height, 24 inches. Width, 10¾ inches.
Dial, 4¼ inches.
French or Rococo Sash, Beveled Glass, Porcelain Dial, Arabic or Roman.
Finished in Barbedienne or Syrian Bronze
List, each, $45 00

**SAPPHO**
Eight-Day, Half-Hour Gong Strike.
Height, 26 inches. Width, 9¼ inches.
Dial, 4¼ inches.
French or Rococo Sash, Beveled Glass, Porcelain Dial, Arabic or Roman.
Finished in Barbedienne or Syrian Bronze
List, each, $35 00

# ANSONIA

*The year entered beside each clock is the year of the catalog from which the illustration was taken. Some clocks may have been produced a few years earlier.*

**176**

### SIBYL AND GLORIA
Eight-Day, Half-Hour Gong Strike.
Height, 27½ inches.   Dial, 4¼ inches.   Width, 10¾ inches.
French or Rococo Sash, Beveled Glass, Porcelain Dial, Arabic or Roman.
Finished in Barbedienne or Syrian Bronze...List, each, $52 50

1901

SIBYL AND GLORIA

**177**

1901

### HEBE AND PIZARRO
Eight-Day. Lever Movement.   Half-Hour Gong Strike.
Height, 25 inches.   Dial, 4¾ inches.   Width, 13½ inches.
French or Rococo Sash, Beveled Glass, Porcelain Dial, Arabic or Roman.
Finished in Barbedienne or Syrian Bronze ...........List, each, $55 00

**178**

1901

### OLYMPIA AND CHLORIS
Eight-Day, Lever Movement.   Half-Hour Gong Strike.
Height, 24¾ inches.   Dial, 4¾ inches.   Width, 15½ inches.
French or Rococo Sash, Beveled Glass, Porcelain Dial, Arabic or Roman.
Finished in Syrian Bronze with Rich Gold Trimmings......List, $55 00
Clock Finished in Syrian Bronze with Rich Gold Trimmings,
  Figure in Rich Gold ................................List,   67 00
Finished all in Rich Gold ................................List,   75 00

# ANSONIA

**EUREKA, WITH GROUP No. 1099**

Eight-Day, Half-Hour Gong Strike.
Height, 23 inches.   Dial, 5 inches.   Width, 21¾ inches.
French or Rococo Sash, Beveled Glass, Porcelain Visible Escapement
Dial, Arabic or Roman.
Finished in Barbedienne or Syrian Bronze. . . . . . .List, complete, $64 75

**HECTOR**

Height, 18 inches.  Width, 9 1/4 inches.
Dial, 4 1/4 inches.
List, each .............................. $31.00

**No. 1053.**
Height, 13½ inches.

**ETRUSCAN.**
Brass or Silver.
EIGHT DAY TIME.
Height, 14½ inches.

**No. 1053.**
Height, 13½ inches.

**APOLLO**

Eight-Day, Half-Hour Gong Strike.
Height, 16¼ inches.   Width, 8¼ inches.
Dial, 4¾ inches.
French Sash, Beveled Glass, Porcelain Dial,
Arabic or Roman.
Finished in Blue, Green or Red.
Gilt Trimmings.
List, each ............................$22 00

# ANSONIA

*For comprehensive listing of Ansonia Clocks, see Tran Duy Ly "Ansonia Clocks - A Guide to Identification & Prices."*

Only clocks in ORIGINAL MINT condition and in good running order will command top prices. A replaced or damaged dial, missing parts, dull finish, worn-off gold gild, re-painted or chipped beveled glass will reduce prices accordingly.

**EUREKA, WITH FIGURE 1085.**
Barbedienne or Silver.
Height, 24 inches.  Width, 22½ inches.
Dial, 4½ inches.   List, each, $80.75.
List, without figure, $61.00.

**SUPERBA AND No. 1181**
Eight-Day, Half-Hour Gong Strike.
Height, 29 inches.  Width, 15 inches.
French or Rococo Sash, Beveled Glass, Porcelain Dial, 4¾ inches, Arabic or Roman.
Brazilian Green Onyx Inlay.
Finished as follows: Art Nouveau, Barbedienne, Queen's Gray, Real Bronze, Syrian.
List Price of Clock.....................$66.00
List Price of Set Complete.............130.00

**LATONIA, WITH FIGURE No. 1012**
Eight-Day, Half-Hour Gong Strike.
Height, 19 inches.  Dial, 5 inches.  Width, 19 inches.
French or Rococo Sash, Beveled Glass, Porcelain Visible Escapement Dial, Arabic or Roman.
Finished in Barbedienne or Syrian Bronze........List, complete, $55.00

# ANSONIA

**MONARCH.**
Bronze Ornaments, French Sash.
Eight Day, Strike.
Dial, 6 inches.
Height........24½ inches.

**TURRET (ANS.)**
WALNUT.   Spring.
Height, 21 inches.   EIGHT DAY, STRIKE.

**TRIUMPH.**
Eight Day, Strike, Silver Cupids, Plate Glass Mirrors,
Bronze Ornaments.
Dial, 7 inches.
Height........................24½ inches.

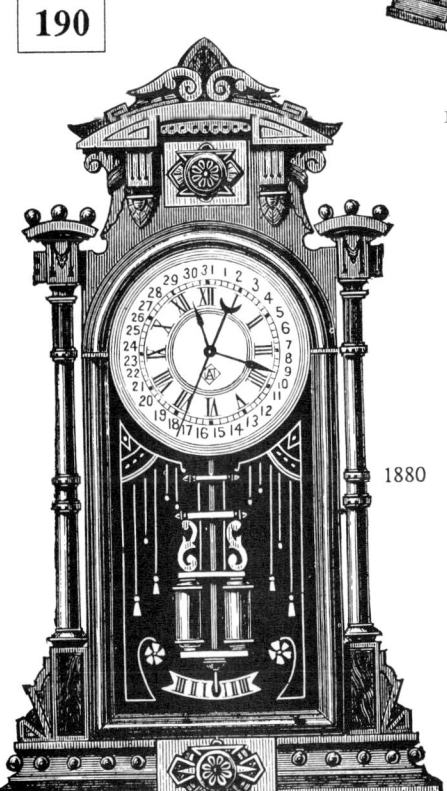

**CARLOS.**
Eight Day, Spring Strike.
Dial, 6 inches.
Height........................22 inches.

**WINDSOR.**
Eight Day, Spring Strike.
Plate Glass Mirrors, Silver Cupids.
Dial, 5 inches.
Height........................21⅜ inches.

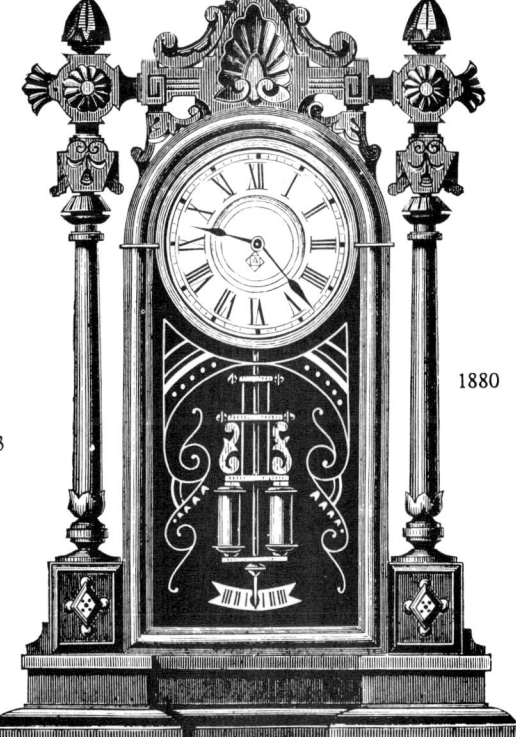

**BELGRADE.**
Eight Day, Spring Strike.
Dial, 6 inches.
Height........................23 inches.

# ANSONIA

### 193

1901

**PAPER WEIGHT**

One-Day Time.
Dial, 1½ inches.
Porcelain Dial, Gilt Centre, Arabic or Roman.
Antique Brass or Silver.
List, each .................................$3 50

### 194

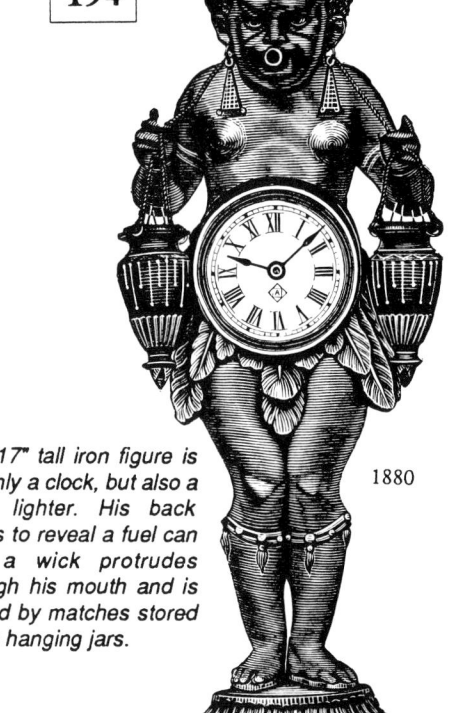

1880

*This 17" tall iron figure is not only a clock, but also a cigar lighter. His back opens to reveal a fuel can and a wick protrudes through his mouth and is lighted by matches stored in the hanging jars.*

**1011.**

Height, - - 17 inches.

### 195

1901

**EAGLE**

One-Day Time.
Watch Movement.
Dial, 1½ inches, Arabic or Roman.
Jeweled Sash.
Height, 8½ inches.
No. 1. All Rich Gold, Green Onyx Base,
List, each, $13.00
No. 3. Rich Gold Bird, Bronze Tree,
Green Onyx Base.........List, each, 11.50

### 196

1883

**SONNET.**

Finished in Bronze, Silver or Gilt, with Hand-painted Decorations.
8 Day, Time.
Height................................... 20 inches.

### 197

1901

**PSYCHE**

One-Day Time.
Watch Movement.
Dial, 1½ inches, Arabic or Roman.
Jeweled Sash.
Height, 7 inches.
Finished in Syrian Bronze, Onyx Base.
List, each, $10.80
Finished in Rich Gold, Onyx Base,
List, each, 12.00

### 198

1901

**HARP**

One-Day Time.
Watch Movement.
Dial, 1½ inches, Arabic or Roman.
Height, 9 inches.
Finished in Rich Gold, Jeweled,
List, each, $9.75

# ANSONIA

Most white metal and brass ornamentals are gilded or silver plated. Great care must be taken when cleaning these surfaces. The thin layer of gilding or silver plating will be easily removed by cleaning solutions and strong rubbing. If you believe cleaning is necessary, hire the services of a professional and avoid the expense of re-gilding or re-plating.

**199** — 1894

**TROTTER.**
Silver Finish.
EIGHT-DAY TIME.
Dial, 2¼ inches.   Height, 6 inches.
List, each, $6.50.

**200**

**BABY**   1901
One-Day Time.
Dial, 1½ inches.   Height, 5¼ inches.
Porcelain Dial, Gilt Centre, Arabic or Roman.
Finished in Antique Brass or Silver,
List, each, $4 85
Finished in Rich Gold..........List, each, 6 25

**201** — 1886

**GOOD LUCK.**
Bronze and Nickel Finish.
ONE DAY TIME.
ONE DAY TIME, ALARM.
Dial, 4 inches.   Height, 6½ inches.

**202** — 1886

**ECHO.**
Bronze and Nickel Finish.
Hand moves automatically and rings Bell.
ONE DAY TIME, ALARM.
ONE DAY, STRIKE.
Dial, 4 inches.   Height, 7½ inches.

**203** — 1894

**BARGE.**
Finished in Silver.
ONE-DAY TIME.
Plain Dial, 2 inches.   Fancy Dial, 2 inches.
List, each, $4.75.   List, each, $5.00.
Height, 9¼ inches.

# ANSONIA

*Most white metal and brass ornamentals are gilded or silver plated. Great care must be taken when cleaning these surfaces. The thin layer of gilding or silver plating will be easily removed by cleaning solutions and strong rubbing. If you believe cleaning is necessary, hire the services of a professional and avoid the expense of re-gilding or re-plating.*

**204**

1895

**BICYCLE.**
Silver Finish.
EIGHT-DAY TIME.
Dial, 2¼ inches.   Height, 9¾ inches.
List, each, $5.00.

*Only clocks in ORIGINAL MINT condition and in good running order will command top prices. A replaced or damaged dial, missing parts, dull finish, worn-off gold gild, re-painted or chipped beveled glass will reduce prices accordingly.*

**205**

1886

**CALENDAR, No. 3056.**
ONE DAY TIME.
Dial, 2 inches.   Height, 11 inches.

**206**

1886

**NOVELTY, No. 44.**
ONE DAY TIME.
Dial, 2 inches.   Height, 7¾ inches.

**207**

1886

**NOVELTY, No. 27.**
Dial, 2 inches.   Height, 7¼ inches.

**208**

**OPERA FAN.**   1895
Silver, Brass or Bronze.
ONE-DAY TIME.
Dial, 2 inches.   Height, 7¾ inches.
List, each, $4.00.   Fancy Dial.   List, each, $4.25.

# ANSONIA

*For comprehensive listing of Ansonia Clocks, see Tran Duy Ly "Ansonia Clocks - A Guide to Identification & Prices."*

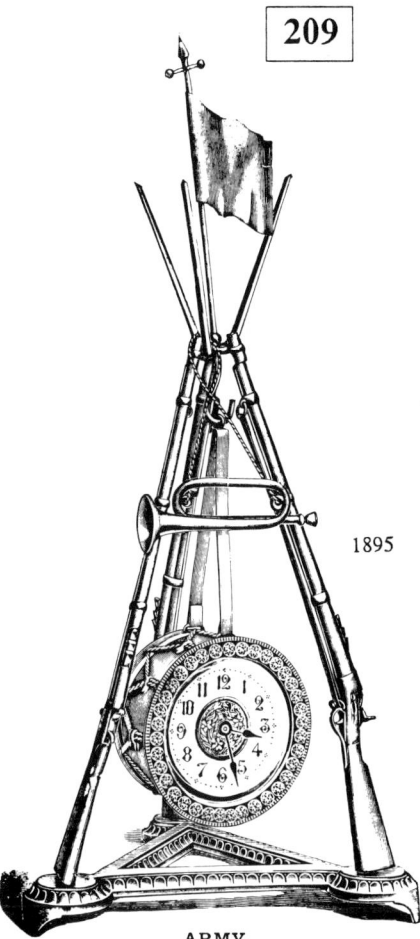

**209**

1895

**ARMY.**
Silver and Gilt.
EIGHT-DAY TIME.
Plain Dial, 2¼ inches. Height, 12½ inches.
List, each, $7.50.
Packed 24 in a case.

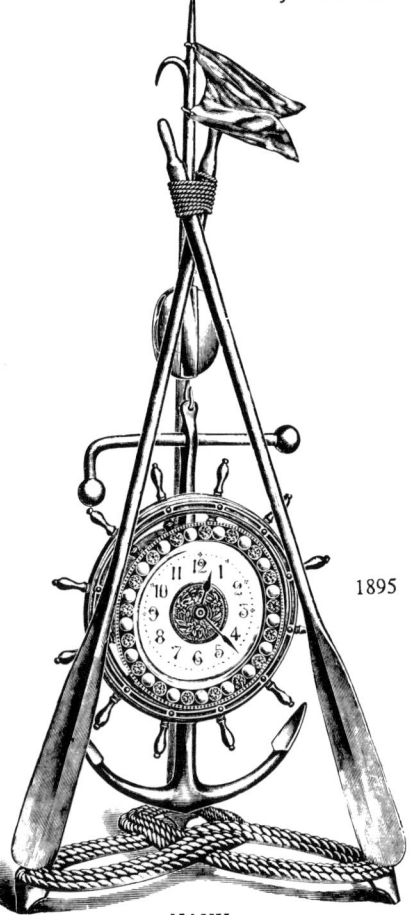

**210**

1895

**NAVY.**
Silver and Gilt.
EIGHT-DAY TIME.
Plain Dial, 2¼ inches. Height, 12½ inches.
List, each, $7.50.
Packed 24 in a case.

**211**

**NIGHT LIGHT** 1901
One-Day Time.
Opaloid Glass Dial, 5 inches. Height, 7 inches.
Finished in Gilt.
List, each ................................$4 75

**212**

1901

**FAIRIES**
Dial, 1½ inches. Height, 6¾ inches.
Finished in Rich Gold.........List, each, $9.25

**213**

1880

## Aladdin Night Light.
EXTRA.
Dial, - - - - 4 inches.

**214**

1880

## GRANDFATHER.
Dial, - - - - 4 inches.

# ANSONIA

Except for those fabulous clocks for which you wouldn't mind cutting a hole in the ceiling, Grandfather or Hanging Clocks under eight feet are the most desirable and practical. Some manufacturers produced certain of these clocks in two styles: one to stand on the floor, and the other to hang on a wall. Of the clocks designed with two styles, the hanging style is considered to be the most valuable.

The year entered beside each clock is the year of the catalog from which the illustration was taken. Some clocks may have been produced a few years earlier.

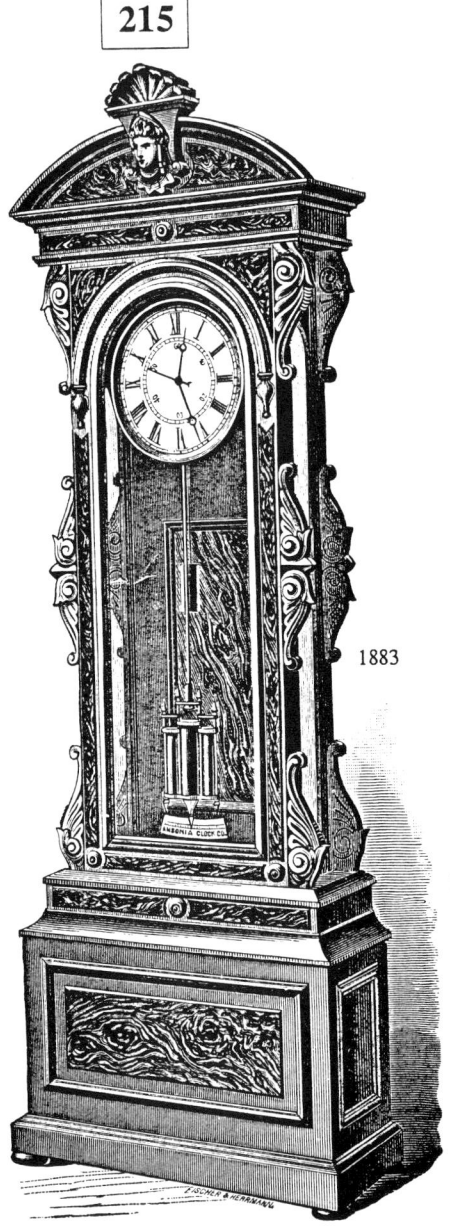

**REGULATOR, No. 11.**
Eight Day, Weight Time.
Dial, 14 inches.
Height....................105 inches.

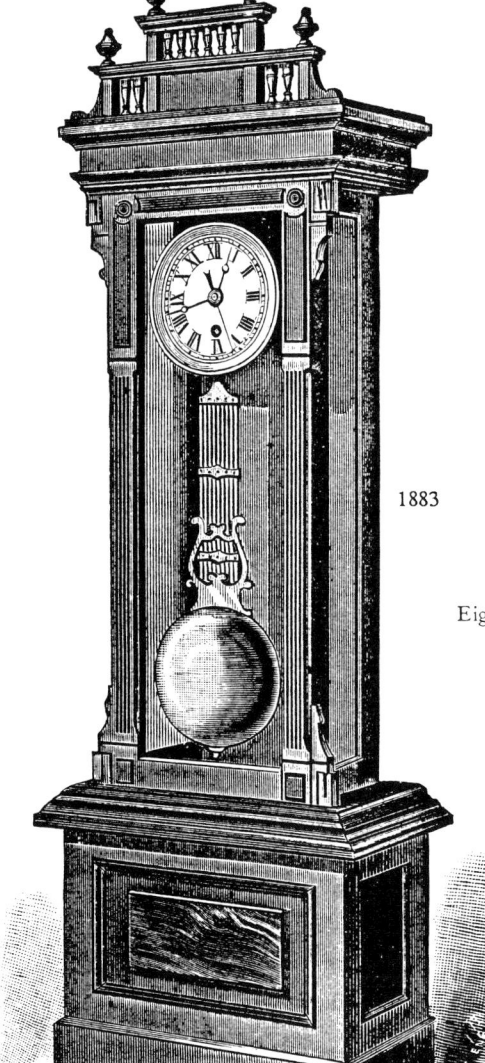

**REGULATOR, NO. 15.**
Eight Day, Weight Time.
Dial, 12 inches.
Height....................101 inches.

REGULATOR No. 15.
Dial, 12 inches. Height, 101 inches.
Eight-Day Weight Time. List, each, $165.00.

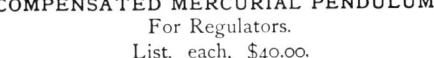

COMPENSATED MERCURIAL PENDULUM.
For Regulators.
List, each, $40.00.

# ANSONIA

**219**

**ANTIQUE STANDING No. 2**

Eight-Day, Half-Hour Gong Strike.
Mahogany or Oak.
Antique Brass Trimmings.
Gilt and Silver Dial, 12 inches.
Height, 104 inches.
List, each .................................. $305 00

**220**

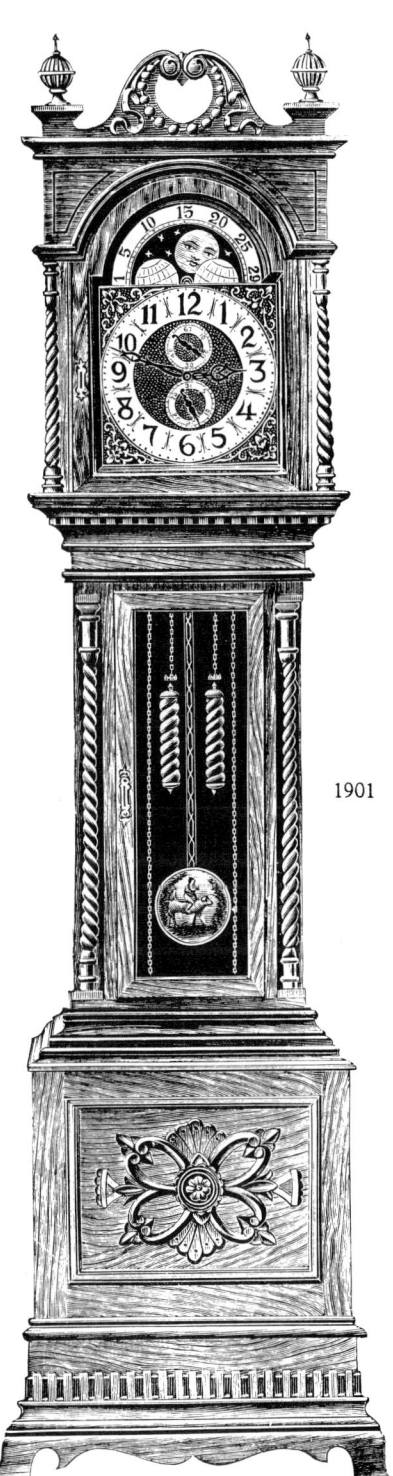

**ANTIQUE STANDING No. 3**

Eight-Day, Half-Hour Gong Strike.
Mahogany or Oak.
Antique Brass Trimmings.
Gilt and Silver Dial, 12 inches.
Height, 100 inches.
List, each .................................. $305 00

**221**

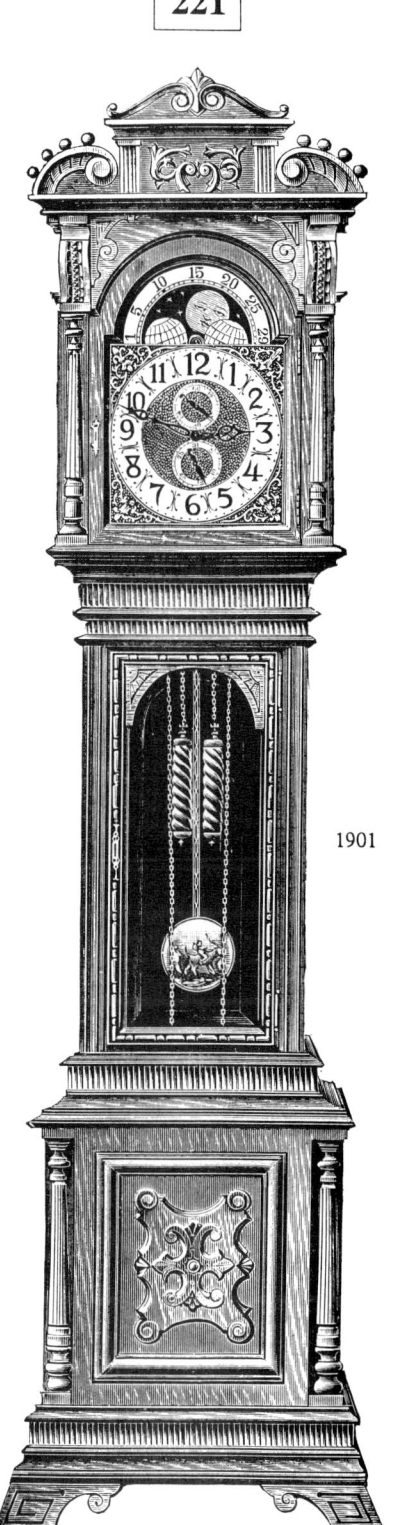

**ANTIQUE STANDING No. 4**

Eight-Day, Half-Hour Gong Strike.
Mahogany or Oak.
Antique Brass Trimmings.
Gilt and Silver Dial, 12 inches.
Height, 102 inches.
List, each .................................. $305 00

# ANSONIA

*Except for those fabulous clocks for which you wouldn't mind cutting a hole in the ceiling, Grandfather or Hanging Clocks under eight feet are the most desirable and practical. Some manufacturers produced certain of these clocks in two styles: one to stand on the floor, and the other to hang on a wall. Of the clocks designed with two styles, the hanging style is considered to be the most valuable.*

### 222

1886

**ANTIQUE HALL CLOCK.**
Finished in Oak or Mahogany, with Antique Brass Trimmings.
EIGHT DAY.
HALF-HOUR, CATHEDRAL GONG, WEIGHT, STRIKE.
Silver Dial. 10 inches.   Height, 7 feet 10 inches.

### 223

1894

**ANTIQUE STANDING, No. 1.**
Oak.
Silver Dial, 12 inches.   Height, 8 feet 8 inches.
List, each, $325.00.

### 224

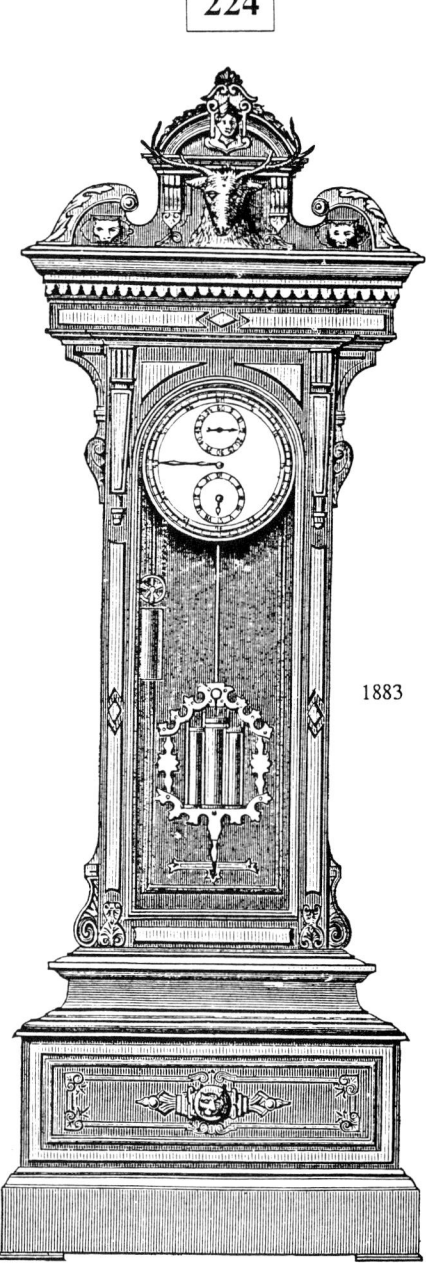

1883

**REGULATOR, No. 8.**

Eight Day, Weight Time.
Dial, 16 inches.
Height....................105 inches.

# ANSONIA

**225**

*For comprehensive listing of Ansonia Clocks, see Tran Duy Ly "Ansonia Clocks - A Guide to Identification & Prices."*

*The year entered beside each clock is the year of the catalog from which the illustration was taken. Some clocks may have been produced a few years earlier.*

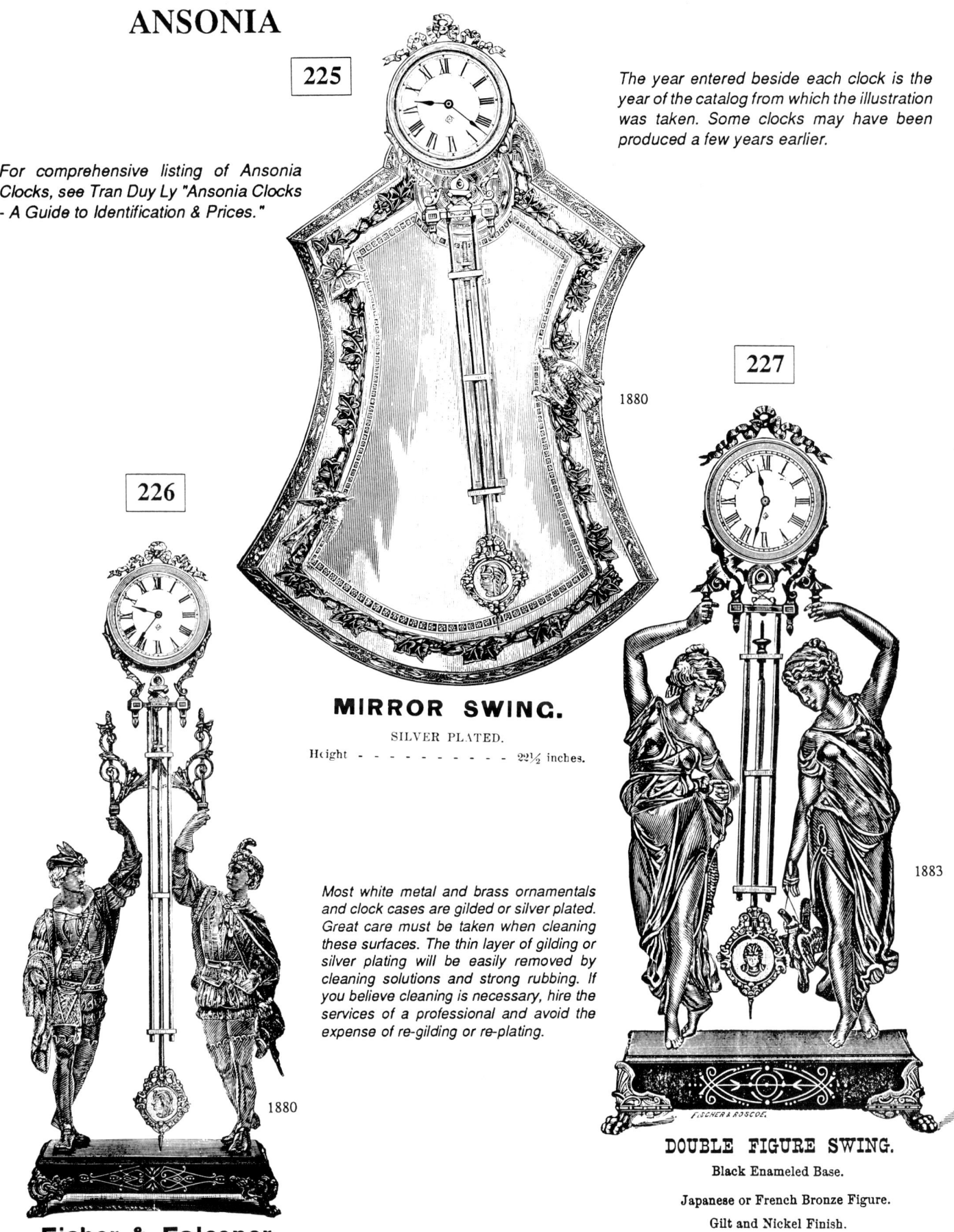

1880

**MIRROR SWING.**
SILVER PLATED.
Height - - - - - - - - - - 22½ inches.

**226**

*Most white metal and brass ornamentals and clock cases are gilded or silver plated. Great care must be taken when cleaning these surfaces. The thin layer of gilding or silver plating will be easily removed by cleaning solutions and strong rubbing. If you believe cleaning is necessary, hire the services of a professional and avoid the expense of re-gilding or re-plating.*

**227**

1883

1880

**Fisher & Falconer Swing.**
Height - - - - - - - 27 inches.

**DOUBLE FIGURE SWING.**
Black Enameled Base.
Japanese or French Bronze Figure.
Gilt and Nickel Finish.
8 Day, Time.
Dial, 4 inches.
Height.................. 25 inches.

# ANSONIA

*Most white metal and brass ornamentals and clock cases are gilded or silver plated. Great care must be taken when cleaning these surfaces. The thin layer of gilding or silver plating will be easily removed by cleaning solutions and strong rubbing. If you believe cleaning is necessary, hire the services of a professional and avoid the expense of re-gilding or re-plating.*

*Only clocks in ORIGINAL MINT condition and in good running order will command top prices. A replaced or damaged dial, missing parts, dull finish, worn-off gold gild, re-painted or chipped beveled glass will reduce prices accordingly.*

**228**

**229**

**230**

1906

1880

1906

**FORTUNA BALL SWING**
Height, 30 inches.
Eight-Day Time, Dial 4 1/2 inches.
Finished in Art Nouveau, Barbedienne or Syrian Bronze, etc.. Ball finished in Blue or Red Enamel; also Barbedienne or Verde Bronze.
List, each............................ $55.00

**FISHERMAN SWING**
Dial, - - - - 4 inches.
Height - - - - 25 inches.

4 inches dial
3 inches dial (rare)

**ARCADIA BALL SWING**
Height, 31 1/2 inches.
Eight-Day Time, Dial 4 1/2 inches.
Finished in Art Nouveau, Barbedienne or Syrian Bronze, etc.. Ball finished in Blue or Red Enamel; also Barbedienne or Verde Bronze.
List, each............................ $50.00

# ANSONIA

*For comprehensive listing of Ansonia Clocks, see Tran Duy Ly "Ansonia Clocks - A Guide to Identification & Prices."*

231

1906

### DIANA BALL SWING
Height, 30 inches.
Eight-Day Time, Dial 4 1/2 inches. Finished in Art Nouveau, Barbedienne or Syrian Bronze, etc.. Ball finished in Blue or Red Enamel; also Barbedienne or Verde Bronze.
List, each............................ $44.00

232

1906

### GLORIA BALL SWING
Height, 28 1/2 inches.
Eight-Day Time, Dial 4 1/2 inches. Finished in Art Nouveau, Barbedienne or Syrian Bronze, etc.. Ball finished in Blue or Red Enamel; also Barbedienne or Verde Bronze.
List, each............................ $44.00

233

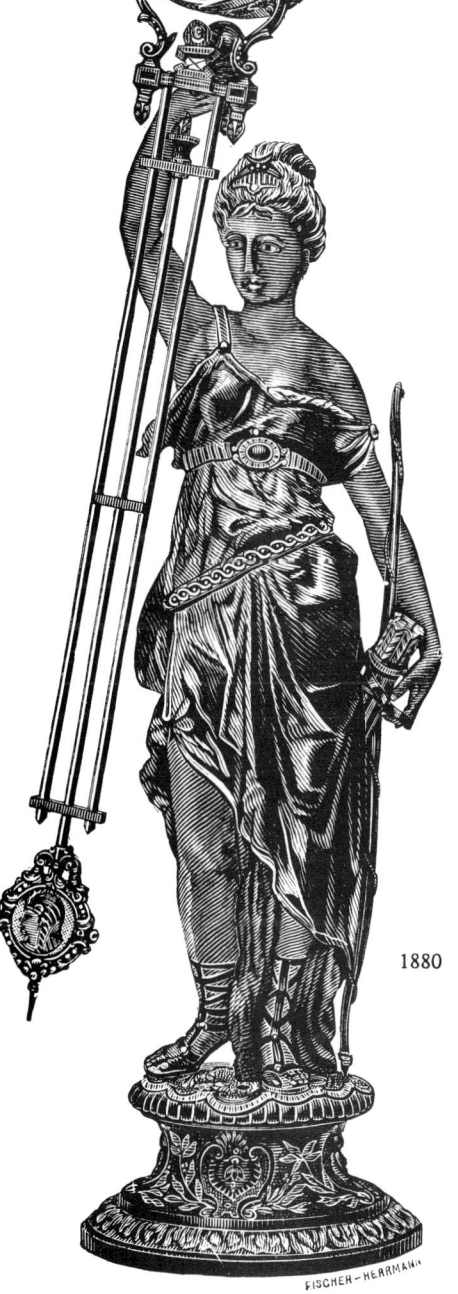

1880

### DIANA SWING.
Dial, - - - - - - - - - 4 inches.
Height, - - - - - - - - - 30 inches.

# ANSONIA

*The year entered beside each clock is the year of the catalog from which the illustration was taken. Some clocks may have been produced a few years earlier.*

### 234

### FISHER OR HUNTER SWING.

Japanese or French Bronze Figure.
8 Day, Time.
Dial, 4 inches.
Black Enameled Base.
Gilt and Nickel Finish
Height............ 24½ inches.

4 inches dial
3 inches dial (rare)

1883

### 235

1901

### JUNO SWING

Eight-Day Time.
Dial, 4 inches.   Height, 28 inches.
Finished in Japanese Bronze..List, each, $21 50
Finished in Syrian Bronze....List, each,  23 65
Boxed singly.

### 236

1906

### HUNTRESS SWING
Or with the Fisher Figure.

Height, 25 inches.
Eight-Day Time, Dial 4 1/2 inches.
Finished in Art Nouveau, Barbedienne or Syrian Bronze, etc.. Ball finished in Blue or Red Enamel; also Barbedienne or Verde Bronze.
List, each............................ $35.00

HUNTER BALL SWING

*Most white metal and brass ornamentals and clock cases are gilded or silver plated. Great care must be taken when cleaning these surfaces. The thin layer of gilding or silver plating will be easily removed by cleaning solutions and strong rubbing. If you believe cleaning is necessary, hire the services of a professional and avoid the expense of re-gilding or re-plating.*

### 237

1906

### JUNO BALL SWING

Height, 28 inches.
Eight-Day Time, Dial 4 1/2 inches.
Finished in Art Nouveau, Barbedienne or Syrian Bronze, etc.. Ball finished in Blue or Red Enamel; also Barbedienne or Verde Bronze.
List, each............................ $36.00

# ANSONIA

73

*For comprehensive listing of Ansonia Clocks, see Tran Duy Ly "Ansonia Clocks - A Guide to Identification & Prices."*

## 𝓜AHOGANY SWINGING 𝓒LOCKS

Eight-Day Time Movements. Pedestals and Swinging Parts Genuine Mahogany, Dull Hand Rubbed. Silver Plated Dials, Three and One-Half Inches Wide. Arabic Numerals.

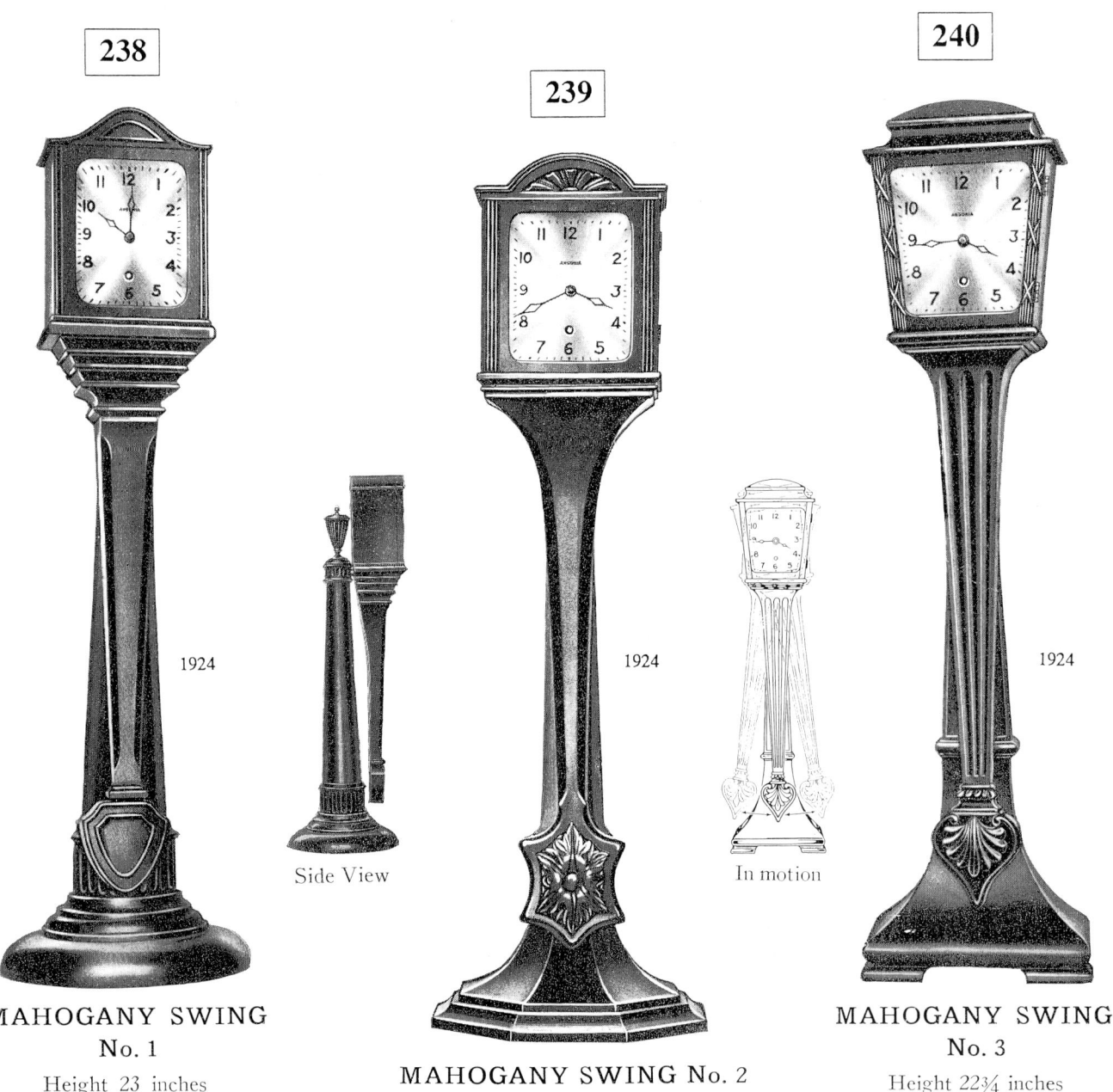

**238** — **239** — **240**

Side View — In motion

**MAHOGANY SWING No. 1**
Height 23 inches
Width 7 inches

**MAHOGANY SWING No. 2**
Height 23¼ inches   Width 7½ inches

**MAHOGANY SWING No. 3**
Height 22¾ inches
Width 6¼ inches

# BOSTON CLOCK COMPANY
## Chelsea, Mass.

The Boston Clock Company was organized May 29, 1884, as successor to the Harvard Clock Company which had been organized October 11, 1880 by James H. Gerry, Joseph H. Eastman and others and primarily produced wall clocks, often similar in style to some of the Howard models, but of lesser quality.

Joseph H. Eastman became manager of the Boston Clock Company and the firm is listed in Boston directories from 1885 through 1894. The firm was a large producer of good grade imitation French "carriage," "crystal regulator" clocks and mantel clocks, often in style and onyx cases. Most of their clocks utilized a good quality movement with a platform or watch-like balance escapement and had tandem winding whereby the time and strike were wound by turning the key in opposite directions in a single keyhole.

By 1890, the firm issued a trade catalog illustrating more than 50 different models. These clocks usually had porcelain dials, and the style and onyx cases used for the mantel clocks, were no doubt imported.

In 1894, the Boston Clock Company failed and Joseph Eastman tried to revive the firm as the Eastman Clock Company the following year. Eastman purchased land on Everett Street in Chelsea on September 13, 1895 and borrowed some $7,000 and commenced building a factory. The Eastman Clock Company was shown in Boston directories only for the year 1896 and Eastman's creditors foreclosed on the firm October 29, 1896.

**241**

1890

**CANDELABRA**
Onyx with Gilt Bronze.
Base, 4 inches.
Height, 18½ inches.

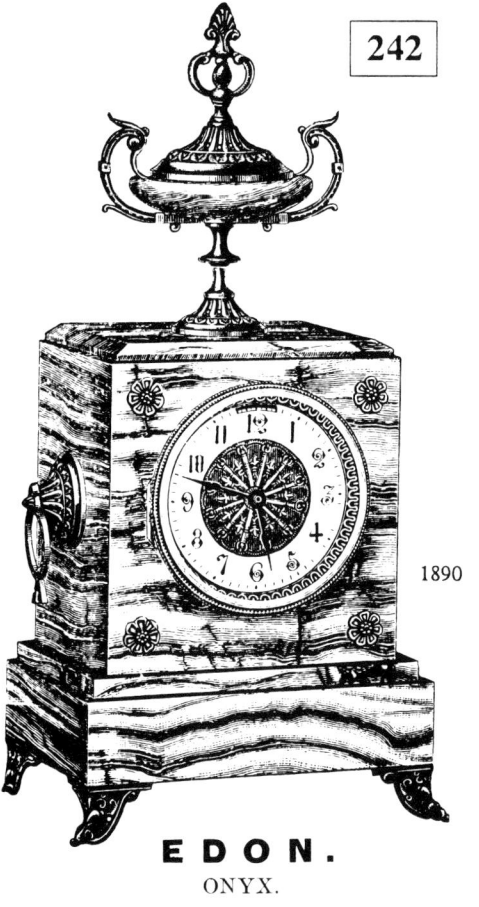

**242**

1890

**E D O N.**
ONYX.
Bronze Gilded Ornaments.
Height, 16½ inches.
Base, 8 inches.
8 Day.   Half-hour Cathedral Strike
Seven Jeweled Movement.

1890

**CANDELABRA**
Onyx with Gilt Bronze.
Base, 4 inches.
Height, 18½ inches.

# BOSTON

**SPARTA.**

## CRYSTAL.

Fine Gold Plated Case.

Seven Jeweled Movement.

Gilded Plates.

Front, Back and Sides, Beveled Glass.

Height, 9¾ inches.   Base, 6½ inches.

8 Day.   Half-hour Cathedral Strike.

### SPARTA.

Fine Gold Plated Case.

Seven Jeweled Movement.

Gilded Plates.

Front, Back and Sides, Beveled Glass.

Height, 6¼ inches.   Base, 3¼ inches.

8 Day Time.

Traveling Case Furnished.

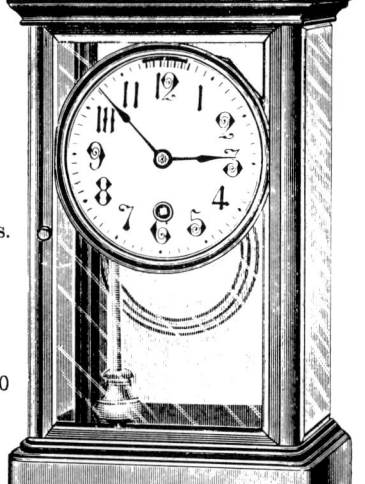

**CRYSTAL.**

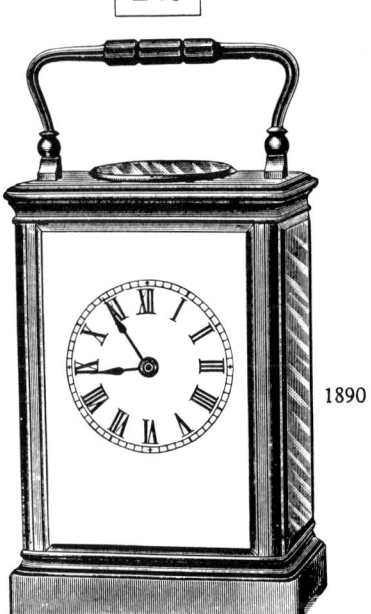

**CYPRUS.**

Fine Gold Plated Case.

Seven Jeweled Movement.

Gilded Plates with Damaskeen Finish.

Front, Back, Sides and Top, Beveled Glass.

Height, 7 inches.   Base, 3¾ inches.

8 Day.   Half-hour Cathedral Strike.

Traveling Case Furnished.

## DELPHUS.

Fine Gold Plated Case.

Eleven Jeweled Movement.

Nickel Plates with Damaskeen Finish.

Front, Back and Sides, Beveled Glass.

Height, 10½ inches.   Base, 7½ inches.

8 Day.   Half-hour Cathedral Strike.

**ATHENS.**

Fine Gold Plated Case.

Seven Jeweled Movement.

Gilded Plates.

Front, Back and Sides, Beveled Glass.

Height, 6½ inches.   Base, 3¼ inches.

8 Day Time.

Traveling Case Furnished.

# BOSTON

**248**

### QUEEN ANNE.
Fine Gold Plated Case.
Seven Jeweled Movement.
Gilded Plates.
Height, 6½ inches.
Base, 3¼ inches.
8 Day Time.
Traveling Case Furnished.
Front, Back and Sides, Beveled Glass.

**QUEEN ANNE.**

**249**

### DELOS.
Fine Gold Plated Case.
Seven Jeweled Movement.
Gilded Plates.
Front, Back and Sides, Beveled Glass.
Base, 3¼ inches.
Height, 6½ inches.
8 Day Time.
Traveling Case Furnished.

**DELOS.**

**250**

**251**

### LOCOMOTIVE.
Heavy Brass Case.
Nickeled or Silvered, if required.
Seven Jeweled Movement.

Dial, 4 inches, silvered.    Diameter, 5½ inches.
Dial, 8   "   "   "   10   "

8 Day Time.

**252**

**JUNO.**
ONYX.
Real Bronze Gilded Ornaments.
Height, 18½ inches.      Base, 11¼ inches.
8 Day. Half-hour Cathedral Strike.
Seven Jeweled Movement.

**ALHAMBRA.**
Fine Gold Plated Case.
Eleven Jeweled Movement.
Nickel Plates with Damaskeen Finish.
Front, Back and Sides, Beveled Glass.
Height, 13¼ inches.      Base, 13½ inches.
8 Day. Half-hour Cathedral Strike.

# CHELSEA CLOCK COMPANY
## Chelsea, Mass.

On October 7, 1896, a group of investors formed a clock company at Kittery, ME called the Boston Clock Company in order to succeed the defunct firm of the same name. The following October 29th, at the time of the foreclosure of Joseph H. Eastman's Eastman Clock Company, one of the principals of the new Boston Clock Company purchased at public auction the newly built factory and boiler of the defunct Eastman Clock Company for $6,300.

A new firm known as the Chelsea Clock Company was formed and on August 4, 1907, the Eastman factory was transferred to the new firm by the new Boston Clock Company. Though this second Boston Clock Company was merely a firm on paper and never commenced manufacturing under that name, the Chelsea Clock Company revived the Boston Clock Company name in 1909 and used it on a line of clocks with lesser grade movements until about the time of the Great Depression. In the 1970's, Chelsea again revived the Boston name for use with a cheaper line utilizing foreign movements.

The Chelsea Clock Company has enjoyed a long, successful business primarily manufacturing superior quality clocks. They were particularly known for their good quality Willard-style banjo clocks and other wall clocks manufactured prior to 1930 and more especially for their marine clocks which they continue to manufacture to the present. By 1975, the firm employed 80.

In the latter part of 1969 the Chelsea Clock Company was sold to Automation Industries of Los Angeles, CA which took over the firm January 2, 1970, and worked under that firm's Kenyon Marine Division of Guilford, CT. In 1972, the firm was taken over by Bunker-Remo of Illinois.

Chelsea remains in business to the present, though perhaps not as financially healthy as in the past. The retail prices of their clocks have skyrocketed in the last decade, some models retailing for more than $1,000. In the 1980's, a disastrous fire caused by an irresponsible neighboring firm destroyed the firm's warehouse and some of the firm's records and most of its repair parts, some dating back to the firm's early years. Prior to this time the firm could supply replacement parts for most of its earlier models.

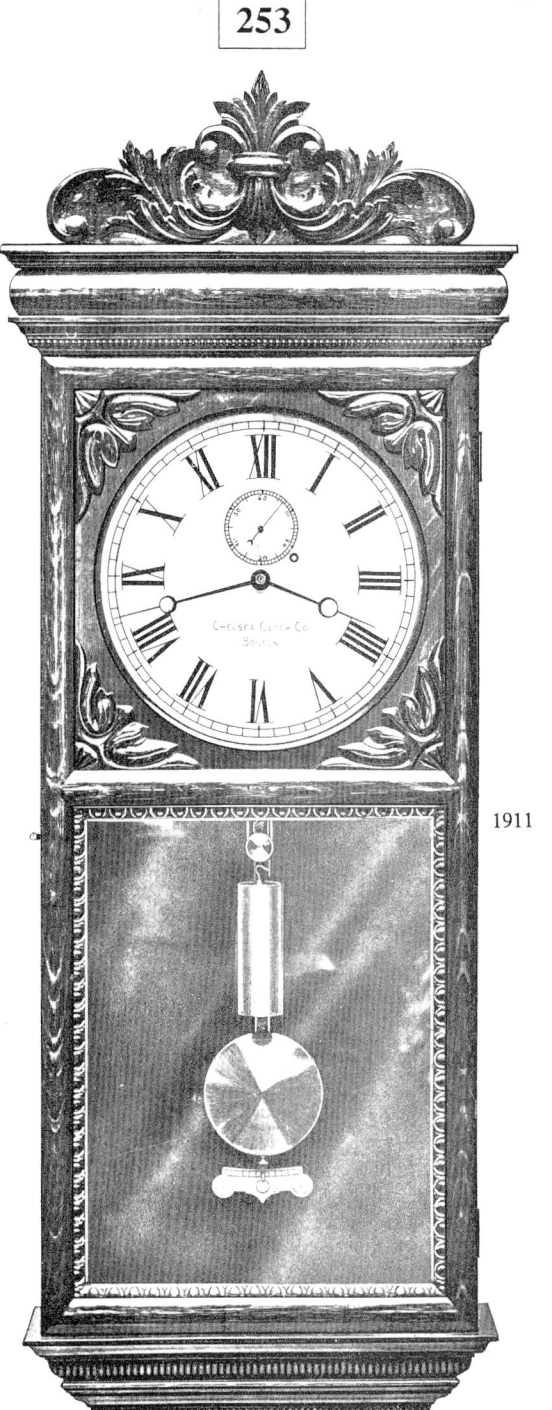

253

1911

## REGULATOR NO. 5

Height, about 4 feet over all — 12-inch White Enameled Non-Crackable Dials — Seconds Hand — Maintaining Power — Exposed Weight — Weight Cord is specially made wire cord, *not* catgut — Cut Polished Steel Pinions — Graham Pallet, Dead Beat Escapement — Extra Heavy Front and Back Plates, insuring rigidity to Movement and good Bearings for Pivots — Hardened and Polished Pivots — Fine workmanship throughout.

**In Oak or Cherry Cases    $45.00**

**In Extra Finely Finished Mahogany Cases . . . .    52.50**

# CHELSEA

**254** — 1911

**NO. 1 PENDULUM**
For SCHOOLS, Offices, Public Buildings, Etc.

☞ Extensively used in Schools in New York, Brooklyn, Boston, etc. Best of its kind on the market.

HARDWOOD CASES of GOLDEN OAK and Light or Dark Colored QUARTERED OAK or CHERRY — 12-inch White Enameled Non-Crackable DIAL — 8-Day — Weight — Weight Cord is specially made wire cord, *not* catgut — Cut Polished Steel Pinions — Recoil Escapement — Hardened and Polished Pivots — Extra Heavy Front and Back Plates, insuring rigidity to movement and good Bearings for Pivots — Fine workmanship throughout.

Height, 34 inches
Price-List, $18

**255** — 1911

**REGULATOR NO. 3**
For Jewelers, Railroad Stations, Offices, Public Buildings, Etc.

☞ BEST of its kind on the market.

HARDWOOD CASES of GOLDEN OAK and Light or Dark Colored QUARTERED OAK or CHERRY — 12-inch White Enameled Non-Crackable DIAL — Seconds Hand — Maintaining Power — 8-Day — Exposed Weight — Weight cord is specially made wire cord, *not* catgut — Cut Polished Steel Pinions — Graham Pallet, Dead Beat Escapement — Extra Heavy Front and Back Plates, insuring rigidity to movement and good Bearings for Pivots — Hardened and Polished Pivots — Fine workmanship throughout.

Height, 37 inches
Price-List, $22

**256** — 1911

**"SPECIAL" AUTO CLOCK**

**257** — 1911

The above illustrates the Inner Clock Case and Clock removed from the Outer Case. Convenient when touring, for using on mantels, bureaus, etc.

| SIZE | "SPECIAL" CAT. NO. | PRICE - LIST |
|---|---|---|
| 2¾ in. | 196 | $36.00 |
| 3½ " | 197 | 45.00 |

*To receive top prices, clocks must be in ORIGINAL MINT condition and in good running order - A replaced or damaged dial, missing parts, will reduce prices accordingly.*

# CHELSEA

The "CHELSEA" 8-Day High-Grade

2¾-inch "Limousine" Auto Clock. (Only furnished in 2¾-inch size.)

| CAT. NO. | PRICE-LIST |
|---|---|
| 193 | $28.50 |

## YACHT WHEEL
### CLOCK

| SIZE | 2¾ INCH | | 3½ INCH | | 4½ INCH | |
|---|---|---|---|---|---|---|
| Size of Dials | Cat. No. | Price | Cat. No. | Price | Cat. No. | Price |
| Plain Silvered Metal | 400 | $46. | 401 | $52. | 402 | $78. |
| "Special" | 407 | 50. | 408 | 57. | 409 | 84. |
| "Special Grand" | 414 | 54. | 415 | 62. | 416 | 90. |

| SIZE | 6 INCH | | 8½ INCH | | 10 INCH | | 12 INCH | |
|---|---|---|---|---|---|---|---|---|
| Size of Dials | Cat. No. | Price | Cat. No. | Price | Cat. No. | Price | Cat. No. | Price |
| Plain Silvered Metal | 403 | $94. | 404 | $116. | 405 | $140. | 406 | $163. |
| "Special" | 410 | 100. | 411 | 125. | 412 | 150. | 413 | 175. |
| "Special Grand" | 417 | 107. | 418 | 134. | 419 | 163. | 420 | 190. |

### Carved Yacht Wheel Clock

| SIZE | 2¾ INCH | | 3½ INCH | | 4½ INCH | | 6 INCH | | 8½ INCH | | 10 INCH | | 12 INCH | |
|---|---|---|---|---|---|---|---|---|---|---|---|---|---|---|
| Style of Dials | Cat. No. | Price | Cat. No. | Price | Cat. No. | Price | Cat. No. | Price | Cat. No. | Price | Cat. No. | Price | Cat. No. | Price |
| Plain Silvered Metal | 900 | $54. | 901 | $61. | 902 | $89. | 903 | $108. | 904 | $134. | 905 | $161. | 906 | $187. |
| "Special" | 910 | 58. | 911 | 66. | 912 | 95. | 913 | 114. | 914 | 143. | 915 | 171. | 916 | 199. |
| "Special Grand" | 920 | 62. | 921 | 71. | 922 | 101. | 923 | 121. | 924 | 152. | 925 | 184. | 926 | 214. |

# CHELSEA

### The "CHELSEA" 8-Day High-Grade
## Clock and Barometer Desk Sets

| SIZE | KIND OF MOVEMENT | CLOCK AND BAROMETER ON BASE | |
|---|---|---|---|
| | | Cat. No. | Price |
| 2¾ inch | Auto (Time only) | 1952 | $ 54.00 |
| 3½ " | Auto (Time only) | 1962 | 60.00 |
| 4½ " | Time | 2877 | 61.00 |
| 4½ " | Hour and Half Hour Striking | 2879 | 77.00 |
| 4½ " | Ship's Bell | 2881 | 82.00 |
| 6 " | Time | 2884 | 75.00 |
| 6 " | Hour and Half Hour Striking | 2886 | 94.00 |
| 6 " | Ship's Bell | 2888 | 99.00 |
| 8½ " | Time | 2891 | 104.00 |
| 8½ " | Hour and Half Hour Striking | 2893 | 121.00 |
| 8½ " | Ship's Bell | 2895 | 126.00 |

## WARDROOM CLOCK

| Size | Plain Silvered Metal Dial |
|---|---|
| 6 inch | $28.00 |
| 8½ " | 36.00 |
| 10 " | 45.00 |
| 12 " | 51.00 |

| SIZE | KIND OF MOVEMENT | CLOCK ON BASE | | BAROMETER ON BASE | | CLOCK AND BAROMETER ON BASE | |
|---|---|---|---|---|---|---|---|
| | | Cat. No. | Price | Cat. No. | Price | Cat. No. | Price |
| 2¾ inch | Auto (Time only) | 2011 | $31.00 | 2002 | $26.00 | 2022 | $51.00 |
| 3½ " | Auto (Time only) | 2041 | 33.00 | 2032 | 29.00 | 2052 | 56.00 |
| 4½ " | Time | 2098 | 31.00 | 2091 | 34.00 | 2113 | 61.00 |
| 4½ " | Hour and Half Hour Striking | 2099 | 47.00 | | | 2114 | 77.00 |
| 4½ " | Ship's Bell | 2100 | 52.00 | | | 2115 | 82.00 |
| 6 " | Time | 2134 | 36.00 | 2127 | 43.00 | 2149 | 75.00 |
| 6 " | Hour and Half Hour Striking | 2135 | 55.00 | | | 2150 | 94.00 |
| 6 " | Ship's Bell | 2136 | 60.00 | | | 2151 | 99.00 |
| 8½ " | Time | 2169 | 48.00 | 2162 | 60.00 | 2184 | 104.00 |
| 8½ " | Hour and Half Hour Striking | 2170 | 65.00 | | | 2185 | 121.00 |
| 8½ " | Ship's Bell | 2171 | 70.00 | | | 2186 | 126.00 |

# CHELSEA

**265**

1911

**CARVED BASE**
CLOCK

Style No. 2

S = Screw Bezel
H = Hinge "

**266**

1911

**CARVED BASE**
CLOCK

Style No. 1

| SIZE (INCH) | Plain Silvered Metal Dials | | | | | | "Special" Dials | | | | | | "Special Grand" Dials | | | | | |
|---|---|---|---|---|---|---|---|---|---|---|---|---|---|---|---|---|---|---|
| | TIME | | Hour and Half Hour STRIKING | | SHIP'S BELL | | TIME | | Hour and Half Hour STRIKING | | SHIP'S BELL | | TIME | | Hour and Half Hour STRIKING | | SHIP'S BELL | |
| | Catalog No. | PRICE | Catalog No. | PRICE | Catalog No. | PRICE | Catalog No. | PRICE | Catalog No. | PRICE | Catalog No. | PRICE | Catalog No. | PRICE | Catalog No. | PRICE | Catalog No. | PRICE |
| 4½ S | 806 | 41 | 814 | $57. | 822 | $62. | 846 | 47. | 854 | $63. | 862 | $68. | 876 | 53. | 884 | $69. | 892 | $74. |
| 6   S | 807 | 47. | 815 | 66. | 823 | 71. | 847 | 53. | 855 | 72. | 863 | 77. | 877 | 60. | 885 | 79. | 893 | 84. |
| 4½ H | 808 | 44. | 816 | 60. | 824 | 65. | 848 | 50. | 856 | 66. | 864 | 71. | 878 | 56. | 886 | 72. | 894 | 77. |
| 6   H | 809 | 50. | 817 | 69. | 825 | 74. | 849 | 56. | 857 | 75. | 865 | 80. | 879 | 63. | 887 | 82. | 895 | 87. |
| 8½ H | 810 | 65. | 818 | 82. | 826 | 87. | 850 | 74. | 858 | 91. | 866 | 96. | 880 | 83. | 888 | 100. | 896 | 105. |
| 10  H | 811 | 87. | 819 | 100. | 827 | 105. | 851 | 97. | 859 | 110. | 867 | 115. | 881 | 110. | 889 | 123. | 897 | 128. |
| 12  H | 812 | 102. | 820 | 114. | 828 | 119. | 852 | 114. | 860 | 126. | 868 | 131. | 882 | 129. | 890 | 141. | 898 | 146. |

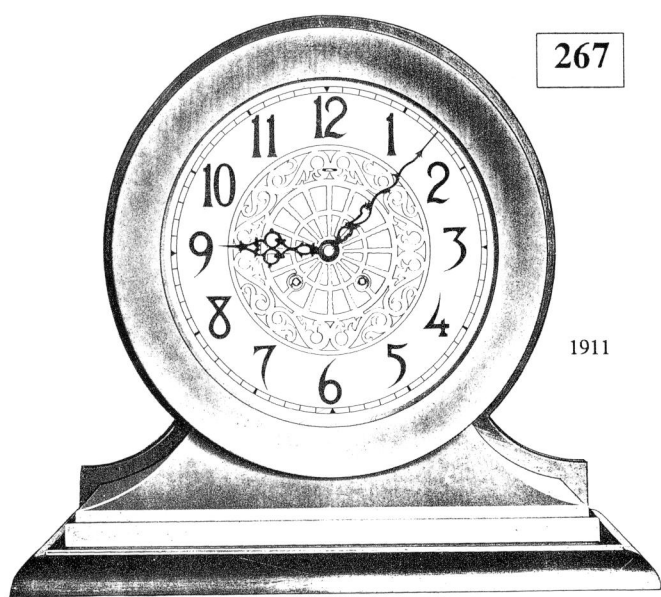

**267**

1911

The "CHELSEA" 8-Day High-Grade

## "SPECIAL GRAND"

SHIP'S BELL CLOCK ON { MAHOGANY and METAL BASE

| SIZE | PRICE - LIST |
|---|---|
| 4½ inch | $ 66.00 |
| 6   " | 75.00 |
| 8½ " | 90.00 |
| 10  " | 110.00 |
| 12  " | 125.00 |

81

# CHELSEA

**268**

1911

### TAMBOUR CLOCK Style No. 2
#### PLAIN SILVERED DIALS

| SIZE | CATALOG NO. | PRICE - LIST |
|---|---|---|
| 4½ inch | 2650 | $ 82.00 |
| 5½ " | 2651 | 93.00 |
| 6½ " | 2652 | 98.00 |
| 8 " | 2653 | 118.00 |
| 10 " | 2654 | 160.00 |

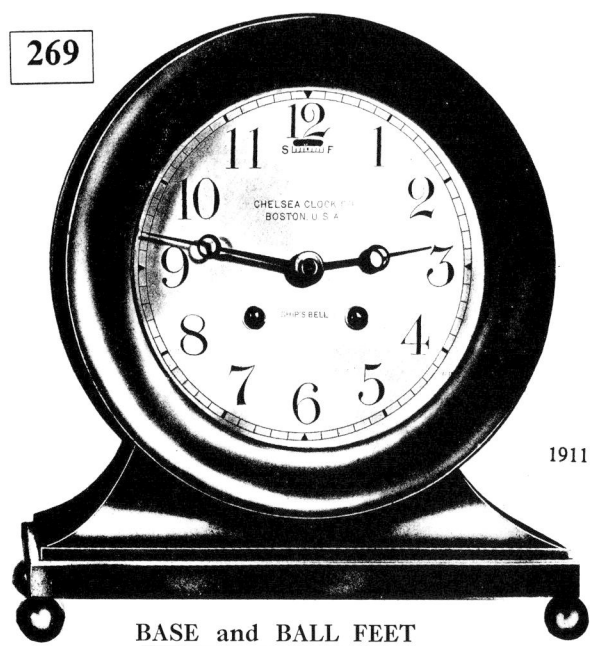

**269**

1911

### BASE and BALL FEET
#### Dials ☞ Plain Silvered Metal Dial

| SIZE | CATALOG NO. | PRICE - LIST |
|---|---|---|
| 4½ inch | 250 | $52.00 |
| 6 " | 251 | 60.00 |
| 8½ " | 252 | 70.00 |
| 10 " | 253 | 85.00 |
| 12 " | 254 | 96.00 |

**270**

1911

### TAMBOUR CLOCK Style No. 1
#### PLAIN SILVERED DIALS

| SIZE | CATALOG NO. | PRICE - LIST |
|---|---|---|
| 4½ inch | 2565 | $ 74.00 |
| 5½ " | 2566 | 84.00 |
| 6½ " | 2567 | 89.00 |
| 8 " | 2568 | 107.00 |
| 10 " | 2569 | 145.00 |

**271**

1911

### MAHOGANY PEDESTAL CLOCK.

| | Dials ☞ | "Hour and Half Hour" | | | "Ship's Bell" | | |
|---|---|---|---|---|---|---|---|
| SIZE | | Plain Silvered Metal | "Special" | "Special Grand" | Plain Silvered Metal | "Special" | "Special Grand" |
| 5½ inch | | $64.00 | $70.00 | $76.00 | $69.00 | $75.00 | $81.00 |
| 6½ " | | 69.00 | 76.00 | 83.00 | 74.00 | 81.00 | 88.00 |
| 8 " | | 74.00 | 82.00 | 90.00 | 79.00 | 87.00 | 95.00 |

¶ These clocks are most frequently ordered with "SPECIAL" style of dial.

☞ All sizes stated are the approximate diameters of dials.

# CHELSEA

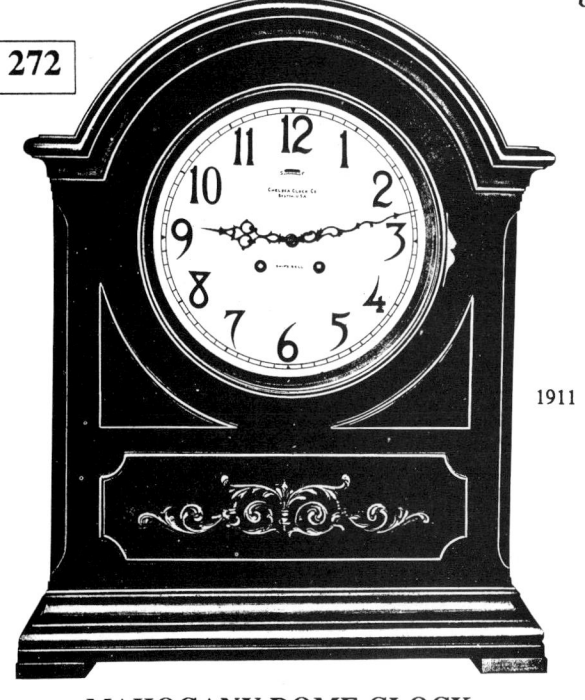

**"Hour and Half Hour"**

| SIZE | Dials → Plain Silvered Metal | "Special" | "Special Grand" |
|---|---|---|---|
| 5½ inch | $73.00 | $79.00 | $85.00 |
| 6½ " | 78.00 | 85.00 | 92.00 |
| 8 " | 83.00 | 91.00 | 99.00 |

**"Ship's Bell"**

| SIZE | Dials → Plain Silvered Metal | "Special" | "Special Grand" |
|---|---|---|---|
| 5½ inch | $78.00 | $84.00 | $ 90.00 |
| 6½ " | 83.00 | 90.00 | 97.00 |
| 8 " | 88.00 | 96.00 | 104.00 |

¶ These Clocks are most frequently ordered with "SPECIAL" style of dial.
☞ All sizes stated are the approximate diameters of dials.

**MAHOGANY DOME CLOCK.**

**WINDSOR CLOCK**

| DIAL | PLAIN SILVERED METAL DIALS | | | |
|---|---|---|---|---|
| Size (Inch) | Hour and Half Hour | | Ship's Bell | |
| | Catalog Number | Price-List | Catalog Number | Price-List |
| 4½ | 2815 | $ 62.00 | 2800 | $ 67.00 |
| 5½ | 2816 | 72.00 | 2801 | 77.00 |
| 6½ | 2817 | 76.00 | 2802 | 81.00 |
| 8 | 2818 | 92.00 | 2803 | 97.00 |
| 10 | 2819 | 126.00 | 2804 | 131.00 |

**MAHOGANY EMPIRE CLOCK.**

| SIZE | Dials → "Hour and Half Hour" | | | "Ship's Bell" | | |
|---|---|---|---|---|---|---|
| | Plain Silvered Metal | "Special" | "Special Grand" | Plain Silvered Metal | "Special" | "Special Grand" |
| 4½ inch | $69.00 | $ 75.00 | $ 81.00 | $ 74.00 | $ 80.00 | $ 86.00 |
| 5½ " | 75.00 | 81.00 | 87.00 | 80.00 | 86.00 | 92.00 |
| 6½ " | 81.00 | 88.00 | 95.00 | 86.00 | 93.00 | 100.00 |
| 8 " | 87.00 | 95.00 | 103.00 | 92.00 | 100.00 | 108.00 |
| 10 " | 99.00 | 109.00 | 122.00 | 104.00 | 114.00 | 127.00 |

# CHELSEA

**275**    **276**

## $2\frac{3}{4}$ AND $3\frac{1}{4}$ INCH BOUDOIR CLOCKS

### TIME (Not striking) CLOCKS in BRONZE METAL CASES

**2¾-INCH**

| DORIC Cat. No. | GOTHIC Cat. No. | STYLE OF DIALS | PRICE - LIST |
|---|---|---|---|
| 2504 | 2500 | White Porcelain | $24.00 |
| 2505 | 2501 | Plain Silvered Metal | 27.00 |
| 2506 | 2502 | "Special" | 31.00 |
| 2507 | 2503 | "Special Grand" | 35.00 |

**3¼-INCH**

| DORIC Cat. No. | GOTHIC Cat. No. | STYLE OF DIALS | PRICE - LIST |
|---|---|---|---|
| 176 | 174 | White Porcelain | $27.00 |
| 541 | 540 | Plain Silvered Metal | 30.00 |
| 543 | 542 | "Special" | 35.00 |
| 545 | 544 | "Special Grand" | 40.00 |

### DORIC
Height of 2¾-inch size, 6⅞ inches  
Height of 3¼-inch size, 8 inches

### GOTHIC
Height of 2¾-inch size, 7⅜ inches  
Height of 3¼-inch size, 8½ inches

**277**    **278**

### DORIC

TIME NOT STRIKING

### LEVER WALL CLOCK

| Size (inch) | WITH WHITE PORCELAIN DIAL | WITH PLAIN SILVERED METAL DIAL | WITH "SPECIAL" DIAL | WITH "SPECIAL GRAND" DIAL |
|---|---|---|---|---|
| 4¼ | $47.00 | $50.00 | $56.00 | $62.00 |
| 5½ |  | 84.00 | 90.00 | 96.00 |
| 6½ |  | 93.00 | 100.00 | 107.00 |
| 8 |  | 140.00 | 148.00 | 156.00 |

IN THREE STYLES OF CASES — viz: No. 1, No. 2, No. 3.

No. 1. The Best, as shown in cut.  
No. 2. Same as No. 1, except omit portion of carved wood at top and bottom of case at points opposite A A.  
No. 3. Same as No. 2, except omit the four carved corner pieces on door, and are not quite as finely finished. A favorite clock for Schools, Offices, etc.  
Otherwise, Styles 2 and 3 are in all respects equal to No. 1.

| CAT. NO. | SIZE | STYLE NO. 1 | STYLE NO. 2 | STYLE NO. 3 |
|---|---|---|---|---|
| 160 | 8 in. | $28.00 | $22.00 | $21.00 |
| 162 | 10 " | 32.00 | 26.00 | 25.00 |
| 164 | 12 " | 36.00 | 30.00 | 29.00 |
| 166 | 18 " | 75.00 | 60.00 | 57.00 |

# WILLIAM L. GILBERT CLOCK COMPANY
Winsted, Connecticut

On July 5, 1871, the William L. Gilbert Clock Company was formed at Winsted, Conn. to succeed the Gilbert Manufacturing Company (1866-1871) which had been dissolved after a fire destroyed the factory. These firms had grown out of the clockmaking operations of William L. Gilbert (1806-1890) who, since 1828, had been involved in various clockmaking partnerships in Bristol, Farmington and Winsted, Connecticut.

In July of 1873, the new factory complex was completed and manufacturing commenced. George B. Owen (1834-1916) had come to Winsted in 1866 as General Manager and ran the firm for nearly 50 years, designing many interesting cases and patenting several clock movement features. Owen also operated a concurrent clock business at Winsted between 1875 and 1894.

In 1897, the Gilbert firm built a four-story building for a new case shop and another building by 1900 for storage and shipping. A three story office building was built in 1902. The recession which began in 1907, along with the financial pressures of their recent expansion headed the firm in a decline that culminated by forcing George B. Owen and his sons to relinquish control in 1914.

Bankruptcy and liquidation were barely avoided in 1914 and a new manager, Charles E. Williams, was appointed and served until his death in 1930, just a few months following the stock market crash. Pressures of the Great Depression and money spent in developing electronic clocks sent the firm into receivership in September, 1932.

On July 20, 1934, a new firm known as the William L. Gilbert Clock Corporation was formed to succeed the earlier company. It was one of the few firms allowed to continue clockmaking during World War II because it was able to manufacture clocks without metal cases, having installed machinery in 1940 to produce cases from moulded papier mache. They had modest profits after the war and in 1954 tooled up to produce an adding machine for General Computing Machines Corporation. In 1957, they were taken over by General. In December, 1964, the clockmaking division was sold to Spartus Corporation of Chicago, not having produced a profit for about 12 years.

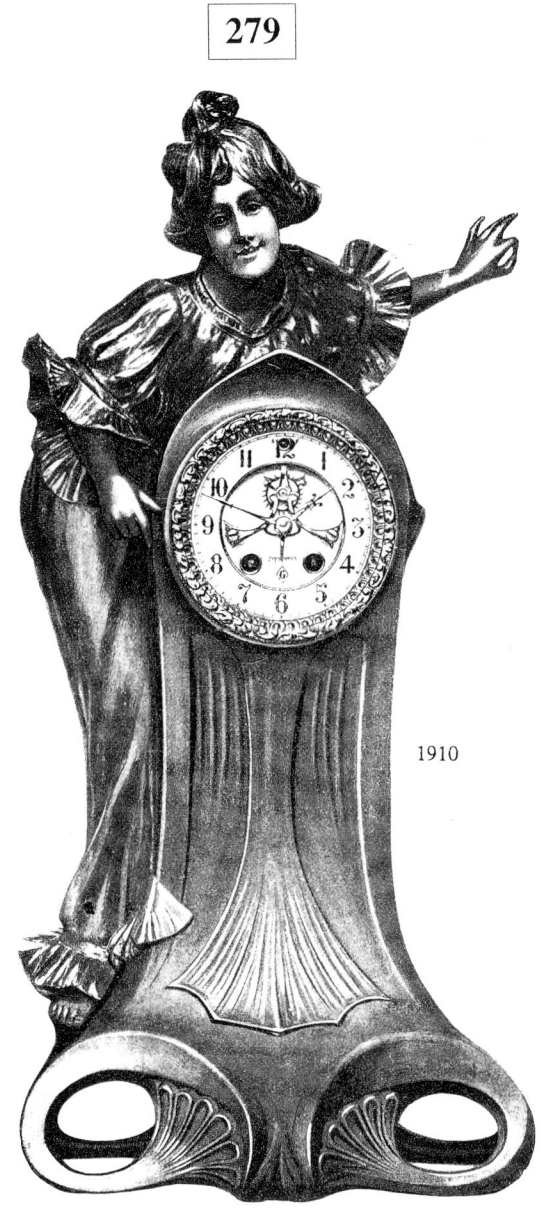

279

1910

**VIOLETTE.**

A distinctive Figure Pedestal Clock in Art Nouveau Design. Height, 19¼ inches. Width, 10 inches. Fitted with Eight-Day Movement, Hour and Half-Hour Strike on Rich Cathedral Gong. 4-inch Ivory Porcelain Dial, with the Gilbert Visible Escapement and Patent Beat Adjuster. Fancy Gilt Rococo Sash, Beveled Glass. Supplied in Rich Ormolu Gold Plate Finish or Violacé Bronze Finish, as listed below:

|  |  | Violacé Bronze Finish. | Ormolu Gold Finish. |
|---|---|---|---|
| VIOLETTE, with Vase No. 1278 | List per set, | $60.00 | $70.00 |
| VIOLETTE (Clock only) | List each, | 45.00 | 52.00 |
| VASE No. 1278 | List per pair, | 15.00 | 18.00 |

# GILBERT

### THALIA.

Eight-Day Gilt Regulator. Height, 11½ inches. Width, 8 inches. The case is Gold Plated and Highly Polished, with heavy raised Ornamental Corner Case Decorations in Rich Ormolu Gold Plate Finish. Beveled Plate Glass Front, Sides and Back. Fitted with Eight-Day Movement, Hour Strike on Rich Cathedral Gong, Half-Hour on Separate Cup Bell. The Movement Plates and all Exposed Parts are Highly Polished. 4½-inch Ivory Porcelain Dial, with the Gilbert Visible Escapement and Patent Beat Adjuster. Visible Pendulum.

THALIA .................................................. List each, $39.00

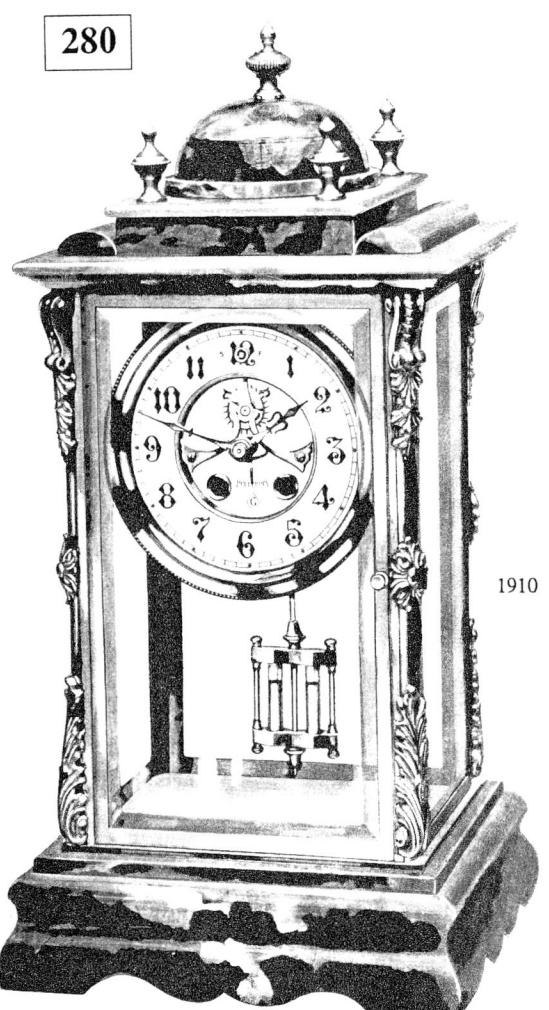

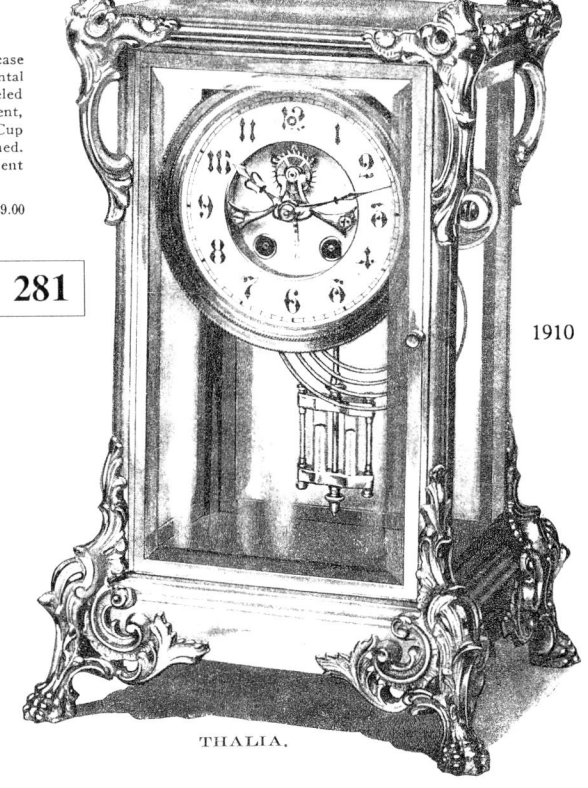

THALIA.

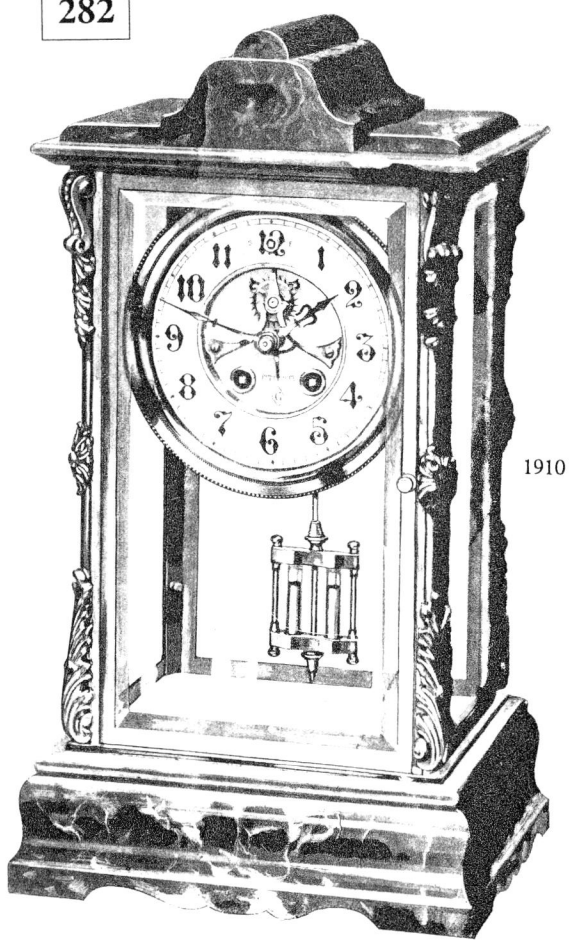

### ALGONQUIN.

Eight-Day Gilt and Onyx Regulator. Height, 14¼ inches. Width, 7½ inches. The Case is made of Rich Brazilian Onyx, with artistically curved Onyx Base and Onyx Dome Top. The Mountings are Gold Plated and Highly Polished, with heavy Ornamental Corner Case Decorations and Top Ornaments in Rich Ormolu Gold Plate Finish. Beveled Plate Glass Front, Sides and Back. Fitted with Eight-Day Movement, Hour Strike on Rich Cathedral Gong, Half-Hour on Separate Cup Bell. The Movement Plates and all Exposed Parts are Highly Polished. 4½-inch Ivory Porcelain Dial, with the Gilbert Visible Escapement and Patent Beat Adjuster. Visible Pendulum.

ALGONQUIN .................................................. List each, $60.00

### ABINGDON.

Eight-Day Gilt and Onyx Regulator. Height, 12½ inches. Width, 7 inches. The Case is made of Rich Brazilian Onyx, with artistically curved Onyx Base and Onyx Top. The Mountings are Gold Plated and Highly Polished, with heavy Ornamental Corner Case Decorations in Rich Ormolu Gold Plate Finish. Beveled Plate Glass Front, Sides and Back. Fitted with Eight-Day Movement, Hour Strike on Rich Cathedral Gong, Half-Hour on Separate Cup Bell. The Movement Plates and all Exposed Parts are Highly Polished. 4½-inch Ivory Porcelain Dial, with the Gilbert Visible Escapement and Patent Beat Adjuster. Visible Pendulum.

ABINGDON .................................................. List each, $52.00

# GILBERT

**MAGDELEINE.**

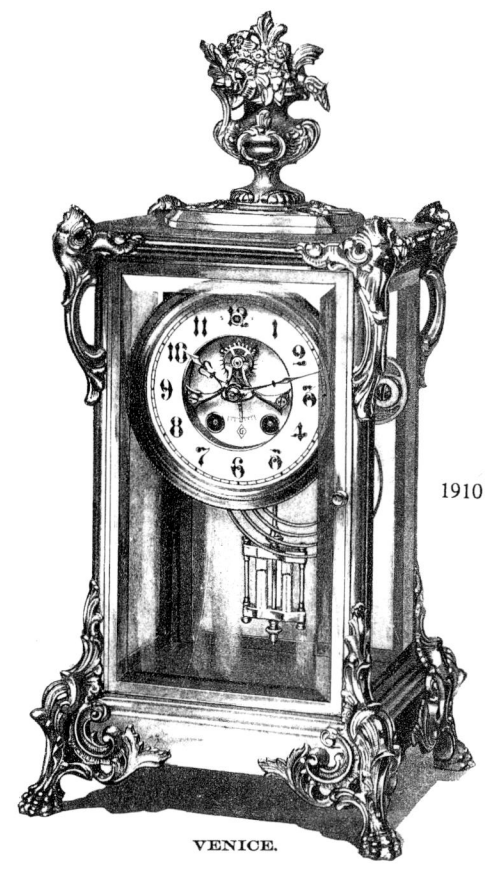

**VENICE.**

Eight-Day Gilt Regulator. Height, 15½ inches. Width, 8 inches. The Case is Gold Plated and Highly Polished, with heavy raised Ornamental Corner and Top Decorations in Rich Ormolu Gold Plate Finish. Beveled Plate Glass Front, Sides and Back. Fitted with Eight-Day Movement, Hour Strike on Rich Cathedral Gong, Half-Hour on Separate Cup Bell. The Movement Plates and all Exposed Parts are Highly Polished. 4½-inch Ivory Porcelain Dial, with the Gilbert Visible Escapement and Patent Beat Adjuster. Visible Pendulum.

VENICE .................................................. List each, $42.00

Eight-Day Gilt and Onyx Regulator. Height, 13⅞ inches. Width, 9⅞ inches. The Case is made of Rich Brazilian Onyx, with Four Round Tapering Onyx Columns. The Mountings are Gold Plated and Highly Polished, with heavy Ornamental Case Decorations in Rich Ormolu Gold Plate Finish. Beveled Plate Glass Front, Sides and Back. Fitted with Eight-Day Movement, Hour Strike on Rich Cathedral Gong, Half-Hour on Separate Cup Bell. The Movement Plates and all Exposed Parts are Highly Polished. 4½-inch Ivory Porcelain Dial, with the Gilbert Visible Escapement and Patent Beat Adjuster. Visible Pendulum.

MAGDELEINE ..................................................List each, $100.00
MAGDELEINE, with Lion Top Ornament ..................List each, 103.00

**ORLEANS.**

Eight-Day Fancy Gilt Regulator. Height, 11 inches. Width, 7 inches. The Case is made of heavy French Crystal Plate Glass, rounded Edges and Top, with heavy Mountings of artistic Scroll and Filigree Design in Rich Ormolu Gold Plate Finish. Fitted with Eight-Day Movement, Hour Strike on Rich Cathedral Gong, Half-Hour on Separate Cup Bell. Visible Pendulum. 4-inch Ivory Porcelain Dial, with the Gilbert Visible Escapement and Patent Beat Adjuster. French Sash, Beveled Glass.

ORLEANS, White Crystal Glass Case .........List each, $48.00
ORLEANS, Green Crystal Glass Case .........List each, 48.00
ORLEANS, Ruby Crystal Glass Case ..........List each, 49.00

**ORLEANS.**

# GILBERT

### TORREON.

Eight-Day Brass Applique Regulator. Height, 14 inches. Width, 8¼ inches. The Case is Solid Brass, artistically ornamented with Heavy Deposit Case Decorations in Pond-Lily Design. Rich Old Brass Finish, Dull Polished. Fitted with Eight-Day Movement, Hour and Half-Hour Strike on Rich Cathedral Gong. 4-inch Ivory Porcelain Dial, with the Gilbert Visible Escapement and Patent Beat Adjuster. French Sash, Beveled Glass.

TORREON ..................................................... List each, $57.00

TORREON.

ISABELLA.

Eight-Day Brass Applique Regulator. Height, 17½ inches. Width, 9¼ inches. The Case is Solid Brass, having Four Heavy Brass Columns supporting Top, and is artistically ornamented with Heavy Deposit Case Decorations in Pond-Lily Design. Rich Old Brass Finish, Dull Polished. Beveled Plate Glass Front, Sides and Back. Fitted with Eight-Day Movement, Hour Strike on Rich Cathedral Gong, Half-Hour on Separate Cup Bell. The Movement Plates and all Exposed Parts are Highly Polished. 4½-inch Ivory Porcelain Dial, with the Gilbert Visible Escapement and Patent Beat Adjuster. Visible Pendulum.

ISABELLA ..................................................... List each, $77.00

### FALMOUTH.

Eight-Day Brass Applique Regulator. Height, 12¼ inches. Width, 7 inches. The Case is Solid Brass, artistically ornamented with Heavy Deposit Case Decorations in Pond-Lily Design. Rich Old Brass Finish, Dull Polished. Beveled Plate Glass Front, Sides and Back. Fitted with Eight-Day Movement, Hour Strike on Rich Cathedral Gong, Half-Hour on Separate Cup Bell. The Movement Plates and all Exposed Parts are Highly Polished. 4½-inch Ivory Porcelain Dial, with the Gilbert Visible Escapement and Patent Beat Adjuster. Visible Pendulum.

FALMOUTH ..................................................... List each, $37.00

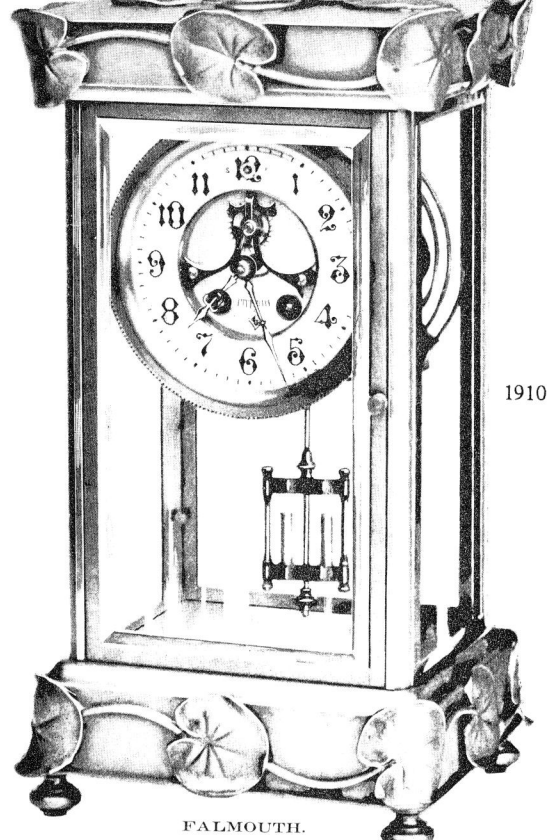

FALMOUTH.

# GILBERT

### 289

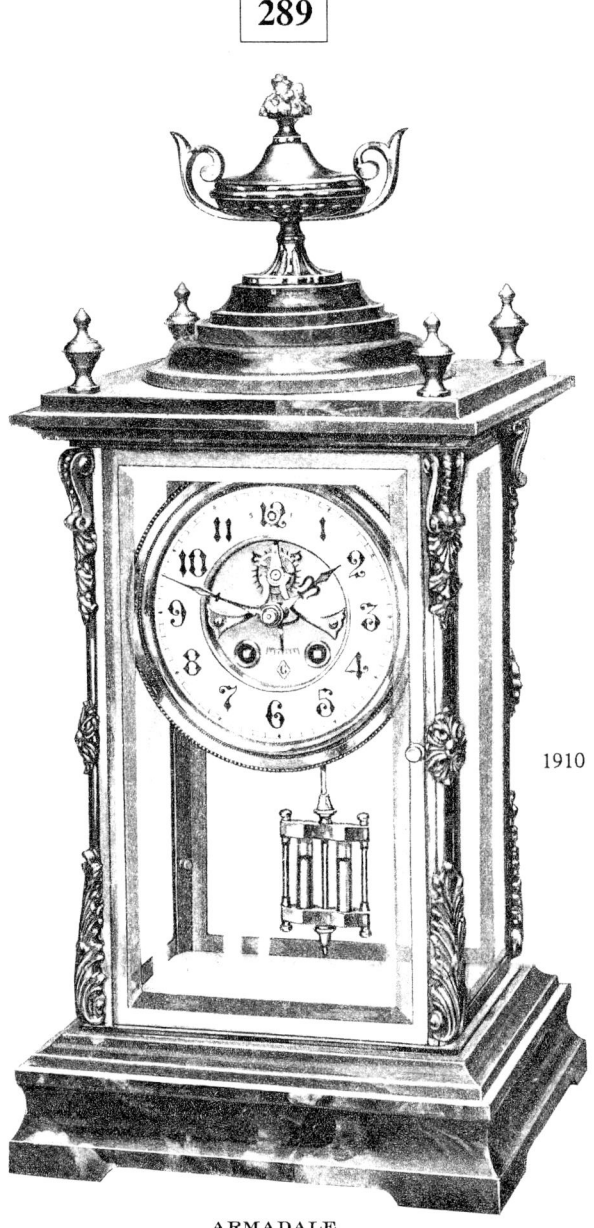

ARMADALE.

Eight-Day Gilt and Onyx Regulator. Height, 15½ inches. Width, 7½ inches. The Case is made of Rich Brazilian Onyx, with artistically curved Onyx Base and Pedestal Top. The Mountings are Gold Plated and Highly Polished with Ornamental Corner Case Decorations, Top Ornaments and Urn in Rich Ormolu Gold Plate Finish. Beveled Plate Glass Front, Sides and Back. Fitted with Eight-Day Movement, Hour Strike on Rich Cathedral Gong, Half-Hour on Separate Cup Bell. The Movement Plates and all Exposed Parts are Highly Polished. 4½-inch Ivory Porcelain Dial, with the Gilbert Visible Escapement and Patent Beat Adjuster. Visible Pendulum.

ARMADALE ................................................. List each, $65.00

### 290

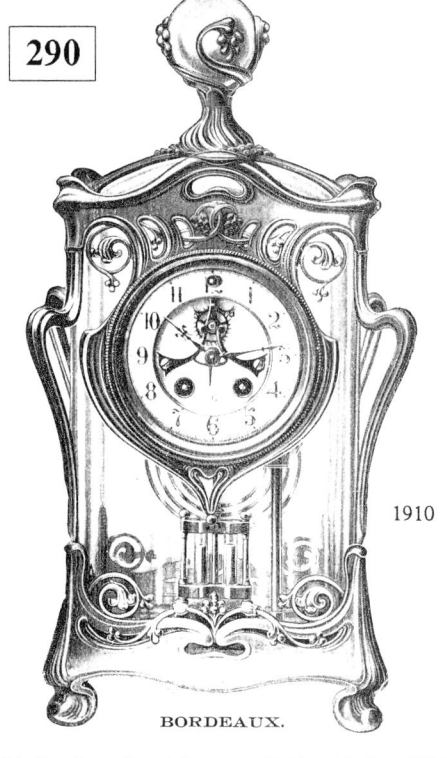

BORDEAUX.

Eight-Day Fancy Crystal Regulator. Height, 14 inches. Width, 8 inches. The Case is made of heavy French Crystal Plate Glass, rounded Edges and Top, with heavy Mountings of artistic Scroll and Filigree Design in Rich Ormolu Gold Plate Finish. Fitted with Eight-Day Movement, Hour Strike on Rich Cathedral Gong, Half-Hour on Separate Cup Bell. Visible Pendulum. The Movement Plates and all Exposed Parts are Highly Polished. 4-inch Ivory Porcelain Dial, with the Gilbert Visible Escapement and Patent Beat Adjuster. French Sash, Beveled Glass.

BORDEAUX, White Crystal Glass Case........List each, $52.00
BORDEAUX, Green Crystal Glass Case........List each, 52.00
BORDEAUX, Ruby Crystal Glass Case........List each, 53.00

### 291

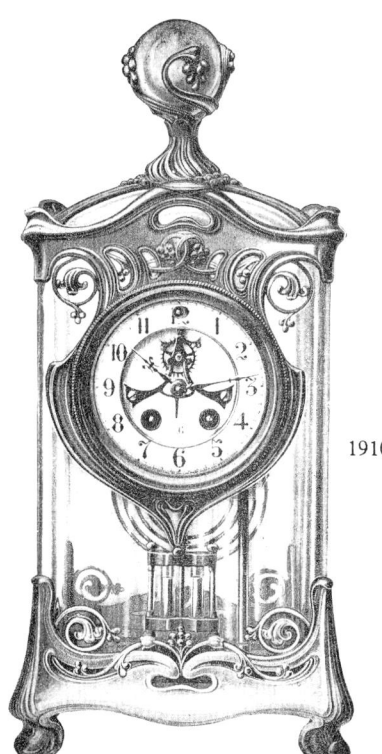

TOURAINE.

Eight-Day Fancy Crystal Regulator. Height, 14 inches. Width, 7 inches. The Case is made of heavy French Crystal Plate Glass, rounded Edges and Top, with heavy Mountings of artistic Scroll and Filigree Design in Rich Ormolu Gold Plate Finish. Fitted with Eight-Day Movement, Hour Strike on Rich Cathedral Gong, Half-Hour on Separate Cup Bell. Visible Pendulum. The Movement Plates and all Exposed Parts are Highly Polished. 4-inch Ivory Porcelain Dial, with the Gilbert Visible Escapement and Patent Beat Adjuster. French Sash, Beveled Glass.

TOURAINE, White Crystal Glass Case........List each, $50.00
TOURAINE, Green Crystal Glass Case........List each, 50.00
TOURAINE, Ruby Crystal Glass Case........List each, 51.00

# GILBERT

Most white metal and brass ornamentals and clock cases are gilded or silver plated. Great care must be taken when cleaning these surfaces. The thin layer of gilding or silver plating will be easily removed by cleaning solutions and strong rubbing. If you believe cleaning is necessary, hire the services of a professional and avoid the expense of re-gilding or re-plating.

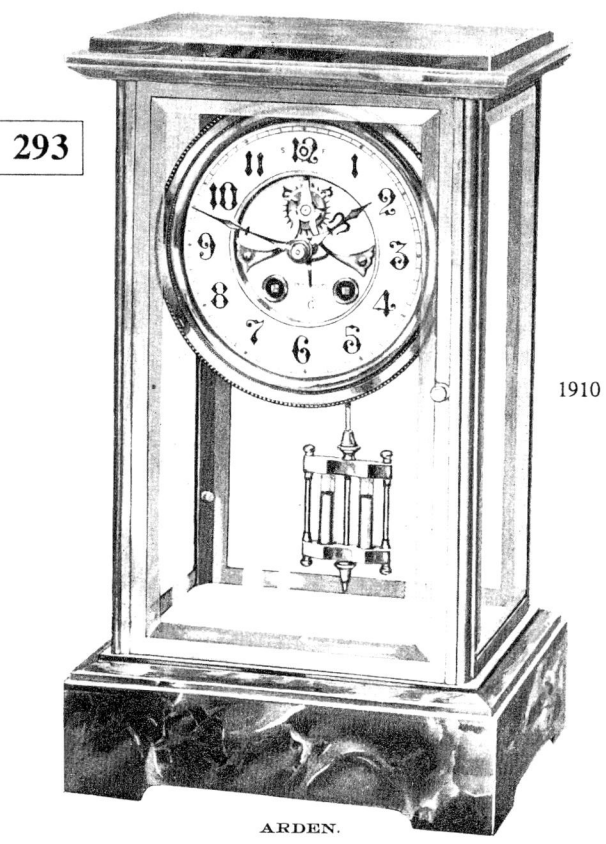

**CHANTELLE.**

Eight-Day Gilt and Onyx Regulator. Height, 12½ inches. Width, 7⅝ inches. The Case is made of Rich Brazilian Onyx, with Fancy Round Edge Onyx Base and Top. The Mountings are Gold Plated and Highly Polished. Beveled Plate Glass Front, Sides and Back. Fitted with Eight-Day Movement, Hour Strike on Rich Cathedral Gong, Half-Hour on Separate Cup Bell. The Movement Plates and all Exposed Parts are Highly Polished. 4½-inch Ivory Porcelain Dial, with the Gilbert Visible Escapement and Patent Beat Adjuster. Visible Pendulum.

CHANTELLE .................................................. List each, $75.00

**ARDEN.**

Eight-Day Gilt and Onyx Regulator. Height, 11¼ inches. Width, 7 inches. The Case is made of Rich Brazilian Onyx, with Fancy Onyx Base and Top. The Mountings are Gold Plated and Highly Polished. Beveled Plate Glass Front, Sides and Back. Fitted with Eight-Day Movement, Hour Strike on Rich Cathedral Gong, Half-Hour on Separate Cup Bell. The Movement Plates and all Exposed Parts are Highly Polished. 4½-inch Ivory Porcelain Dial, with the Gilbert Visible Escapement and Patent Beat Adjuster. Visible Pendulum.

ARDEN ......................................................... List each, $44.00

**PIERRE.**

Eight-Day Fancy Crystal Regulator. Height, 14½ inches. Width, 8 inches. The Case is made of heavy French Crystal Plate Glass, rounded Edges and Top, with heavy Mountings of artistic Scroll and Filigree Design in Rich Ormolu Gold Plate Finish. Fitted with Eight-Day Movement, Hour Strike on Rich Cathedral Gong, Half-Hour on Separate Cup Bell. Visible Pendulum. The Movement Plates and all Exposed Parts are Highly Polished. 4-inch Ivory Porcelain Dial, with the Gilbert Visible Escapement and Patent Beat Adjuster. French Sash, Beveled Glass.

PIERRE, White Crystal Glass Case............List each, $52.00
PIERRE, Green Crystal Glass Case............List each, 52.00
PIERRE, Ruby Crystal Glass Case.............List each, 53.00

**PIERRE.**

# GILBERT

### RUSKIN.

Eight-Day Brass Applique Regulator. Height, 14 inches. Width, 8¼ inches. The Case is Solid Brass, artistically ornamented with Heavy Deposit Case Decorations in Pond-Lily Design. Rich Old Brass Finish, Dull Polished. Beveled Plate Glass Front, Sides and Back. Fitted with Eight-Day Movement, Hour Strike on Rich Cathedral Gong, Half-Hour on Separate Cup Bell. The Movement Plates and all Exposed Parts are Highly Polished. 4½-inch Ivory Porcelain Dial, with the Gilbert Visible Escapement and Patent Beat Adjuster. Visible Pendulum.

RUSKIN .................................................... List each, $50.00

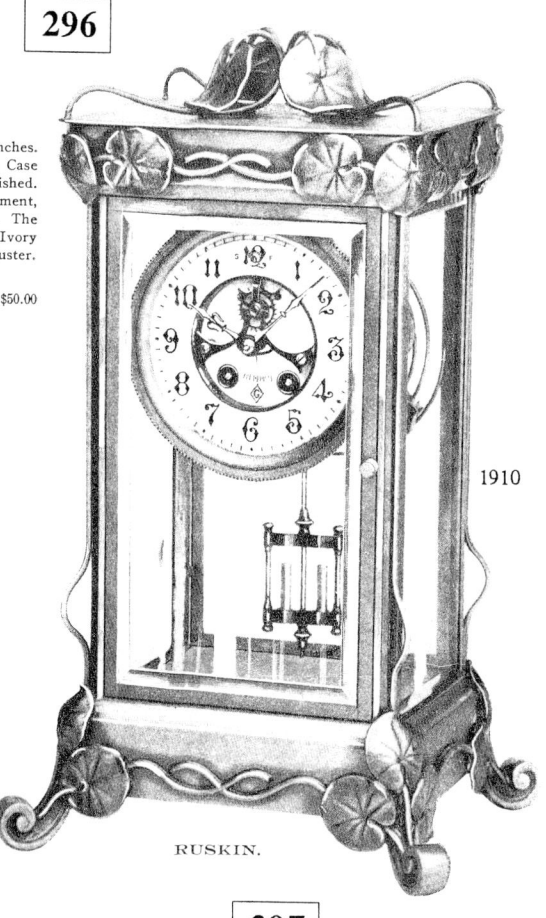

RUSKIN.

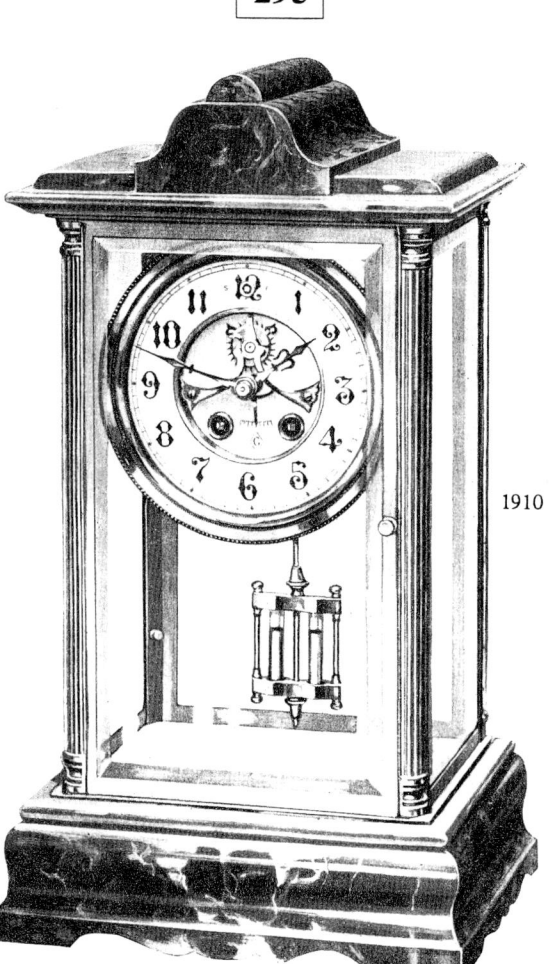

ANNESLEY.

### ANNESLEY.

Eight-Day Gilt and Onyx Regulator. Height, 12½ inches. Width, 7 inches. The case is made of Rich Brazilian Onyx, with artistically curved Onyx Base and Onyx Top. The Mountings and Fluted Corner Case Decorations are Gold Plated and Highly Polished. Beveled Plate Glass Front, Sides and Back. Fitted with Eight-Day Movement, Hour Strike on Rich Cathedral Gong, Half-Hour on Separate Cup Bell. The Movement Plates and all Exposed Parts are Highly Polished. 4½-inch Ivory Porcelain Dial, with the Gilbert Visible Escapement and Patent Beat Adjuster. Visible Pendulum.

ANNESLEY .................................................. List each, $50.00

### AVALON.

Eight-Day Brass Applique Regulator. Height, 10½ inches. Width, 6½ inches. The Case is Solid Brass, artistically ornamented with Heavy Deposit Case Decorations in Water-Lily Design. Rich Old Brass Finish, Dull Polished. Fitted with Eight-Day Movement, Hour and Half-Hour Strike on Rich Cathedral Gong. 4-inch Ivory Porcelain Dial, with the Gilbert Visible Escapement and Patent Beat Adjuster. French Sash, Beveled Glass.

AVALON .................................................... List each, $43.00

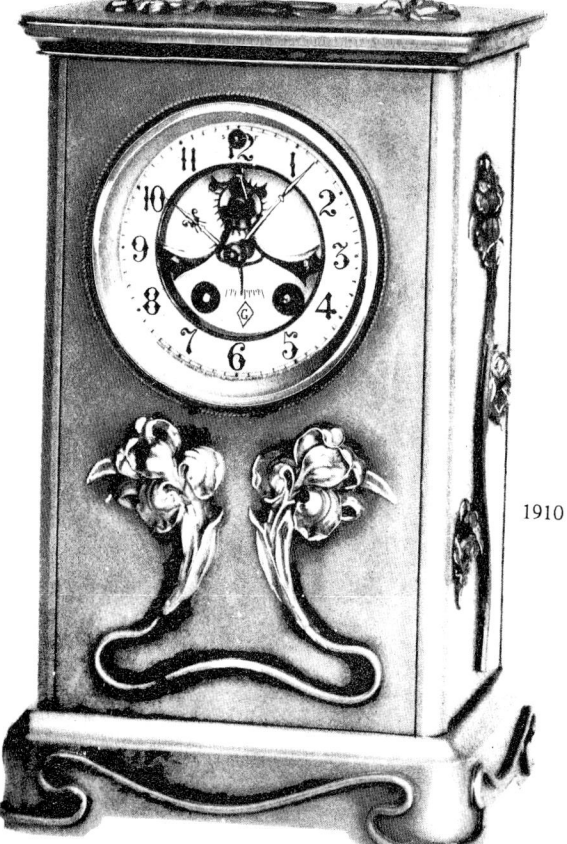

AVALON.

# GILBERT

**ORIENTAL—Oak or Walnut.**
HEIGHT, 27 INCHES. WIDTH, 15 INCHES.
8 INCH DIAL.
Eight Day Gong, Calendar only,
List each, $11 00
Not fitted with Alarm.

1901

**SHARON—Oak or Walnut.**
HEIGHT, 37 INCHES. 8 INCH DIAL.
Eight Day Calendar Gong only.
List each, . . . . $16 00

**ELBERON—Oak or Walnut.**
HEIGHT, 28 INCHES. WIDTH, 16 INCHES.
8 INCH DIAL.
Eight Day Gong, Calendar only,
List each, $11 50
Not fitted with Alarm.

1901

*The year entered beside each clock is the year of the catalog from which the illustration was taken. Some clocks may have been produced a few years earlier.*

*Only clocks in ORIGINAL MINT condition and in good running order will command top prices. A replaced or damaged dial, missing parts, tablets which are either cracked, flacking, redone will reduce prices accordingly.*

# GILBERT

*For comprehensive listing of Calendar Clocks, see Tran Duy Ly "Calendar Clocks - A Guide to Identification & Prices."*

**301**

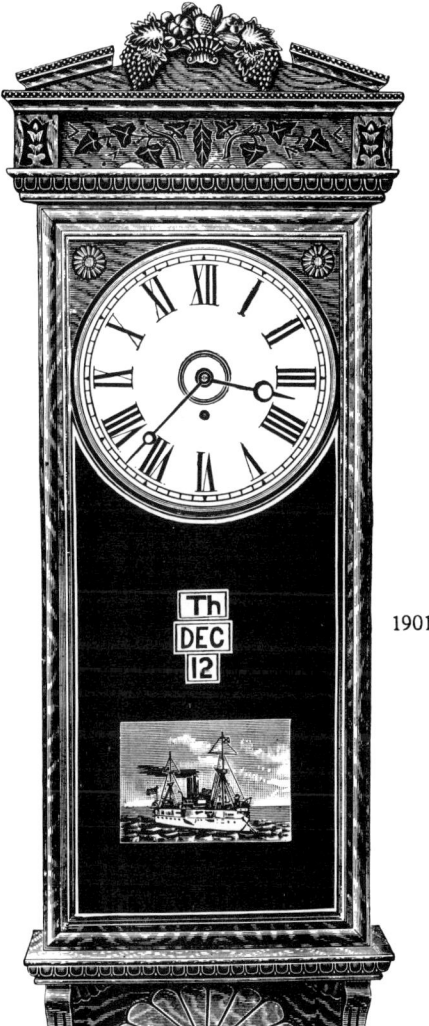

**MAINE—Oak Only.**
HEIGHT, 49½ INCHES. WIDTH, 20 INCHES.
12 INCH DIAL.
Eight Day Strike Gong Calendar.
List each, . . . . . $19 00

**302**

**OFFICE DROP CALENDAR—Walnut.**
*(Maranville Patent.)*
HEIGHT, 34 INCHES. 15 INCH DIAL.
Eight Day Time Calendar,
List each, $12 00
Eight Day Strike Wire Bell Calendar, . . . List each. 13 50
**Also made with Spanish or Portuguese Calendar.**

**303**

**BERKSHIRE—Oak or Walnut.**
HEIGHT, 38 INCHES. 8 INCH DIAL
Eight Day Calendar Gong only.
List each, . . . . . $15 50

**304**

**LENOX—Oak or Walnut.**
HEIGHT, 35½ INCHES. 8 INCH DIAL.
Eight Day Calendar Gong only.
List each, . . . . . $15 00

# GILBERT

**OSSA.**
4½ INCH DIAL. HEIGHT, 6 INCHES.
**One Day Nickel, Time, Calendar.**
List each, . . . . . $1 90
**One Day, Nickel, Alarm, Calendar.**
List each, . . . . . $2 10

**REGULATOR—B—Oak Only.**
HEIGHT, 29 INCHES. 12 INCH DIAL.

| | | |
|---|---|---|
| Eight Day Time, | List each, | $6 50 |
| Eight Day Strike Wire Bell, | List each, | 7 50 |
| Eight Day Time Calendar, | List each, | 7 00 |
| Eight Day Strike Wire Bell Calendar, | List each, | 8 00 |

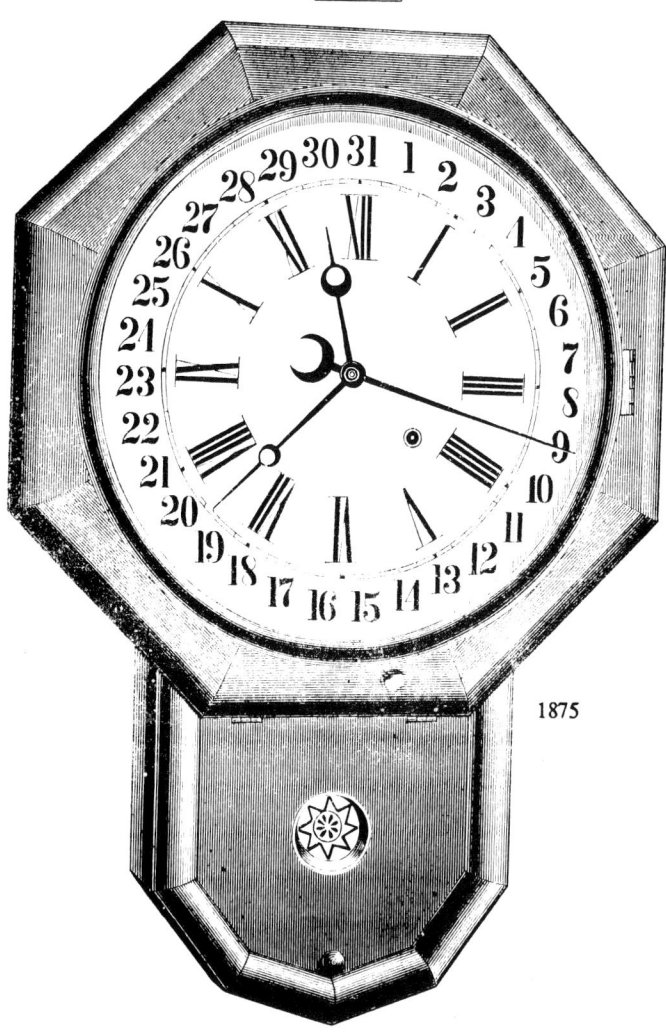

**ADMIRAL—Oak Only.**
HEIGHT, 26¾ INCHES. 12 INCH DIAL.

| | | |
|---|---|---|
| Eight Day Time Calendar, | List each, | 5 90 |
| Eight Day Strike Wire Bell Calendar, | List each, | 6 90 |

**DROP OCTAGON CALENDAR**
15½ wide, 22½ high.
EIGHT DAY SPRING. STRIKE OR TIME.

# GILBERT

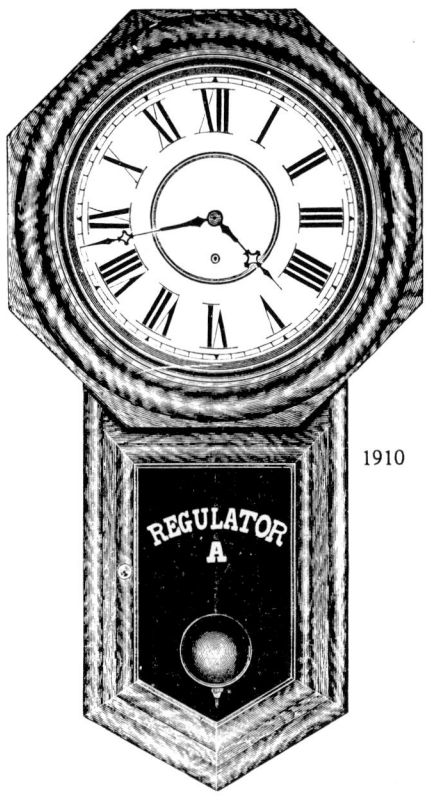

**REGULATOR G.**
Eight-Day Hanging Spring Clock. Height, 31 inches. 12-inch Dial. Fitted with Eight-Day Time or Strike Movement, as listed below. Oak Finish only
REGULATOR G, Time .............. List each, $9.90
REGULATOR G, Wire Bell Strike.. List each, 10.90
REGULATOR G, Cathedral Gong Strike,
　　　　　　　　　　　　　List each, 11.40

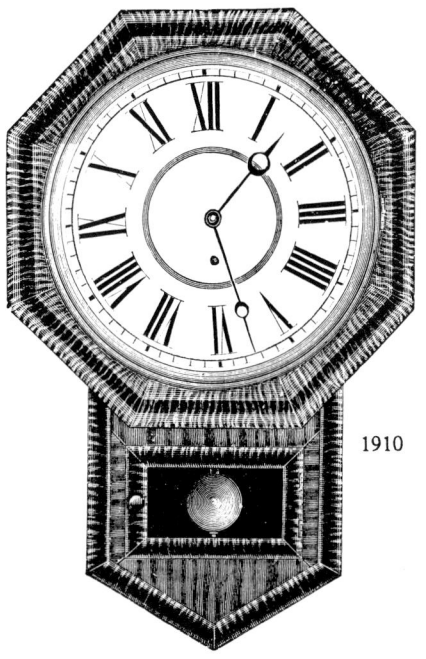

**12-INCH OCTAGON DROP.**
Rosewood Veneered, or Solid Oak.
Height, 25½ inches.　　　　12-inch Dial.
Eight-Day Time ...................... List each, $6.50
Eight-Day Strike Wire Bell ......... List each, 7.50
Eight-Day Time Calendar ........... List each, 7.00
Eight-Day Strike Wire Bell Calendar. List each, 8.00

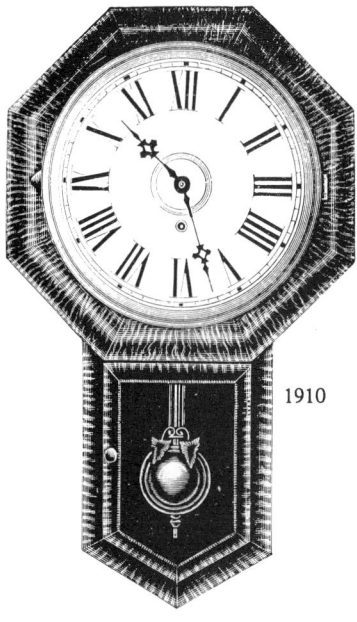

**10-INCH OCTAGON DROP.**
Rosewood Veneered, or Solid Oak.
Height, 23½ inches.　　　　10-inch Dial.
Eight-Day Time ....................... List each, $6.10
Eight-Day Strike Wire Bell ......... List each, 7.10

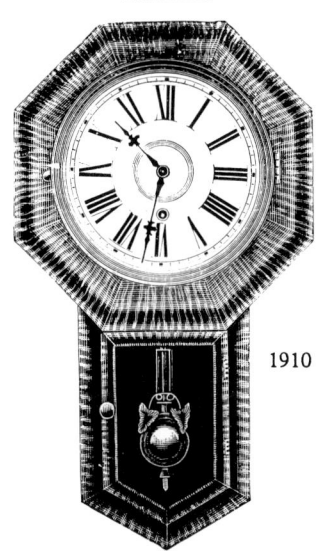

**8-INCH OCTAGON DROP.**
Rosewood Veneered, or Solid Oak.
Height, 20½ inches.　　　　8-inch Dial.
Eight-Day Time ..................... List each, $5.75
Eight-Day Strike Wire Bell ......... List each, 6.75

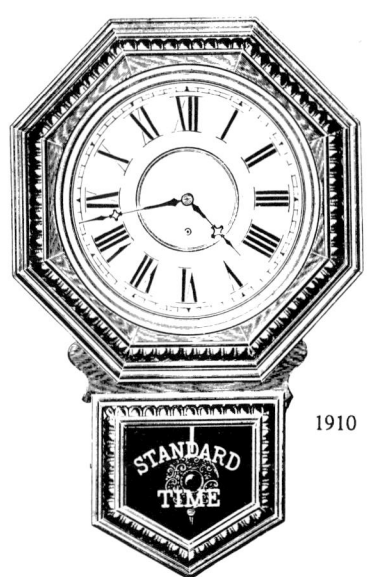

**ADMIRAL.**
Eight-Day Hanging Spring Clock. Height, 26¾ inches. 12-inch Dial. Fitted with Eight-Day Time or Strike Movement, as listed below. Oak Finish only.
ADMIRAL, Time .................... List each, $5.55
ADMIRAL, Wire Bell Strike ........ List each, 6.55

**CONSORT—Oak Only.**
HEIGHT, 31 INCHES.　12 INCH DIAL.
Eight Day Time, . . . . . . List each, $6 00
Eight Day Strike Wire Bell, . . List each, 7 00
Eight Day Time Calendar, . . List each, 6 50
Eight Day Strike Wire Bell Calendar, . List each, 7 50

# GILBERT

### 315

**JANEIRO.**—Mosaic Veneered.

Height, 22 inches.   10-inch Dial.
Eight-Day Time .......................List each, $6.25
Eight-Day Time Calendar ............List each, 6.75
Eight-Day Strike Wire Bell..........List each, 7.25
Eight-Day Strike Wire Bell Calendar.List each, 7.75

### 316
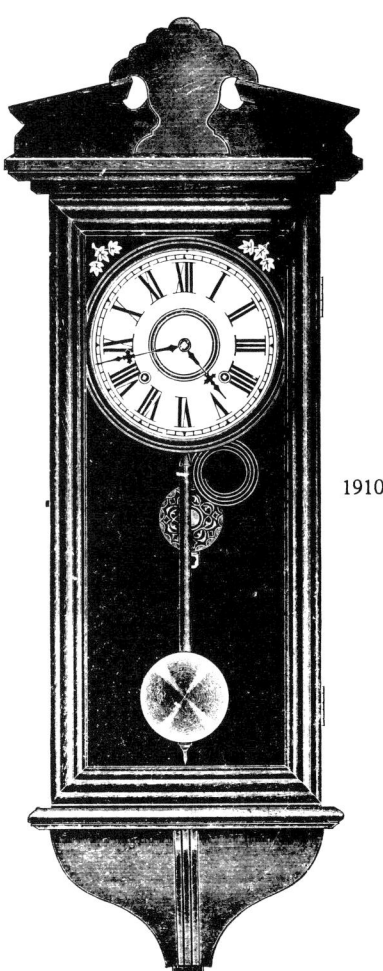

**BARKSDALE.**

Eight-Day Hanging Spring Clock. Height, 37 inches. 8-inch Dial. The Case is a neat Design, with Ornamental Mouldings. Fitted with Eight-Day Time or Strike Movement, as listed below. Oak or Cherry (Mahogany Finish).
BARKSDALE, Time .................List each, $8.50
BARKSDALE, Wire Bell Strike.....List each, 9.50
BARKSDALE, Cathedral Gong Strike.List each, 10.00

### 317

**JANEIRO.**—Rosewood Veneered.

Height, 22 inches.   10-inch Dial.
Eight-Day Time .......................List each, $6.25
Eight-Day Time Calendar ............List each, 6.75
Eight-Day Strike Wire Bell ..........List each, 7.25
Eight-Day Strike Wire Bell Calendar.List each, 7.75

### 318
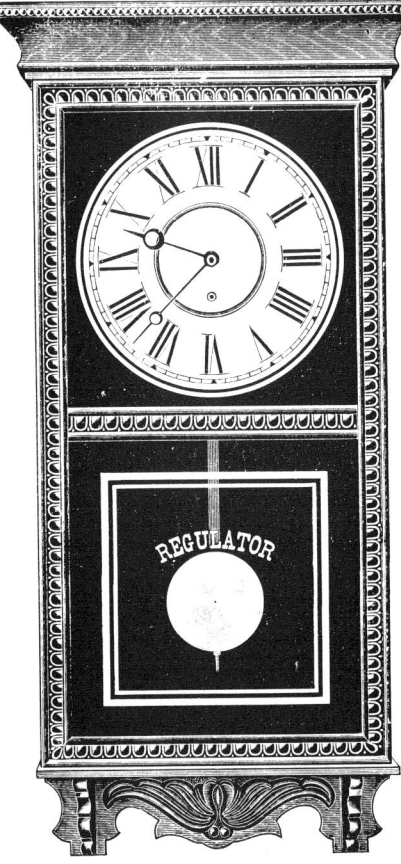

### 319

**OBSERVATORY.**

Eight-Day Hanging Spring Clock. Height, 37 inches. Width, 15¾ inches. 12-inch Dial. Fitted with Eight-Day Time or Strike Movement, as listed below. Oak Finish only.
OBSERVATORY, Time .............List each, $6.50
OBSERVATORY, Wire Bell Strike..List each, 7.50

**WASHINGTON.**

Eight-Day Hanging Spring Clock. Height, 38 inches. Width, 19 inches. 12-inch Dial. Fitted with Eight-Day Time or Strike Movement, as listed below. Oak Finish only.
WASHINGTON, Time ..............List each, $6.70
WASHINGTON, Wire Bell Strike...List each, 7.70

# GILBERT

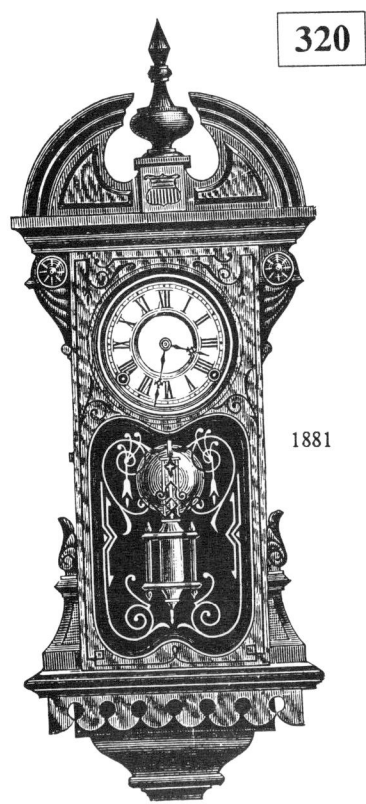

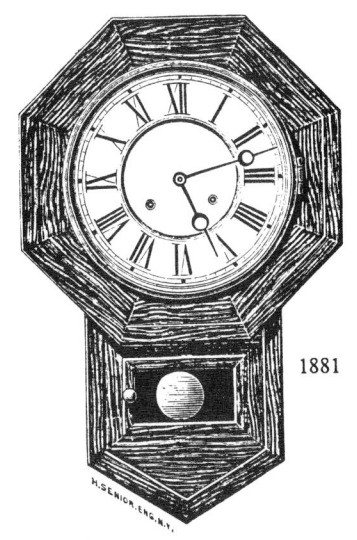

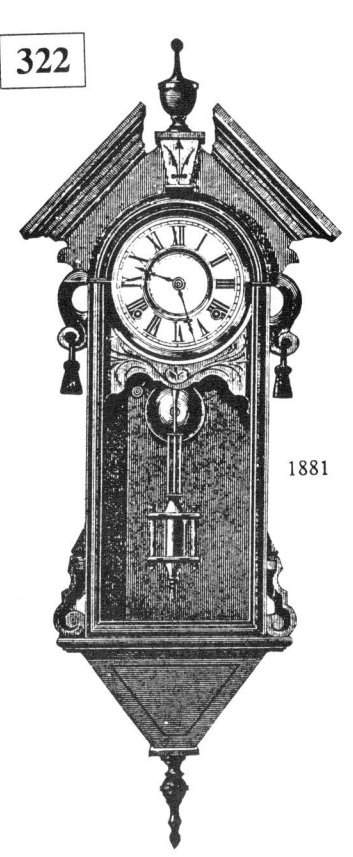

1881

**SHIELD.**
Eight-day Spring Strike.

Height, 29 inches; 6 inch Dial.

**OCTAGON DROP.**
Eight-day Time. Eight-day Strike, Spring.

Height, 25½ inches; 12 inch Dial.

**AMERICA.**
Eight-day Time.
Eight-day Strike, Spring.

Height, 32 inches; 6 inch Dial.

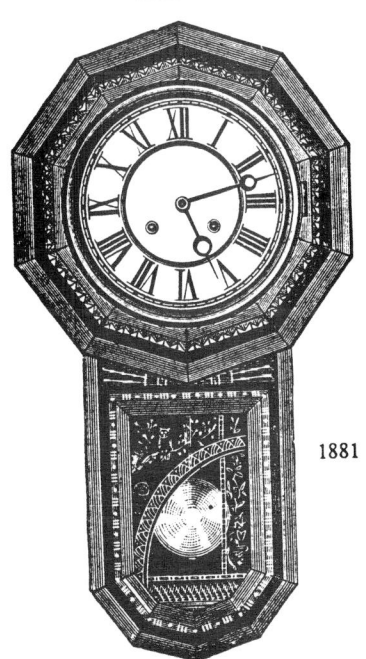

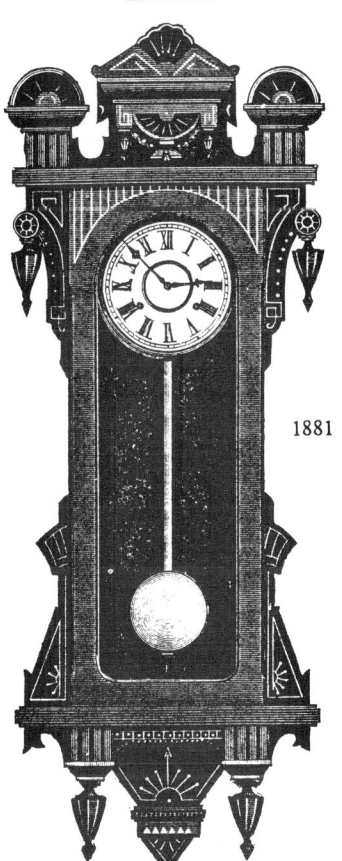

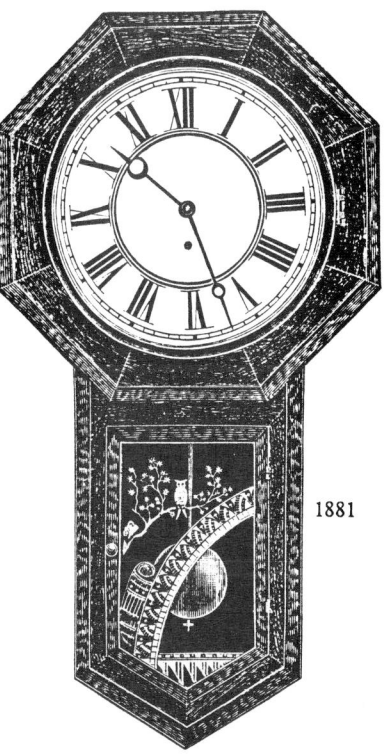

**STAR DROP.**
(WALNUT.)

Eight-day Time. Eight-day Strike, Spring.
Height, 32½ inches; 12 inch Dial.

**ITASCA.**
Eight-day Time. Eight-day Strike, Spring

Height, 46 inches; 8 inch Dial.

**RESOLUTE.**
Eight-day Time. Eight-day Strike, Spring.

Height, 31½ inches; 12 inch Dial.

# GILBERT

**326**

**GIRARD.**

Eight-Day Hanging Clock. Height, 29 inches. 6-inch Dial. Fitted with Eight-Day Half-Hour Strike Movement. Oak or Walnut Finish.

| | | |
|---|---|---|
| GIRARD, Wire Bell Strike | List each, | $5.00 |
| GIRARD, Wire Bell Alarm | List each, | 5.50 |
| GIRARD, Gong Strike | List each, | 5.50 |
| GIRARD, Gong Alarm | List each, | 6.00 |

**327**

**CAMBRIDGE.**

Eight-Day Hanging Clock. Height, 29 inches. 6-inch Dial. Fitted with Eight-Day Half-Hour Strike Movement. Oak or Walnut Finish.

| | | |
|---|---|---|
| CAMBRIDGE, Wire Bell Strike | List each, | $5.00 |
| CAMBRIDGE, Wire Bell Alarm | List each, | 5.50 |
| CAMBRIDGE, Gong Strike | List each, | 5.50 |
| CAMBRIDGE, Gong Alarm | List each, | 6.00 |

*Only clocks in ORIGINAL MINT condition and in good running order will command top prices. A replaced or damaged dial, missing parts, tablets which are either cracked, flacking, redone will reduce prices accordingly.*

**328**

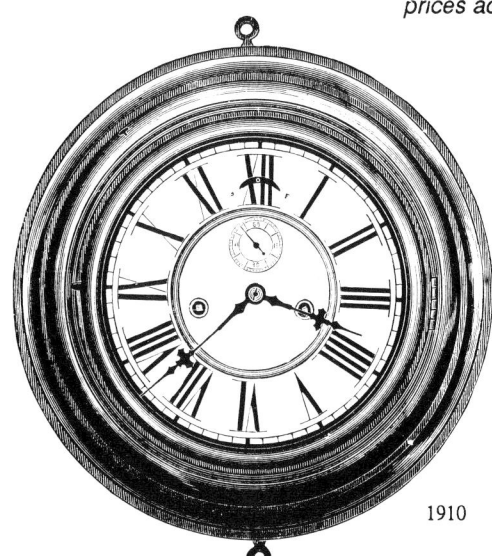

**BRASS LEVER.**—One Day.

| | | |
|---|---|---|
| One-Day, 6-inch, Brass Lever, Time | List each, | $3.90 |
| One-Day, 6-inch, Brass Lever, Strike | List each, | 4.75 |
| One-Day, 8-inch, Brass Lever, Time | List each, | 4.65 |
| One-Day, 8-inch, Brass Lever, Strike | List each, | 5.50 |

**329**

**OXFORD.**

Eight-Day Hanging Clock. Height, 28 inches. 6-inch Dial. Fitted with Eight-Day Half-Hour Strike Movement. Oak or Walnut Finish.

| | | |
|---|---|---|
| OXFORD, Wire Bell Strike | List each, | $5.00 |
| OXFORD, Wire Bell Alarm | List each, | 5.50 |
| OXFORD, Gong Strike | List each, | 5.50 |
| OXFORD, Gong Alarm | List each, | 6.00 |

# GILBERT

**330**

1910

**DREXEL.**

Eight-Day Hanging Clock. Height, 24 inches. 6-inch Dial. Fitted with Eight-Day Half-Hour Strike Movement. Oak Finish only.

DREXEL, Wire Bell Strike .........List each, $4.25
DREXEL, Wire Bell Alarm .........List each, 4.75
DREXEL, Gong Strike ............List each, 4.75
DREXEL, Gong Alarm ............List each, 5.25

**ROUND CORNER LEVER.—Eight Day.**

Eight-Day, 6-inch, Round Corner, Time .............. List each, $5.10
Eight-Day, 8-inch, Round Corner, Time .............. List each, 5.75
Eight-Day, 8-inch, Round Corner, Strike ............. List each, 7.70
Eight-Day, 12-inch, Round Corner, Time ............. List each, 6.75
Eight-Day, 12-inch, Round Corner, Strike ........... List each, 8.50

**331** **332**

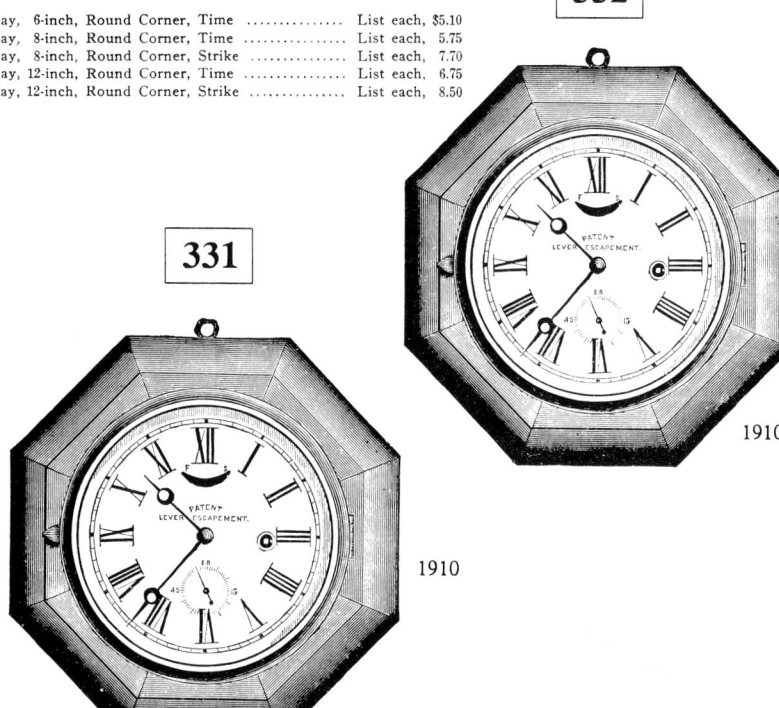

1910

**ROUND CORNER LEVER.—One Day.**

One-Day, 4-inch, Round Corner, Time ................. List each, $2.75
One-Day, 6-inch, Round Corner, Time ................. List each, 3.90
One-Day, 6-inch, Round Corner, Strike ............... List each, 4.75
One-Day, 8-inch, Round Corner, Time ................. List each, 4.65
One-Day, 8-inch, Round Corner, Strike ............... List each, 5.50

**333**

**DUNDEE.**

Eight-Day Hanging Clock. Height, 24 inches. 6-inch Dial. Fitted with Eight-Day Half-Hour Strike Movement. Oak Finish only.

DUNDEE, Wire Bell Strike .........List each, $4.25
DUNDEE, Wire Bell Alarm .........List each, 4.75
DUNDEE, Gong Strike ............List each, 4.75
DUNDEE, Gong Alarm ............List each, 5.25

1910

**334**

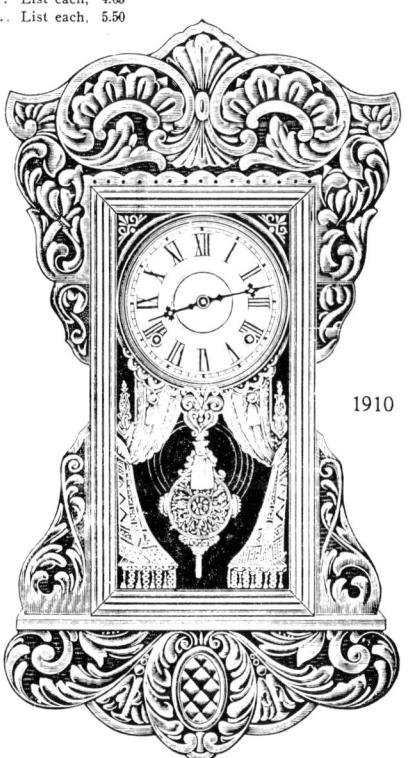

1910

**DURHAM.**

Eight-Day Hanging Clock. Height, 24 inches. 6-inch Dial. Fitted with Eight-Day Half-Hour Strike Movement. Oak Finish only.

DURHAM, Wire Bell Strike .........List each, $4.25
DURHAM, Wire Bell Alarm .........List each, 4.75
DURHAM, Gong Strike ............List each, 4.75
DURHAM, Gong Alarm ............List each, 5.25

# GILBERT

*The year entered beside each clock is the year of the catalog from which the illustration was taken. Some clocks may have been produced a few years earlier.*

**335**

**WALNUT HANGING WEIGHT.**

Thirty Hour Weight. Strike.

47 inches high.

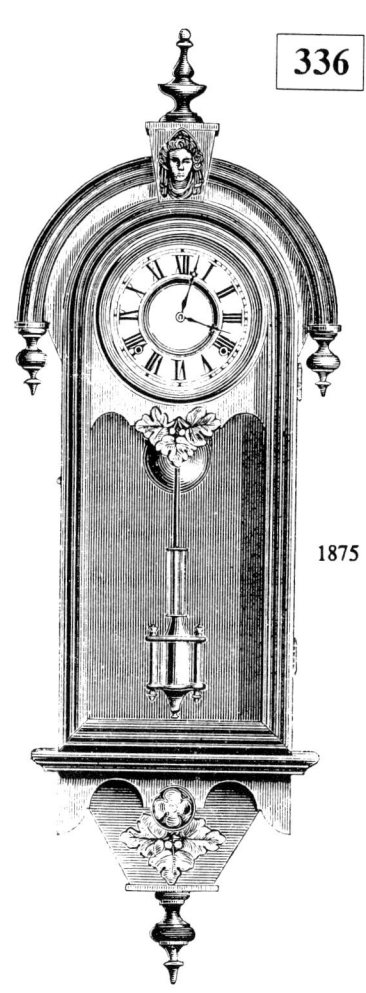

**336**

1875

**WALNUT PENDANT.**

35 inches high.

Eight Day Spring. Strike.

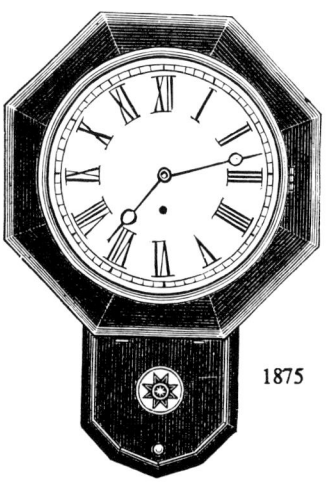

**338**

1875

**OCTAGON DROP.**

15½ wide, 22½ high.

Eight Day Spring. Strike or Time.

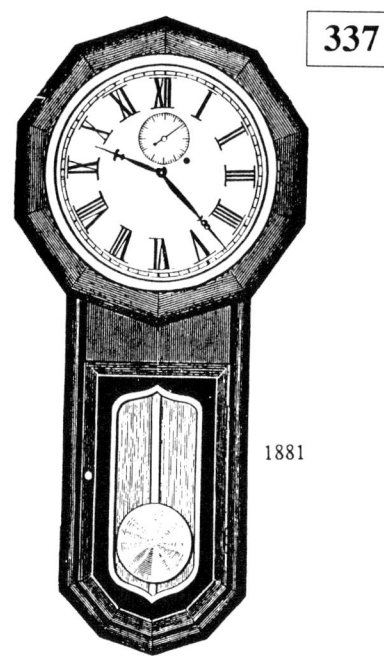

**337**

1881

**REGULATOR No. 2.**

Eight-day Time. Weight.

Dead-Beat Escapement.

Height, 33¼ inches; 12 inch Dial.

**339**

1910

**REGULATOR B.**

Eight-Day Hanging Spring Clock. Height, 29 inches. 12-inch Dial. Fitted with Eight-Day Time or Strike Movement, as listed below. Oak Finish only.

REGULATOR B, Time ..............List each, $6.50
REGULATOR B, Time Calendar ....List each, 7.00
REGULATOR B, Wire Bell Strike..List each, 7.50
REGULATOR B, Wire Bell Strike Calendar,
List each, 8.00

# GILBERT

To receive top prices, clocks must be in ORIGINAL MINT condition and in good running order - A replaced or damaged dial, missing parts, will reduce prices accordingly.

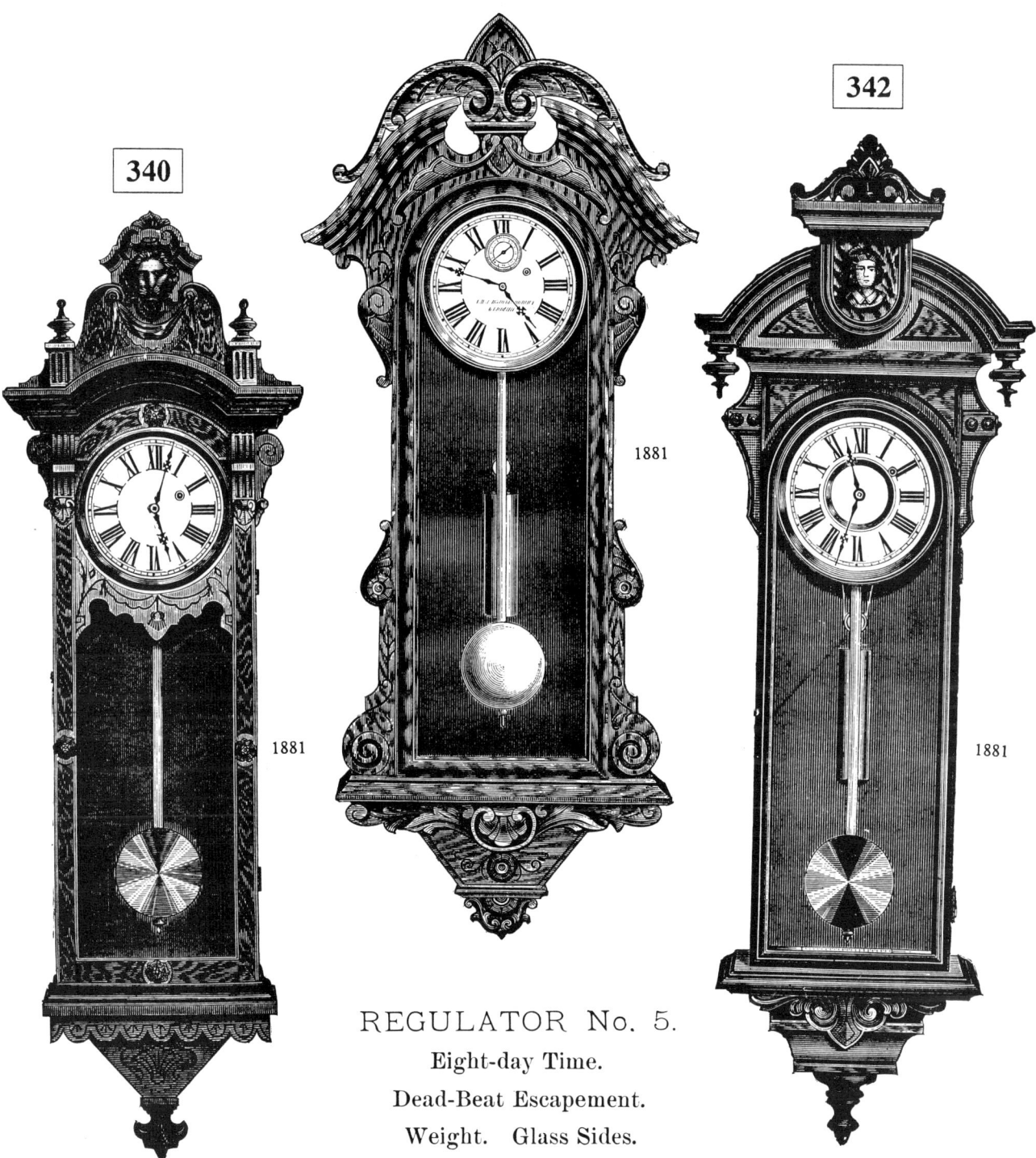

**340**

REGULATOR No. 6.
Weight.   Glass Sides.
Eight-day Time.
Dead-Beat Escapement.

Height, 52 inches ;  8 inch Dial.

**341**

REGULATOR No. 5.
Eight-day Time.
Dead-Beat Escapement.
Weight.   Glass Sides.

Height, 46 inches ;  8 inch Dial.

**342**

REGULATOR No. 4.
Weight.   Glass Sides.
Dead-Beat Escapement.
Eight-day Time.

8 inch Dial.

# GILBERT

**PARACHUTE.**
**WALNUT**
15¼ inches high.
Eight Day Time.

**ANCHOR LEVER.**
One-day Time.
Height, 14 inches; 3½ inch Dial.

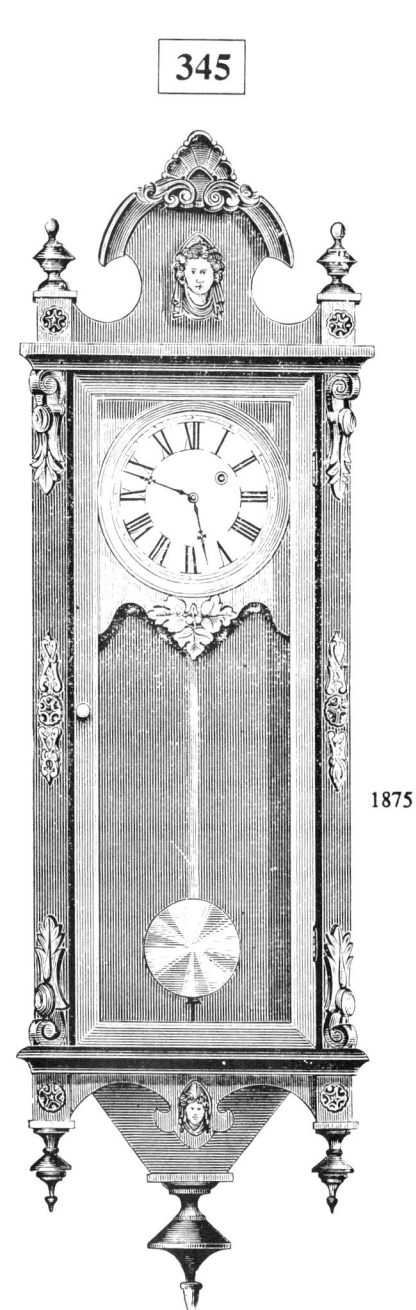

**No. 3 REGULATOR.**
Eight Day Time.
Dead Beat Escapement.
50 inches high.

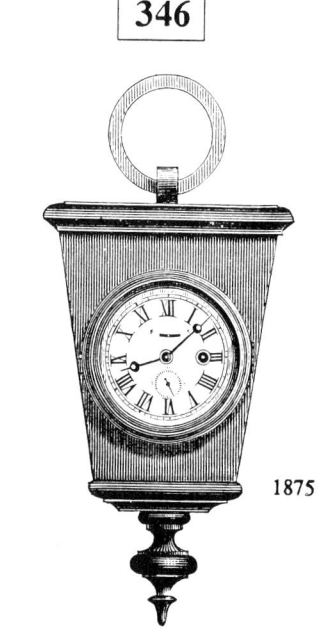

**MASONIC LEVER.**
**WALNUT**
14½ inches high.
Thirty Hour, 4 inch. Time.

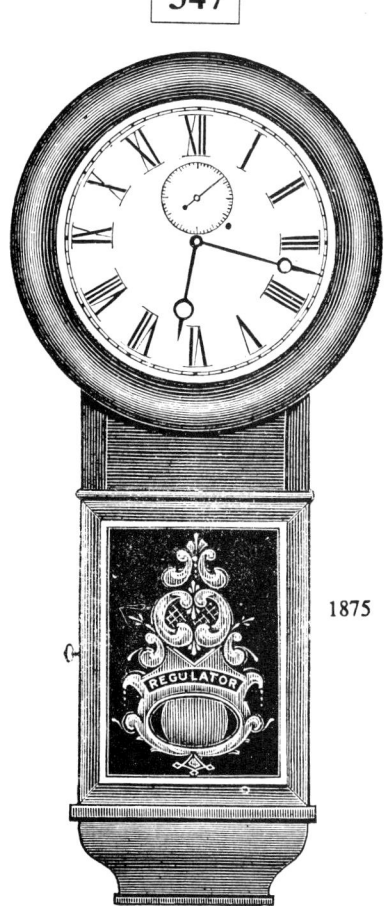

**No. 1 REGULATOR.**
33½ inches high.
Eight Day Time.
Dead-Beat Escapement.

# GILBERT

**348**

### GALLERY.
#### 12-INCH DIAL.

Eight-Day Hanging Spring Gallery Clock. Back Measurement of Case, 18 inches. Depth of Case, 5 inches. Fitted with Eight-Day Pendulum Movement, Time only. Oak Case, Highly Polished Cabinet Finish.

GALLERY, Time, 12-inch Dial .....List each, $14.00

**349**

**350**

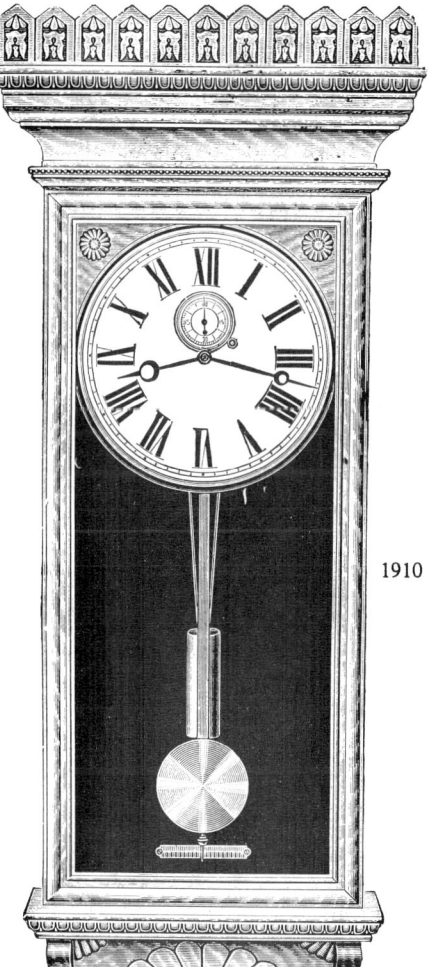

**351**

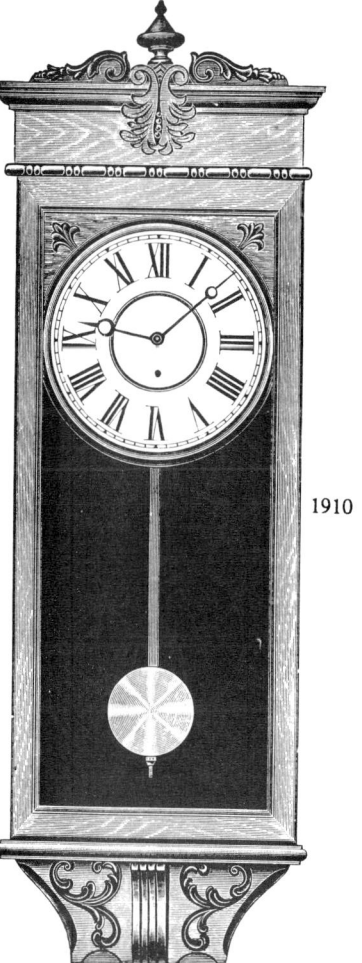

### DEFENDER.

Height, 49½ inches. Width, 20 inches. 12-inch Dial. The Case is handsomely carved and finished in Oak only. Fitted with Eight-Day Spring Pendulum Movement, Time or Strike, as listed below.

DEFENDER, Time ................List each, $14.90
DEFENDER, Cathedral Gong Strike.List each, 16.40

### REGULATOR No. 14.

Height, 50 inches. Width, 20 inches. 12-inch Dial, with Second Hand. The Case is handsomely carved and finished in Oak only. Fitted with Eight-Day Weight Movement, Time only. Dead Beat Escapement and Retaining Power.

REGULATOR No. 14 .............List each, $24.00

### CONSTITUTION.

Height, 49¼ inches. Width, 17¼ inches. 12-inch Dial. The Case is handsomely carved and finished in Oak only. Fitted with Eight-Day Spring Pendulum Movement, Time or Strike, as listed below.

CONSTITUTION, Time ............List each, $16.50
CONSTITUTION, Cathedral Gong Strike,
List each, 18.00

# GILBERT

### 352

**SARATOGA—Oak or Walnut.**
HEIGHT, 37 INCHES. 8 INCH DIAL.
Eight Day Time, . List each, $11 00
Eight Day Cathedral Gong only,
  List each, . . . 12 50
Eight Day Time Calendar,
  List each, . . . 11 50
Eight Day Cathedral Gong
  Calendar, . List each, 13 00

### 353

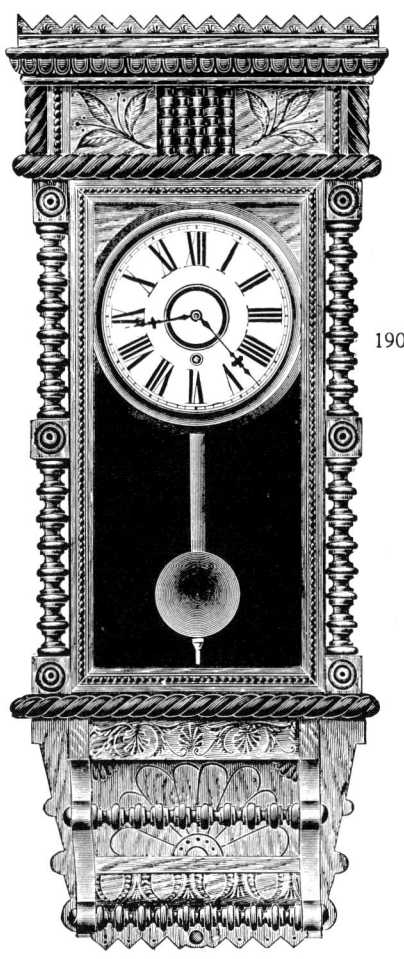

**BRIGHTON—Oak or Walnut.**
HEIGHT, 38 INCHES. 8 INCH DIAL.
Eight Day Time, . List each, $10 00
Eight Day Cathedral Gong only,
  List each, . . 11 50

### 354

**NEWPORT—Oak or Walnut.**
HEIGHT, 35½ INCHES. 8 INCH DIAL.
Eight Day Time, . List each, $9 00
Eight Day Cathedral Gong only,
  List each, . . . 10 50

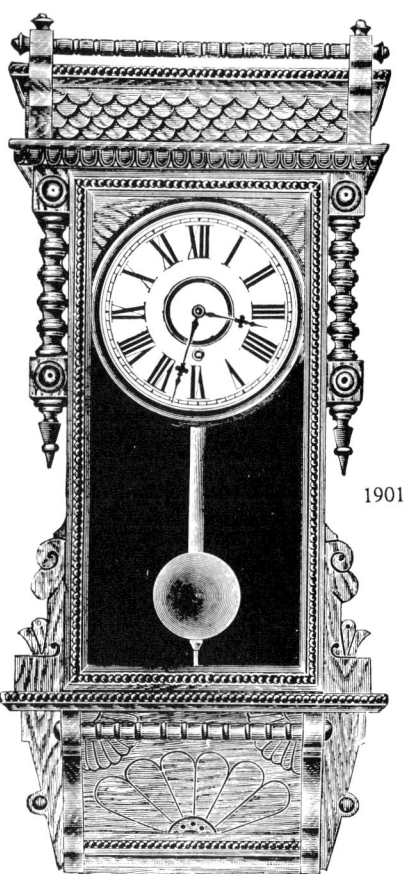

### 355

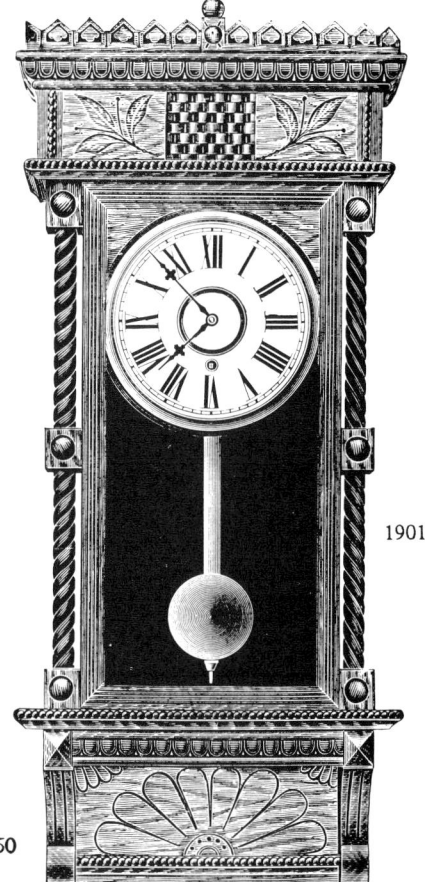

**ASBURY—Oak or Walnut.**
HEIGHT, 37 INCHES. 8 INCH DIAL.
Eight Day Time, . List each, $9 50
Eight Day Cathedral Gong only,
  List each, . . . 11 00

# GILBERT

*Except for those fabulous clocks for which you wouldn't mind cutting a hole in the ceiling, Grandfather or Hanging Clocks under eight feet are the most desirable and practical. Some manufacturers produced certain of these clocks in two styles: one to stand on the floor, and the other to hang on a wall. Of the clocks designed with two styles, the hanging style is considered to be the most valuable.*

356      357      358

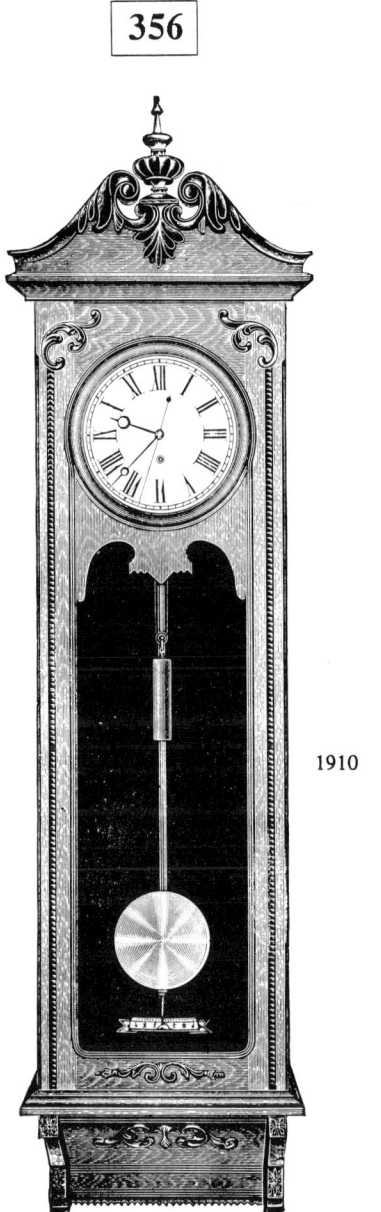

1910

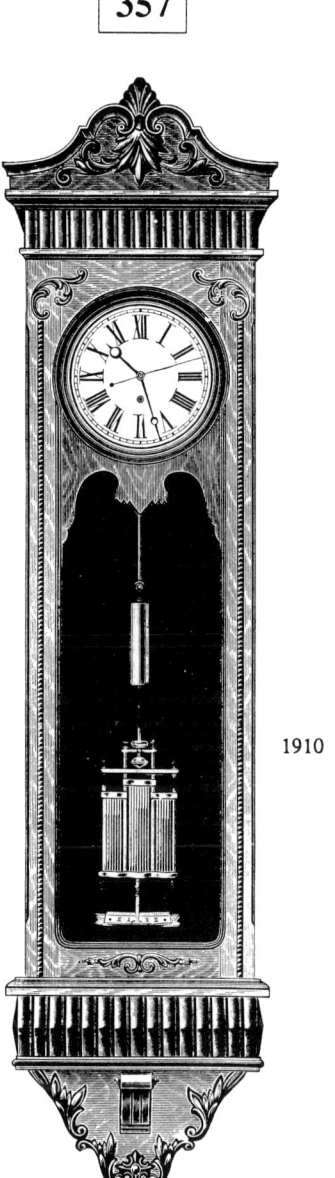

1910

1910

**REGULATOR No. 20, HANGING.**
Wood Rod.
Oak or Cherry (Mahogany Finish).
Height, 6 feet, 11 inches. Width, 20½ inches. 12-inch Porcelain Dial, with Sweep Second Hand. The Case is handsomely carved and has a Fine Cabinet Finish. Fitted with Eight-Day Weight Movement, Dead Beat Escapement and Retaining Power. Wood Rod, Brass Pendulum Ball.
REGULATOR No. 20, HANGING, Wood Rod,
List each, $75.00

**REGULATOR No. 22, HANGING**
Oak or Cherry (Mahogany Finish).
Height, 7 feet, 3½ inches. Width, 21¾ inches. 12-inch Porcelain Dial, with Sweep Second Hand. The Case is handsomely carved and has a Fine Cabinet Finish. Fitted with Eight-Day Weight Movement, Dead Beat Escapement and Retaining Power. Mercurial Compensating Pendulum, with Mercury in Three Cut-glass Jars. Also supplied with Gridiron Pendulum, and Wood Rod with Brass Ball Pendulum, as listed below.
REGULATOR No. 22, HANGING, Mercurial
Pendulum ..................List each, $125.00
REGULATOR No. 22, HANGING, Gridiron
Pendulum ..................List each, 105.00
REGULATOR No. 22, HANGING, Wood Rod,
List each, 95.00

**REGULATOR No. 20, HANGING.**
Oak or Cherry (Mahogany Finish).
Height, 6 feet, 11 inches. Width, 20½ inches. 12-inch Porcelain Dial, with Sweep Second Hand. The Case is handsomely carved and has a Fine Cabinet Finish. Fitted with Eight-Day Weight Movement, Dead Beat Escapement and Retaining Power. Mercurial Compensating Pendulum, with Mercury in Three Cut-glass Jars. Also supplied with Gridiron Pendulum, as listed below.
REGULATOR No. 20, HANGING, Mercurial
Pendulum ..................List each, $105.00
REGULATOR No. 20, HANGING, Gridiron
Pendulum ..................List each, 85.00

# GILBERT

*The year entered beside each clock is the year of the catalog from which the illustration was taken. Some clocks may have been produced a few years earlier.*

**359**

UNA DROP

Eight-day Spring, Strike.
Height, 28¼ inches; 6 inch Dial.

1881

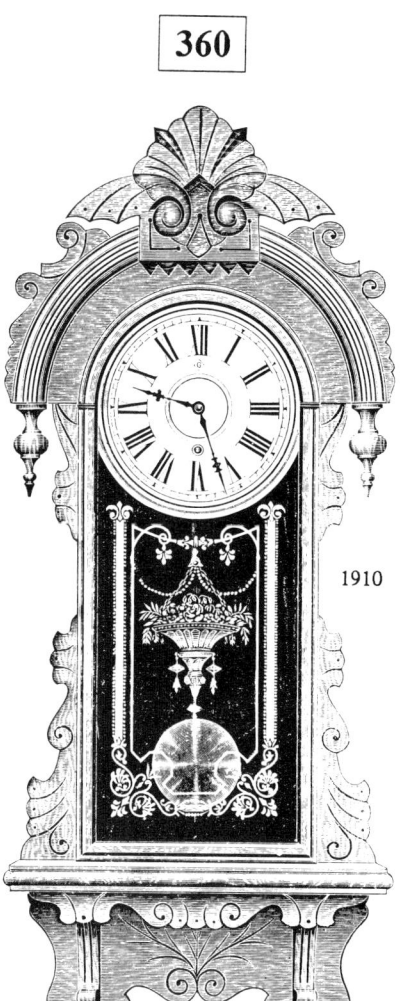

**360**

1910

### STOCKWELL.

Eight-Day Hanging Spring Clock. Height, 36 inches. 8-inch Dial. The Case is handsomely designed, with Ornamental Case Decorations in Carved and Turned Wood. Fitted with Eight-Day Time or Strike Movement, as listed below. Oak or Walnut Finish.
STOCKWELL, Time .................List each, $10.00
STOCKWELL, Wire Bell Strike....List each, 11.00
STOCKWELL, Cathedral Gong Strike,
List each, 11.50

**361**

1910

### GLENWOOD.

Eight-Day Hanging Spring Clock. Height, 36 inches. 8-inch Dial. The Case is handsomely designed, with Ornamental Case Decorations in Carved and Turned Wood. Fitted with Eight-Day Time or Strike Movement, as listed below. Oak or Walnut Finish.
GLENWOOD, Time .................List each, $11.00
GLENWOOD, Wire Bell Strike.....List each, 12.00
GLENWOOD, Cathedral Gong Strike,
List each, 12.50

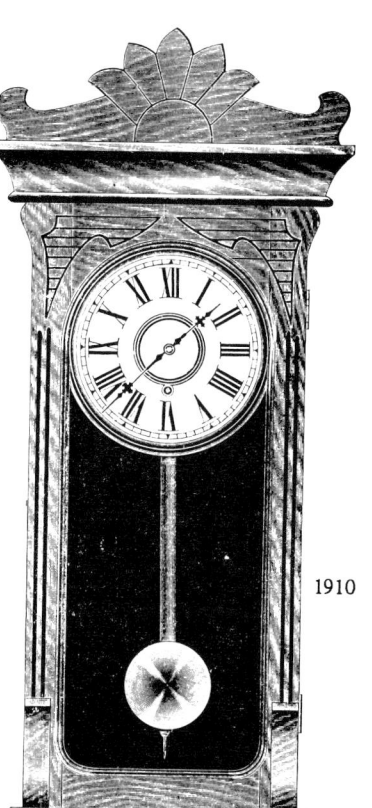

**362**

1910

### BANCROFT.

Eight-Day Hanging Spring Clock. Height, 37 inches. 8-inch Dial. The Case is a neat Design, with Ornamental Mouldings. Fitted with Eight-Day Time or Strike Movement, as listed below. Oak or Cherry (Mahogany Finish).
BANCROFT, Time .................List each, $8.75
BANCROFT, Wire Bell Strike........List each, 9.75
BANCROFT, Cathedral Gong Strike.List each, 10.25

# GILBERT

*Except for those fabulous clocks for which you wouldn't mind cutting a hole in the ceiling, Grandfather or Hanging Clocks under eight feet are the most desirable and practical. Some manufacturers produced certain of these clocks in two styles: one to stand on the floor, and the other to hang on a wall. Of the clocks designed with two styles, the hanging style is considered to be the most valuable.*

**363**

### GALLERY.
#### 18-INCH DIAL.

Eight-Day Hanging Spring Gallery Clock. Back Measurement of Case, 24 inches. Depth of Case, 5 inches. Fitted with Eight-Day Pendulum Movement, Time only. Oak Case, Highly Polished Cabinet Finish.

GALLERY, Time, 18-inch Dial..................List each, $21.00

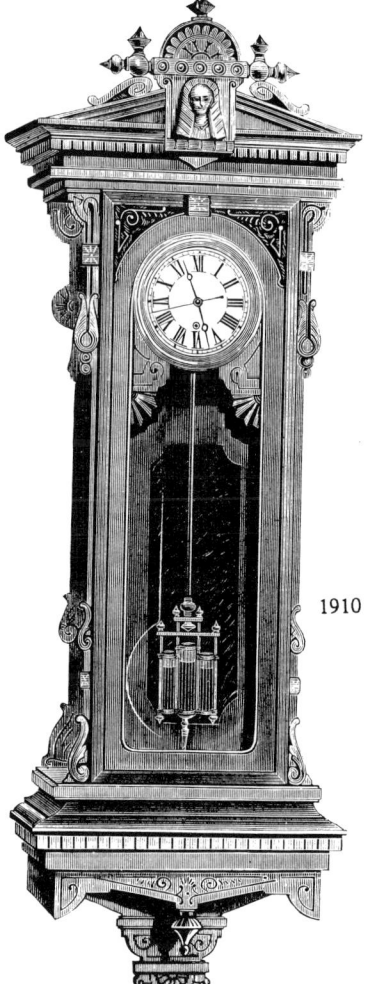

**364**

### REGULATOR, No. 8—(Hanging).
#### Walnut, Cherry (Mahogany Finish) or Oak.
HEIGHT, 8 FEET 7 INCHES. WIDTH, 2 FEET 9 INCHES.

Cabinet Finish, Glass Sides, 12 inch Porcelain Dial, Dead Beat Escapement, Retaining Power, Brass Weight, Sweep Second. Gridiron Pendulum.

List each, . . . . $110 00

**365**

### REGULATOR, No. 16—(Hanging).
#### Walnut, Cherry (Mahogany Finish) or Oak.
HEIGHT, 8 FEET 7 INCHES. WIDTH, 2 FEET 9 INCHES.

Cabinet Finish, Glass Sides, 12 inch Porcelain Dial, Dead Beat Escapement, Sweep Second, Brass Weight, Retaining Power. Mercurial Compensating Pendulum, with Mercury in three cut glass jars.

List each, . . . . $130 00

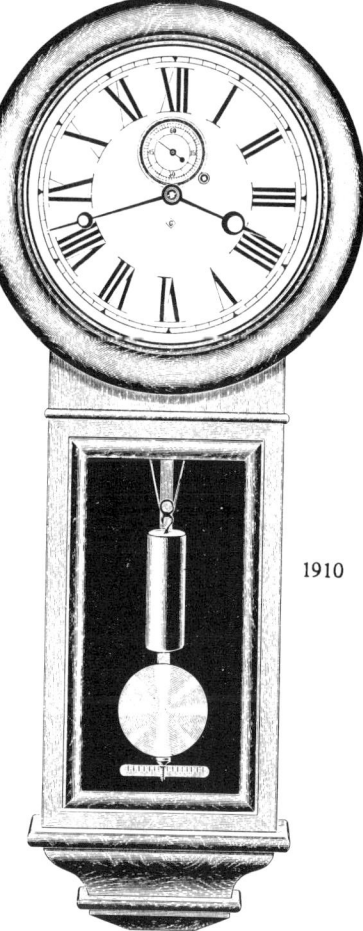

**366**

### REGULATOR No. 66, HANGING.
#### Oak or Cherry (Mahogany Finish).

Height, 38 inches. Width, 15¼ inches. 12-inch Dial, with Second Hand. The Case is handsomely designed and has a Fine Cabinet Finish. Fitted with Eight-Day Weight Movement, Time only. Dead Beat Escapement and Retaining Power. Wood Rod, Brass Pendulum Ball.

REGULATOR No. 66, Hanging.....List each, $25.00

# GILBERT

108

Only clocks in ORIGINAL MINT condition and in good running order will command top prices. A replaced or damaged dial, missing parts, tablets which are either cracked, flacking, redone will reduce prices accordingly.

### 369

**LINDEN.**

Eight-Day Hanging Spring Clock. Height, 37 inches. 8-inch Dial. The Case is handsomely carved and well finished. Fitted with Eight-Day Time or Strike Movement, as listed below. Oak or Walnut Finish.

LINDEN, Time ........................List each, $8.75
LINDEN, Wire Bell Strike .........List each, 9.75
LINDEN, Cathedral Gong Strike....List each, 10.25

### 368

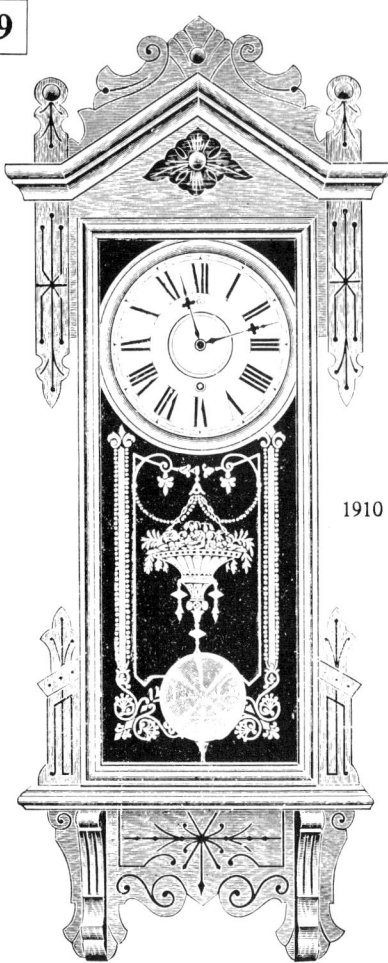
1910

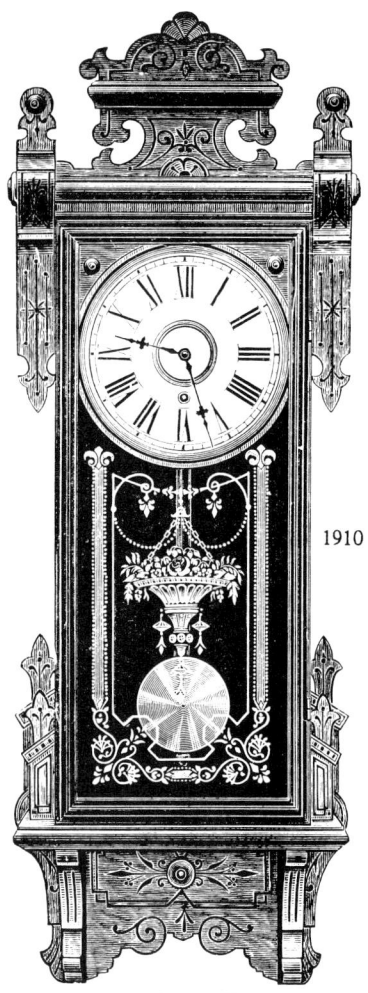
1910

### 367

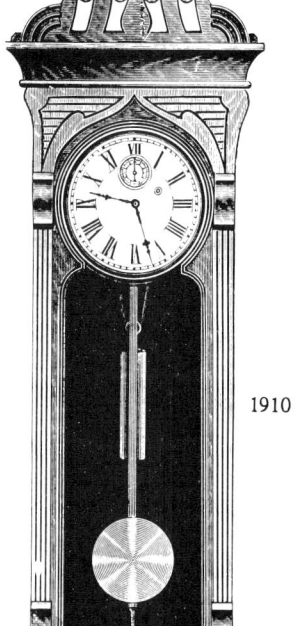
1910

### 370

1910

**THESPIAN.**

Eight-Day Hanging Spring Clock. Height, 37 inches. 8-inch Dial. The Case is handsomely carved and well finished. Fitted with Eight-Day Time or Strike Movement, as listed below. Oak or Walnut Finish.

THESPIAN, Time ....................List each, $8.75
THESPIAN, Wire Bell Strike.......List each, 9.75
THESPIAN, Cathedral Gong Strike..List each, 10.25

**REGULATOR No. 21.**

Oak or Cherry (Mahogany Finish).

Height, 48½ inches. Width, 18 inches. 12-inch Dial, with Second Hand. The Case is handsomely carved and well finished. Fitted with Eight-Day Weight Movement, Time only. Dead Beat Escapement and Retaining Power. Wood Rod, Brass Pendulum Ball.

REGULATOR No. 21..............List each, $25.00

**REGULATOR No. 64.**

Oak or Cherry (Mahogany Finish).

Height, 42½ inches. Width, 14¼ inches. 8-inch Dial, with Second Hand. The Case is handsomely carved and well finished. Fitted with Eight-Day Weight Movement, Time only. Dead Beat Escapement and Retaining Power. Wood Rod, Brass Pendulum Ball.

REGULATOR No. 64...............List each, $25.00

# GILBERT

### 371

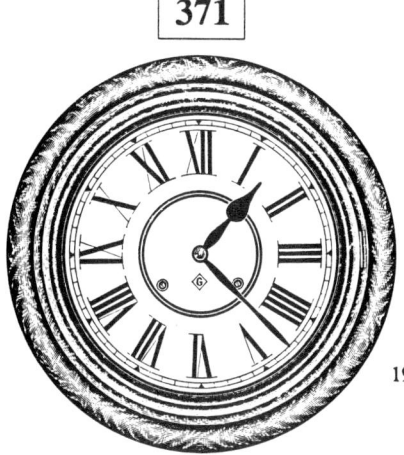

**LOBBY.**
**12-INCH DIAL.**

Eight-Day Hanging Spring Clock. Diameter of Case, 16 inches. 12-inch Dial. Fitted with Eight-Day Pendulum Movement, Time or Strike, as listed below. Oak or Mahogany, Highly Polished Cabinet Finish.

LOBBY, Time .......................List each, $6.60
LOBBY, Cathedral Gong Strike .....List each, 7.60

### 372

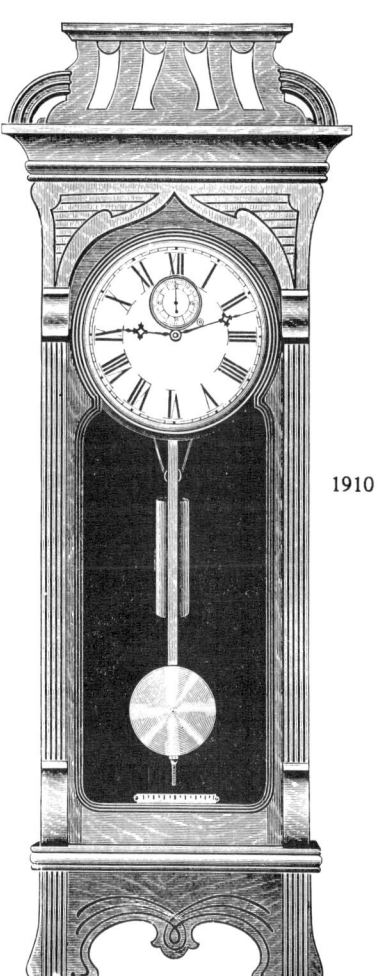

1910

**REGULATOR No. 65.**
Oak or Cherry (Mahogany Finish).

Height, 50¼ inches. Width, 18 inches. 10-inch Dial, with Second Hand. The Case is handsomely carved and well finished. Fitted with Eight-Day Weight Movement, Time only. Dead Beat Escapement and Retaining Power. Wood Rod, Brass Pendulum Ball.

REGULATOR No. 65..............List each, $26.50

### 373

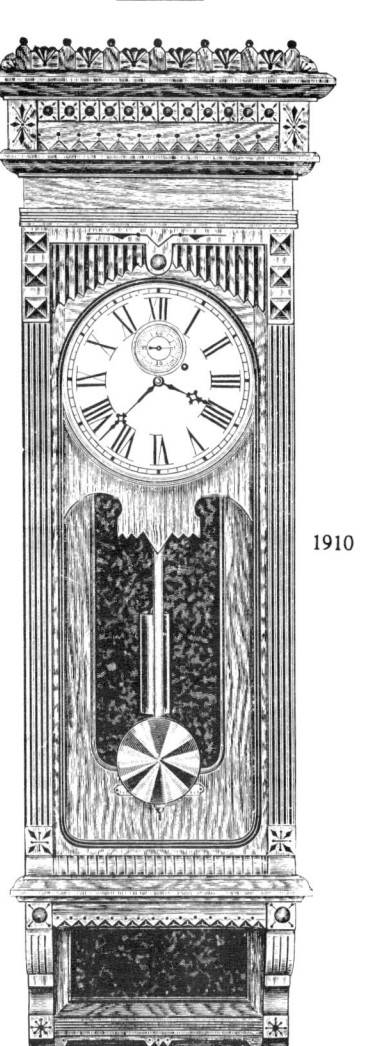

1910

**REGULATOR No. 10, HANGING.**
Oak only.

Height, 53 inches. Width, 18½ inches. 10-inch Dial, with Second Hand. The Case is handsomely carved and has a Fine Cabinet Finish. Glass Sides. Fitted with Eight-Day Weight Movement, Dead Beat Escapement and Retaining Power. Wood Rod, Brass Pendulum Ball.

REGULATOR No. 10, HANGING, Wood Rod,
List each, $32.00

### 374

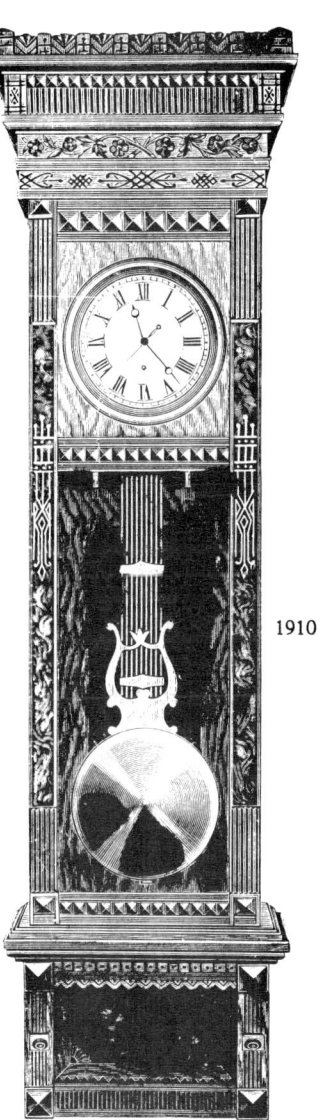

1910

**REGULATOR No. 12, HANGING.**
Oak or Cherry (Mahogany Finish).

Height, 7 feet, 8 inches. Width, 30 inches. 12-inch Porcelain Dial, with Sweep Second Hand. The Case is handsomely carved and has a Fine Cabinet Finish. Glass Sides. Fitted with Eight-Day Weight Movement, Dead Beat Escapement and Retaining Power. Gridiron Pendulum. Also supplied with Mercurial Compensating Pendulum, with Mercury in Three Cut-glass Jars, as listed below.

REGULATOR No. 12, HANGING, Gridiron
Pendulum ......................List each, $100.00
REGULATOR No. 12, HANGING, Mercurial
Pendulum ......................List each, 120.00

# GILBERT

Most white metal and brass ornamentals and clock cases are gilded or silver plated. Great care must be taken when cleaning these surfaces. The thin layer of gilding or silver plating will be easily removed by cleaning solutions and strong rubbing. If you believe cleaning is necessary, hire the services of a professional and avoid the expense of re-gilding or re-plating.

**CURFEW.**
Eight Day Half Hour Strike on Chime Bell.
HEIGHT, 17 INCHES. WIDTH, 16 INCHES. 5½ INCH DIAL.
Gilt Arch and Bell, Gilt Feet, Marbleized Sides
White or Cream Dial, . . . . . List each, $10 75
Gilt Perforated Dial, . . . . . List each, 10 75

CURFEW

**CURFEW.**—Visible Escapement.
Illustration shows Black Enamel Finish, with Scotch Granite Marbleized Sides, and Gilt Dome Top with Curfew Bell.
Eight-Day Enameled Wood Parlor Clock. Height, 17⅜ inches. Width, 16½ inches. 5½-inch Dial, with the Gilbert Visible Escapement and Patent Beat Adjuster. Fancy Gilt Rococo Sash. Fitted with Eight-Day Movement, Hour and Half-Hour Strike on Heavy Chime Bell. Black Enamel Finish, with Scotch Granite Marbleized Sides. The Case is Gilt Engraved and has heavy Cast Gilt Ornaments. Solid Metal Bell Dome, Gilt Finish, with Polished Gilt Bell.
CURFEW, Visible Escapement and Patent Beat Adjuster..................List each, $12.00
CURFEW, White or Gilt Perforated Dial (Not V. E. and P. B. A.)..................List each, 11.00

**MELODY.**
Eight Day, Half Hour Strike on Chime Bell.
HEIGHT, 18 INCHES. WIDTH, 11 INCHES. 5½ INCH DIAL.
Gilt Arch Bell and Ornaments, Marbleized, Assorted Colors.
White or Cream Dial, . . . List each, $9 50
Gilt Perforated Dial, . . . List each, 9 50

# GILBERT

### 378

1910

**PORCELAIN No. 419.**
Eight-Day Porcelain Clock in Rich Color Decorations. Height, 11⅝ inches. Width, 9¼ inches. 5-inch Dial. Rococo Sash. Fitted with Eight-Day Movement, Hour Strike on Rich Cathedral Gong, Half-Hour on separate Cup Bell. Supplied in Rich Color Decorations only, Gold Striping. Colors: Cobalt Blue, Georgia Green, Ruby and Peacock Blue.

PORCELAIN No. 419, White or Gilt Perforated
  Dial .................................. List each, $9.00
PORCELAIN No. 419, Ivory Porcelain Dial,
                                     List each, 9.50

### 379

1901

**PORCELAIN, No. 416. Rich Color.**
Eight Day Half Hour Strike Gong.
HEIGHT, 11½ INCHES. WIDTH, 9 INCHES. 5 INCH DIAL.
Rich Color Decorations—Cobalt Blue, Bronze Green, Violet, Rich Pink
Gilt Perforated or Cream Dial. . . List each, $9 50
Rococo Sash, Ivory Dial, . . . . List each, 10 00

### 380

1901

**PORCELAIN, No. 414. Rich Color.**
Eight Day Half Hour Strike Gong.
HEIGHT, 10¼ INCHES. WIDTH, 8¾ INCHES. 5 INCH DIAL.
Rich Color Decorations—Cobalt Blue, Bronze Green, Violet, Rich Pink.
Gilt Perforated or Cream Dial. . . List each, $8 75
Rococo Sash, Ivory Dial, . . . List each, 9 25

### 381

1901

**PORCELAIN, No. 244.**
Eight Day Half Hour Strike Gong.
HEIGHT, 11⅝ INCHES. WIDTH, 12 INCHES.
Rich Decorations—Cobalt Blue and Ruby.
Gilt Perforated or Cream Dial. . . List each, $12 00
Rococo Sash, Ivory Dial, . . . List each, 12 50

# GILBERT

### 382

1901

**PORCELAIN, No. 422.**
Eight Day Half Hour Strike Gong.
HEIGHT, 14 INCHES. WIDTH, 12 INCHES. 5 INCH DIAL.
Rich Color Decorations—Cobalt Blue, Georgia Green,
Violet and Rich Pink.
Gilt Perforated or Cream Dial . . . List each, $12 00
Rococo Sash, Ivory Dial, . . . List each, 12 50

### 383

1910

**SALVIA.—Alarm.**
Rich Porcelain.
A large and handsomely modeled Design in Rich Porcelain, with Bright Burnished Gold Decorations. Height, 8 inches. Width, 6¼ inches. Fitted with 30-Hour Alarm Movement. 3½-inch Dial, with Alarm Indicator on Dial. French Sash, Beveled Glass.
SALVIA . . . . . . . . . . . . . . . . List each, $3.75

### 384

1910

**GOLF.**
Rich Porcelain.
A large and handsomely modeled Design in Rich Porcelain, with beautifully colored Golf Scene and Bright Burnished Gold Striping. Height, 9 inches. Width, 8½ inches. Fitted with 30-Hour Time Movement. 2-inch Dial, with Second Hand. Convexed Beveled Glass.
GOLF . . . . . . . . . . . . . . . . List each, $4.00

### 385

1901

**PORCELAIN, No. 423.**
Eight Day Half Hour Strike Gong.
HEIGHT, 12¼ INCHES. WIDTH, 12 INCHES.
Rich Color Decorations—Cobalt Blue, Georgia Green,
Violet and Rich Pink.
Gilt Perforated or Cream Dial . . . List each, $12 00
Rococo Sash, Ivory Dial, . . . List each, 12 50

# GILBERT

*In the case of Porcelain clocks, color is the most important factor in estimating value. Bright reds, blues and frosted pinks are usually considered to be the most desirable. In order to receive top prices, clocks must be in ORIGINAL MINT condition and in good running order. Damaged dial, chipped porcelain will substantially reduce prices.*

### 386

1910

**PORCELAIN No. 434.**

Eight-Day Porcelain Clock in Rich Color Decorations. Height, 10⅝ inches. Width, 12½ inches. 5-inch Dial. Rococo Sash. Fitted with Eight-Day Movement, Hour Strike on Rich Cathedral Gong, Half-Hour on Separate Cup Bell. Supplied in Rich Color Decorations only, Gold Striping. Colors: Cobalt Blue, Georgia Green, Ruby and Water Green.

PORCELAIN No. 434, White or Gilt Perforated Dial.........List each, $11.50
PORCELAIN No. 434, Ivory Porcelain Dial...................List each, 12.00

### 387

1910

**COACHING.**
Rich Porcelain.

A large and handsomely modeled Design in Rich Porcelain, with beautifully colored Coaching Scene and Bright Burnished Gold Striping. Height, 9 inches. Width, 8½ inches. Fitted with 30-Hour Time Movement. 2-inch Dial, with Second Hand. Convexed Beveled Glass.

COACHING .........................List each, $4.00

### 388

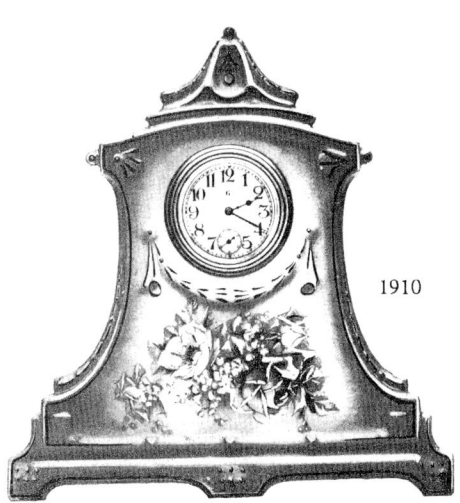

1910

**FLORETTA.**
Rich Porcelain.

A large and handsomely modeled Design in Rich Porcelain, with beautifully colored Floral Decorations and Bright Burnished Gold Striping. Height, 9 inches. Width, 8½ inches. Fitted with 30-Hour Time Movement. 2-inch Dial, with Second Hand. Convexed Beveled Glass.

FLORETTA .......................List each, $4.00

### 389

1910

**PORCELAIN No. 437.—Visible Escapement.**

Eight-Day Porcelain Clock in Rich Color Decorations. Height, 11¾ inches. Width, 13⅜ inches. 5-inch Gilt Perforated Dial, with the Gilbert Visible Escapement and Patent Beat Adjuster. Rococo Sash. Fitted with Eight-Day Movement, Hour Strike on Rich Cathedral Gong, Half-Hour on Separate Cup Bell. Supplied in Rich Color Decorations only, Gold Striping. Colors: Cobalt Blue, Ruby, Rich Pink, Peacock Blue, Water Green and Georgia Green.

PORCELAIN No. 437, With Visible Escapement and Patent Beat Adjuster,
List each, $13.50

# GILBERT

*The year entered beside each clock is the year of the catalog from which the illustration was taken. Some clocks may have been produced a few years earlier.*

### 390

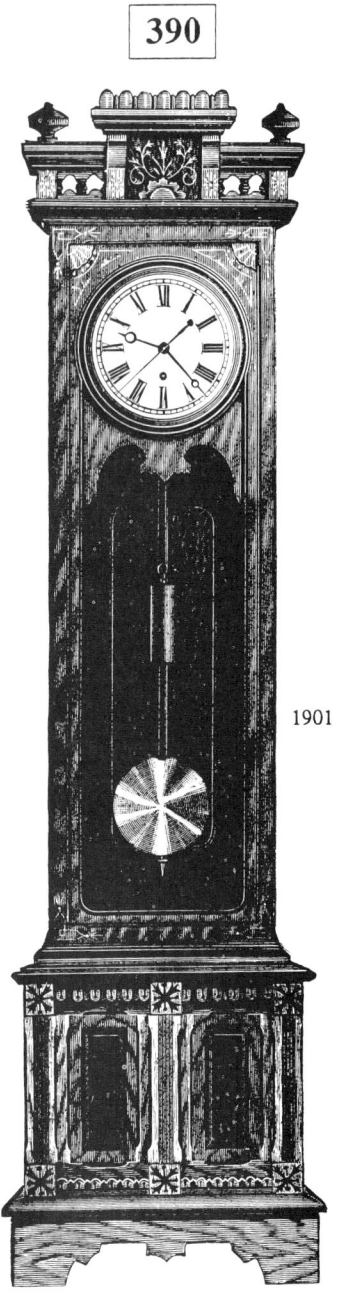

1901

**REGULATOR No. 9—(Standing).**
Walnut, Oak or Cherry
(Mahogany Finish).
HEIGHT, 7 FEET 7 INCHES.   WIDTH, 2 FEET 1 INCH.
Cabinet Finish, Glass Sides, 12 inch Porcelain Dial, Dead Beat Escapement, Sweep Second, Brass Weight, Retaining Power. Wood Rod Brass Ball,   .   .   List each, $90 00

### 391

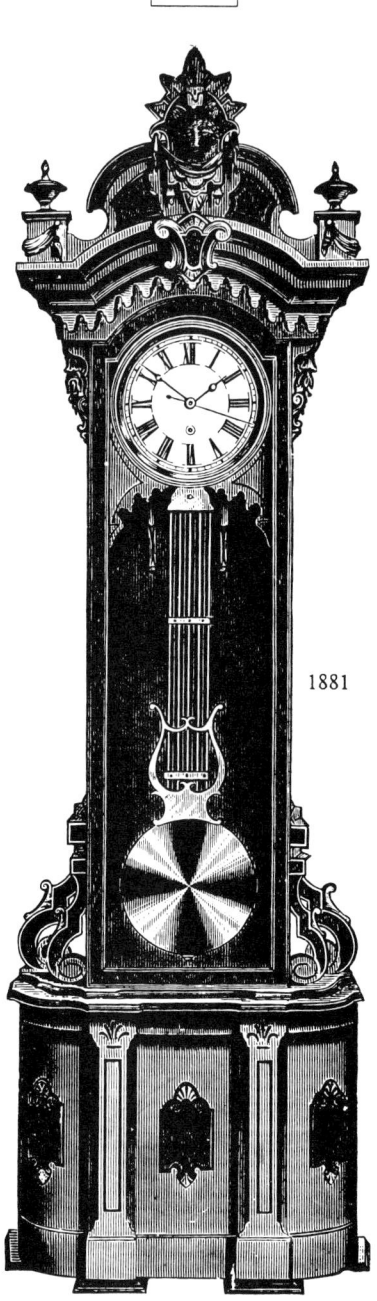

1881

## REGULATOR No. 7.
Eight-day Time.   Weight.
Glass Sides.   Swiss Movement.

Height, 8½ feet; 12 inch Dial.

### 392

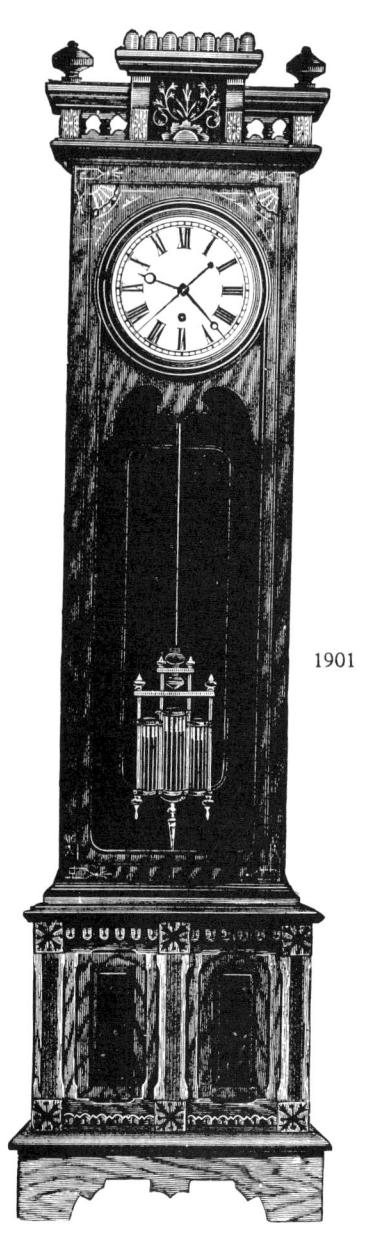

1901

**REGULATOR, No. 18—(Standing).**
Walnut, Oak, Cherry or Cherry Mahogany Finish.
HEIGHT, 7 FEET 7 INCHES.   WIDTH 2 FEET 1 INCH.
Cabinet Finish, Glass Sides, 12 inch Porcelain Dial, Dead Beat Escapement, Sweep Second, Brass Weight, Retaining Power. Mercurial Compensating Pendulum, with Mercury in three cut glass jars.
List each.   .   .   .   .   $120 00
Gridiron Pendulum,   List each,   100 00

# GILBERT

*Except for those fabulous clocks for which you wouldn't mind cutting a hole in the ceiling, Grandfather or Hanging Clocks under eight feet are the most desirable and practical. Some manufacturers produced certain of these clocks in two styles: one to stand on the floor, and the other to hang on a wall. Of the clocks designed with two styles, the hanging style is considered to be the most valuable.*

*To receive top prices, clocks must be in ORIGINAL MINT condition and in good running order - A replaced or damaged dial, missing parts, will reduce prices accordingly.*

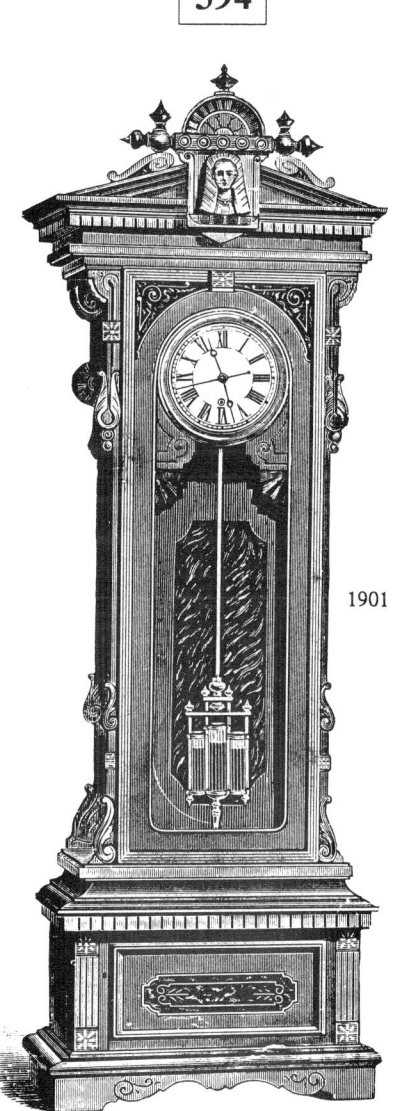

1901

**REGULATOR, No. 8—(Standing).**
Walnut, Cherry (Mahogany Finish) or Oak.
HEIGHT, 8 FEET 10 INCHES.   WIDTH, 2 FEET 10 INCHES.
Cabinet Finish, Glass Sides, 12 inch Porcelain Dial, Dead Beat Escapement, Sweep Second Brass Weight, Retaining Power. Gridiron Pendulum.
List each, . . . $125.00

1901

**REGULATOR, No. 16—(Standing).**
Walnut, Cherry (Mahogany Finish) or Oak.
HEIGHT, 8 FEET 10 INCHES.   WIDTH, 2 FEET 10 INCHES.
Cabinet Finish, Glass Sides, 12 inch Porcelain Dial, Dead Beat Escapement, Sweep Second, Brass Weight, Retaining Power. Mercurial Compensating Pendulum, with Mercury in three cut glass jars.
List each, . . . $145.00

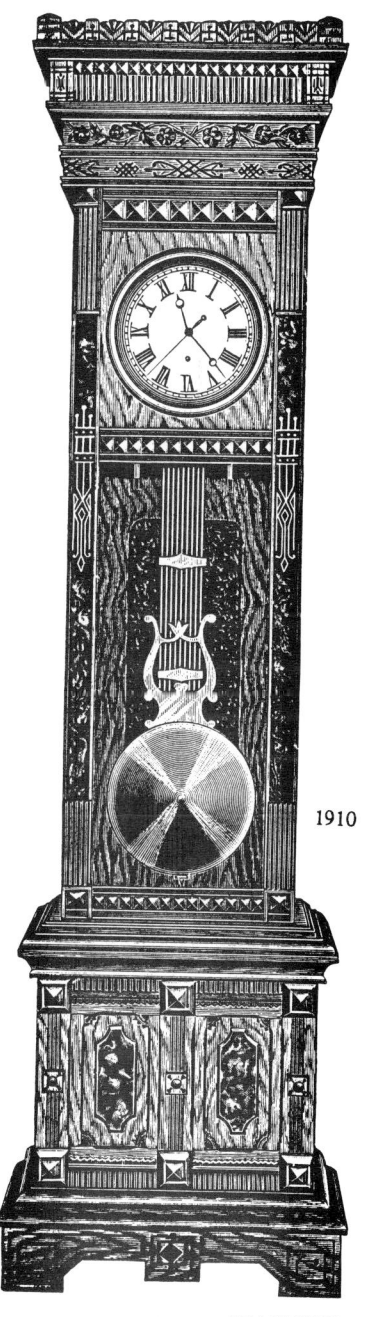

1910

**REGULATOR No. 12, STANDING.**
Oak or Cherry (Mahogany Finish).
Height, 8 feet, 6 inches. Width, 33 inches. 12-inch Porcelain Dial, with Sweep Second Hand. The Case is handsomely carved and has a Fine Cabinet Finish. Glass Sides. Fitted with Eight-Day Weight Movement, Dead Beat Escapement and Retaining Power. Gridiron Pendulum. Also supplied with Mercurial Compensating Pendulum, with Mercury in Three Cut-glass Jars, as listed below.
REGULATOR No. 12, STANDING, Gridiron Pendulum ................List each, $110.00
REGULATOR No. 12, STANDING, Mercurial Pendulum ................List each, 130.00

# E. HOWARD WATCH & CLOCK COMPANY
Boston, Mass.

The E. Howard Watch & Clock Company was formed as a joint stock corporation on December 1, 1881 to succeed an earlier firm of similar name founded by Edward Howard (1813-1904). Howard, a clockmaking apprentice of Aaron Willard, Jr. had commenced business with David P. Davis, manufacturing high-grade wall clocks under the name of Howard & Davis in 1842. They also became known for their manufacture of sewing machines, fire engines and precision balances. About 1843, with a third partner, Luther Stephenson, they began to also manufacture tower clocks.

In 1857, David P. Davis left the firm and Howard & Davis was dissolved and was succeeded by E. Howard & Company. Both Howard and Davis had also been involved in watch manufacturing, somewhat unsuccessfully, since 1850. In 1857, Edward Howard began a new watch company and on March 24, 1861 combined his clock and watch businesses into one joint stock corporation, the Howard Clock & Watch Company, which failed in 1863. Thereafter, Howard formed a new company called the Howard Watch & Clock Company (transposing clock & watch) on October 1, 1863 which was successful for some years but was reorganized in 1881 after financial setbacks of a few years previous.

In 1881, Edward Howard sold out his personal interests and retired, leaving the firm to new management. This firm continued the manufacture of many clock styles, primarily weight-driven wall timepieces and regulators of fine quality. Only a couple common wall models, #5 and #10, were produced as stock items, all others being manufactured by special order. Clocks were manufactured at Roxbury, a part of Boston, but in 1934 the operation was moved to Waltham, MA.

A new firm known as Howard Clock Products was formed November 5, 1934 to succeed the earlier firm. Clock production was on the wane, but precision gear cutting business kept the firm profitable, particularly from government contract work. Production of smaller clocks ceased in 1958 and the last tower clock was produced in 1964. The production of Self-Winding master clocks stopped in 1940, although secondary clocks were still made and sold until 1956, mostly for existing clock systems.

However, in 1975, Dana J. Blackwell, as a new Vice-president of the firm, revived clock production, reintroducing several of the more popular models to the market. Movements in these later clocks maintained the high standards the Howard firm had become famous for and cases were made to very strict specifications.

Sadly, the older owners of the firm sold the business to a young seemingly successful businessman in August of 1977. This new owner eventually fired most of the firm's knowledgeable management and proceeded to drain it financially. By 1980, when the firm was at the verge of bankruptcy, he was caught with a bunch of hired thugs in an attempt to blow up the factory building. After a lengthy trial he was convicted, though never served any time in jail.

At the time of owner's arrest, the Federal Government stepped in and the Howard firm was placed under Chapter 11 of the bankruptcy code. A shrewd manager was brought in by the bankruptcy court and after creditors were satisfied, the firm sold the clockmaking portion of the business to private investors who continue to offer fine Howard clocks.

# HOWARD

*The year entered beside each clock is the year of the catalog from which the illustration was taken. Some clocks may have been produced a few years earlier.*

No. 4 is 2 feet 8 inches long, with 8 inch dial.
No. 5 is 2 feet 5 inches long, with 7 inch dial.

Both recoil escapements, and accurate time-keepers. The cases are from well-seasoned hard wood, stained in imitation of rosewood, and polished. For use in dwelling-houses, offices and rooms, they are well adapted.

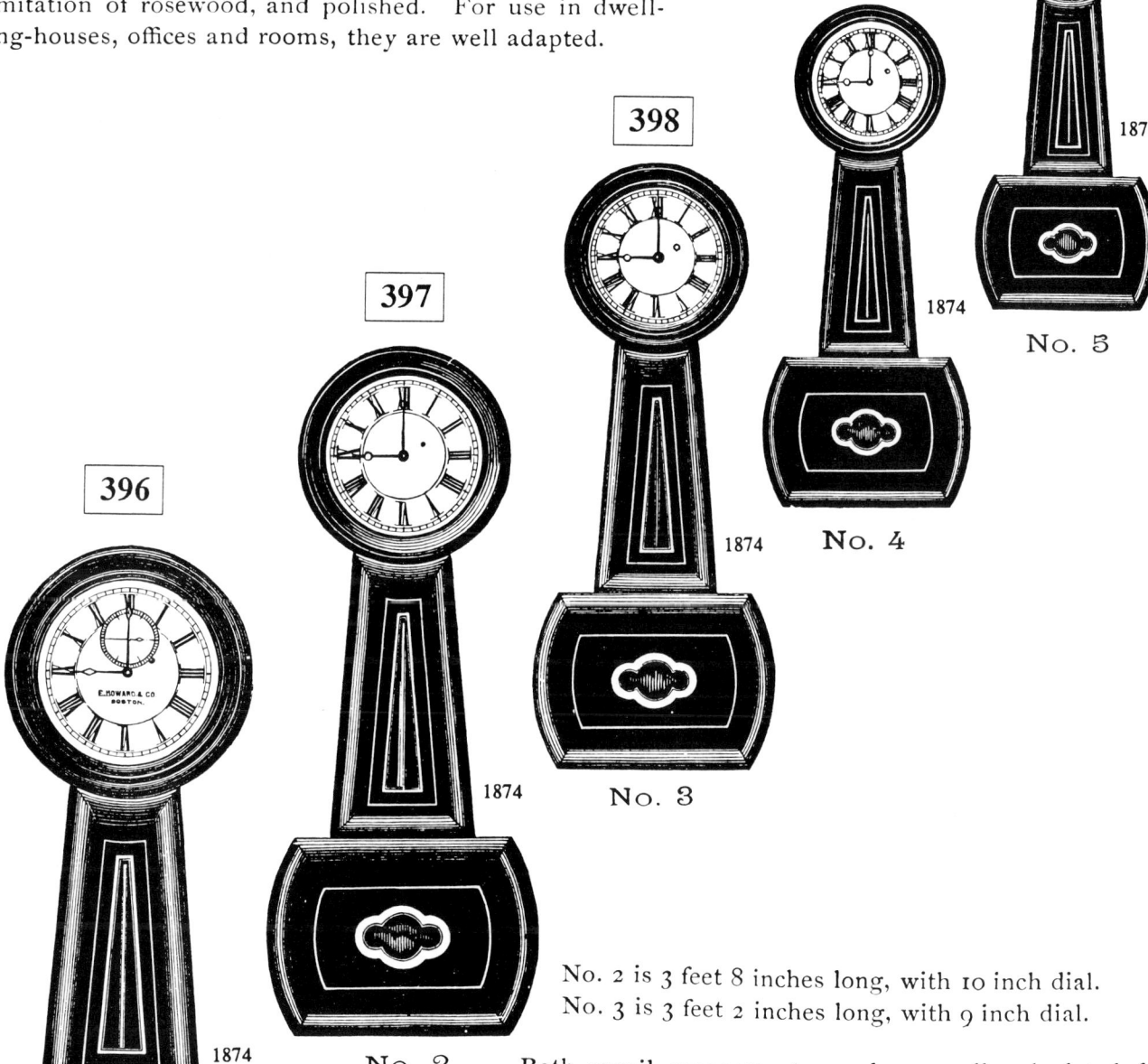

No. 2 is 3 feet 8 inches long, with 10 inch dial.
No. 3 is 3 feet 2 inches long, with 9 inch dial.

Both recoil escapements, and are well calculated for office or school use, and may be fully relied upon for accurate time-keeping. The cases are of seasoned hard wood, stained rosewood, varnished and polished.

No. 1 Regulator.
Movement with Graham Dead Beat Escapement. Enameled Metal Dial, 12 inches diameter. Case of Cherry, stained Rosewood, 4 feet 2 inches long.

# HOWARD

No. 9 is 3 feet 1 inch long, with 9 inch dial.
No. 10 is 2 feet 9 inches long, with 8 inch dial.

Both recoil escapements and accurate time-keepers. The cases are made from well-seasoned black walnut wood, and polished. These clocks are well adapted for use in dwelling-houses.

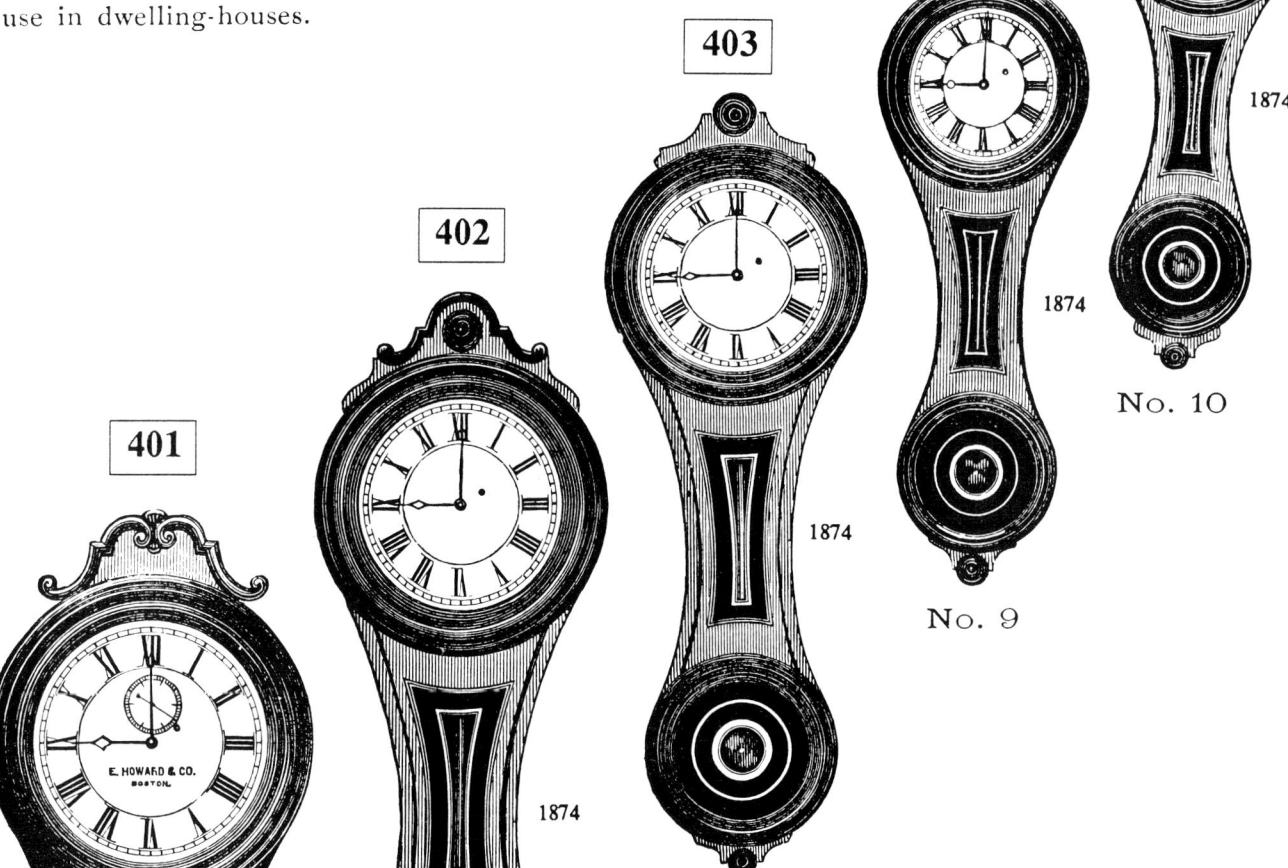

*Only clocks in ORIGINAL MINT condition and in good running order will command top prices. A replaced or damaged dial, missing parts, tablets which are either cracked, flacking, redone will reduce prices accordingly.*

No. 7 is 4 feet 2 inches long, with 12 inch dial.
No. 8 is 3 feet 8 inches long, with 11 inch dial.

Both recoil escapements, and accurate time-keepers. The cases are made from well-seasoned black walnut wood, polished.

No. 6 Regulator.
Movement with Graham Dead Beat Escapement. Enameled Metal Dial, 14 inches diameter. Case of Black Walnut, 4 feet 10 inches long.

# HOWARD

No. 14 is 3 feet 6 inches long, with 10 inch dial.
No. 11 is 2 feet 7 inches long, with 10 inch dial.

Both recoil escapements and accurate time-keepers. No. 14 case is made from well-seasoned black walnut wood, and well polished.

No. 11 case is made from cherry wood, and stained imitation of rosewood. It is particularly adapted for rooms where a large dial is required, and the cost low.

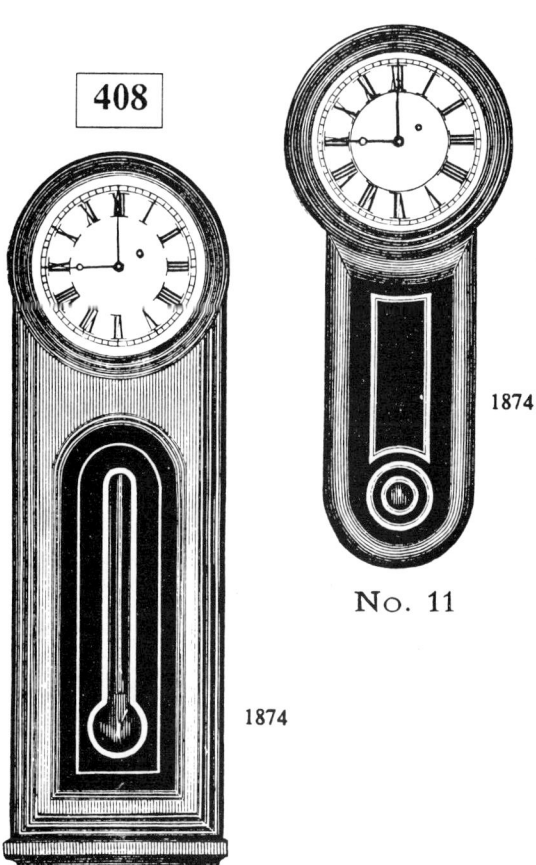

No. 11

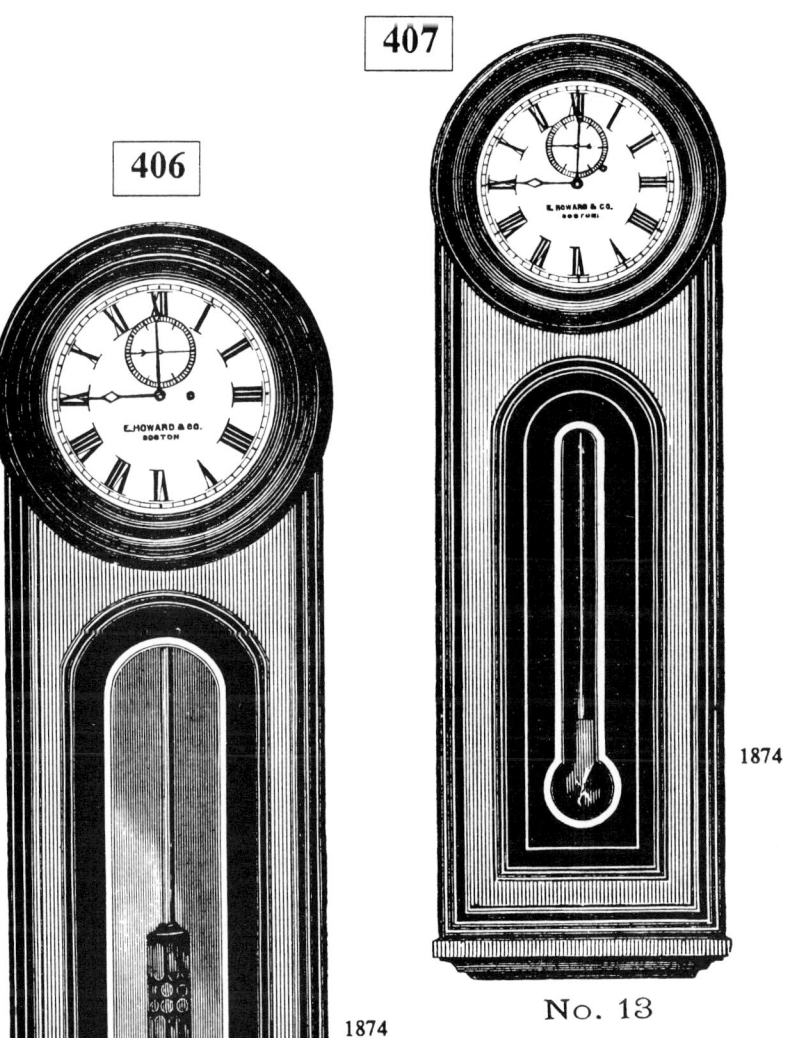

No. 14

No. 13

No. 12

No. 13 Regulator. It has full-length pendulum, with 12 inch dial, 4 feet 8 inches long, beats seconds, has maintaining power, with dead-beat escapement, and case made of black walnut, highly polished. For accuracy of performance it is equal to anything but a compensated pendulum.

No. 12 Regulator. It has full-length pendulum, 14 inch dial, 5 feet long, beats seconds, has maintaining power, with dead-beat escapement, pendulum rod of wood, and loaded glass jar for pendulum ball. Is finely finished in every respect. For accuracy of performance it is equal to anything but a compensated pendulum, while its appearance is solid and ornamental This can be arranged with a Compensated Pendulum if desired; also, jewelled if required.

# HOWARD

No. 38

This represents No. 38 Clock, black walnut case, 14 inch dial, 6 feet long. It has a full-length pendulum, seconds-dial, with dead-beat escapement and maintaining power.

No. 39
Black walnut, 12 inch dial, 5 feet long.

No. 40
Black walnut, 11 inch dial, 4 feet 6 inches long.

No. 41
Black walnut, 9 inch dial, 4 feet long.

No. 42
Black walnut, 8 inch dial, 3 feet 8 inches long.

# HOWARD

**415**

**416**

**417**

*The year entered beside each clock is the year of the catalog from which the illustration was taken. Some clocks may have been produced a few years earlier.*

1889

1889

1889

**No. 75.**

**No. 75.**

Case of Mahogany.

12-inch dial, case 2 feet 10 inches long.

**No. 85.**

## REGULATOR.

Case of Mahogany.

14-inch dial, case 5 feet long.

Movement has a full-length pendulum.

Movement with Graham Dead Beat Escapement.
Enameled Zinc Dial, 14 inches diameter.
Case of Mahogany.
5 feet long.

# HOWARD

### No. 100.
14 inch Enameled Metal Dial. Case of Oak, Cherry or Mahogany, 3 feet 6 inches long.

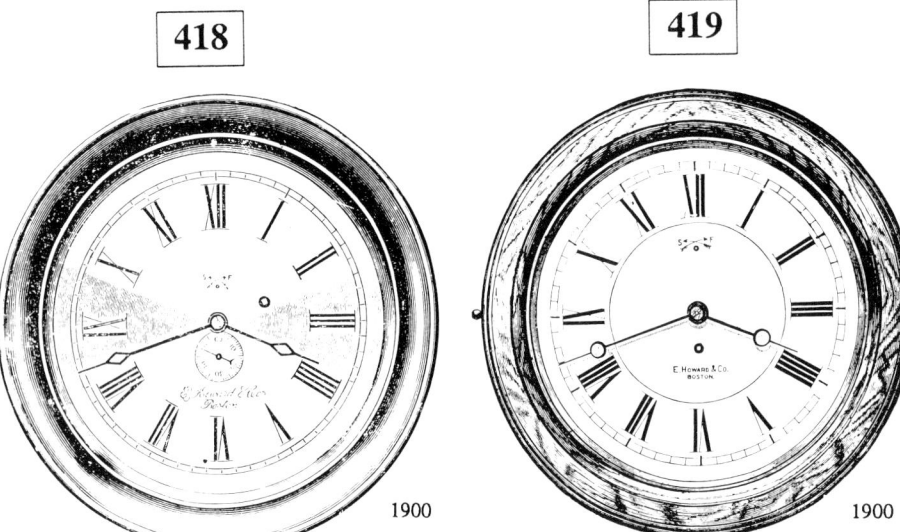

No. 69.    No. 123A.

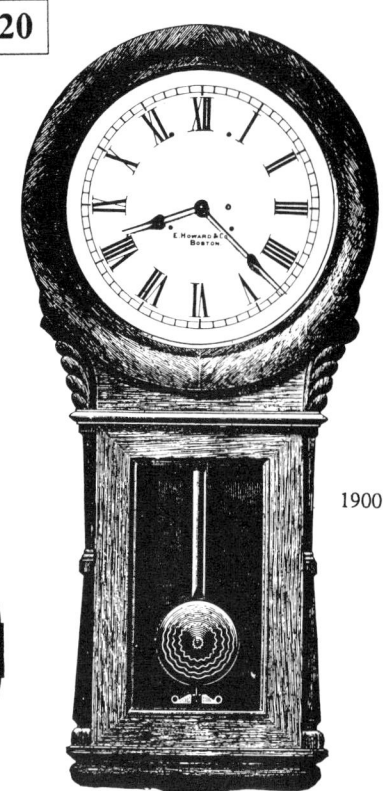

No. 100.

Movements finely finished throughout with jeweled escapement and chronometer balance. They may be adjusted upon order, and will then fulfil all requirements of a chronometer. No. 69 and No. 123A are made with dials 4½, 6, 8, 10 and 12 inches in diameter. The 4½ in., 6 in. and 8 in. have porcelain enameled dials. The 10 in. and 12 in. have silvered or white enameled metal dials.

The No. 69 cases are made of heavy metal finished in polished brass, copper or nickel; the No. 123A cases are made of wood to match other furniture.

No. 99.

Designed and especially appropriate for cabins of yachts. 6, 8 or 10 inch dial of Onyx, with raised figures and dots finished in gold, silver or bronze, with hands to match. A white enameled metal dial can be used on this clock if desired.

### No. 72.
### REGULATOR.

Movement with Graham Dead Beat Escapement.
Enameled Zinc Dial, 12 and 14 inches diameter.
Cases of Black Walnut.
14-inch, case 5 feet 5 inches long.
12-inch, case 5 feet 4 inches long.

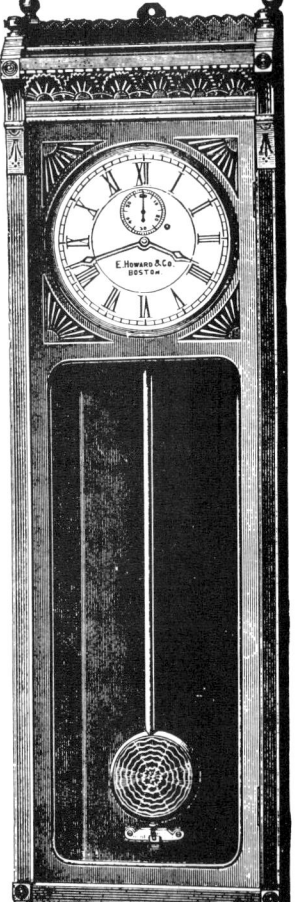

No. 72 Regulator.

# HOWARD

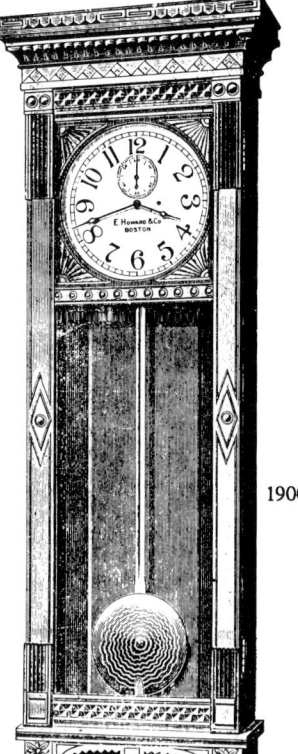

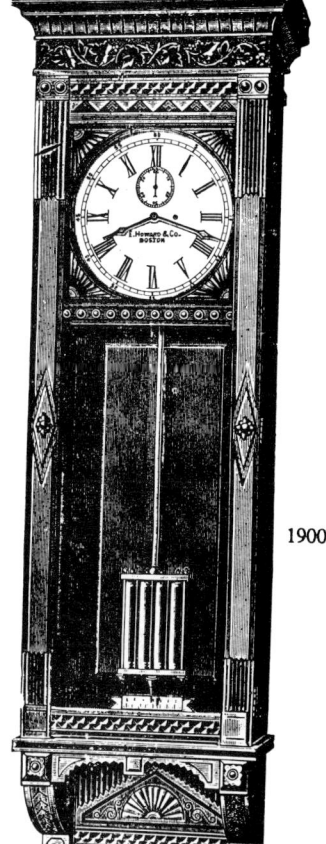

**No. 86a Regulator.**

The design of our No. 86 Regulator has been received with so much favor that we have recently made it in a smaller size (N86A) with less expensive movement to meet the wants of Jewelers having small but elegant stores, and desiring a handsome but comparatively low-priced Regulator.

The movement is strongly made and well finished throughout, has Graham Dead Beat Escapement and pendulum rod of perfectly seasoned, straight-grained cherry with heavy bob of brass filled with zinc for compensation.

Case of oak or cherry, **5 feet 3 inches** long. **12-inch** silvered or white enameled metal dial.

**No. 86 Regulator.**

No. 3 Movement, Graham Dead Beat Escapement. 14 inch Silvered or White Enameled Metal Dial. Case of Oak or Cherry, 6 feet 4 inches long.

**No. 86 Regulator.**
With Mercurial Pendulum.

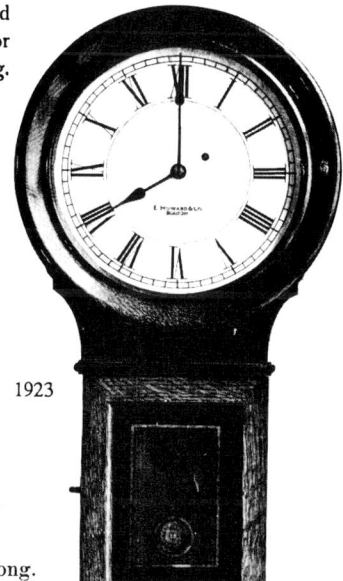

**No. 58.**

Case of Walnut, Oak or Cherry, 12 inch Dial, case 5 feet 4 inches long. 10 inch Dial, case 4 feet 3 inches long. 8 inch Dial, case 3 feet 6 inches long.

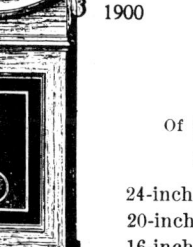

No. 70

**No. 70.**

Of this pattern there are five sizes.

Cases of Black Walnut.

24-inch dial, case 4 feet 8 inches long.
20-inch dial, case 4 feet long.
16-inch dial, case 3 feet 5 inches long.
14-inch dial, case 3 feet long.
12-inch dial, case 2 feet 8 inches long.

No. 70.

# HOWARD

*Only clocks in ORIGINAL MINT condition and in good running order will command top prices. A replaced or damaged dial, missing parts, tablets which are either cracked, flacking, redone will reduce prices accordingly.*

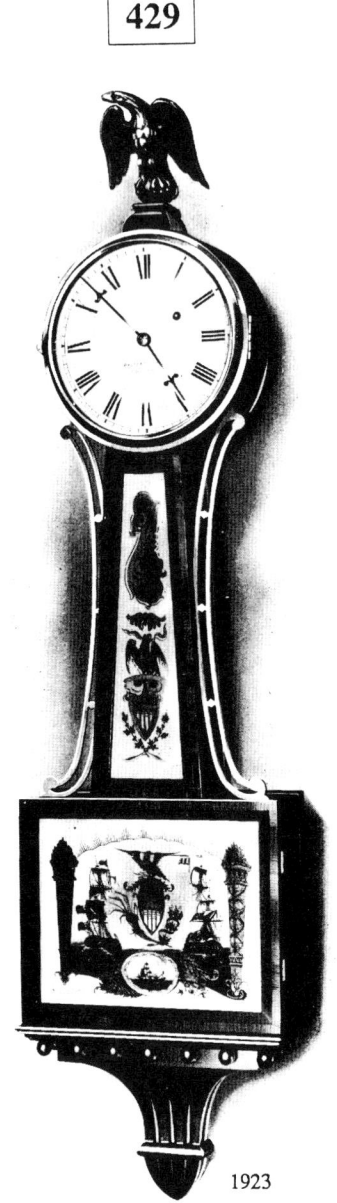
**429**
1923

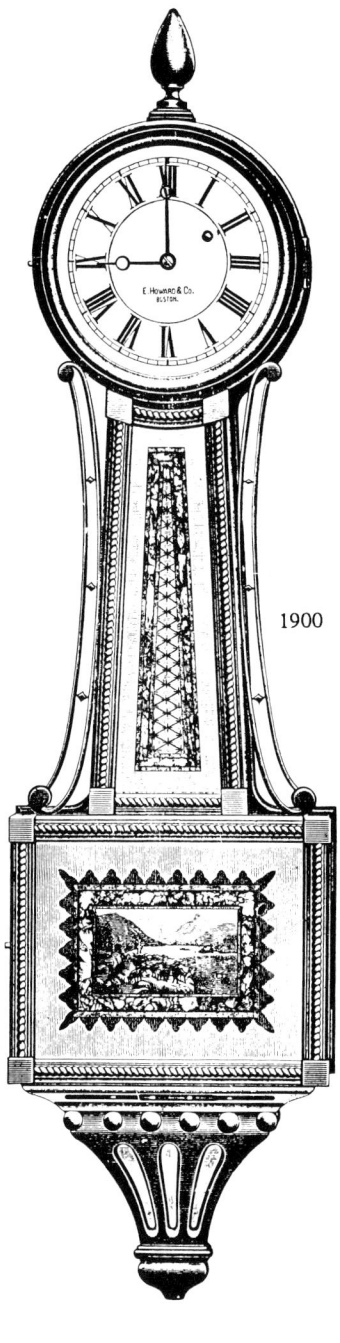
**430**
1900

No. 95

Case of mahogany with brass ornaments, white enamel dial.
Convex glass set in cast brass bezel.

Dial, 8 inches diameter.   Case, 3 feet 4 inches long.

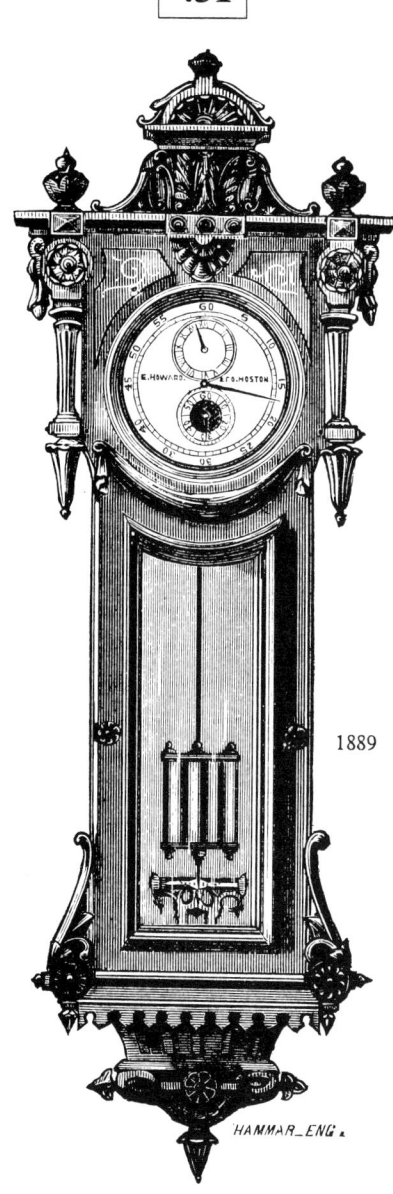
**431**
1889

No. 49

**Regulator.**

14 inch dial.   7 feet 6 inches long.

# HOWARD

*The year entered beside each clock is the year of the catalog from which the illustration was taken. Some clocks may have been produced a few years earlier.*

**432** — 1889 — No. 57.

**REGULATOR.**
No. 3 Movement.
Graham Dead Beat Escapement.
Silvered Electro-plate Dial, 14 inches diameter.
Case of Black Walnut.
6 feet 2 inches long.

**433** — 1900 — No. 67 Regulator.
No. 3 Movement, Graham Dead Beat Escapement. 15 inch Silvered Dial. Case Walnut or Oak, 7 feet 9 inches long

**434** — 1889 — No. 67.

**REGULATOR.**
No. 3 Movement; Graham Dead Beat Escapement; 16-inch Dials, White Enameled Zinc; 14-inch Dials, Silvered Electro-plate; Cases, Black Walnut.

16-inch, standing 10 feet 3 inches high.
16-inch, hanging 9 feet 3 inches long.
14-inch, standing 8 feet 9 inches high.
14-inch, hanging 7 feet 9 inches long.

# HOWARD

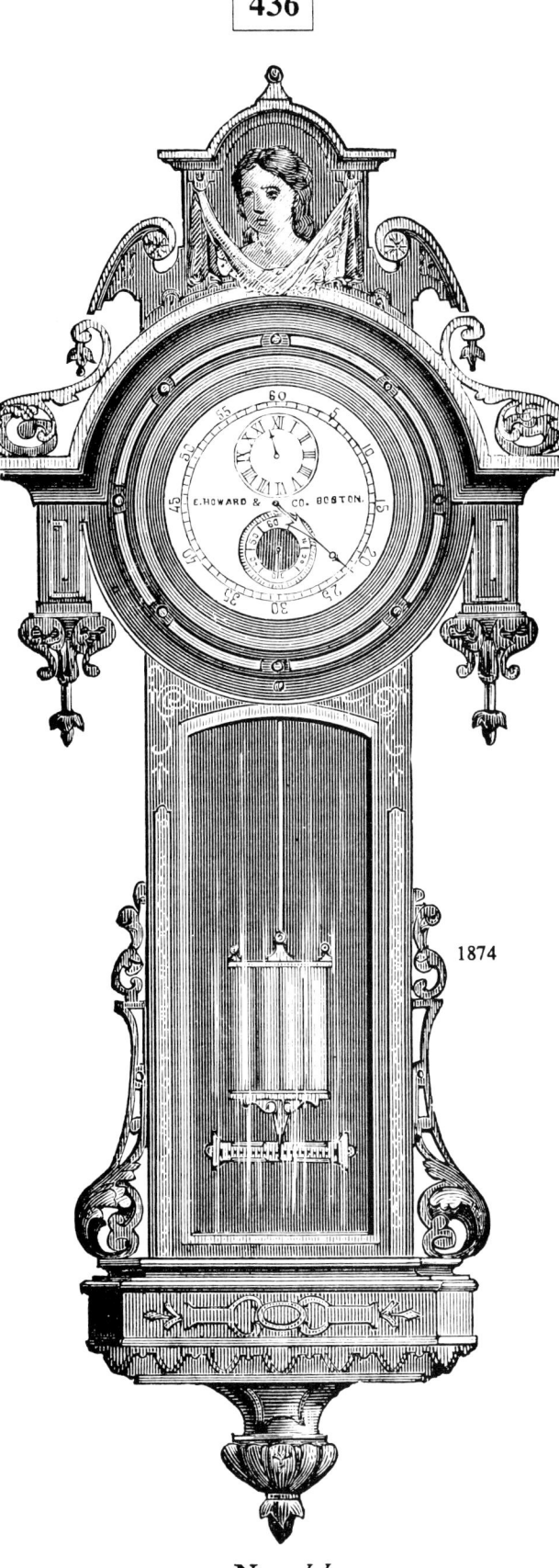

## REGULATOR.
### No. 67.
We make of this pattern four sizes.

16 inch dial, standing case, 9 feet 3 inches high, 38 inches wide.
14 " " " 8 " 9 " " 34½ " "
16 " " hanging " 10 " 3 " " 38 " "
14 " " " 7 " 9 " " 34½ " "

### No. 44

14 inch dial.   7 feet 2 inches long.

# HOWARD

*Except for those fabulous clocks for which you wouldn't mind cutting a hole in the ceiling, Grandfather or Hanging Clocks under eight feet are the most desirable and practical. Some manufacturers produced certain of these clocks in two styles: one to stand on the floor, and the other to hang on a wall. Of the clocks designed with two styles, the hanging style is considered to be the most valuable.*

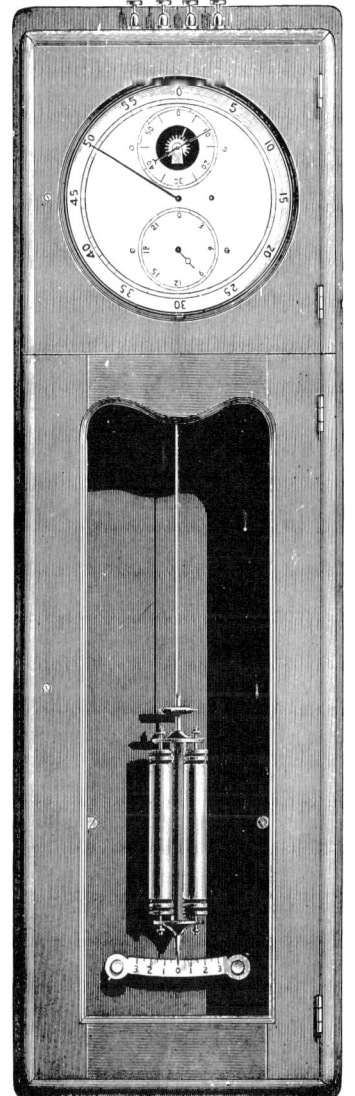

437

1900

438

1900

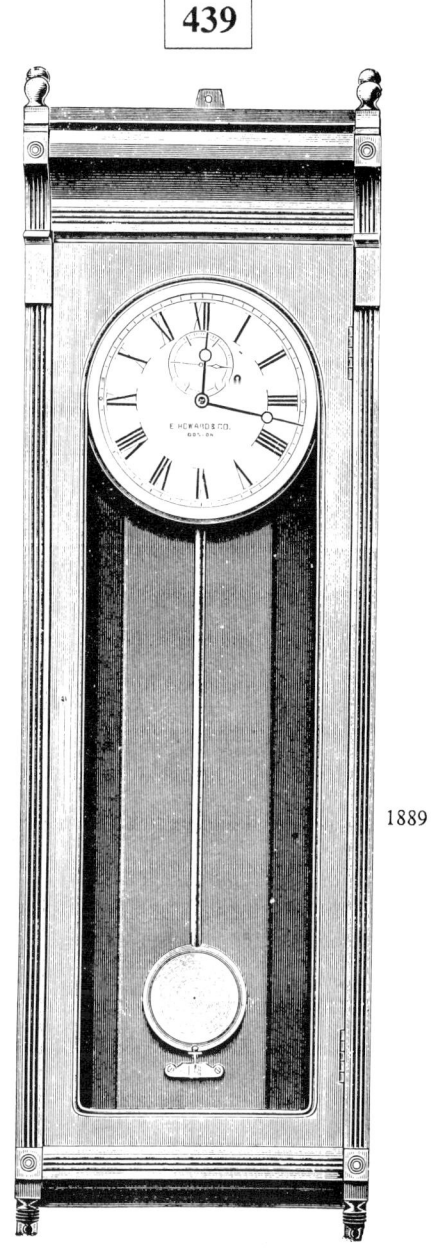

439

1889

**No. 89.**
**REGULATOR.**

Movement with Graham Dead Beat Escapement.
Enameled Zinc Dial, 12 inches diameter.
Case of Oak, 5 feet 5 inches long.

## Astronomical Regulator, No. 74.

The back, on which the movement, pendulum, weight, and case are mounted, is cast in one piece, and has four bosses cast upon it, on which the back is to be seated when mounted, and through which screws and bolts are to pass in securing it into place.

Dimensions of back, 5 feet long, 19 inches wide. The Pendulum is constructed with a bob of four jars clustered around the rod, to render the mercury columns as nearly as possible, as susceptible to the changes of temperature as the rod.

We make three grades of movements for this Regulator.

# HOWARD

*Only clocks in ORIGINAL MINT condition and in good running order will command top prices. A replaced or damaged dial, missing parts, tablets which are either cracked, flacking, redone will reduce prices accordingly.*

### 440 — 1880

No. 36

This cut represents our No. 36 Regulator. It has full-length pendulum, 14 inch dial, 6 feet long, beats seconds, has maintaining power, with dead-beat escapement, pendulum rod of wood, and loaded glass jar for pendulum ball. Is finely finished in every respect. For accuracy of performance it is equal to anything but a compensated pendulum, while its appearance is solid and ornamental. This can be arranged with a compensated pendulum. Also, jeweled if required.

### 441 — 1880

No. 59.

Of the above pattern we make four sizes.

No. 59, 3 feet long, 6-inch Dial.
No. 59, 3 feet 10 inches long, 8-inch Dial.
No. 59, 4 feet 8 inches long, 10-inch Dial.
No. 59, 5 feet 10 inches long, 12-inch Dial.

### 442 — 1888

No. 71.

**REGULATOR.**
Movement with Graham Dead Beat Escapement.
Enameled Zinc Dial, 12 inches diameter.
Case of Black Walnut.
5 feet 10 inches long.

# HOWARD

*The year entered beside each clock is the year of the catalog from which the illustration was taken. Some clocks may have been produced a few years earlier.*

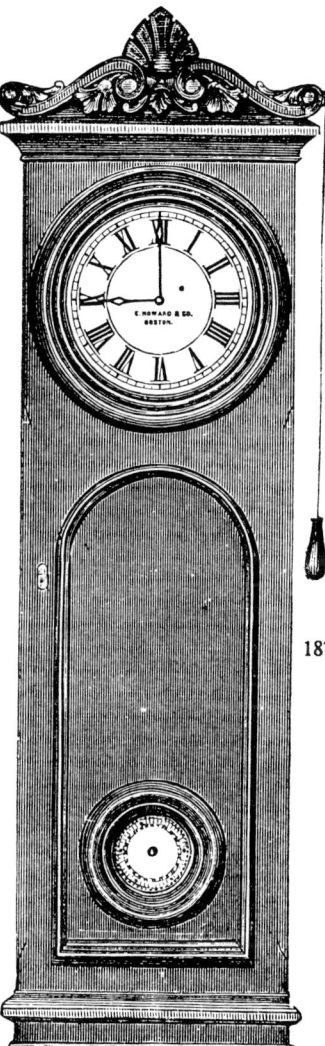

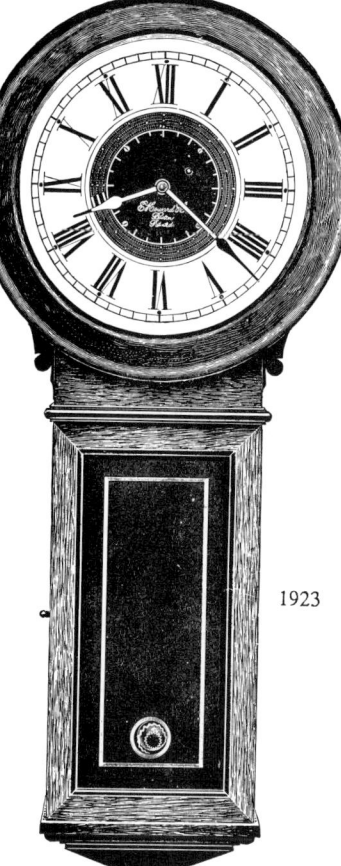

1923

1874

1923

### No. 70.
### PROGRAM CONTACT CLOCK

Arranged to ring a system of bells in Schools, Colleges, Factories, or other buildings requiring simultaneous signals at stated times in any number of rooms.

We show a cut of our No. 70 clock arranged with Program Contact.

The Dial is made in such a manner that pins can be inserted to make contacts at any five minutes of the day or night, and transmit electric currents for ringing bells or making other signals.

This Program Contact can be attached to other patterns of our clocks with dials from 12 to 16 inches in diameter.

### Watch Clock.

### No. 101.
### MULTIPLE ELECTRIC PROGRAM CLOCK
#### FOR
SCHOOLS, COLLEGES, OR FACTORIES WHERE MORE THAN ONE PROGRAMME IS NECESSARY

Arranged to ring two or three sets of bells, each set containing any number of bells desired. All bells of a set to be rung simultaneously at as many periods (five minutes apart) as may be required through the day or night. Programs easily changed by altering locations of contact pins in clock dial.

No. 37 Clock, for the detection of delinquent watchmen on night duty in factories, railroad stations and public buildings. This clock has been lately improved, and is the only one that gives a record of the watchman's duty two nights in succession, giving on Monday the record of Saturday and Sunday nights previous, thereby avoiding the necessity of visiting the clock on Sunday. Black walnut case, 11 inch dial, 5 feet 6 inches long.

# HOWARD

## Marble Dial Clocks

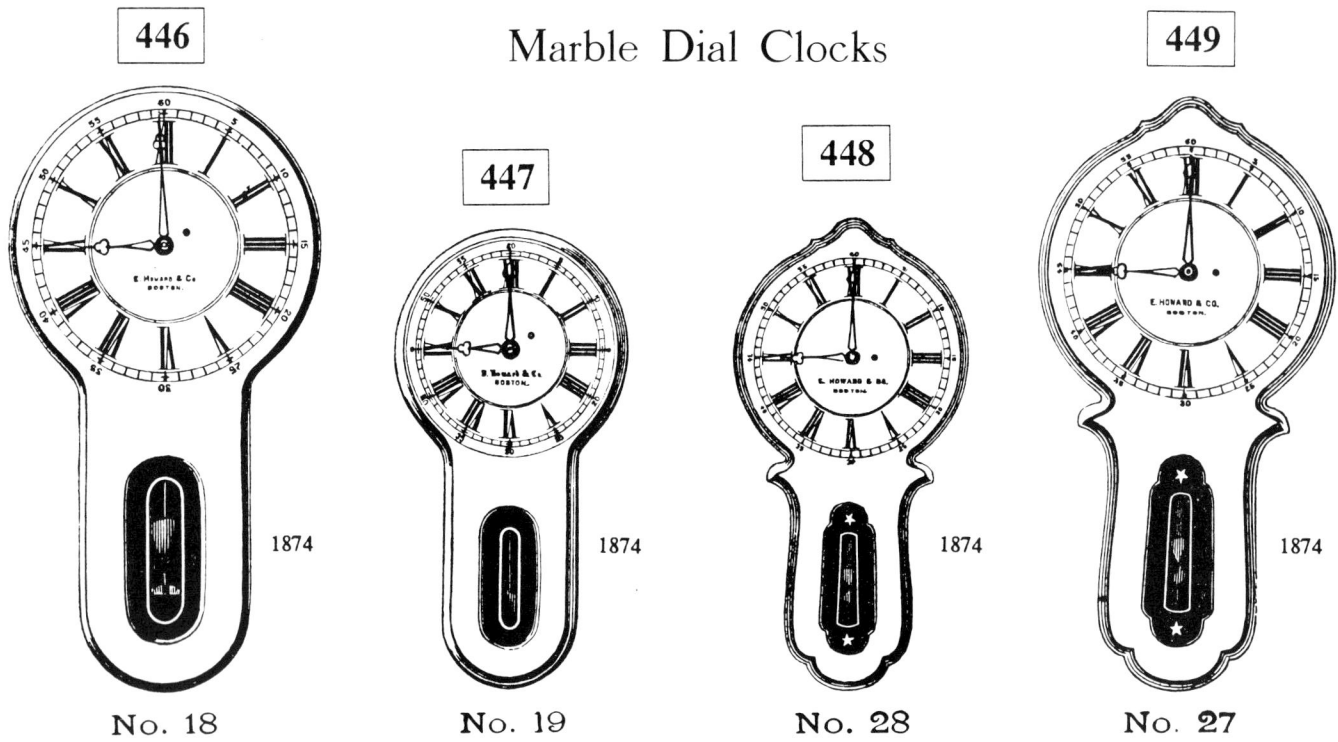

No. 18    2 feet 10 inches long, with 18 inch dial.
No. 19    2 feet 2 inches long, with 14 inch dial.

No. 27    2 feet 11 inches long, with 18 inch dial.
No. 28    2 feet 4 inches long, with 14 inch dial.

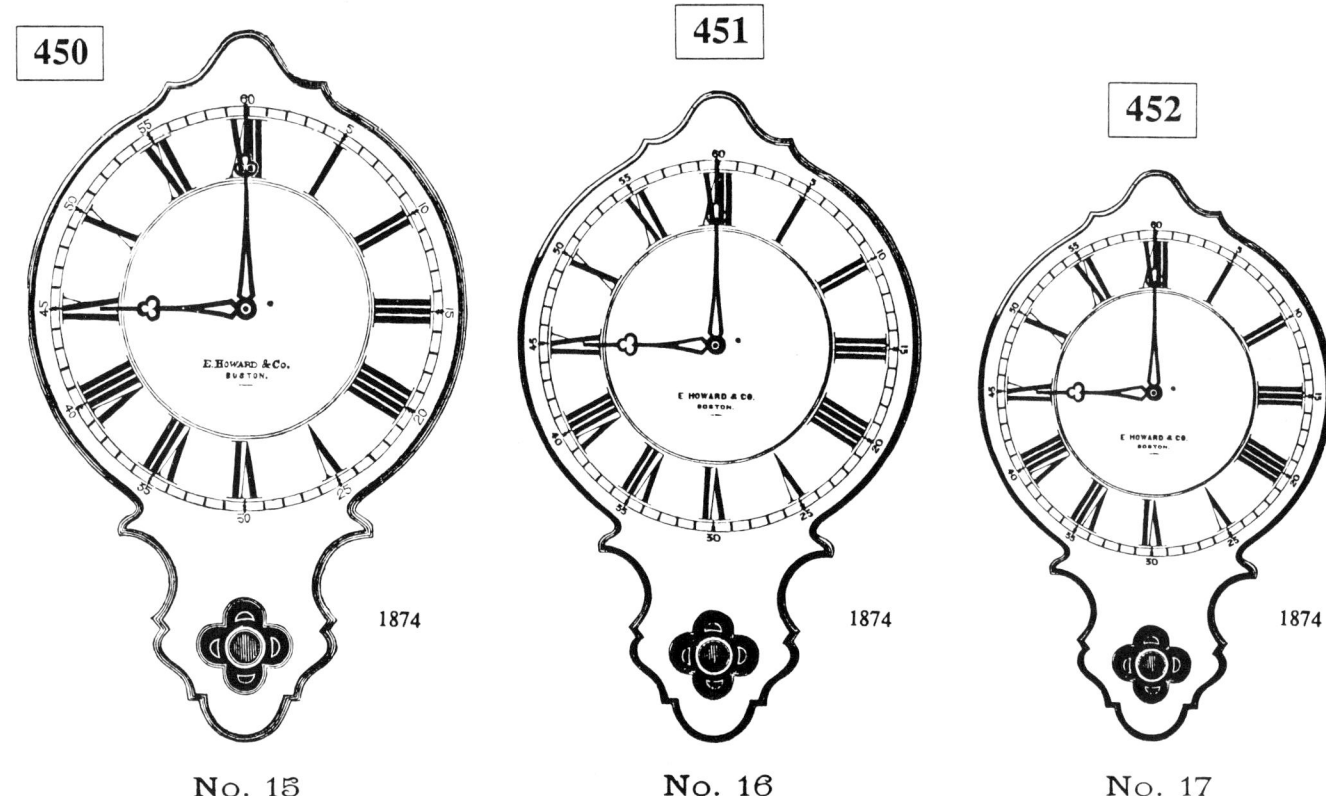

No. 15 Clock.    4 feet 9 inches long, with 36 inch dial.
No. 16 Clock.    4 feet long, with 30 inch dial.
No. 17 Clock.    3 feet 6 inches long, with 24 inch dial.

# HOWARD — Marble Dial Clocks

**453**

1874

**No. 33 Clock.** 4 feet 2 inches wide, 5 feet 9 inches long. This clock has a marble dial 2 feet in diameter. Case made of black walnut, and finely finished.

**454**

1874

The above represents No. 34 Clock. Black walnut case, 24 inch marble dial, 5 feet 8 inches long.

**455**

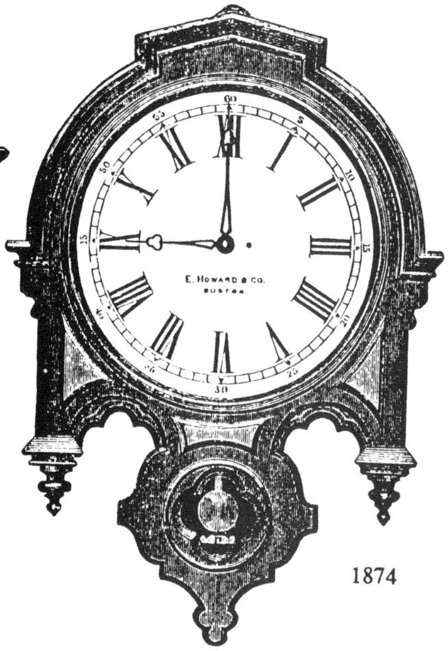

1874

**No. 35**

Black walnut case, 24 inch marble dial, 4 feet long.

**456**

1900

**No. 63.
Designed for
Church or Chapel.**

Case of Oak, Walnut or Mahogany.
Height, 5 feet 2 inches. Width, 3 feet 9 inches.
Dial of Italian Marble, 24 inches in diameter.

**457**

1923

**"Bayonne"**

This clock can also be furnished with a glass dial for illumination when a secondary movement is used.

Heighth 43½ inches.   Extreme width 51 inches.

Bronze case with marble dial and bronze numerals.

8-day, weight and pendulum movement or secondary electric movement operated from a Master clock.

# HOWARD

## Marble Dial Clocks

**458**

No. 138

24 inch dial.

Fine 8-day, weight and pendulum movement.

Height 44½ inches.
Width 30½ inches.
Depth 5 inches.

**459**

No. 21

**460**

No. 20

Extreme Dimensions
24 inch dial, 28x28 inches   30 inch dial, 36x36 inches
36 inch dial, 42x42 inches

No. 20

No. 21

Dials, 24 ins., 30 ins., and 36 ins. in diameter when mechanical movements are used.

Fine 8-day weight and pendulum movements all confined within the case, which is four inches deep.

Dials, 12 ins., 18 ins., 24 ins., 30 ins. and 36 ins. when secondary electric movements are used. With the latter a Master clock is necessary.

**461**

No. 140

| 24 inch dial. | 5 feet 4 inches long. | 37 inches high. |
| 30 inch dial. | 6 feet 8 inches long. | 48 inches high. |

**462**

No. 29

| 24 inch dial. | 5 feet 3 inches long. | 38 inches high. |
| 30 inch dial. | 6 feet 6 inches long. | 48 inches high. |

Cases of walnut, oak, mahogany or white enamel and gold.

8-day, weight and pendulum movements or secondary electric movements operated from a Master clock.

# HOWARD

*Except for those fabulous clocks for which you wouldn't mind cutting a hole in the ceiling, Grandfather or Hanging Clocks under eight feet are the most desirable and practical. Some manufacturers produced certain of these clocks in two styles: one to stand on the floor, and the other to hang on a wall. Of the clocks designed with two styles, the hanging style is considered to be the most valuable.*

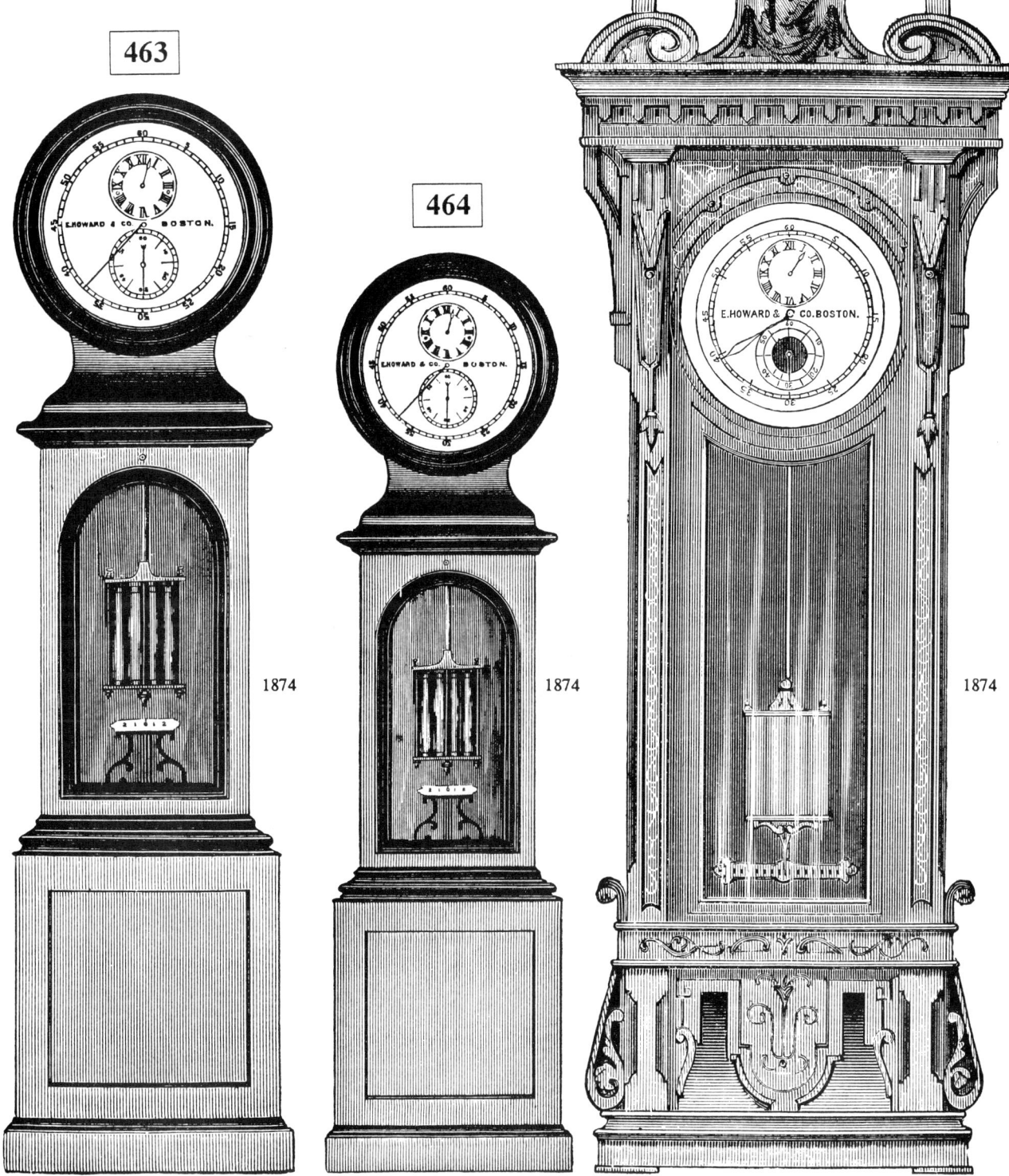

**Regulator, No. 23,**
16 inch dial.
7 feet 3 inches long.

**Regulator, No. 25.**
14 inch dial.
6 feet 9 inches long.

**Regulator, No. 43,**
14 inch dial.
8 feet 8 inches long.

# HOWARD

*Except for those fabulous clocks for which you wouldn't mind cutting a hole in the ceiling, Grandfather or Hanging Clocks under eight feet are the most desirable and practical. Some manufacturers produced certain of these clocks in two styles: one to stand on the floor, and the other to hang on a wall. Of the clocks designed with two styles, the hanging style is considered to be the most valuable.*

**466** — 1874 — **Regulator.** No. 48, 14 inch dial. 9 feet 6 inches long.

**467** — 1874 — **Regulator.** No. 47, 14 inch dial. 9 feet 6 inches long.

**468** — 1900 — No. 63 Regulator. No. 3 Movement, Graham Dead Beat Escapement. 14 inch Silvered Dial. Case Walnut or Oak, 8 feet 9 inches high.

# HOWARD

*To receive top prices, clocks must be in ORIGINAL MINT condition and in good running order - A replaced or damaged dial, missing parts, will reduce prices accordingly.*

**No. 61 Regulator**
No. 3 Movement, Graham Dead Beat Escapement. 14 inch Silvered or White Enameled Metal Dial. Case of Oak, Walnut or Cherry, 7 feet 10 inches long.

**Regulator. No. 45,**
14 inch dial.
8 feet 2 inches long.

**Regulator. No. 46,**
18 inch dial.
10 feet 6 inches long.

# HOWARD

### No. 77.

14-inch round dial. Height, 8 feet, 10 inches. Mahogany. Beveled glass. Wire gong. Hours strike. Graham Dead Beat escapement.
Price: $260.00

### No. 79.

15-inch arched with moon and calendar. Height, 10 feet 2 inches. Mahogany, Graham Dead Beat escapement, Hours strike on Cathedral gong. Westminster Chimes on wire gongs or the Cambridge chime on saucer gongs.
Price: $1,200.00

### No. 80.

Height, 8 feet, 11 inches. Mahogany, Graham Dead Beat escapement. Hours strike on Cathedral gong. Westminster chimes on Wire gongs or Cambridge chimes on saucer gongs.
Price: $760.00

# HOWARD

### 475

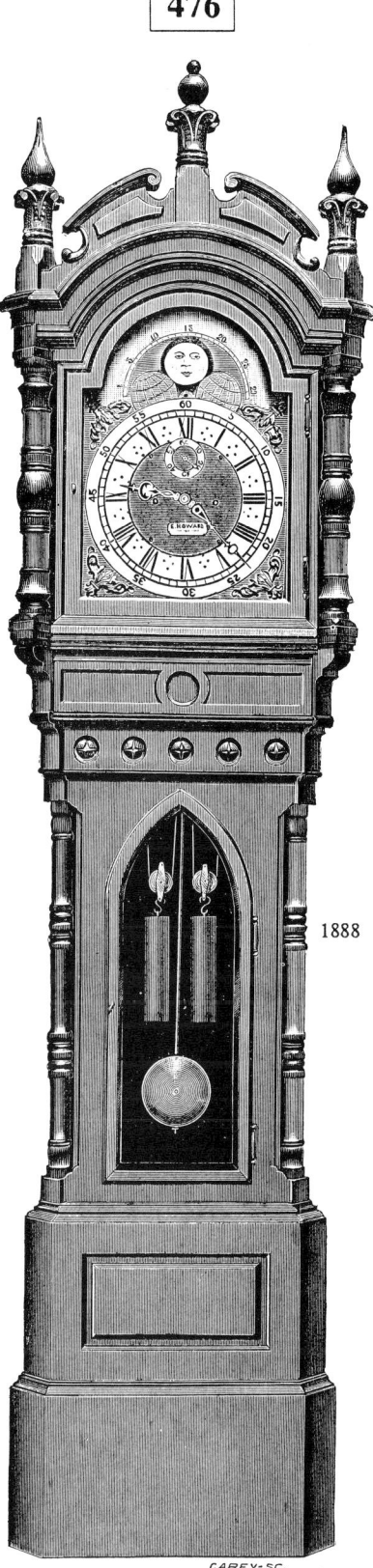

### 476

### 477

137

## No. 81.

13-inch arched dial. Height, 8 feet, 2 inches. Mahogany. Brass ornaments. Hours & half hours strike on saucer gong. Graham Dead Beat escapement.
Price: $275.00

## No. 82.—13 in.

13 inch arched dial. Height, 8 feet, 3 inches. Graham Dead Beat escapement. Hours & half hours strike on Cathedral gong.
Price: $360.00

## No. 82.—15 in.

15-inch arched dial, moon & calendars. Height, 9 feet, 3 inches. Mahogany or oak. Graham Dead Beat escapement. Hours strike on Catheral gong. Westminster chimes on wire gong.
Price: $825.00

# HOWARD

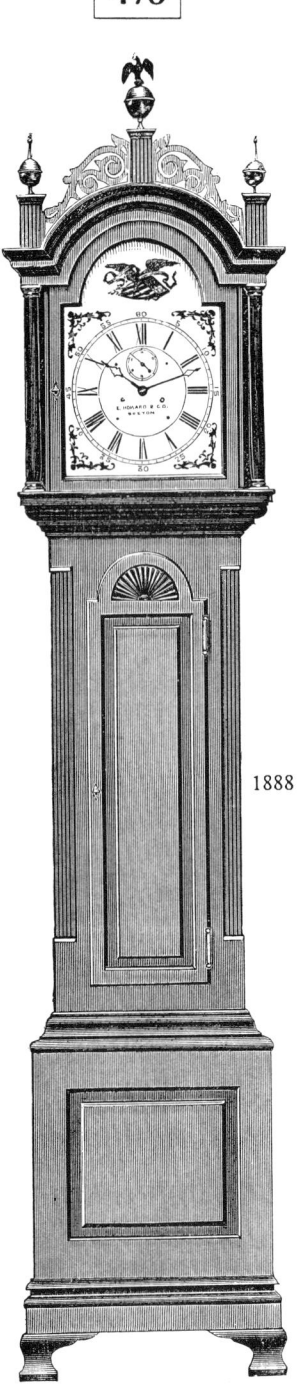

**478**

### No. 83.
12-inch arched dial. Height, 7 feet, 9 1/2 inches. Hours strike on saucer gong. Dead Beat escapement.
Price: $170.00

**479**

### No. 84.
14-inch round silvered dial. Height, 8 feet, 4 inches. Mahogany. Hours strike on Wire gong. Graham Dead Beat escapement.
Price: $250.00

**480**

### No. 87.
14-inch round silver-faced dial. Height, 8 feet, 7 inches. Mahogany or oak. Cathedral gong. Hours & half hours strike. Graham Dead Beat escapement.
Price: $375.00

# HOWARD

*The year entered beside each clock is the year of the catalog from which the illustration was taken. Some clocks may have been produced a few years earlier.*

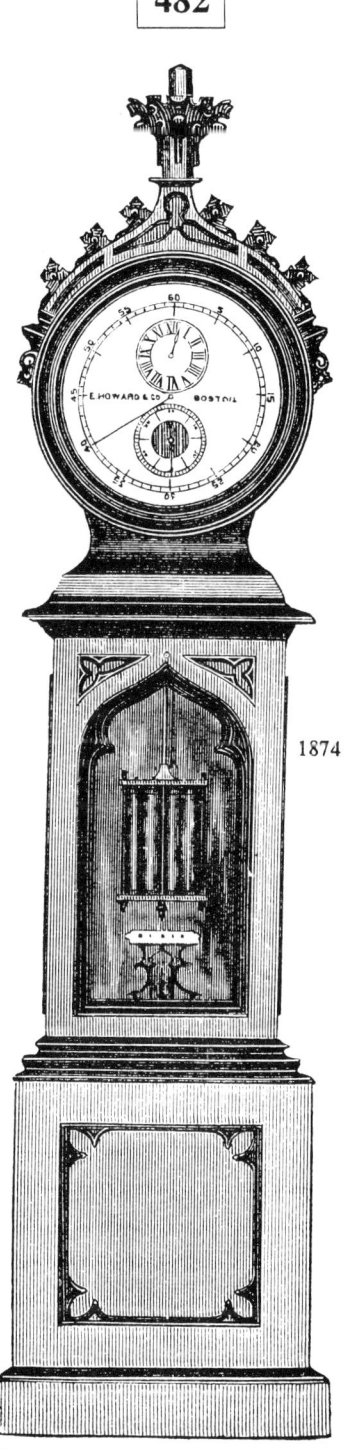

**481** — **Regulator. No. 22,** 16 inch dial. 8 feet 9 inches long. *1874*

**482** — **Regulator. No. 24,** 14 inch dial. 8 feet long. *1874*

**483** — **No. 88.** 14-inch round silver-faced dial. Height, 8 feet, 10 inches. Mahogany or oak. Cathedral gong. Hours & half hours strike. Graham Dead Beat escapement. Price: $315.00 *1888*

# E. INGRAHAM
## & COMPANY
### Bristol, Connecticut

E. Ingraham & Company was formed in 1860, succeeding several earlier clock manufacturing firms in which casemaker Elias Ingraham had been involved, notably Brewster & Ingrahams (1843-1852), E. & A. Ingrahams (1852-1856) and Elias Ingraham & Company (1857-1860). The firm originally rented, and later purchased, a shop on Birge's Pond in Bristol which had been used by a number of clockmaking firms since 1820.

Having originally purchased their movements from various sources, in 1865 the firm decided to establish their own movement making facility. A hardware shop was moved onto a piece of land owned by the firm and veteran clockmaker Anson L. Atwood set up and managed the movement department for Ingraham for some years.

Elias Ingraham (1805-1885) designed a variety of popular cases and case features for the firm, receiving 17 patents between 1857 and 1873. Many of his cases utilized an unusual figure "8" door design for which he had received a patent in 1857. Rosewood veneered case models with names such as "Doric", "Venetian", and "Ionic" were often made in several sizes and held their popularity with the public for many years.

Elias Ingraham's son Edward Ingraham (1830-1892) succeeded his father as head of the business in 1885. Edward had also received an important patent in 1884 for a method of applying black enamel paint (Japan) to wooden clock cases. Using this method to produce cheaper imitations of French marble mantel clocks, was a great success. Though the process was soon imitated by most other clock manufacturers, the Ingraham firm became a leading maker of "black mantel" clocks, introducing 221 models plus special order styles in the following three decades.

1805 - 1885

In 1887, the firm had its first great expansion with the erection of a 300 foot long, 4 story case shop. A new office building and movement shop was built between 1902 and 1904. In 1913, they began to manufacture a non-jewelled pocket watch and added wrist watch models to the line in 1932, producing more than 65 million pockets watches and 15 million wrist watches by the time this production ceased in the mid-1960's.

Ingraham's clock and watchmaking ceased totally during World War II and pendulum clocks production did not resume after the war. After the war, electric clocks, added to the line about 1930, were then a major part of their product line as were watches, alarm clocks, fuses and timers (the latter two were established during war-time production).

In 1964, a modern and much smaller factory was constructed in the southern part of Bristol and the old complex was abandoned and later demolished. Little if any clock production was done at the new factory as it was almost totally devoted to manufacture of more profitable fuses. The firm was sold to McGraw-Edison, a conglomerate, in 1967 and the Bristol factory presently produces Bussman fuses. Production of electric clocks with the Ingraham trademark continues at a plant which the firm built at Laurinburg, North Carolina in 1959.

# INGRAHAM

*For comprehensive listing of Calendar Clocks, see Tran Duy Ly "Calendar Clocks - A Guide to Identification & Prices."*

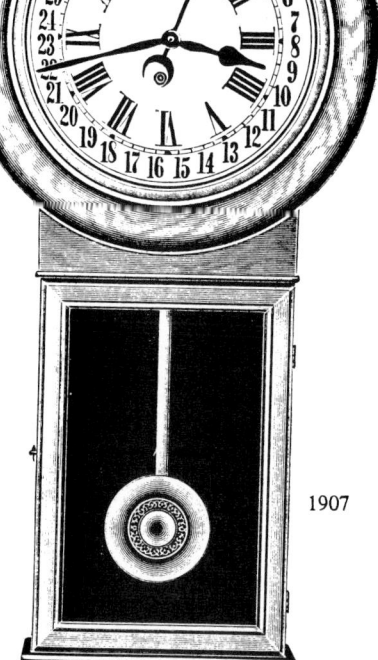

## ORMOND
### SOLID OAK—HAND POLISHED

Dial, 12 inches

Height, 35½ inches

| | |
|---|---|
| Eight Day, Time | $11.70 |
| Eight Day, Strike | 12.70 |
| Eight Day, Time, Calendar | 12.10 |
| Eight Day, Strike, Calendar | 13.10 |

1907

**ORMOND**
SOLID OAK—HAND POLISHED

## RELIANCE—Solid Oak

Dial, 12 inches

Height, 38¼ inches

| | |
|---|---|
| Eight Day, Time | $ 8.00 |
| Eight Day, Strike | 9.00 |
| Eight Day, Time, Calendar | 8.40 |
| Eight Day, Strike, Calendar | 9.40 |

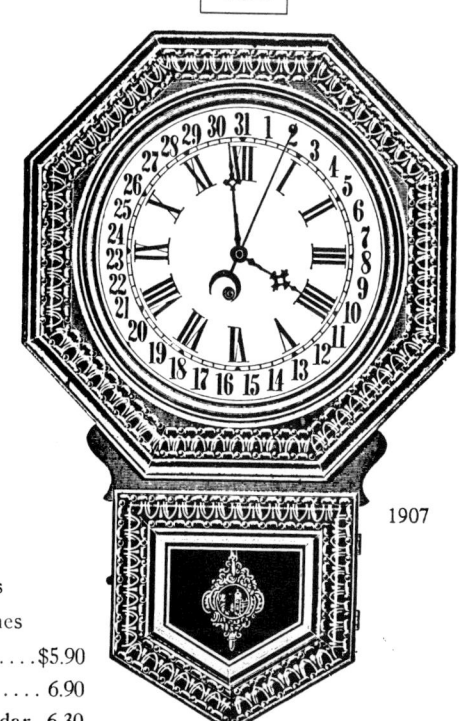

## LYRIC
### SOLID OAK

Dial, 12 inches
Height, 27 inches

| | |
|---|---|
| Eight Day, Time | $5.90 |
| Eight Day, Strike | 6.90 |
| Eight Day, Time, Calendar | 6.30 |
| Eight Day, Strike, Calendar | 7.30 |

**LYRIC—Solid Oak**

# INGRAHAM

*For comprehensive listing of Calendar Clocks, see Tran Duy Ly "Calendar Clocks - A Guide to Identification & Prices."*

### DEW DROP

**IMITATION ROSEWOOD OR OAK FINISH**

Dial, 12 inches

Height, 23½ inches

| | |
|---|---|
| Eight Day, Time | $ 5.85 |
| Eight Day, Strike | 6.85 |
| Eight Day, Time, Calendar | 6.25 |
| Eight Day, Strike, Calendar | 7.25 |

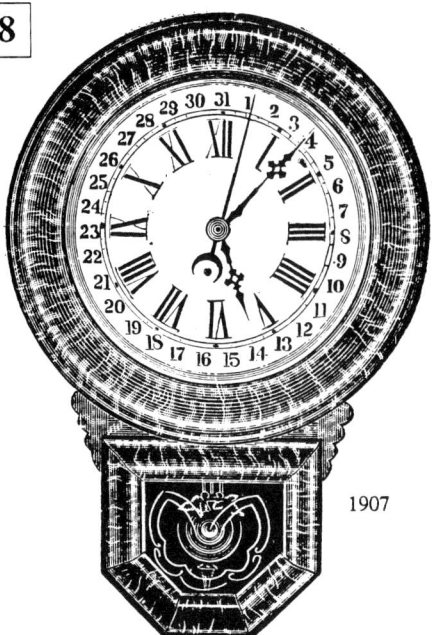

1907

**DEW DROP**

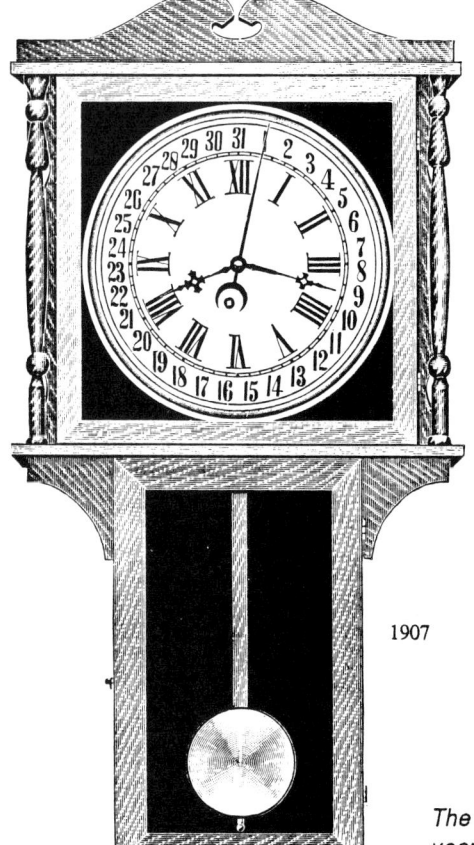

1907

*The year entered beside each clock is the year of the catalog from which the illustration was taken. Some clocks may have been produced a few years earlier.*

### HIGHLAND

**SOLID OAK—RUBBED FINISH**

Dial, 12 inches

Height, 37¾ inches

| | |
|---|---|
| Eight Day, Time | $ 8.00 |
| Eight Day, Strike | 9.00 |
| Eight Day, Time, Calendar | 8.40 |
| Eight Day, Strike, Calendar | 9.40 |

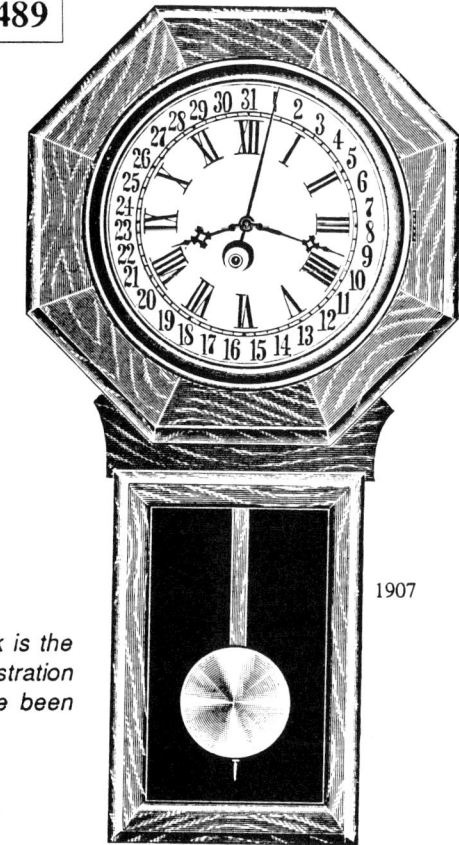

1907

### HERALD

**SOLID OAK—RUBBED FINISH**

Dial, 12 inches     Height, 33 inches

| | |
|---|---|
| Eight Day, Time | $ 6.90 |
| Eight Day, Strike | 7.90 |
| Eight Day, Time, Calendar | 7.30 |
| Eight Day, Strike, Calendar | 8.30 |

# INGRAHAM

**490**

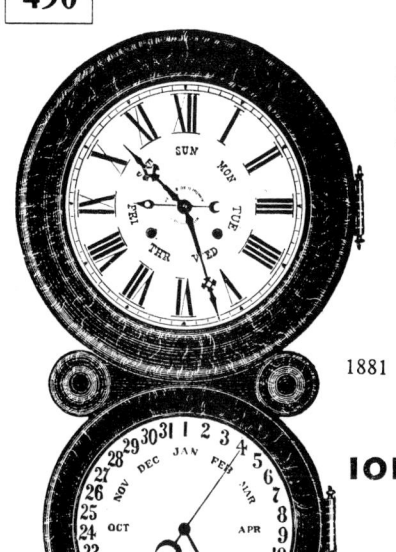
1881

For comprehensive listing of Calendar Clocks, see Tran Duy Ly "Calendar Clocks - A Guide to Identification & Prices."

### IONIC CALENDAR.

8 Day Time......... $13.00
8 Day Strike........ $14.00
Time Dial, 9 inch.
Calendar Dial, 5 1/2 inch.
Height, 22 inches.

### IONIC CALENDAR.

ROSEWOOD.
Upper Dial, 12 inches.
Height, 29½ inches.

| | | |
|---|---|---|
| 8 Day Ionic Time, Calendar | | 15.35 |
| 8 " " " Spanish Dial. | | 15.35 |
| 8 " " Strike, " | | 16.85 |
| 8 " " " Spanish Dial. | | 16.85 |

**IONIC CALENDAR.**

**491**

1881

### IONIC CALENDAR.

**492**

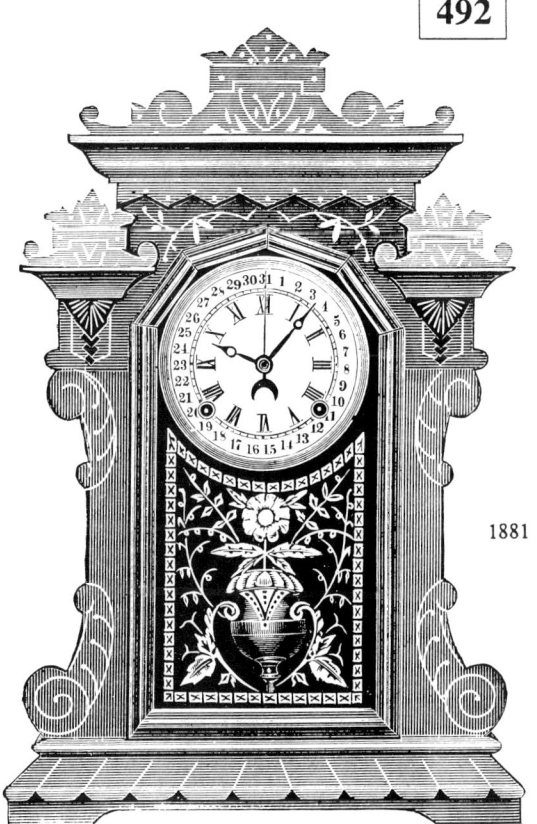
1881

### INDEX CALENDAR.
(Walnut.)

Eight Day, Strike, . . . . . $5.10

6-inch Dial.   Height, 21½ inches.

We are prepared to fit all Eight Day Clocks having *6-inch Dial* and without Alarm, with a cheap and reliable Calendar, showing Day of the Month, which cannot get out of order; hands can be turned forward or backward without injury to the Clock, at 60 cents additional list.

**493**

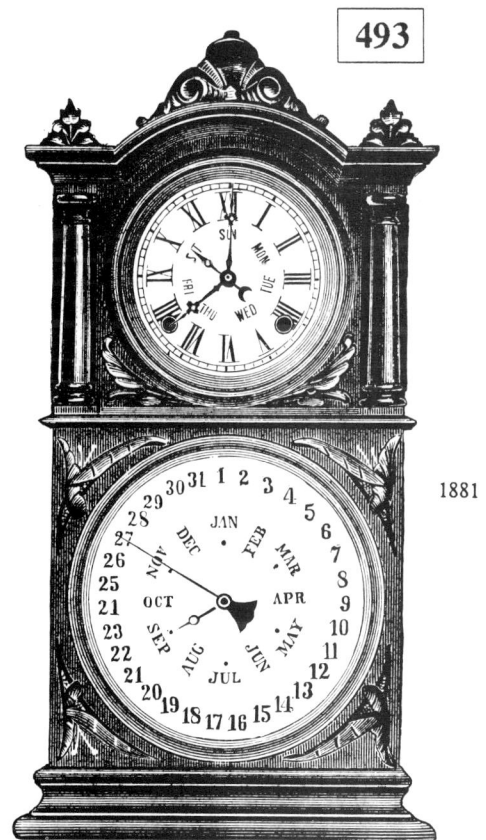
1881

### INGRAHAM CALENDAR.
(Lewis Patent.)

Time Dial, 6 inches ; Calendar Dial, 8 inches.
Height, 21½ inches.

Eight Day, Time....... $12.50
Eight Day, Strike...... 14.00

# INGRAHAM

**494**

1880

### REFLECTOR.
Spring.

Height, 29½ inches.
Dial, 12 inches.
8 Day Time.
8 Day Strike.

**495**

### LUSTRE
SOLID OAK

Dial, 12 inches
Height, 27 inches
Eight Day, Time.........$5.90
Eight Day, Strike........6.90
Eight Day, Time, Calendar, 6.30
Eight Day, Strike, Calendar, 7.30

1908

**LUSTRE—Solid Oak**

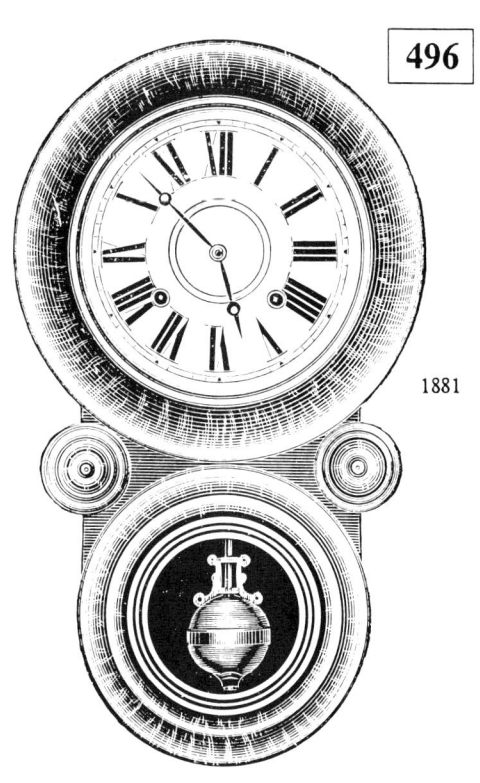

**496**

1881

### IOTA.
8-inch Dial.
Height, 19 inches.
Mosaic, 15 cents additional. Gilt, 55 cents.

Eight Day, Time, . . $4.05

Eight Day, Strike, . 4.75

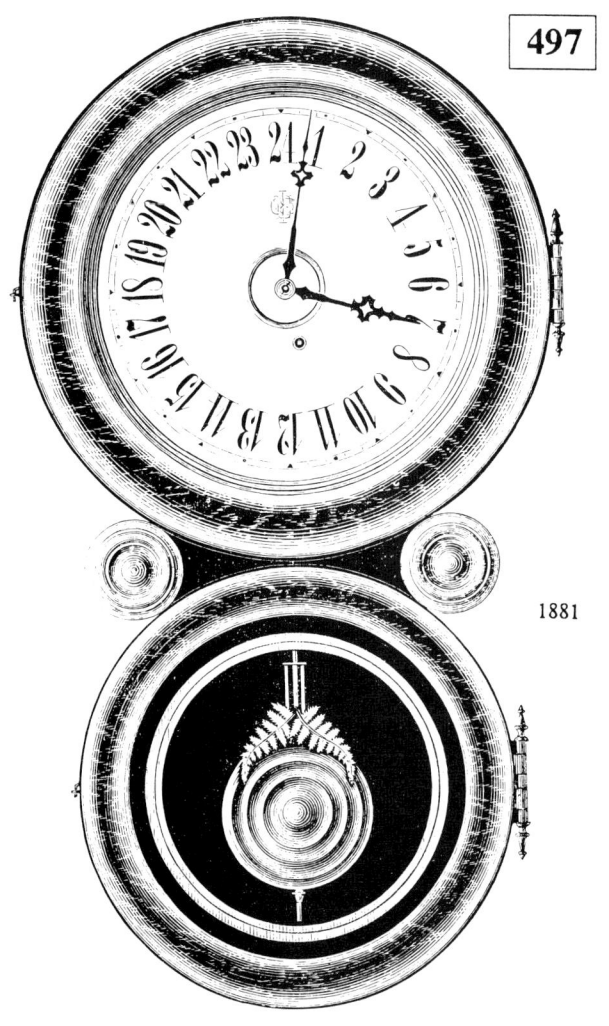

**497**

1881

### REFLECTOR. 24-Hour Dial.

Eight Day, Time, . . $5.80

# ITHACA CALENDAR CLOCK COMPANY
### Ithaca, New York

The Ithaca Calendar Clock Company was formed in 1865 to manufacture clocks with a calendar mechanism for which a patent had been granted on April 18, 1865 to Henry B. Horton and for an improvement patented on August 28, 1866. With practically no capital, they began operations in 1865.

Some of the earliest of the firm's clocks were installed in iron front cases, patented by Horton in 1866, which were cast by a local foundry. Wooden cases in many styles were later manufactured by the company. The upper clock movements were manufactured in Connecticut, primarily by the firms of E. N. Welch and Laporte Hubbell prior to 1890. The calendar mechanism was manufactured at Ithaca and the clocks were assembled, tested and marketed from there. Henry Horton and Merritt Wood had patented a device for testing the calendar mechanism on June 11, 1867 and supposedly as many as 108 clocks could be tested in one day.

In 1866, the firm moved to another facility and a joint stock corporation was formed. The business grew so well that in 1874, they commenced a new three-story building which was occupied the following June. This factory was totally destroyed by fire on February 12, 1876 but a new factory was thereafter erected which exists to the present.

No doubt the firm's greatest prosperity was from about 1875 to 1900. In 1898 the firm added a non-calendar floor standing clock to the line. Ithaca "grandfather" clocks with inexpensive spring or weight-driven Connecticut movements were manufactured for nearly 20 years.

Literally several dozen different standard calendar models were manufactured over a 50 year period, ranging in size from less than 16 to 72 inches tall in both mantel and hanging styles. Some of the models were redesigned during this 50 year period, so there are some model variations which range from minor changes in dimensions to instances where the same models have a totally different appearance. As the cases were manufactured by the firm, a number of special order models were produced, some in floor standing cases.

By 1917, the firm was bankrupt and on March 14, 1917, the real estate and personal property of the firm were sold at public auction to settle its liabilities.

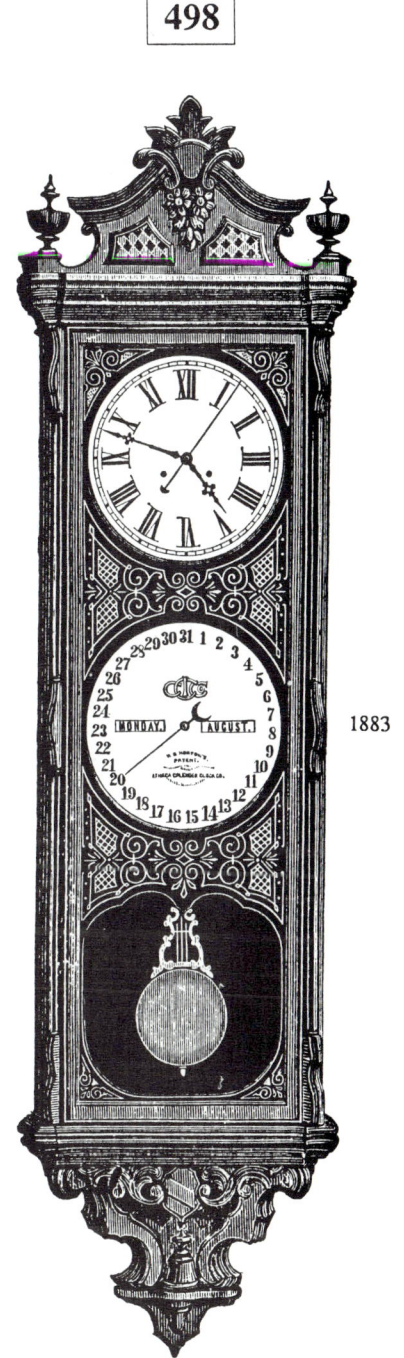

498

1883

NO. 1 REGULATOR.
WALNUT.—SWEEP SECOND.
Eight Day—Weight—Time.
Thirty-nine inch Wood Rod.
Height, 72 inches.
12 inch Calendar Dial.
12 inch Time Dial.

146

# ITHACA

*For comprehensive listing of Calendar Clocks, see Tran Duy Ly "Calendar Clocks - A Guide to Identification & Prices."*

**499** 1883

### NO. 3 VIENNA.
WALNUT.

Thirty Day—Double Spring—Time.

Twenty-six inch Wood Rod.
8 inch Time Dial.   8 inch Calendar Dial.
Height, 52 inches.

**500** 1883

### No. 2 BANK.
WALNUT.

Eight Day—Weight—Time.

Thirty-one inch Wood Rod.
12 inch Time Dial.   12 inch Calendar Dial.
Height, 61 inches.

**501** 1883

### No. 0 BANK.
WALNUT.

Eight Day—Weight—Time.

Thirty-one inch Wood Rod.
12 inch Time Dial.   12 inch Calendar Dial.
Height, 61 inches.

# ITHACA

## No. 15 MELROSE
CHERRY. EBONIZED. POLISHED.
Heavy Cast Brash Sash. Beveled French Crystals.
Eight Day, Half Hour, Slow Strike Gong
Perpetual Calendar
6-inch Time Dial, 8-inch Calendar Dial,
Black and Gold Trimmings
Height, 22 inches

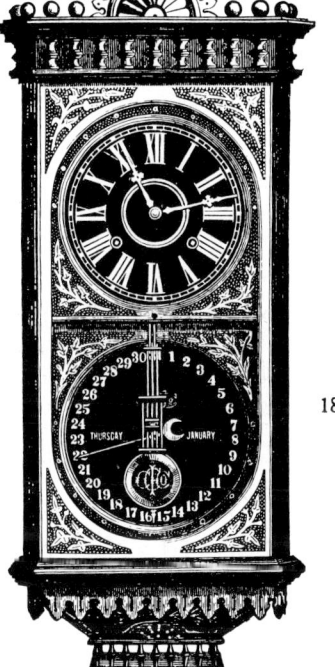

No. 15 MELROSE.

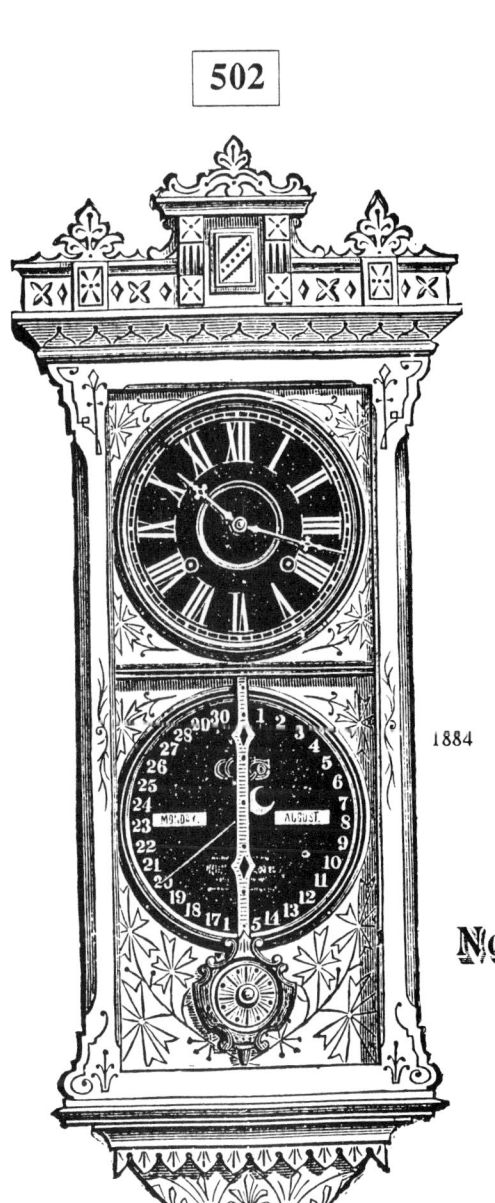

### No. 5½ HANGING BELGRADE.
WALNUT OR ASH.

Walnut, with White Dials.
Ash, with Black Dials.

Eight Day—Half-hour slow Strike—Gong. $32 50.
Thirty Day—Time—Double Spring.
Time Dial 8 Inches. Calendar Dial 8 Inches.
Height 37½ Inches.

### No. 12 Hanging Kildare.
MAHOGANY.
Eight Day—Spring—Strike—$30 00.

Thirty Day—Spring—Time—$31 00.

8 inch Time Dial. 8 inch Calendar Dial.
Height, 32 1-2 inches.

### No. 3½ PARLOR.
WALNUT CASE, WITH EBONY
TRIMMINGS AND ELABORATE
CARVINGS, NICKEL-PLATED SASH, BELL,
PENDULUM ROD, AND CALENDAR.
With Glass Ball.

Fancy Bracket for Hanging, extra.

5 inch Time Dial.   8 inch Calendar Dial (Glass).
Height, 20 inches.

Eight Day—Spring—Strike.
Thirty Day—Double Spring—Time.

# ITHACA

*The year entered beside each clock is the year of the catalog from which the illustration was taken. Some clocks may have been produced a few years earlier.*

For comprehensive listing of Calendar Clocks, see Tran Duy Ly "Calendar Clocks - A Guide to Identification & Prices."

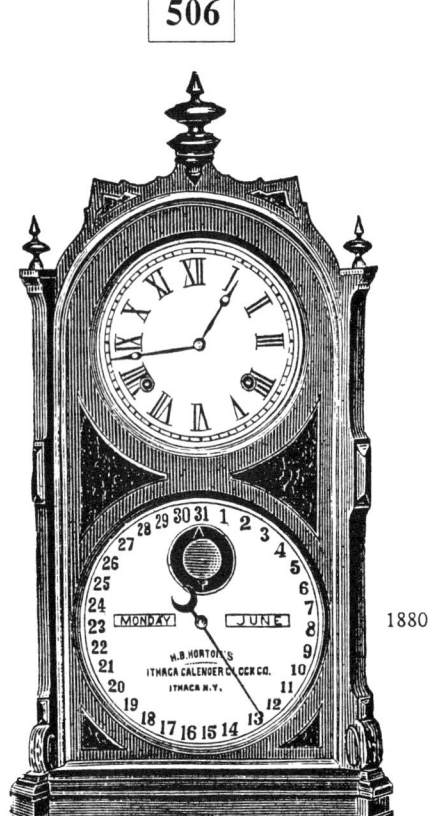

506

1880

### NO. 9 SHELF COTTAGE.

WALNUT.

Eight Day—Spring—Strike—$15 00.

Eight Day—Spring—Strike—Alarm—$16 00.

7 inch Time Dial. 8 inch Calendar Dial.
Height, 23 inches.

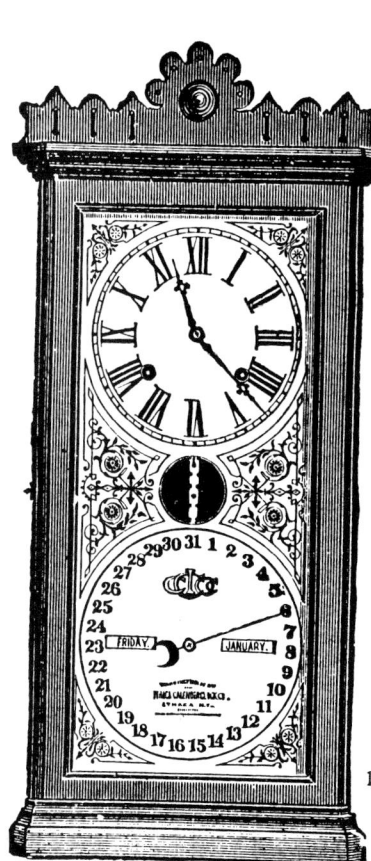

507

1885

### No. 14 GRANGER.
WALNUT.

Eight Day—Half-hour Strike. $14 30.
Eight Day—Half-hour Slow Strike—Gong. $14 93.
Eight Day—Half-hour Strike—Alarm. $15 60.
Time Dial 8 Inches. Calendar Dial
8 Inches. Height 25 Inches.

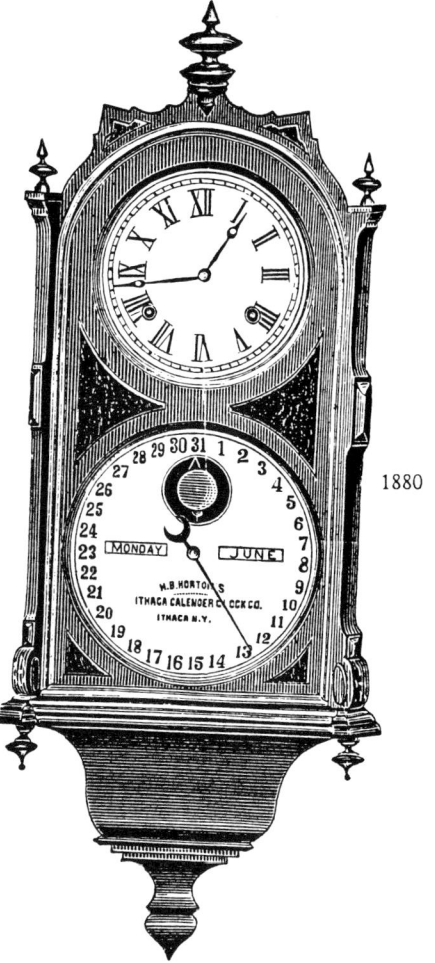

508

1880

### NO. 7 HANGING COTTAGE.

WALNUT.

Eight Day—Spring—Strike—$16 00.

Eight Day—Spring Strike—Alarm—$17 00.

7 inch Time Dial. 8 inch Calendar Dial.
Height, 29 inches.

# THE
# F. KROEBER CLOCK CO.

New York, N. Y.

Florenz Friederick Martin Kroeber was born in 1840 in the city of Paderborn, Westphalia, Germany. His early years were spent in the city of Cologne, but his parents brought their family to America about the spring of 1850, settling in New York City. By the age of 19, "Florence" Kroeber began to work in the clock store of Owen & Clark at John Street as a bookkeeper.

In 1861, Owen & Clark was dissolved but one of the partners, George B. Owen, continued the operation under his own name. When Owen went to Winsted, Conn. to become general manager of W. L. Gilbert & Company in 1864, Florence Kroeber took over the Owen business. This was strictly a marketing operation, both of domestic and imported clocks, though labels with F. Kroeber were often applied.

In 1868, Kroeber went into partnership with Nicholas Mueller, a German immigrant who ran a business of producing bronzed cast metal figurines and figured metal clock case fronts. Two years later Kroeber married Mueller's daughter. Though the partnership lasted only about a year, the two families had close business ties and even rented adjoining New York stores for many years.

Florence Kroeber's photograph which accompanied his obituary in the May 24, 1911 edition of **The Jewelers' Circular-Weekly.**

About 1870 Kroeber began to manufacture some cases of his own design and contracted with Connecticut manufacturers for movements, some to his own specifications. For over 25 years the operation was successfull. It was incorporated as the F. Kroeber Clock Company in 1887 and that year a second New York store was opened in midtown Manhattan. Their 1888 catalog of 208 pages illustrated 286 clocks and 43 figurines, over 90% of American manufacture.

Business began to decline with the depression of 1893 and in 1895 the midtown store was closed. By 1898, their catalog of 115 pages offered only 182 clocks and 31 figurines, with only about 80% being American made. In 1899 the corporation went into receivership. Kroeber moved into a smaller store on Maiden Lane and spent about a year settling the accounts of the company.

Kroeber continued marketing clocks as a private venture under his own name, most being purchased from Connecticut. Except for cuckoo clocks, foreign made clocks were no longer offered. This business ended in bankruptcy in January, 1904 and because it was not a corporation, Kroeber was personally ruined. For the next seven years he worked as an employee in clock and watch departments of various New York firms. He died May 16, 1911 of tuberculosis.

# KROEBER

*The year entered beside each clock is the year of the catalog from which the illustration was taken. Some clocks may have been produced a few years earlier.*

**510** — MILFORD. 12-Inch Dial. Eight-Day Pendulum. Time or Strike. Oak or Walnut. 1898

**509** — VIENNA No. 26. Weight. Length, 51 Inches. 7-Inch Porcelain Dial. Eight-Day Time. Oak. 1898

**511** — BRISTOL. 12-Inch Dial. Length, 32 Inches. Eight-Day Spring. Time or Strike. 1898

**512** — VIENNA No. 27. Weight. Length, 53 Inches. 7-Inch Porcelain Dial. Eight-Day Time. Walnut. 1898

# KROEBER

**513**

BRASS LEVER—
For Locomotives.
6-INCH DIAL.
ONE-DAY TIME.
1898

**514** 
1898

**515**

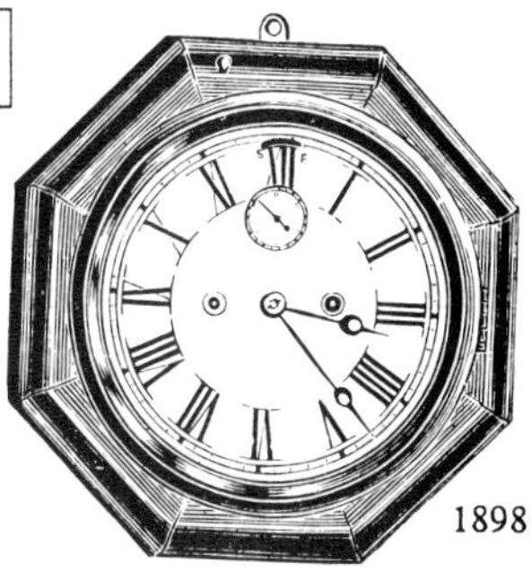

ROUND CORNER
OCT. LEVER.
4 TO 10-INCH DIAL.
ONE-DAY TIME OR STRIKE.
1898

**516** 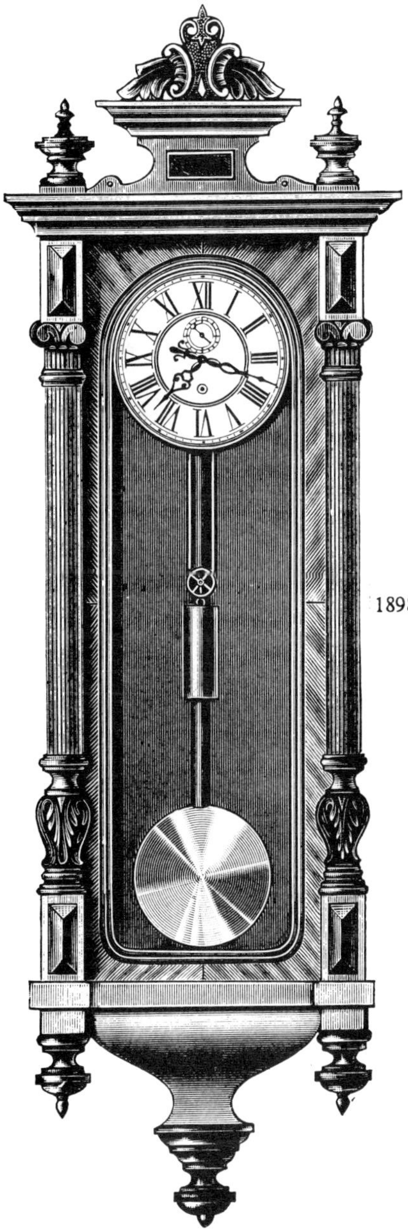
1898

HARTFORD—SOLID OAK.

**517**
1898

**518**
1898

VIENNA No. 1.
WEIGHT.
LENGTH, 49 INCHES.
7-INCH PORCELAIN DIAL.
EIGHT-DAY TIME.
WALNUT.

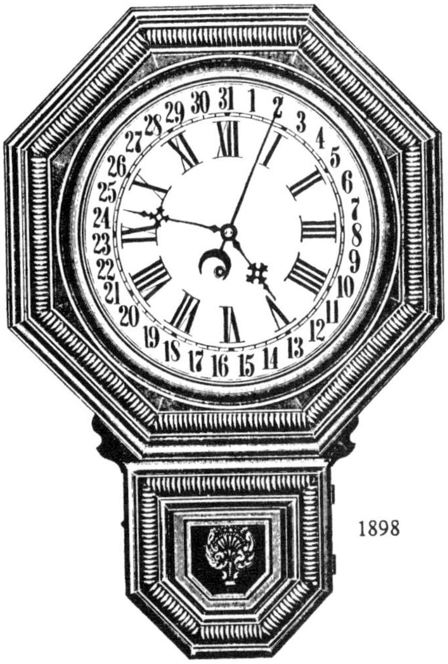

DROP OCTAGON—OAK.
10-INCH DIAL. HEIGHT, 21 INCHES.
12-INCH DIAL. HEIGHT, 24 INCHES.
EIGHT-DAY SPRING. TIME OR STRIKE.

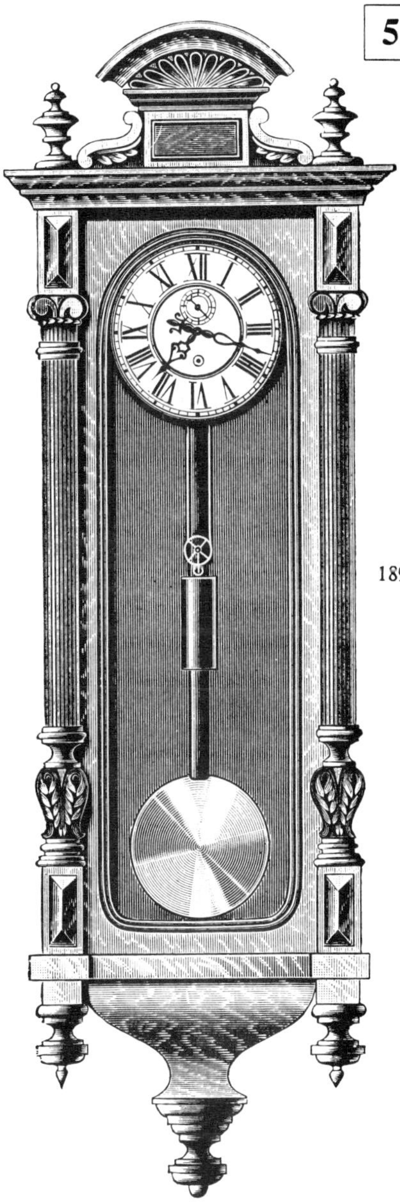

VIENNA No. 2.
WEIGHT.
LENGTH, 49 INCHES.
7 INCH PORCELAIN DIAL.
EIGHT-DAY TIME.
OAK.

# KROEBER

*To receive top prices, clocks must be in ORIGINAL MINT condition and in good running order - A replaced or damaged dial, missing parts, will reduce prices accordingly.*

**519**

SOLID OAK LEVER.
4 OR 6-INCH DIAL.
ONE-DAY TIME.

1898

**520**

SOLID OAK LEVER.
3-INCH PORCELAIN DIAL.
ONE-DAY TIME.

1898

**521**

**522**

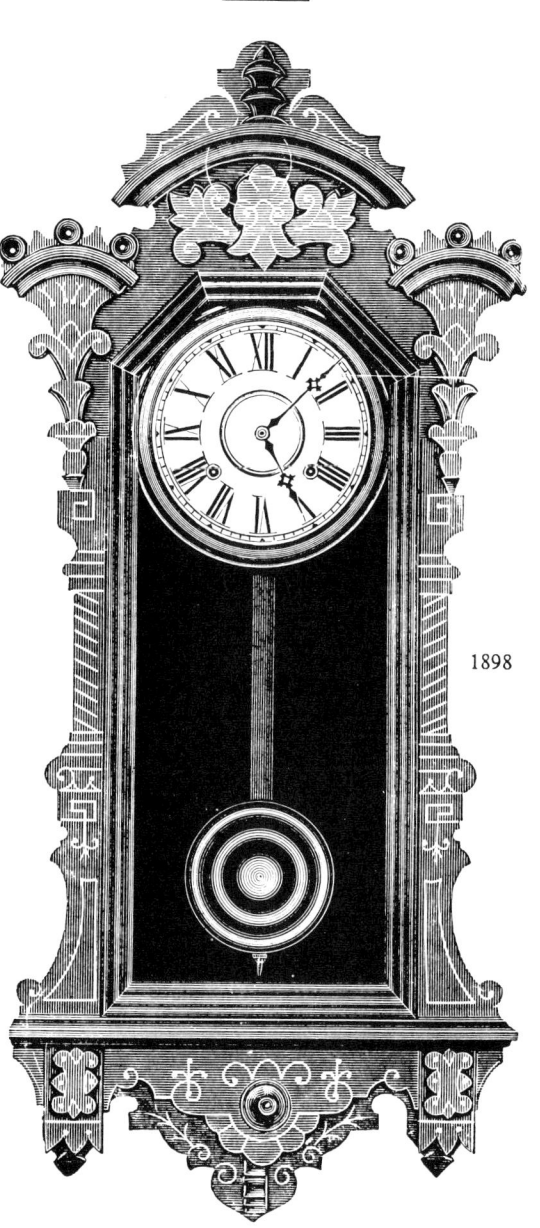

1898

**ARCTIC**—OAK.

DIAL, 8 INCHES.
HEIGHT, 37 INCHES.

**PACIFIC**—OAK.

DIAL, 8 INCHES.
HEIGHT, 37 INCHES.

# KROEBER

*The year entered beside each clock is the year of the catalog from which the illustration was taken. Some clocks may have been produced a few years earlier.*

**523**

**SCYTHIA.**
EBONY.
With Real Marquetries and Beveled Plate Glass Mirror in Panel.
8 Day Strike.   5 inch Porcelain Dial,
Height, 27 inches.
Net Price............$9.00

1870

**SCYTHIA.**

**524**

1870

**SCYTHIA.**
EBONY.
With Real Marquetries and Beveled Plate Glass Mirror in Panel.
15 Day Strike.   Visible Escapement.   5 inch Porcelain Dial.
Height, 27 inches.
Net Price,............$11.00

**SCYTHIA.**

**525**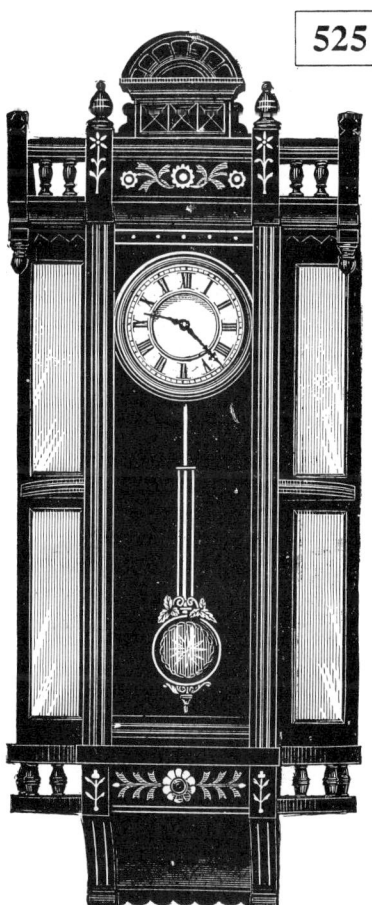

1870

**526**

1870

**PARTHIA.**
EBONY.
With Real Marquetries, Beveled Plate Glass Mirrors on Side Brackets, and Drawers at Bottom.
8 Day Strike.   5 inch Porcelain Dial.
Height, 32½ inches.
Net Price........................$15.50

**PARTHIA.**

**PARTHIA.**
EBONY.
With Real Marquetries, Beveled Plate Glass Mirrors on Side Brackets, and Drawer at Bottom.
15 Day Strike.   Visible Escapement.   5 inch Porcelain Dial.
Height, 32½ inches.
Net Price..................$17.50

**PARTHIA.**

# KROEBER

*The year entered beside each clock is the year of the catalog from which the illustration was taken. Some clocks may have been produced a few years earlier.*

**527**

1898

**WATERMILL.**
Time. Alarm.
Dial, 4 Inches. Dial, 4 Inches.
With Movable Waterwheel.
Nickel.

**528**

1898

**WINDMILL.**
Time. Alarm.
Dial, 4 Inches. Dial, 4 Inches.
With Movable Wind Wings.
Nickel.

**529**

1898

**CLUB NIGHT**—Alarm.
Dial, 4 Inches.
Nickel.

**530**

1898

**TETE A TETE**—Alarm.
Dial, 4 Inches.
Nickel.

**531**

1898

**BALLET DANCER**—Alarm.
Dial, 4 Inches.
Nickel.
with movable figures.

# KROEBER

### 532

**VENTURA.**
Height, 13½ Inches.  Width, 17 Inches.
EIGHT-DAY GONG STRIKE,
WITH ECLIPSE MOVEMENT.

### 533

**ALABAMA.**
Height, 21 Inches.  Width, 24 Inches.
EIGHT-DAY GONG STRIKE, WITH ECLIPSE MOVEMENT.
FOR DIALS, CONSULT PAGES 4 AND 5.

### 534

**ARIZONA.**
Height, 20 Inches.  Width, 18 Inches.
EIGHT-DAY GONG STRIKE, WITH ECLIPSE MOVEMENT.

### 535

**CALIFORNIA.**
Height, 18 Inches.  Width, 17 Inches.
EIGHT-DAY GONG STRIKE, WITH ECLIPSE MOVEMENT.

# KROEBER

**536**

**SANTA BARBARA.**

Height, 18 Inches.   Width, 17 Inches.

EIGHT-DAY GONG STRIKE, WITH ECLIPSE MOVEMENT.

**537**

**MONTANA.**

Height, 22½ Inches.   Width, 18 Inches.

EIGHT-DAY GONG STRIKE, WITH ECLIPSE MOVEMENT.

**538**

**INDIAN**—Bronze.
8 Day, French White or Gilt Dial.
20½ Inches High.

**539**

**EASEL, No. 1**—Alarm.
Height, 13 Inches.

**540**

**BRONZE No. 280.**
Mexican Onyx Base.
French White or Gilt Dial.
15 Inches High.

# KROEBER

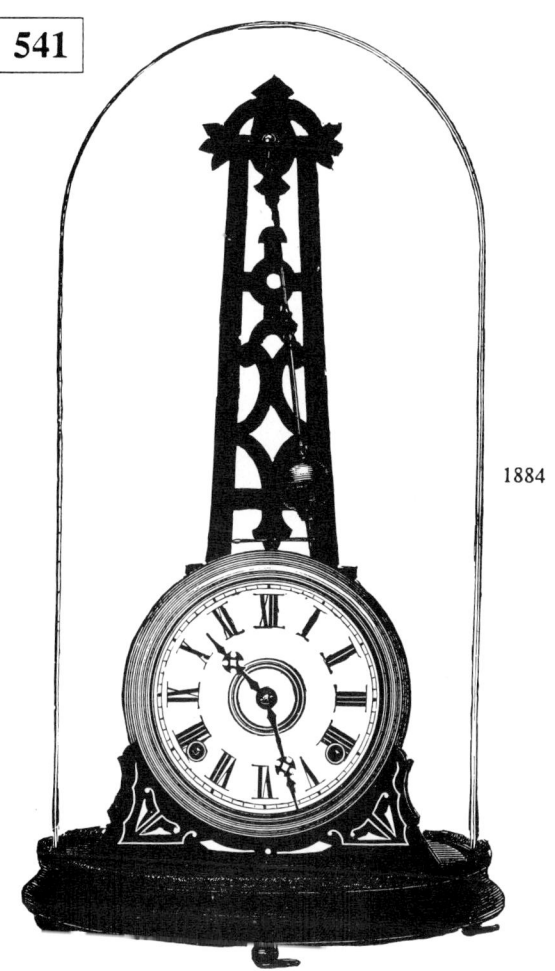

1884

**NOISELESS ROTARY, No. 1.**

8 Day, Strike........................ $10 13
Attractive and useful.
Totally noiseless in the Escapement.
Ebony Case, Nickel Pendulum.
With Glass Shade. Total Height, 20½ inches.

1898

**MOUNTAIN BOY—ALARM.**
HEIGHT, 7 INCHES.
POLISHED WOOD CASE. DECORATED BISQUE FIGURE.

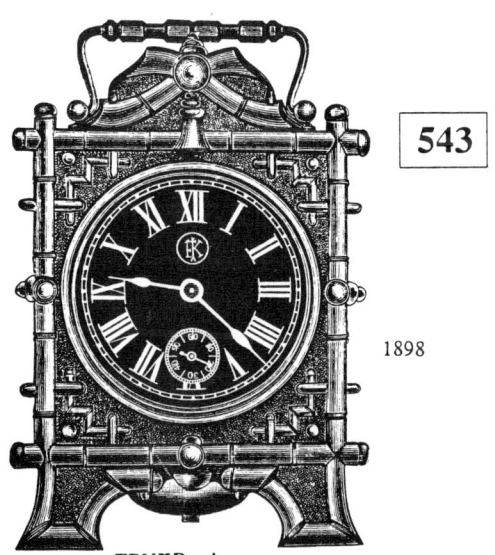

1898

**TRUMP—ALARM.**
BLACK DIAL    HEIGHT, 6½ INCHES
NICKEL.

1884

**NOISELESS ROTARY, No. 3.**

8 Day, Strike........................ $16 88
Ebony Case, Bronze Figure, Nickel Pendulum.
Attractive and useful. Totally noiseless in the Escapement.
No. 2. Similar Style to No. 3, but Smaller. Height, 22 inches, $13 13
Total Height, 22½ inches.

# KROEBER

**545**

**CONQUEST.**
HEIGHT, 21½ INCHES.   5-INCH DIAL.
EIGHT-DAY GONG STRIKE.
WALNUT.
ALSO WITH ILLUMINATING DIAL.

**546**

**BERMUDA.**
HEIGHT, 19 INCHES.   6-INCH WHITE
OR BLACK DIAL.
EIGHT-DAY GONG STRIKE.
ASH.

**547**

**CABINET No. 58.**
HEIGHT, 16 INCHES.
ASH.
EIGHT-DAY GONG STRIKE,
WITH ECLIPSE MOVEMENT.

**548**

**TEXAS.**
HEIGHT, 23 INCHES.   6-INCH DIAL.
EIGHT-DAY GONG STRIKE.
WALNUT.

**549**

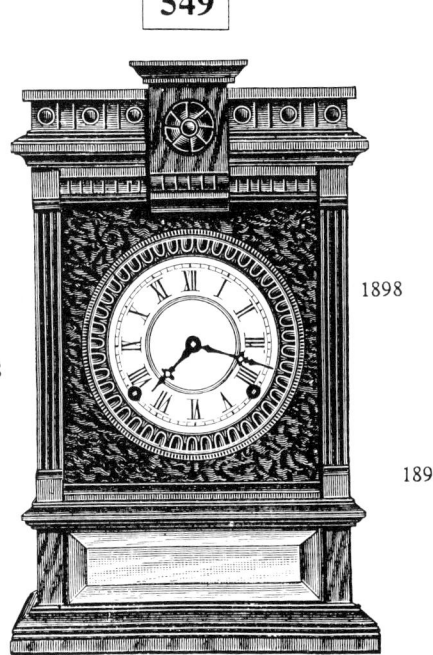

**CALCUTTA.**
HEIGHT, 15 INCHES.
WALNUT—CABINET FINISH.
EIGHT-DAY GONG STRIKE,
WITH ECLIPSE MOVEMENT.

**550**

**POLARIS.**
HEIGHT, 20 INCHES.   6-INCH DIAL.
EIGHT-DAY GONG STRIKE.
WALNUT.

# KROEBER

**551**

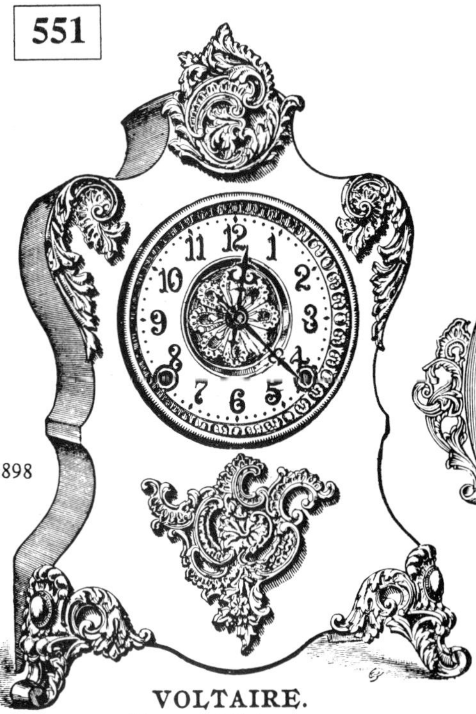

**VOLTAIRE.**
HEIGHT, 13¾ INCHES.
EIGHT-DAY GONG STRIKE, WITH ECLIPSE MOVEMENT.
FURNISHED IN BLACK AND COLORS.

1898

**552**

**CONDÉ.**
HEIGHT, 14¾ INCHES.
EIGHT-DAY GONG STRIKE, WITH ECLIPSE MOVEMENT.
FURNISHED IN BLACK AND COLORS.

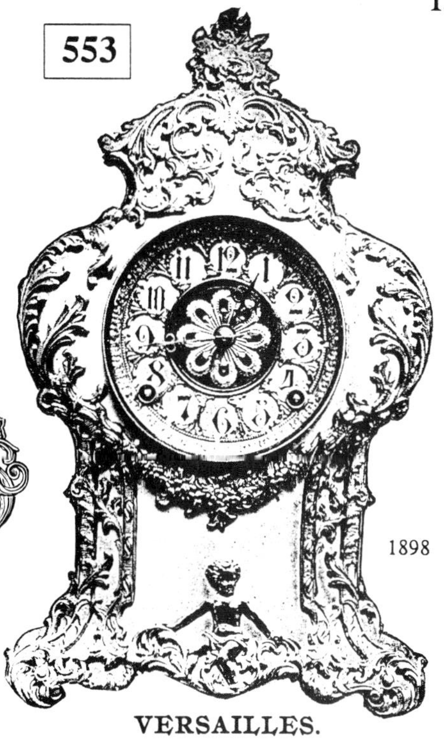

1898

**553**

**VERSAILLES.**
HEIGHT, 15 INCHES.
EIGHT-DAY GONG STRIKE, WITH ECLIPSE MOVEMENT.
FURNISHED IN BLACK AND COLORS.

1898

**554**

**555**

**POMPADOUR.**
HEIGHT, 18¼ INCHES.
EIGHT-DAY GONG STRIKE, WITH ECLIPSE MOVEMENT.
FURNISHED IN BLACK AND COLORS.

1898

*The year entered beside each clock is the year of the catalog from which the illustration was taken. Some clocks may have been produced a few years earlier.*

# KROEBER

*In the case of Porcelain clocks, color is the most important factor in estimating value. Bright reds, blues and frosted pinks are usually considered to be the most desirable. In order to receive top prices, clocks must be in ORIGINAL MINT condition and in good running order. Damaged dial, chipped porcelain will substantially reduce prices.*

**556**

CHINA No. 12.
8 Day American Dial.
French Decoration.
11¼ inches high.
With Eclipse Movement.

**557** — 1898

CHINA No. 22.
8 Day American Dial.
Blue Decore.   French Decore.
13½ inches high.
With Eclipse Movement.

**558** — 1898

CHINA No. 11.
8 Day American Dial.
Delft.
11¼ inches high.
With Eclipse Movement.

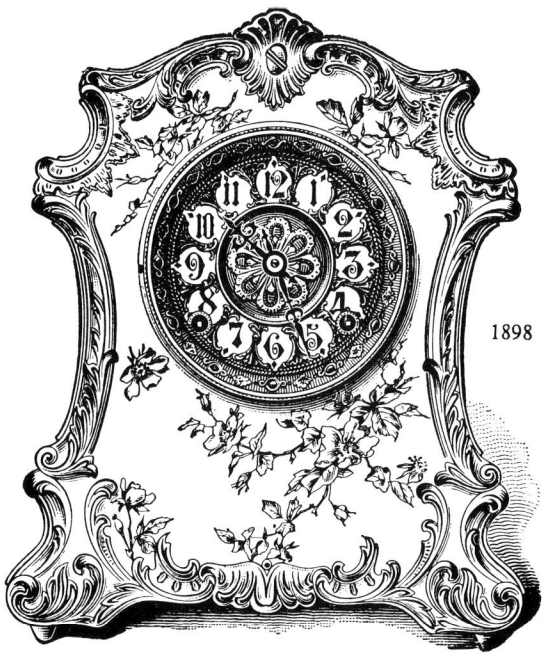

**559** — 1898

CHINA No. 14.
8 Day American Dial.
French Decorations.
12¼ inches high.
With Eclipse Movement.

# KROEBER

*The year entered beside each clock is the year of the catalog from which the illustration was taken. Some clocks may have been produced a few years earlier.*

**561** — 1898

**CHINA No. 20.**
8 Day American Dial.
FRENCH DECORATION.
10½ inches high.
With Eclipse Movement.

**560** — 1898

**CHINA No. 13.**
8 Day American Dial.
DELFT.
12¼ inches high.
With Eclipse Movement.

**562** — 1898

**CHINA No. 23. ORANGE.**
"    "   24. BLUE.
"    "   25. PINK.
9½ inches high.
With Eclipse Movement.

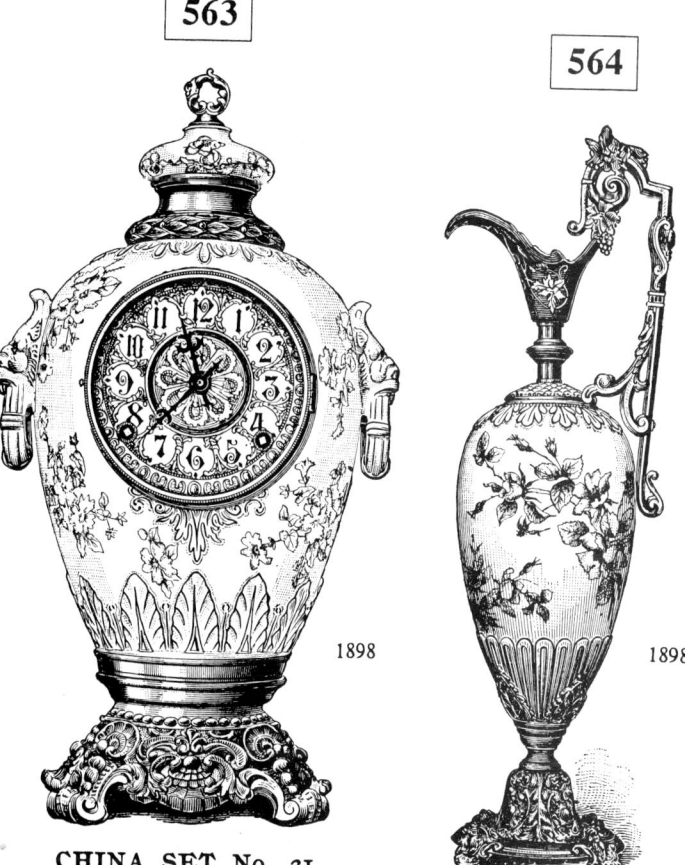

**563** — 1898   **564** — 1898

**CHINA SET No. 31.**
CHINA CLOCK No. 31.
16½ inches high.

VASES No. 3.
18 inches high.

FURNISHED IN COBALT (ROYAL) BLUE OR PINK DECORATION.
CHINA WITH GILT METAL TRIMMINGS.

# KROEBER

*The year entered beside each clock is the year of the catalog from which the illustration was taken. Some clocks may have been produced a few years earlier.*

**565** — 1898
**CHINA No. 18.**
8 Day American Dial.
french decoration.
11½ inches high.
With Eclipse Movement.

**566** — 1898
**LAURA.**
1 Day. french or delft.
8½ inches high.

**567** — 1898
**CHINA No. 15.**
8 Day American Dial.
french decore.
12½ inches high.
With Eclipse Movement.

**568** — 1898
**CHINA No. 17.**
8 Day American Dial.
french decoration.
10½ inches high.
With Eclipse Movement.

**569** — 1898
**CHINA No. 21.**
8 Day American Dial.
blue decore, french decore.
13½ inches high.
With Eclipse Movement.

**570** — 1898
**CHINA No. 19.**
8 Day American Dial.
french decoration.
12½ inches high.
With Eclipse Movement.

# KROEBER

571

*In the case of Porcelain clocks, color is the most important factor in estimating value. Bright reds, blues and frosted pinks are usually considered to be the most desirable. In order to receive top prices, clocks must be in ORIGINAL MINT condition and in good running order. Damaged dial, chipped porcelain will substantially reduce prices.*

1898

**CHINA No. 16.**
8 Day American Dial.
DELFT.
10½ inches high.
With Eclipse Movement.

572  572 A

1898   1898

**RUTH.**
1 Day. Delft.
8½ inches high.

**RUTH.**
1 Day. French.
8½ inches high.

573

1898

**LILLIAN**
FRENCH DECORE.
6 inches high.

574

1898

**BLANCHE.**
FRENCH DECORE.
6 inches high.

575    576

1898    1898    1898

**CANDELABRA No. 4.**
18 inches high.

**CHINA SET No. 32.**
CHINA CLOCK, No. 32.
17 inches high.

**CANDELABRA No. 4.**
18 inches high.

FURNISHED IN COBALT (ROYAL) BLUE OR LIGHT BLUE DECORATION.    CHINA WITH GILT METAL TRIMMINGS.

# NEW HAVEN CLOCK COMPANY
### New Haven, Connecticut

Incorporated February 7, 1853, the New Haven Clock Company was formed by clockmaker Hiram Camp and others to supply clock movements to the Jerome Manufacturing Company, then the largest clockmaking operation in the world. Three years later, the Jerome firm went bankrupt and in April of 1856 the New Haven Clock Company raised an additional $20,000 and purchased the Jerome operation.

By 1860, the firm employed 300 men and 15 women and was producing about 170,000 clocks a year. In 1866, the old Jerome factory was destroyed by fire, but a new brick factory was soon built which survives today with many additions. Their working force had increased to 460 men, 52 women and 88 children by 1880 and nearly half a million dollars worth of clocks were produced that year. Non-jewelled pocket watches were added to the line that year and were offered until the 1950's.

New Haven had trade connections with Jerome & Co. Ltd., an English sales firm which continued business independently after the bankruptcy of the Jerome Manufacturing Company in 1856. By the 1870's the firm was using the trademark "Jerome & Co." on some of their products and purchased the entire English operation in 1904.

For years the Directors of the New Haven operation had drained the resources of the company with large dividends and by 1890 they were in serious financial condition. Hiram Camp (1811-1892), founder and long-time president, resigned as president in September, 1891 and his successor Samuel A. Galpin struggled to keep the firm afloat. They nearly went into bankruptcy in 1894 but raised enough money to continue until March of 1897 when the firm was reorganized.

In 1902 leadership of the firm passed to Walter C. Camp who is better known as the "father of American football." However, Camp modernized the watch manufacturing department in 1904, resulting in cheaper production costs, and had wrist watches added to the line in 1915. Camp was succeeded in 1923 by Edwin P. Root.

When Richard H. Whitehead succeeded Root as president of the firm in February, 1929, it was again in financial trouble. This worsened when the Great Depression hit in November of that year but Whitehead's able leadership kept the firm operating with growing profitability through the time of World War II. From 1943 through 1945, the firm was involved almost totally in manufacturing war products.

In March of 1946, the firm resumed clock and watch manufacture and was reorganized as the New Haven Clock and Watch Company. That year it fell into the control of foreign inventors involved with Swiss watchmaking interests. Whitehead resigned as president and thereafter the firm deteriorated steadily. In 1956, it was reorganized under Chapter X of the bankruptcy act and in 1960 the operation ceased and the clock manufacturing facilities were sold on March 22-24, 1960 at public auction and by private negotiation.

**577**

FLYING PENDULUM CLOCK.
Cases are finished in Oak, Mahogany and Ebony.
The Latest Novelty in Clock Escapements. The Flying Ball takes the place of the Pendulum, and keeps good time.
The best Show Window Attraction ever made. Will draw a crowd wherever exhibited.

No. 5107, 1 Day, Time..........................$5 18

1884

# NEW HAVEN

165

**578** — 1880

**579** — 1880

**580** — 1880

## REGULATOR, D. R.
Spring.

Height, 31 inches.  8 Day Time.
Dial, 12 inches.  8 Day Strike.
Also with Alarm.

*The year entered beside each clock is the year of the catalog from which the illustration was taken. Some clocks may have been produced a few years earlier.*

## CAMBRIA (SPRING).
Walnut.

Height, 42 inches.  8 Day Time.
Dial, 8 inches.  8 Day Strike.

## REGULATOR, B. B. (SPRING.)
Walnut.

Height, 41 inches.  8 Day Time.
Dial, 8 inches.  8 Day Strike.

# NEW HAVEN

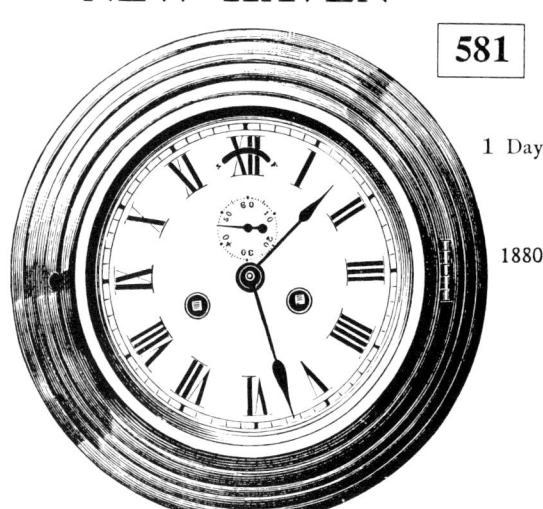

### 581
### CANTON.
Nickel or Brass.
6 and 8 inch Dial.
1 Day or 8 Day Time, Time Alarm or Strike.

CANTON

### 582
**EAGLE.**

CARVED OAK OR CHERRY.

Height, 40 inches.    Dial, 17 inches.

| Code Word. | | | | Price. |
|---|---|---|---|---|
| (Maid) | 8 Day Time, | Oak | ................... | $49.50 |
| (Major) | 8 " " | Cherry | ............... | 49.50 |
| (Taint) | 8 " " | Oak, 24 Hour Dial | ...... | 52.00 |
| (Tainture) | 8 " " | Cherry, 24 " | ...... | 52.00 |

Showing both 12 and 24 Hour System of Time.

### 583
### BAROMETER REGULATOR.
Spring.

Height, 37 inches.    8 Day Time.
Dial, 12 inches.    8 Day Strike.

# NEW HAVEN

*The year entered beside each clock is the year of the catalog from which the illustration was taken. Some clocks may have been produced a few years earlier.*

### 584

1886

### 585

1886

### 586
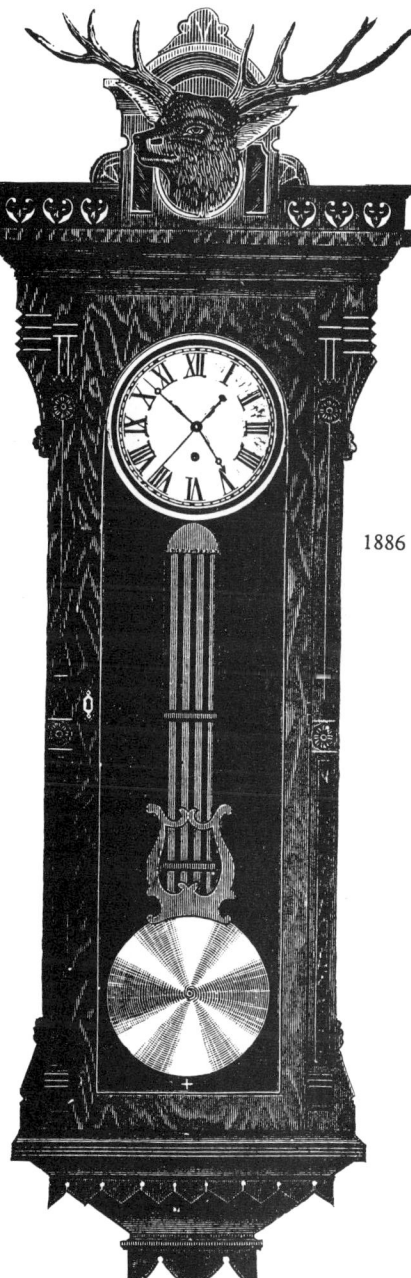
1886

## STANDARD TIME No. 1.

WALNUT, OAK, OR MAHOGANY.

Height, 42 inches.   Dial, 14 inches.

## STANDARD TIME No. 2.

WALNUT, OAK, OR MAHOGANY.

Height, 45 inches.   Dial, 12 inches.

Weight or Spring Movement.

Wooden Rod, Brass Ball.

The Movements in the Standard Time Clocks are specially made and Regulated for them; Screw Pillars, Ornamented Plates, and Stop Work.

An extra charge of $3.30 list will be made for the above Clocks in Mahogany. Walnut Cases will be sent in every instance, unless style of wood is mentioned.

| | | | | | |
|---|---|---|---|---|---|
| 35 | 110 | 10 ft. 11 in. | Purse | 8 Day Standard Time, No. 1, Spring | $27.50 |
| 20 | 72  | 6 ft. 4 in.   | Quail | 8   "      "      "    "  2,  "    | 19.00 |
| 33 | 85  | 6 ft. 4 in.   | Quaker| 8   "      "      "    "  2, Weight | 22.00 |

## REGULATOR No. 3.

WALNUT.

Code Word.   Height, 7 feet 2 inches.  Width at top, 27 inches.  Width at base, 25 inches.   Price.

(Gag) 8 Day Time..................................................................$120.00

Finished with glass sides, movement of best Swiss pattern, sweep second hand, dead-beat pin escapement, 12-inch porcelain dial, steel and brass pendulum rods, adjusting to heat and cold.

We will make this also in mahogany or ash, if so ordered.

# NEW HAVEN

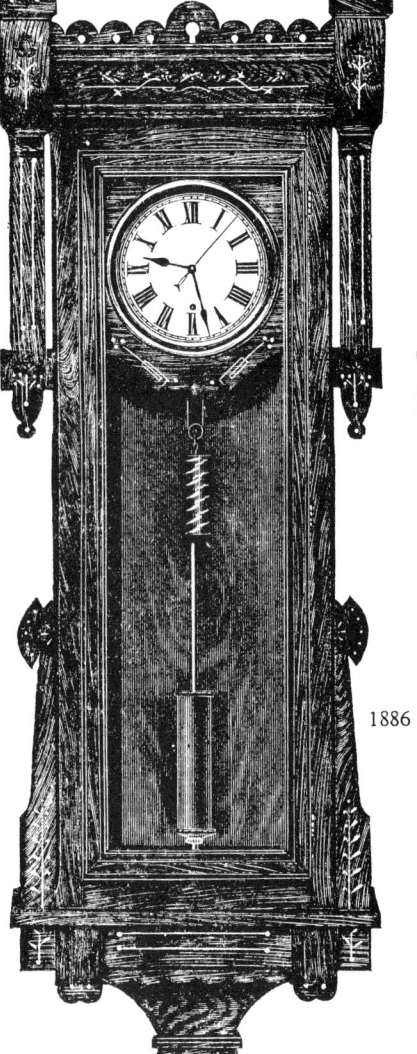

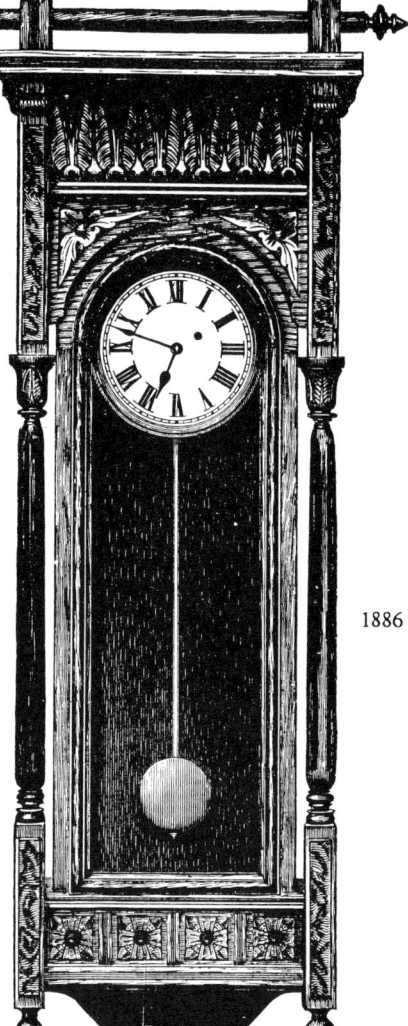

## REGULATOR No. 0.

**BLACK WALNUT.**

Length, 84 inches.   Width, 31 inches.

| | | | | |
|---|---|---|---|---|
| 8 Day Weight Time, | | Walnut | | $90.00 |
| 8 " " " | | Ash | | 90.00 |
| 8 " " " | | Mahogany | | 102.00 |
| 8 " " " | Strike, | Walnut | | 105.00 |
| 8 " " " | " | Ash | | 105.00 |
| 8 " " " | " | Mahogany | | 117.00 |

Dial, 12 inches, with deep Brass Matting.

Sweep Second Hand.   Mercurial Ball, if ordered.

## REGULATOR No. 2.

**WALNUT.**

Code Word.   Height, 7 feet 4 inches.   Width at top, 29 inches.   Width at base, 24½ inches.   Price.
(Gaelic)   8 Day Time ................................................................. $120.00

Finished with Glass Sides, Movement of best Swiss pattern, Sweep Second Hand, Dead-beat Pin Escapement, 12-inch Porcelain Dial, Steel and Brass Pendulum Rods, adjusting to heat and cold.
We will make this also in Mahogany or Ash, if so ordered.

## REGULATOR No. 00.

OAK OR MAHOGANY.

WOOD PENDULUM ROD,

POLISHED BRASS WEIGHTS.

Weight.   Cut Pinion Movement.

Height, 63 inches.

Dial, 10 inches, with heavy Brass Matting.

| | | | Price. |
|---|---|---|---|
| 8 Day Time, | Mahogany | | $90.00 |
| 8 " " | " | Sweep Seconds | 97.50 |
| 8 " " | Oak | | 82.50 |
| 8 " " | " | Sweep Seconds | 90.00 |

## REGULATOR No. 00. No. 2.

Height, 77 inches.   Dial, 10 inches.

8 Day Time, Sweep Seconds ... $130.00

# NEW HAVEN

*To receive top prices, clocks must be in ORIGINAL MINT condition and in good running order - A replaced or damaged dial, missing parts, will reduce prices accordingly.*

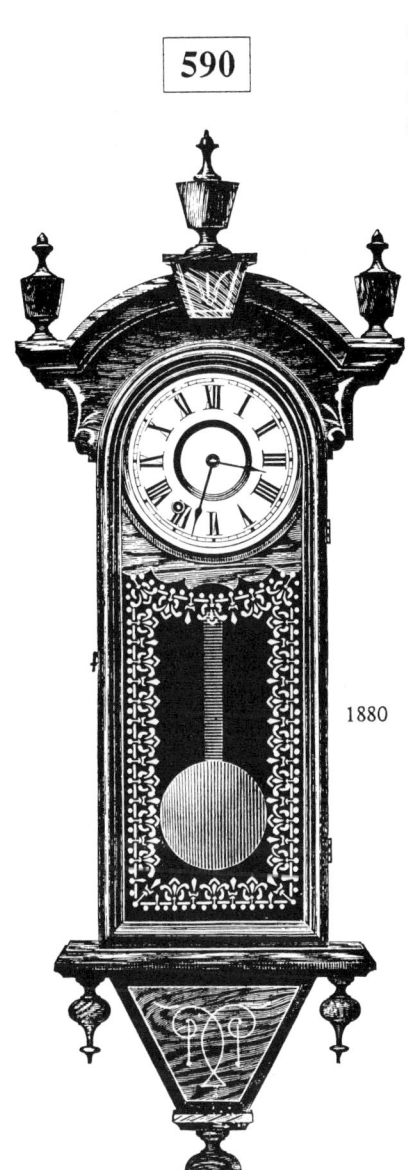

**590**

1880

## WINNIPEG REGULATOR (SPRING).

Walnut.

Height, 35 inches.      8 Day Time.
Dial, 6 inches.         8 Day Strike.

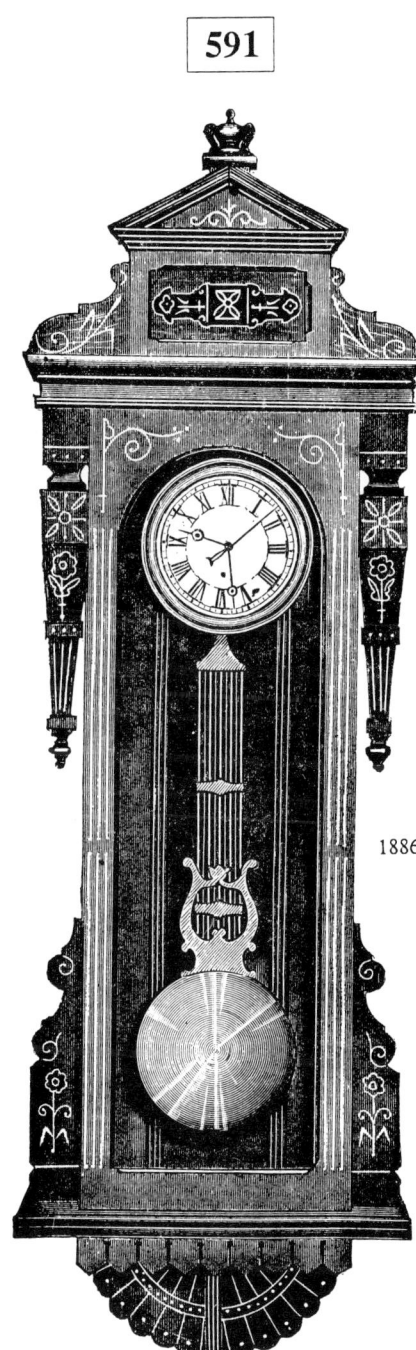

**591**

1886

## REGULATOR No. 9.

WALNUT.

(Gaiter) 8 Day Time ............................................. $90.00
    With Base ..............................................net, extra  7.50

Finished with Glass Sides, Movement of best Swiss Pattern, Sweep Second Hand, Dead-beat Escapement, 12-inch Porcelain Dial, Steel and Brass Pendulum Rods, adjusting to heat and cold.

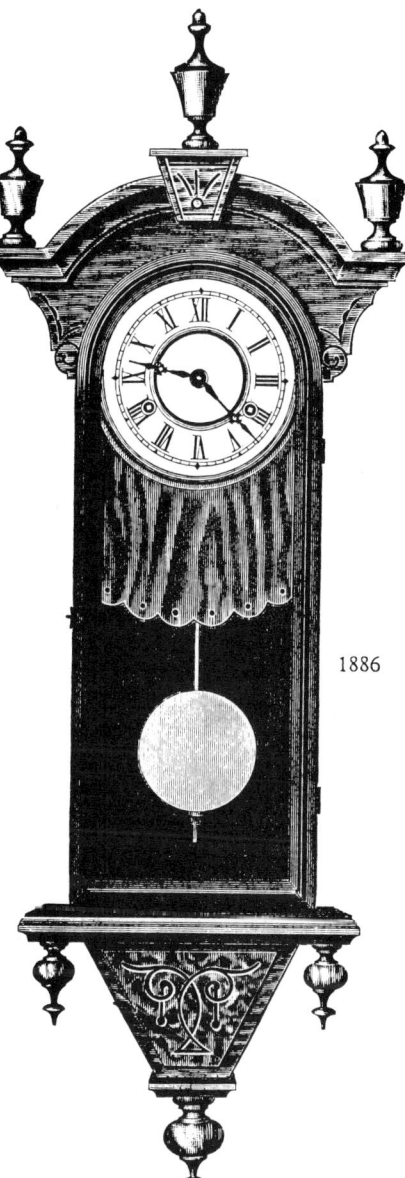

**592**

1886

## WINNIPEG (Spring).

WALNUT.

| Net Weight. | Gross Weight. | Cubic Meas't. |
|---|---|---|
| 8¼ lbs. | 23 lbs. | 1 ft. 3 in. |

Height, 35 inches.   Dial, 6 inches.

| Code Word. | Price. |
|---|---|
| (Hope) 8 Day Time | $ 8.50 |
| (Hone) 8 Day Strike | 10.00 |

# NEW HAVEN

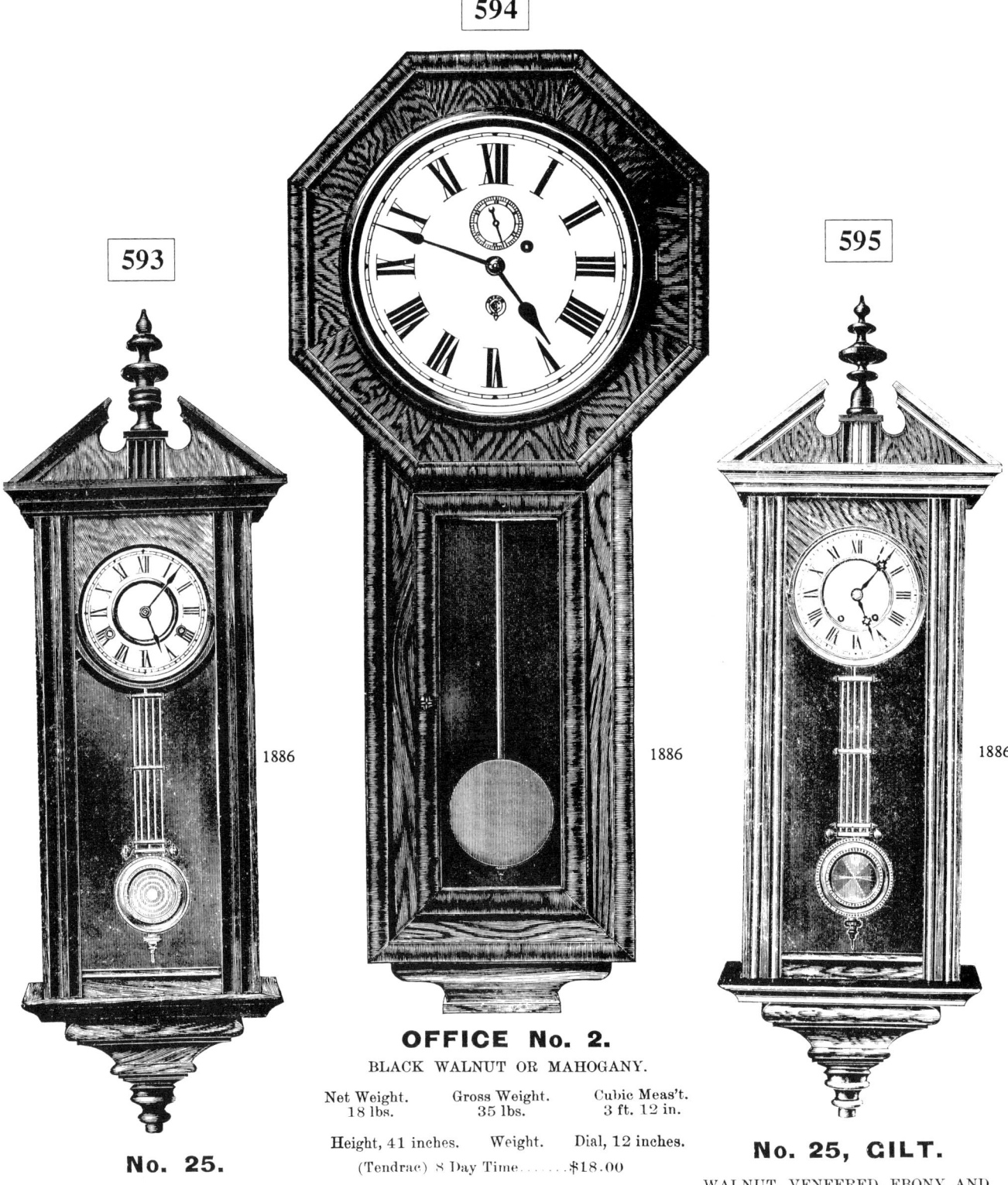

### 593

### 594

### 595

**No. 25.**

WALNUT, VENEERED, EBONY TRIMMINGS.

| No. of Clocks in a Case. | Net Weight. | Gross Weight. | Cubic Meas't. |
|---|---|---|---|
| 1 | 9½ lbs. | 25 lbs. | 2 ft. |

Height, 33 inches.   Dial, 6 inches.

| Code Word. | | Price. |
|---|---|---|
| (Tenrec) | 8 Day Time | $8.50 |
| (Tennon) | 8 "  " Porcelain Dial | 9.00 |
| (Tensome) | 8 "  Strike | 10.00 |
| (Tenter) | 8 "  " Porcelain Dial | 11.00 |

**OFFICE No. 2.**

BLACK WALNUT OR MAHOGANY.

| Net Weight. | Gross Weight. | Cubic Meas't. |
|---|---|---|
| 18 lbs. | 35 lbs. | 3 ft. 12 in. |

Height, 41 inches.   Weight.   Dial, 12 inches.

(Tendrac) 8 Day Time........$18.00

**No. 25, GILT.**

WALNUT, VENEERED, EBONY AND GILT TRIMMINGS.

| No. of Clocks in a Case. | Net Weight. | Gross Weight. | Cubic Meas't. |
|---|---|---|---|
| 1 | 9½ lbs. | 25 lbs. | 2 ft. |

Height, 33 inches.   Dial, 6 inches.

| Code Word. | | Price. |
|---|---|---|
| (Bid) | 8 Day Time | $10.00 |
| (Bide) | 8 "  " Porcelain Dial | 10.50 |
| (Cabas) | 8 "  Strike | 11.50 |
| (Cablet) | 8 "  " Porcelain Dial | 12.50 |

# NEW HAVEN

**596** 

**597**

**598**

1886

## No. 26.

WALNUT, VENEERED, EBONY TRIMMINGS.

| No. of Clocks in a Case. | Net Weight. | Gross Weight. | Cubic Meas't. |
|---|---|---|---|
| 1 | 9 lbs. | 26 lbs. | 2 ft. |

Height, 33 inches.   Dial, 6 inches.

| Code Word. | | | Price. |
|---|---|---|---|
| (Tenure) | 8 Day Time | | $9.50 |
| (Tepefy) | 8 " | " Porcelain Dial | 10.50 |
| (Ternate) | 8 " | " Strike | 11.00 |
| (Termer) | 8 " | " Porcelain Dial | 12.00 |

## No. 26, GILT.

WALNUT, VENEERED, EBONY AND GILT TRIMMINGS.

| No. of Clocks in a Case. | Net Weight. | Gross Weight. | Cubic Meas't. |
|---|---|---|---|
| 1 | 9 lbs. | 26 lbs. | 2 ft. |

Height, 33 inches.   Dial, 6 inches.

| Code Word. | | | Price. |
|---|---|---|---|
| (Benzant) | 8 Day Time | | $11.50 |
| (Bias) | 8 " | " Porcelain Dial | 12.50 |
| (Bib) | 8 " | " Strike | 13.00 |
| (Bible) | 8 " | " Porcelain Dial | 14.00 |

## OFFICE No. 1.

BLACK WALNUT OR MAHOGANY.

| Net Weight. | Gross Weight. |
|---|---|
| 18 lbs. | 35 lbs. |

Height, 42 inches.   Weight.   Dial, 12 inches.

8 Day Time .......... $21.00

Movement has Cut and Polished Steel Pinions,
Solid Frames, Screw Pillars, and Maintaining Power.
Fitted with Standard Time Weight Movement.

# NEW HAVEN

*Except for those fabulous clocks for which you wouldn't mind cutting a hole in the ceiling, Grandfather or Hanging Clocks under eight feet are the most desirable and practical. Some manufacturers produced certain of these clocks in two styles: one to stand on the floor, and the other to hang on a wall. Of the clocks designed with two styles, the hanging style is considered to be the most valuable.*

**599** — 1886

**GAMBIA.**

ASH, WALNUT, OR MAHOGANY.

| No. of Clocks in a Case. | Net Weight. | Gross Weight. | Cubic Meas't. |
|---|---|---|---|
| 1 | 18 lbs. | 30 lbs. | 5 ft. 1 in. |

Height, 49 inches.   Dial, 8 inches.

| Code Word. | | Price. |
|---|---|---|
| (Powder) | 8 Day Time | $16.50 |
| (Power) | 8  "  Strike | 18.00 |
| (Potler) | 8  "   "   Cathedral Gong | 18.50 |

**600** — 1886

**REGULATOR No. 10.**

OAK, BLACK WALNUT, AND MAHOGANY.

Height, 77½ inches.   Dial, 12 inches.

| Code Word. | | Price. |
|---|---|---|
| (Temporist) | 8 Day Time, Sweep Second | $ 70.00 |
| (Tenable) | 8  "   "    "    "   Mercurial Ball | 130.00 |

**601** — 1886

**OBI.**

ASH, WALNUT, OR MAHOGANY.

| No. of Clocks in a Case. | Net Weight. | Gross Weight. | Cubic Meas't. |
|---|---|---|---|
| 1 | 21 lbs. | 59 lbs. | 4 ft. 2 in. |

Height, 49 inches.   Dial, 8 inches.

| Code Word. | | Price. |
|---|---|---|
| (Print) | 8 Day Time | $16.15 |
| (Prize) | 8  "  Strike | 17.65 |
| (Pound) | 8  "   "   Cathedral Gong | 18.15 |

# NEW HAVEN

**602**

**THORNTON — Hanging.**

Height, 74 inches.   Width, 22 inches.   Twelve-inch Porcelain Dial.

Oak or Cherry.

Eight-day.   Fitted with Swiss Movement, Gridiron Pendulum, or Our Own Fine Regulator Movement, Wooden Pendulum and Cylindrical Ball.

Dead-beat Escapement, Sweep Second, Retaining Power.

List price, $103.00

**603**

**ADMIRAL**

Ash Burl, Polished Veneered.

Height, 62 inches.   Dial, 18 inches.

List price, Eight-day Time, $50.00

**604**

**STANDISH — Oak.**

Height, 26¼ inches.   Width, 13 inches.
Dial, 6 inches.

| | | |
|---|---|---|
| List price, | Eight-day, Strike, | $6.00 |
| " " | Eight-day, Strike Alarm, | 6.50 |
| " " | Eight-day, Gong, | 6.40 |
| " " | Eight-day, Gong Alarm, | 6.90 |

**605**

**SALEM — Oak.**

Height, 26½ inches.   Width, 12¾ inches.
Dial, 6 inches.

| | | |
|---|---|---|
| List price, | Eight-day, Strike, | $6.00 |
| " " | Eight-day, Strike Alarm, | 6.50 |
| " " | Eight-day, Gong, | 6.40 |
| " " | Eight-day, Gong Alarm, | 6.90 |

# NEW HAVEN

**606**

**VAMOOSE—Solid Oak.**

Height, 45 inches.   Dial, 10 inches.

List price, Eight-day Time, . . . $11.25
"   "  Eight-day Strike, Gong,   12.65
"   "  Eight-Day Time, Calendar, 11.75
"   "  Thirty-Day Time,    "     13.25

**607**

**NEW YORK — Solid Oak.**

Height, 97 inches.   Twelve-inch Porcelain Dial.

Eight-day.  Fitted with Swiss Movement, Gridiron Pendulum, or Our Own Fine Regulator Movement, Wooden Pendulum and Cylindrical Ball.

Dead-beat Escapement, Sweep Second, Retaining Power.

List price, $114.00

Fitted with Three Jar Mercurial Pendulum, Extra list, $40.00

**608**

**GRECIAN—Oak.**

Height, 51 inches.   Width, 20 inches.
Dial, 10 inches.

List price, Eight-day Time, . . . $16.75
"   "  Eight-day Strike, Gong,   18.25
"   "  Eight-day Time, Weight
Movement, . . . .   20.25

# NEW HAVEN

**COLUMBIA**—Solid Oak or Cherry.

Height, 48½ inches.   Dial, 8 inches.
List price, Eight-day Time,   . $14.25
List price, Eight-day Time, Half-
 hour Strike, Gong, . . . 15.75

**MAYWOOD**—Oak.

Height, 43½ inches.   Width, 17 inches.
Dial, 8 inches.

List price, Eight-day Time,   $10.50
 "   "   Eight-day Time, Calendar, 11.00
 "   "   Eight-day Strike, Gong, 12.00
 "   "   Eight-day Strike, Gong,
Calendar, . . . 12.50

**TROJAN**—Solid Oak.

Height, 44 inches.   Dial, 8 inches.
List price, Eight-day Time,   $12.00
List price, Eight-day Time, Half-
 hour Strike, Gong, . . . 13.50

# NEW HAVEN

*The year entered beside each clock is the year of the catalog from which the illustration was taken. Some clocks may have been produced a few years earlier.*

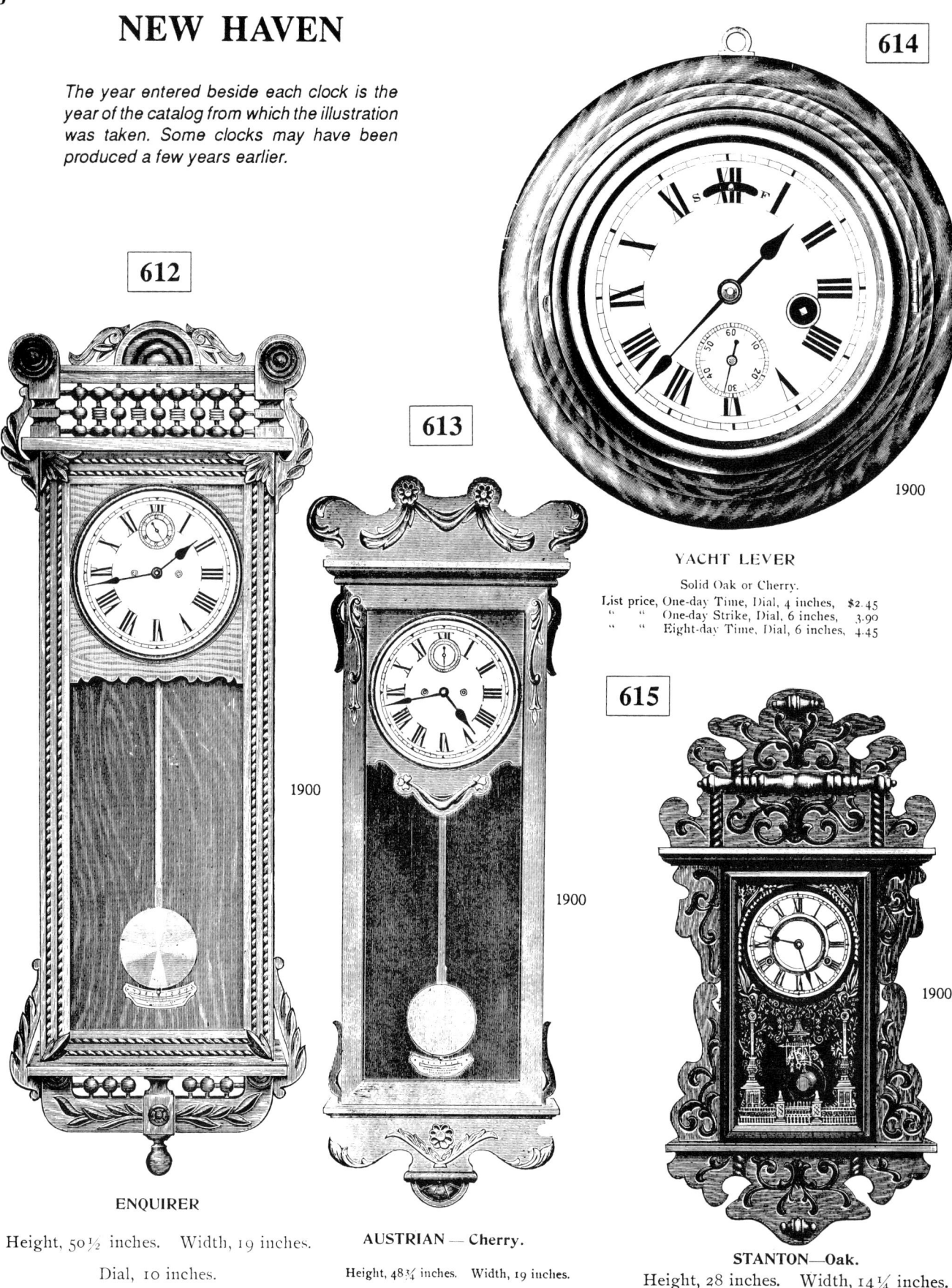

**612** — 1900

**ENQUIRER**

Height, 50½ inches.  Width, 19 inches.
Dial, 10 inches.

List price, Eight-day Time, . $18.50
" " Eight-day Strike, Gong, 20.00
" " Eight-day Time, Weight
Movement, . . . 22.00

**613** — 1900

**AUSTRIAN — Cherry.**

Height, 48¾ inches.  Width, 19 inches.
Dial, 10 inches.

List price, Eight-day Time, . $16.75
" " Eight-day Strike, Gong, 18.25
" " Eight-day Time, Weight
Movement, . . . 20.25

**614** — 1900

**YACHT LEVER**

Solid Oak or Cherry.
List price, One-day Time, Dial, 4 inches, $2.45
" " One-day Strike, Dial, 6 inches, 3.90
" " Eight-day Time, Dial, 6 inches, 4.45

**615** — 1900

**STANTON — Oak.**

Height, 28 inches.  Width, 14¼ inches.
Dial, 6 inches.

List price, Eight-day, Strike, $6.50
" " Eight-day, Strike Alarm, 7.00
" " Eight-day, Gong, 6.90
" " Eight-day, Gong Alarm, 7.40

# NEW HAVEN

**616** — 1886

**617** — 1905

**DULCE**

Height, 24 inches.  Width, 16 inches.
12 inch Wood Dial.  Solid Oak, Flemish Finish.
Cast Brass Hands and Numerals.

| | | |
|---|---|---|
| Eight-day Time | List Price, | $6.45 |
| Eight-day Strike | " " | 7.45 |

**618** — 1905

**TAMPA—Oak**

Height, 38 inches.  Width, 15½ inches.
Dial, 12 inches.

| | | |
|---|---|---|
| Eight-day Time | List Price, | $6.70 |
| Eight-day, Strike | " " | 7.70 |
| Eight-day Time, Calendar | " " | 7.20 |
| Eight-day, Strike, Calendar | " " | 8.20 |

**619** — 1900

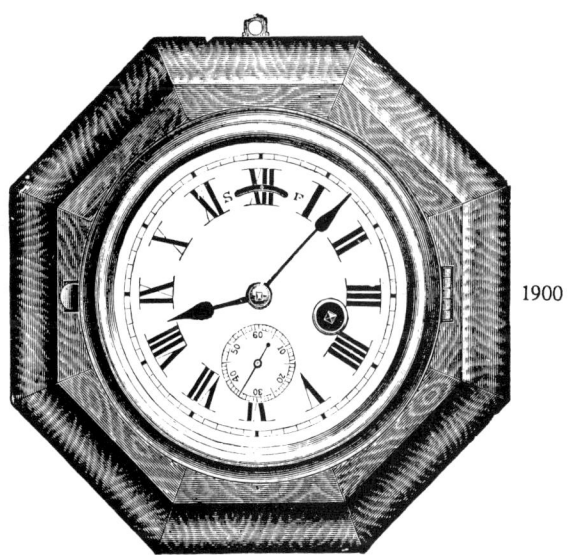

**WOOD LEVER**

## REGULATOR, C. C.

### WALNUT.

| Net Weight. | Gross Weight. | Cubic Meas't. |
|---|---|---|
| 11 lbs. | 34 lbs. | 2 ft. 10 in. |

Height, 37½ inches.  Dial, 8 inches.

| Code Word. | | Price. |
|---|---|---|
| (Finger) | 8 Day Time | $10.50 |
| (Fire) | 8 Day Strike | 12.00 |

| | | | | | |
|---|---|---|---|---|---|
| One-day, | 4-inch, Time, | $2.45 | One-day, | 10-inch, Strike, | $5.25 |
| " | 6-inch, Time, | 3.40 | Eight-day, | 6-inch, Time, | 4.45 |
| " | 6-inch, Time, Alarm, | 3.90 | " | 8-inch, Time, | 5.00 |
| " | 6-inch, Strike, | 3.90 | " | 8-inch, Strike, | 7.00 |
| " | 8-inch, Time, | 4.00 | " | 10-inch, Time, | 5.80 |
| " | 8-inch, Time, Alarm, | 4.50 | " | 10-inch, Strike, | 7.80 |
| " | 8-inch, Strike, | 4.50 | " | 12-inch, Time, | 6.55 |
| " | 10-inch, Time, | 4.75 | " | 12-inch, Strike, | 8.55 |

# NEW HAVEN

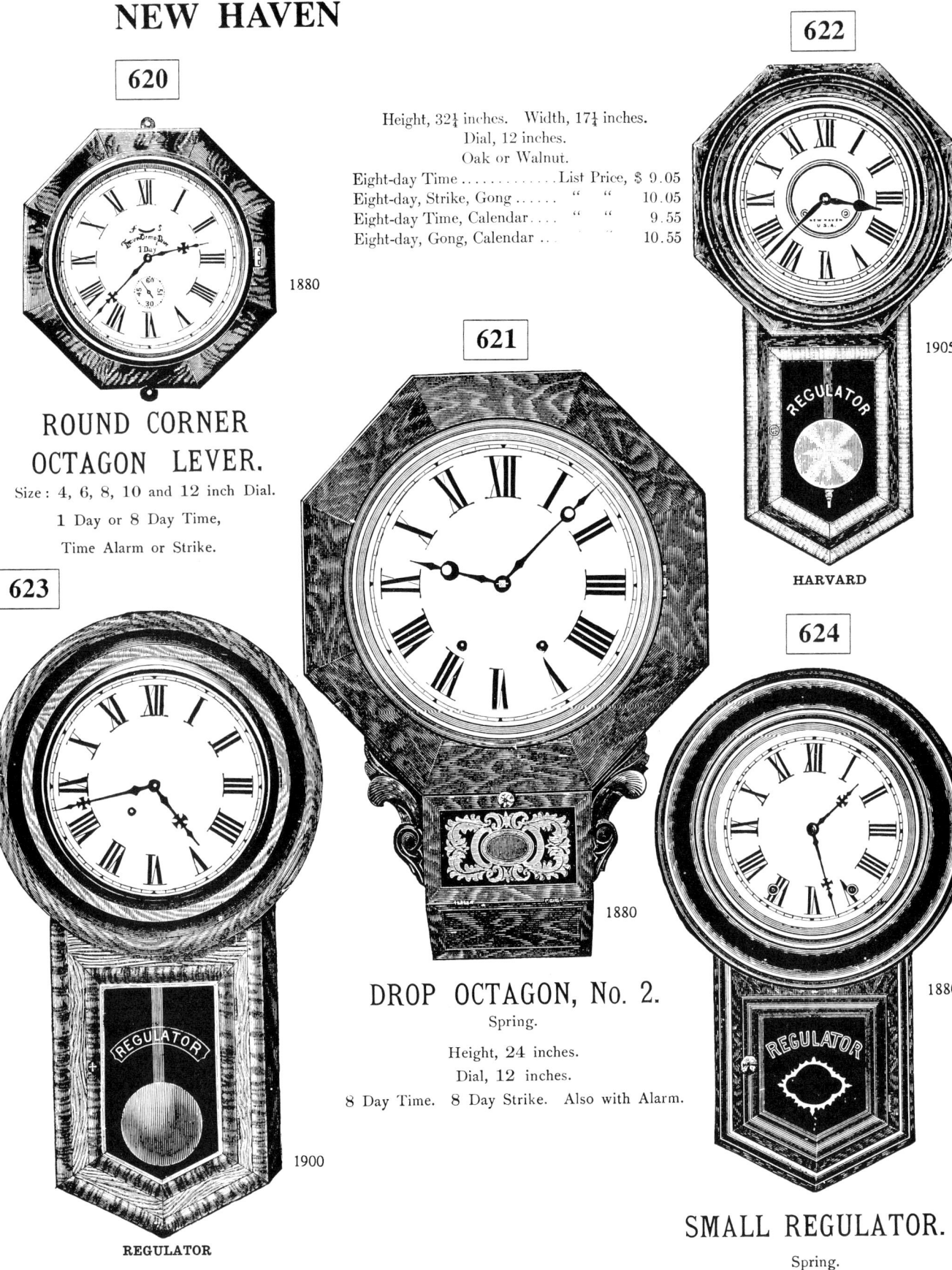

**620**

### ROUND CORNER OCTAGON LEVER.
Size: 4, 6, 8, 10 and 12 inch Dial.
1 Day or 8 Day Time,
Time Alarm or Strike.

Height, 32¼ inches.  Width, 17¼ inches.
Dial, 12 inches.
Oak or Walnut.
Eight-day Time ............ List Price, $ 9.05
Eight-day, Strike, Gong ......  "   "   10.05
Eight-day Time, Calendar ....  "   "    9.55
Eight-day, Gong, Calendar ...            10.55

**621**

**622**

**HARVARD**

**623**

**REGULATOR**

Height, 32 inches.   Dial, 12 inches.
Oak or Walnut Veneer, Solid Oak or
Walnut Circle
Eight-day Time ............ List Price, $ 9.05
Eight-day, Strike, Gong ......  "   "   10.05

### DROP OCTAGON, No. 2.
Spring.
Height, 24 inches.
Dial, 12 inches.
8 Day Time.  8 Day Strike.  Also with Alarm.

**624**

### SMALL REGULATOR.
Spring.
Height, 21 inches.
Dial, 8 inches.
8 Day Time.  8 Day Strike.  Also with Alarm.
Also made with Gilt Edge Molding.

# NEW HAVEN

**625**

**ELFRIDA**

Height, 49 inches.  Width, 19 inches.

Dial, 10 inches.

List price, Eight-day Time, . . $17.50
" " Eight-day Strike, Gong, 19.00
" " Eight-day Time, Weight Movement, . . . . 21.00

**626**

THE BANK REGULATOR—Oak.

Height, 33 inches.  Width, 18 inches.
Dial, 12 inches.

List price, Eight-day Time, . . $6.55
" " Eight-day Strike, . . 7.55
" " Eight-day Time, Calendar, 7.05
" " Eight-day Strike, Calendar, 8.05

**627**

BLAKE — Solid Oak.

Height, 27 inches.    Dial, 12 inches.

List price, Eight-day Time, $5.75
" " Eight-day Strike, 6.75
" " Thirty-day Time, 7.75

**628**

BRADDOCK — Solid Oak.

Height, 27 inches.    Dial, 12 inches.

List price, Eight-day Time, $6.00
" " Eight-day Strike, 7.00
" " Thirty-day Time, 8.00

# NEW HAVEN

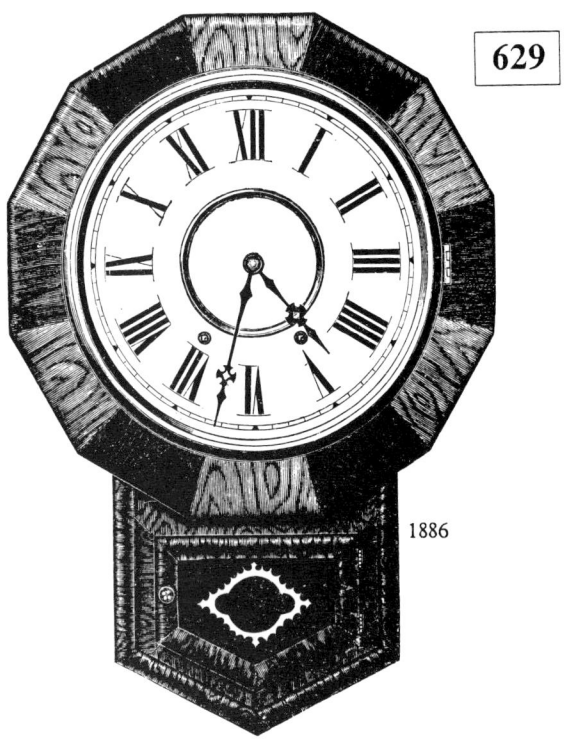

**629**

1886

## MOSAIC DROP.

Height, 24 inches.   Dial, 12 inches.

| Code Word. | | | | | | Price. |
|---|---|---|---|---|---|---|
| (Halt) | 8 Day | 12 inch | Time | | | $6.60 |
| (Hail) | 8 " | 12 " | " | Alarm | | 7.20 |
| (Halo) | 8 " | 12 " | Strike | | | 8.10 |
| (Halse) | 8 " | 12 " | " | Alarm | | 8.70 |

**631**

1900

### 10-inch DROP OCTAGON—Brass Bands.

Veneered Rosewood, or Zebra Solid Oak.

Height, 21 inches.   Dial, 10 inches.

List price, Eight-day Time, . . $6.00
" " Eight-day Strike, . . 7.00
" " Eight-day Time, Calendar, 6.50
" " Eight-day Strike, Calendar, 7.50

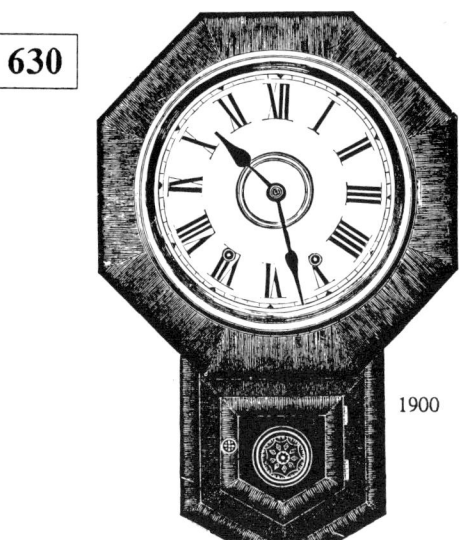

**630**

1900

### TEN-INCH DROP OCTAGON

Polished Veneered Rosewood or Zebra or Solid Oak.

Height, 21 inches.

List price, Eight-day Time, $5.50
" " Eight-day Strike, 6.50

**632**

1886

## DROP OCTAGON

ROSEWOOD OR ZEBRA.   BRASS BAND.

Height, 24 inches.   Dial, 12 inches.

| Betrust | 8 Day | 12 inch | Time | | $6.25 |
|---|---|---|---|---|---|
| Bevel | 8 " | 12 " | " Alarm | | 6.85 |
| Beware | 8 " | 12 " | " Strike | | 7.75 |
| Bey | 8 " | 12 " | " Alarm | | 8.35 |
| Beylic | 8 " | 12 " | Time, Calendar | | 7.25 |
| Bezan | 8 " | 12 " | Strike, " | | 8.75 |

# NEW HAVEN

633

1886

## SMALL DROP OCTAGON
ROSEWOOD AND ZEBRA.
GILT AND GILT BUTTONS.

Height, 21 inches.  Dial, 10 inches.

| | | | | |
|---|---|---|---|---|
| Ivy...... | 8 Day | 10 in. | Time................ | $6.60 |
| Jaguar... | 8 " | 10 " | " Alarm............ | 7.20 |
| Than.... | 8 " | 10 " | Strike.............. | 8.10 |
| Thar.... | 8 " | 10 " | " Alarm............ | 8.70 |
| Junk.... | 8 " | 10 " | Time, Calendar..... | 7.60 |
| King.... | 8 " | 10 " | Strike " | 9.10 |

634

1886

## DROP OCTAGON
ROSEWOOD OR ZEBRA. GILT AND GILT BUTTONS.

Height, 24 inches.  Dial, 12 inches.

| | | | | |
|---|---|---|---|---|
| Dame..... | 8 Day Drop Oct. R. C. | Time, Gilt........ | $6.55 |
| Eclipse... | 8 " " " | " " Alarm.. | 7.15 |
| Damsel... | 8 " " " | Strike, " ...... | 8.05 |
| Edict..... | 8 " " " | " " Alarm.. | 8.65 |
| Damper.. | 8 " " " | Time, "Calendar | 7.55 |
| Dale...... | 8 " " " | Strike, " ...... | 9.05 |

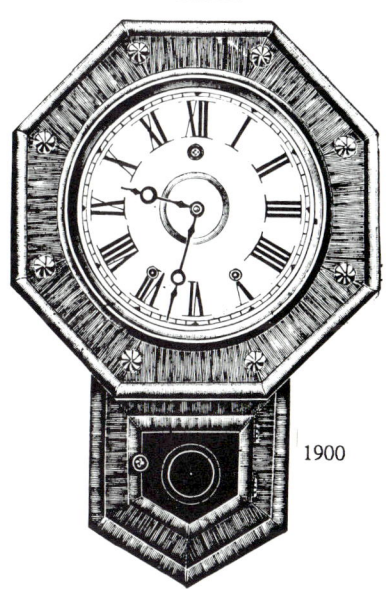

635

1900

**10-inch DROP OCTAGON—Gilt and Gilt Buttons.**

Veneered Rosewood, or Zebra Solid Oak.
Height, 21 inches.  Dial, 10 inches.

List price, Eight-day Time, . . $6.00
"     "    Eight-day Strike, . . 7.00
"     "    Eight-day Time, Calendar, 6.50
"     "    Eight-day Strike, Calendar, 7.50

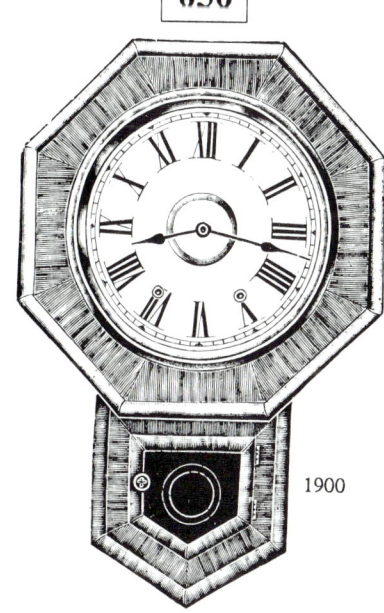

636

1900

**10-inch DROP OCTAGON, GILT**

Veneered Rosewood, or Zebra Solid Oak.
Height, 21 inches.  Dial, 10 inches.

List price, Eight-day Time, . . $5.90
"     "    Eight-day Strike, . . 6.90
"     "    Eight-day Time, Calendar, 6.40
"     "    Eight-day Strike, Calendar, 7.40

# NEW HAVEN

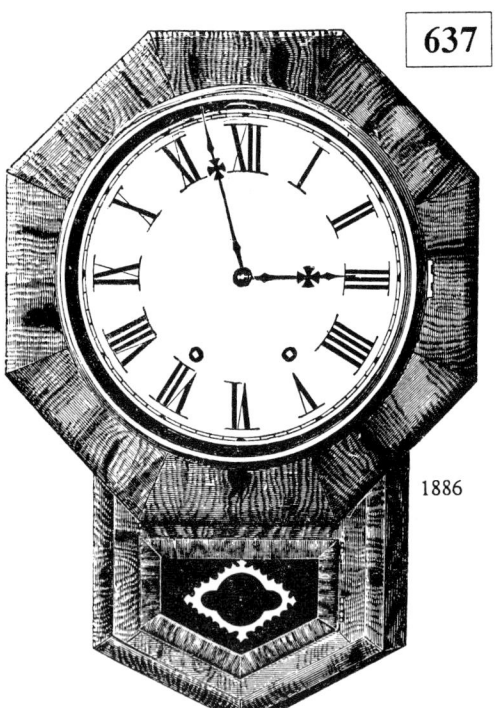

**637**

**DROP OCTAGON, R. C.**

Spring.

Height, 24 inches.
8 Day Time.
8 Day Strike.

Dial, 12 inches.
Also with Alarm.

Also made with Gilt Edge Molding.

**DROP OCTAGON, R. C.**

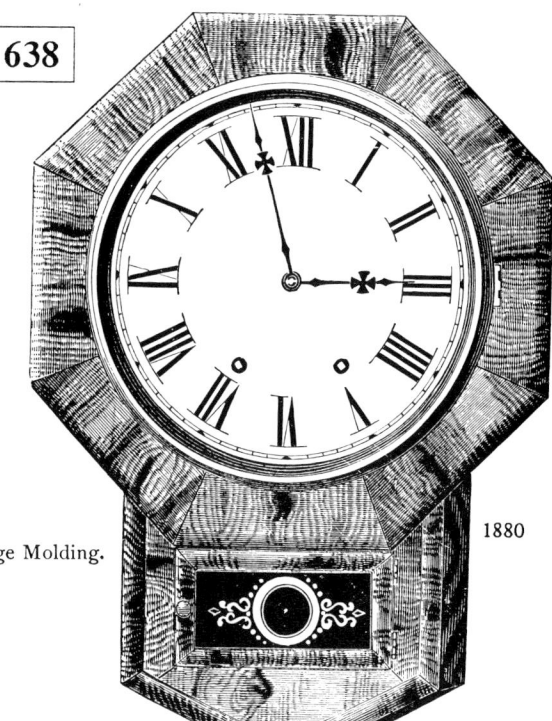

**638**

**DROP OCTAGON, R. C.**

**639**

**640**

**641**

**SMALL DROP OCTAGON, R. C.**

Spring.

Height, 17 inches.

Dial, 8 inches.

8 Day Time.  8 Day Strike.  Also with Alarm.

Also made with Gilt Edge Molding.

**SMALL DROP OCTAGON**

ROSEWOOD OR ZEBRA.

Height, 21 inches.     Dial, 10 inches.

| | | | | | |
|---|---|---|---|---|---|
| 8 Day 10 in. Drop Octagon Time | | | | | $5.70 |
| 8 " 10 " " " Alarm | | | | | 6.30 |
| 8 " 10 " " " Strike | | | | | 7.20 |
| 8 " 10 " " " " Alarm | | | | | 7.80 |
| 8 " 10 " " " Time, Gilt | | | | | 6.30 |
| 8 " 10 " " " " Alm. | | | | | 6.90 |

**SMALL DROP OCTAGON GILT**

ROSEWOOD OR ZEBRA.

Height, 21 inches.     Dial, 10 inches.

| | | | | | |
|---|---|---|---|---|---|
| Paw | 8 Day 10 in. Drop Oct. Strike, Gilt | | | | $7.80 |
| Pool | 8 " 10 " " " Alm. | | | | 8.40 |
| Pea | 8 " 10 " " Time, Calendar | | | | 6.70 |
| Pear | 8 " 10 " " Strike, " | | | | 8.20 |
| Pearl | 8 " 10 " " Time, Cal. Gilt | | | | 7.30 |
| Pebble | 8 " 10 " " Strike, " | | | | 8.80 |

# NEW HAVEN

**642** — 1887

### CORSAIR.
BLACK WALNUT.
Height, 24 inches.  Dial, 6 inches.

| Code Word. | | Price. |
|---|---|---|
| (Earnest) | 8 Day Strike | $10.00 |
| (Eaves) | 8 Day Strike, Alarm | 10.60 |
| (Edge) | 8 Day Strike, Cathedral Gong | 10.50 |
| (Eddy) | 8 Day Strike, Cathedral Gong, Alarm | 11.10 |

The above Clock has ¾ in. Beveled Glass in Door and Solid Metal Ornaments.

**643** — 1880

### OCCIDENTAL.
Walnut.
Height, 24 inches.
Dial, 6 inches.
8 Day Strike.
8 Day Strike, Alarm.

**644** — 1880

### APOLLO.
Walnut.
Height, 20 inches.
Dial, 5½ inches.
8 Day Strike.
8 Day Strike, Alarm.

**645** — 1900

### OAKDALE
Oak with Silver Ornaments.    Walnut with Gilt Ornaments.
Height, 24 inches.  Width, 15¼ inches.  Dial, 6 inches.

| List price, | Eight-day, Strike, | $8.65 |
|---|---|---|
| " " | Eight-day, Strike Alarm, | 9.15 |
| " " | Eight-day, Gong, | 9.05 |
| " " | Eight-day, Gong Alarm, | 9.55 |

# NEW HAVEN

646

1880

### ELBE.
Walnut.

Height, 24 inches.
Dial, 6 inches.
8 Day Strike.
8 Day Strike, Alarm.

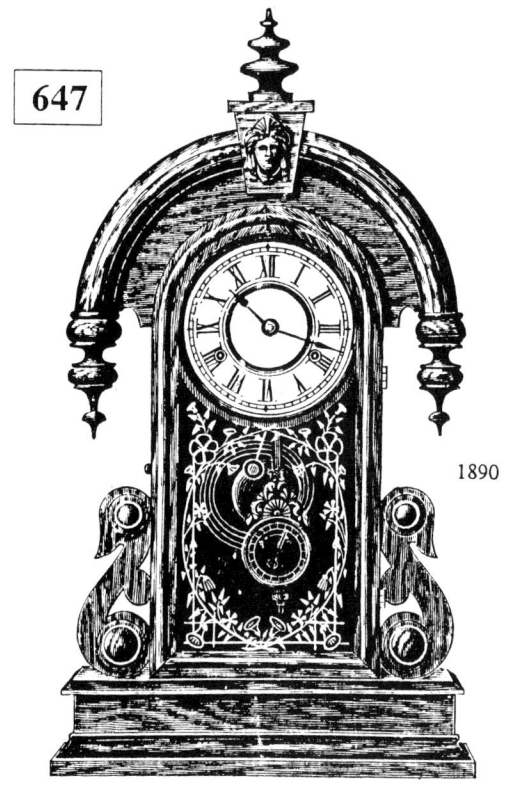

647

1890

### PARISIAN.
BLACK WALNUT.

Height, 24 inches.   Dial, 6 inches.

| | |
|---|---|
| 8 Day Strike | $7.20 |
| 8 Day Strike, Alarm | 7.80 |
| 8 Day Strike, Cathedral Gong | 7.70 |
| 8 Day Strike, Cathedral Gong, Al'm | 8.30 |

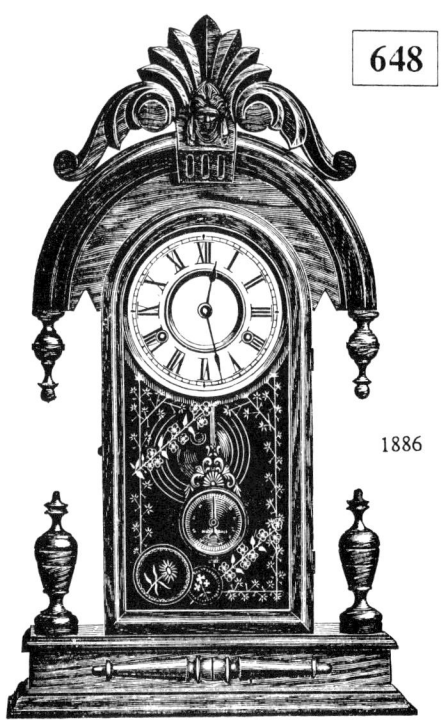

648

1886

### No. 503.
BLACK WALNUT.

Height, 24 inches.   Dial, 6 inches.

| Code Word. | | | Price. |
|---|---|---|---|
| (Ferret) | 8 Day Strike | | $7.20 |
| (Ferrule) | 8 " | " Alarm | 7.80 |
| (Neck) | 8 " | " Cathedral Gong | 7.70 |
| (Ware) | 8 " | " " Alarm | 8.30 |

649

1880

### DANUBE.
Walnut.

Height, 24 inches.
Dial, 6 inches.
8 Day Strike.
8 Day Strike, Alarm.

# NEW HAVEN

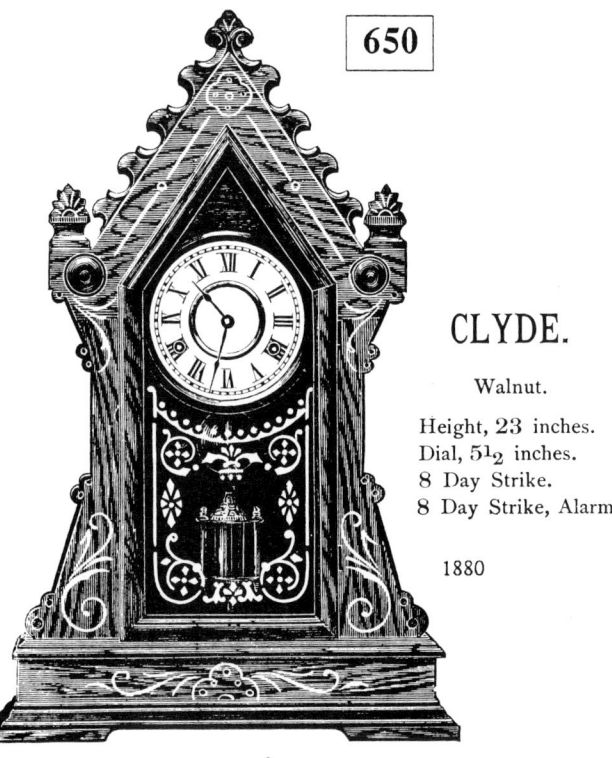

### CLYDE.
Walnut.

Height, 23 inches.
Dial, 5½ inches.
8 Day Strike.
8 Day Strike, Alarm.

1880

1887

### COUNTESS.
BLACK WALNUT.

Height, 26 inches.  Dial, 6 inches.

| Code Word. | | Price. |
|---|---|---|
| (Economy) | 8 Day Strike | $11.00 |
| (Electric) | 8 Day Strike, Alarm | 11.60 |
| (Elevate) | 8 Day Strike, Cathedral Gong | 11.50 |
| (Elevator) | 8 Day Strike, Cathedral Gong, Alarm | 12.10 |

The above Clock has 12 Mirrors and Solid Metal Head.

### CHAMPION.
Walnut.

8 Day Strike.
8 Day Strike, Alarm.
Height, 25 inches.
Dial, 6 inches.

1880

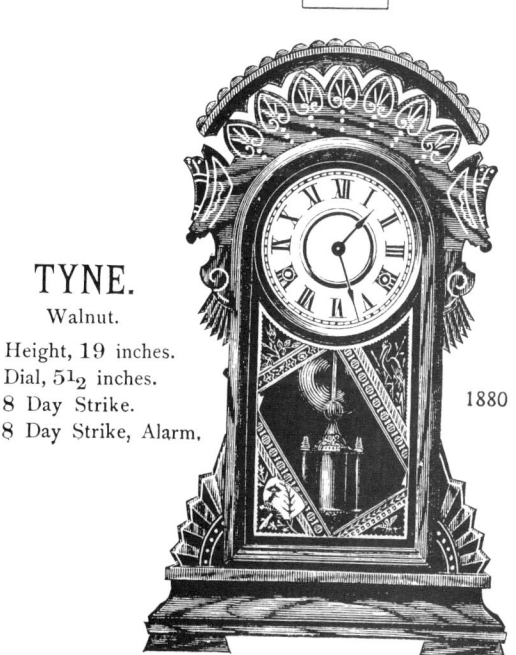

### TYNE.
Walnut.

Height, 19 inches.
Dial, 5½ inches.
8 Day Strike.
8 Day Strike, Alarm.

1880

# NEW HAVEN

### EIGHT-BELL CHIME No. 1.
Case of Solid Mahogany with best French Satin Gilt Trimmings, or of Oak with Antique Silver Trimmings.

Height, 18 inches.

Eight-day Movement with Westminster Chime of Eight Bells, chiming every quarter. The hour is struck on a rich Cathedral Gong. Seven-inch Metal Dial with Etched Rim, raised Numerals and Ornaments to match Trimmings on Case.

List price for either finish, $41.75

1900

1900

### FOUR-BELL WESTMINSTER CHIME, No. 5.
Height, 22¼ inches.   Width, 12¼ inches.
Light Oak with Oxidized Silver or Bronze Trimmings; or Mahogany with French Satin Gold Trimmings.
Six-inch Metal Dial with Fancy Etched Center, finished to match Trimmings.
List price, $36.25

### WESTMINSTER CHIME
These Clocks are fitted with an Eight-day Strike Movement of the Standard Cabinet pattern and a Chime Movement which plays upon four Cathedral Gongs the same tune as that of the Chimes in Westminster Abbey. The music for each quarter is shown on this page. These Gongs are tuned with great care and are played by mechanism made under Willcock patent of August 28, 1894. The richness of tone is secured by the Gunther Resonator, patented March 27, 1896. The hour is struck on a fifth gong.

1900

### FOUR-BELL WESTMINSTER CHIME, No. 3.
Height, 18½ inches.   Width, 12 inches.
Light Oak with Oxidized Silver or Bronze Trimmings; or Mahogany with French Satin Gold Trimmings.
Six-inch Metal Dial with Fancy Etched Center, finished to match Trimmings.
List price, $55.75

# NEW HAVEN

**657**

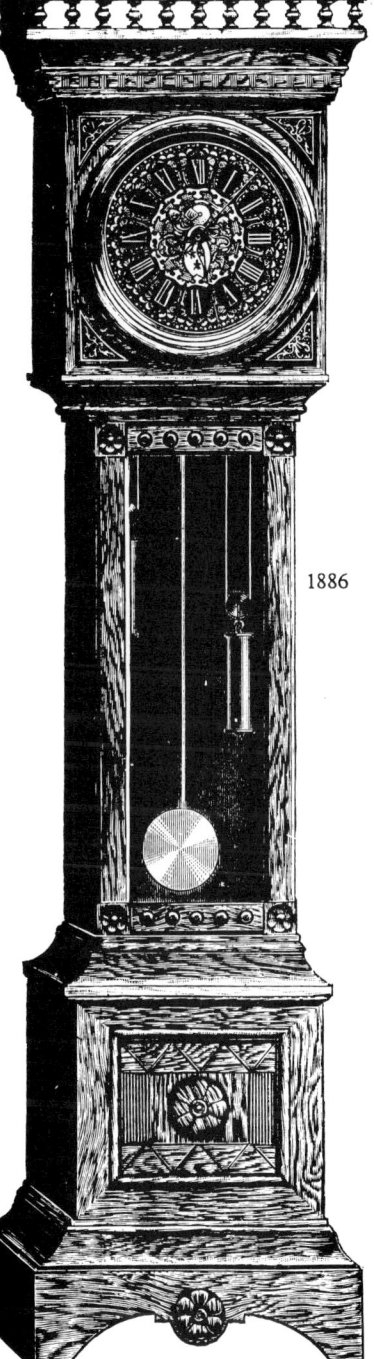

1886

**HALL CLOCK No. 1.**

OAK, MAHOGANY, OR WALNUT, HAND CARVED, CABINET FINISHED.

| No. of Clocks in a Case. | Net Weight. | Gross Weight. | Cubic Meas't. |
|---|---|---|---|
| 1 | 100 lbs. | 236 lbs. | 19 ft. 10 in. |

Code Word.    Height, 84 inches.    Alabaster Dial, 12 inches, Hand Painted.    Price.
(Slave)   8 Day Strike on Gong, Oak ................................. Net. $65.00
(Sleigh)   8   "    "    "    "   Ebony ................................. Net. 65.00
(Slice)   8   "    "    "    "   Walnut ................................. Net. 65.00

Seconds Pendulum, with cut and polished steel pinions, and highly finished movement.

**658**

1900

**CHIPPENDALE.**

Antique Oak, Mahogany or Walnut.
Silver Dial.
Height, 8 feet.    Dial, 12 inches.
Fitted with the Finest Weight Movement.
Eight-day Strike, Cathedral Gong.

List price, $190.00

**659**

1886

**REGULATOR    No. 8.**

Height, 10 feet 6 inches.    WALNUT.

Finished with Glass Sides,

Movement

of best Swiss pattern,

Sweep Second Hand,

Dead-beat Escapement,

12-inch Porcelain Dial,

Steel and Brass

Pendulum Rods,

adjusting to heat and cold.

Code Word.    Price.
(Gaiety) 8 Day Time, $165.

# NEW HAVEN

*To receive top prices, clocks must be in ORIGINAL MINT condition and in good running order - A replaced or damaged dial, missing parts, will reduce prices accordingly.*

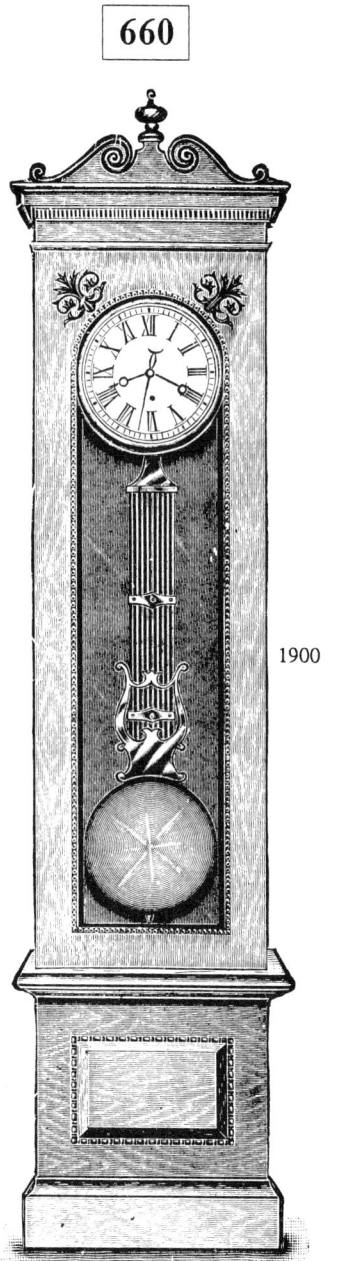

**660**

1900

**THORNTON — Solid Oak or Cherry.**
Standing or Hanging.
Height, 7 feet 10 inches.   Dial, 12 inches.
Dead Beat Pin Escapement and Retaining Power with Sweep Second.
List price, with our own Movement, Round Pendulum Ball and Wooden Rod, $120.00
List price, with Swiss Movement, Gridiron Pendulum, as illustrated, $150.00

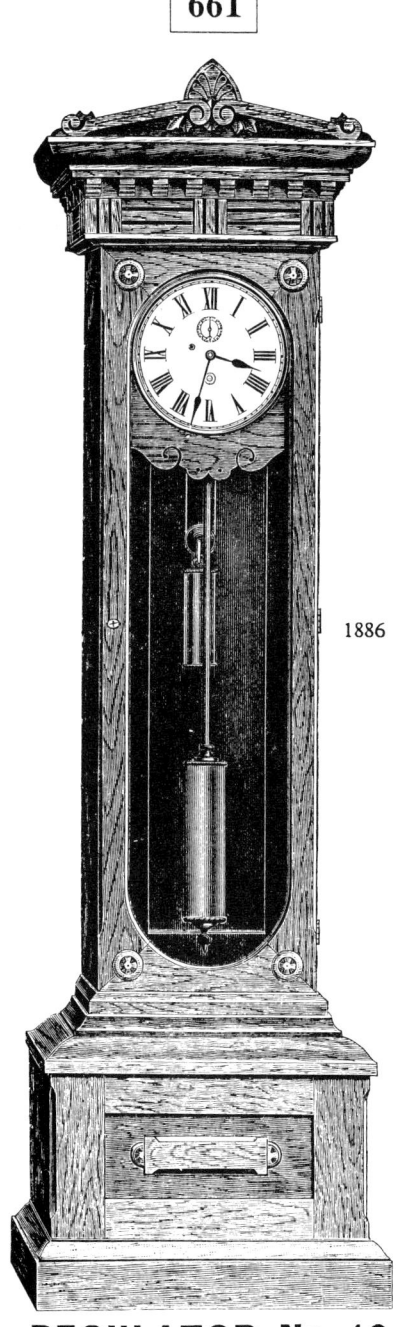

**661**

1886

**REGULATOR No. 10. STANDING**
WITH BASE.
OAK, BLACK WALNUT, AND MAHOGANY.
Height, 90 inches.
Dial, 12 inches.
8 Day Time ...$82.00

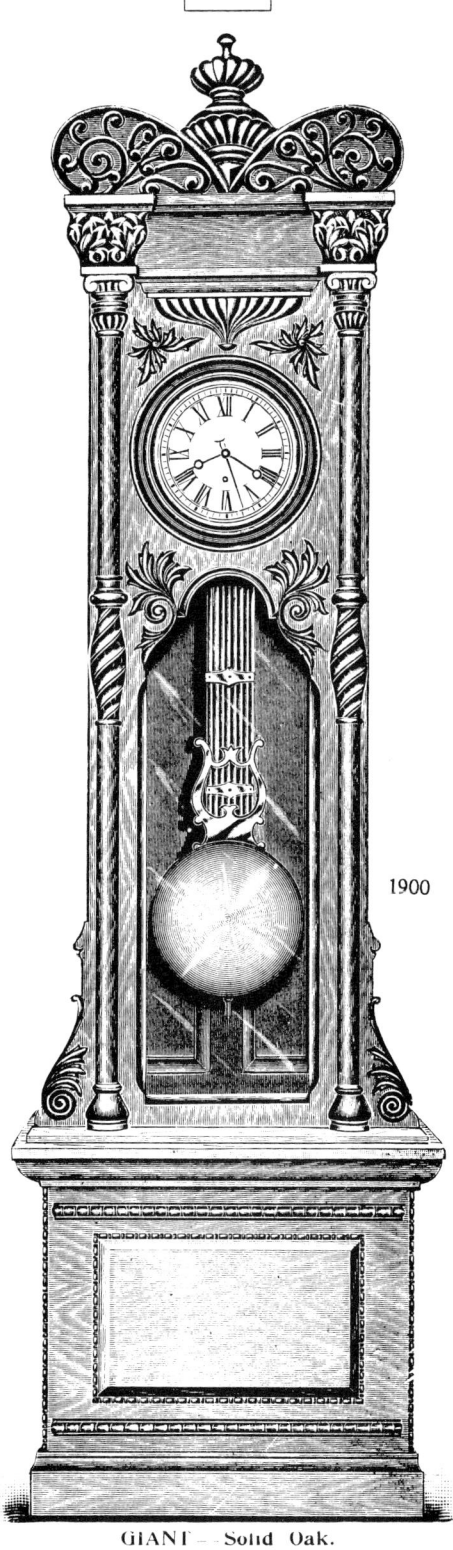

**662**

1900

**GIANT — Solid Oak.**
Height, 10 feet 3 inches.   Dial, 12 inches.
Width of Base, 3 feet.
Dead Beat Pin Escapement and Retaining Power with Sweep Second.
List price, with our own Movement, Round Pendulum Ball and Wooden Rod, $160.00
List price, with Swiss Movement, Gridiron Pendulum, as illustrated, $190.00
Fitted with Three-Jar Mercurial Pendulum, Extra list, $40.00

# NEW YORK STANDARD WATCH COMPANY
New York, N. Y.

On August 13, 1885, the New York Standard Watch Company was organized and watches were placed on the market in 1888. About 1896, the firm began to manufacture an electrically wound pendulum clock whose movement was patented February 25, 1896 by Sigismund Fisher, a Russian emigrant then living at Brooklyn, NY. Also having 1/10 interest in Fisher's patent was Victor D. Brenner of New York, no doubt the same man well known as the designer of the Lincoln head penny.

Most clocks produced by New York Standard are large pendulum regulator clocks in walnut or oak cases. Although Fisher's patent also gave specifications for a striking mechanism, most if not all of the clocks manufactured were timepieces. If not before, production of these electric clocks ceased when the New York Standard Watch Company was purchased by the Keystone Watch Case Company in 1903.

1899

**No. 16** .................. **$41 25**
With or Without Independent Second. Pendulum 12 Beats, Ebonized Wood Rod, 2 lb. Ball.

Oak, Walnut or Mahogany Veneer.
Height, 37 inches. Width, 18 inches.
Dial, 12 inches.

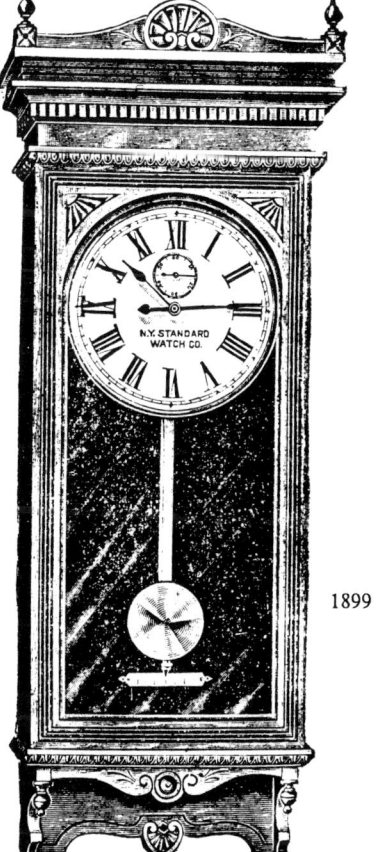

1899

**No. 1** ................ **$103 75**
Sweep Second, With Fine "Dead Beat" Effect, Easily Read for Timing Watches. Pendulum Beats Seconds, Extra Heavy Ball. Has Auxiliary Device for Very Fine Regulation.

Quartered Oak, Cherry and Mahogany.
Beautifully Finished.
Height, 85 inches. Width, 22 inches.
Dial, 12 inches.

1899

**No. 10** ................ **$51 88**
With or Without Independent Second, Pendulum 76 Beats, Ebonized Wood Rod, 2 lb. Ball.

Solid Quartered Oak, Cherry (Natural) and Mahogany Finish.
Height, 49 inches. Width, 19 inches.
Dial, 12 inches.

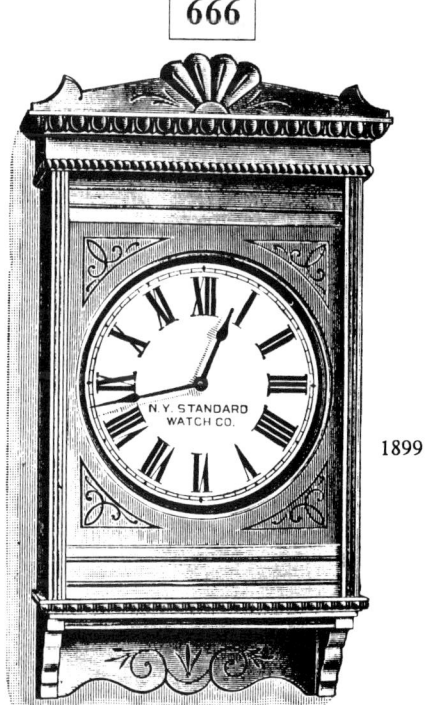

1899

**No. 20** ................ **$46 25**
With or Without Independent Second. Pendulum, 120 Beats, Wood Rod, 2 lb. Ball.

Solid Quartered Oak and Mahogany.
Height, 32 inches. Width, 20½ inches. Dial 14 inches.

# PARKER & WHIPPLE MANUFACTURING COMPANY
# THE PARKER CLOCK COMPANY
### Meriden, Conn.

In 1868, the Parker & Whipple Manufacturing Company was organized by John E. Parker, H. J. P. Whipple and other manufacturing investors at Meriden, CT to succeed an earlier firm called Parker & Whipple which had been manufacturing locks since 1895. This new firm continued to manufacture locks until 1880.

In 1880, Parker & Whipple obtained the right to manufacture novelty timepiece and alarm clocks under the patents of Arthur E. Hotchkiss of Cheshire, CT. That year they built a new clock factory at West Meriden. Manager of the Parker & Whipple factory during this period was Almeron Lane whose brother, Frederick A. Lane, was superintendent of the Yale Clock Company at New Haven, CT. which also manufactured A. E. Hotchkiss' patent movements.

In 1879, Charles E. Parker (b. 1809) of Meriden became an investor in Parker & Whipple. Parker was one of the wealthiest men in the area at that time, having been involved in the manufacturing of rifles, shot guns, locks, Britannia ware, German silver, machine tools, sewing machines, printing presses, coffee grinders, waffle irons, wood screws and many other related products. In 1893, he bought out Parker & Whipple, changing the name to the Parker Clock Company.

The Parker Clock Company continued the manufacture of novelty timepieces and alarms, adding a line of larger drum cased alarm clocks. In 1919, they expanded their facility, adding a 2 1/2 story building 60 x 40 feet. The Meriden papers of May 28, 1926 reported the firm had closed its doors, though it was apparently reopened thereafter. The Parker Clock Company was finally dissolved in 1934.

**667** — 1907 — No. 11. One-half Size. SOLID BRASS.

**668** — 1907 — No. 2. (Calendar.) Two-third Size. GILT OR NICKEL.

**669** — 1907 — No. 5.—Without Calendar. No. 6.—With Calendar.

# THE PARKER CLOCK CO.

**670**

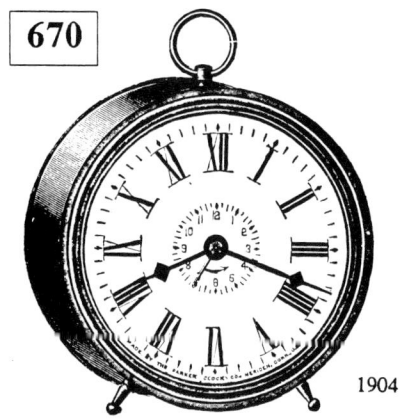

1904

### Rotary Hammer Alarm.

No.  106.

**HEAVY NICKEL-PLATED CASE, DUST PROOF.**

Price, each.................................$1 44
In case lots of 50 clocks, each................. 1 40
Printing name or business card on dial, in lots of 50 clocks, extra............................. 1 60

**673**

1910

### Rotary Hammer Alarm.
### No. 63

**Heavy Brass Nickel Plated Case, Dust-Proof.**

Cut half size.
Dial 3 inches. Height 6½ inches.

This clock is similar to our No. 61, but with alarm on back of clock instead of under the base. Somewhat cheaper in price, but of same good quality.
Price, each .........................$2.20

### No. 43

One day time. Same as Nos. 63, but without alarm.
Price, each........ ................$1.80

**671**

1910

### Rotary Hammer Alarm.

No.  61

**One Day Time Alarm, Heavy Brass Nickel Plated Case, with Alarm Shut-off Latch.**

Cut half size.
Dial 3½ inches. Height 6 inches.

Movement is without doubt perfect, because it has quick-beat watch escapement with mirror polished steel pallet faces. Cut steel pinions throughout. Mainsprings can be removed without disturbing train or escapement. Will run and keep time over 48 hours.
The alarm movement is separate from the time movement, and will ring loud and long. Is equipped with alarm shut-off latch, making it the most serviceable and least liable to get out of order of any clock on the market.
Price, each ............................$2.40

**672**

1905

**No. 604. $4.00**
Nickel case, cut steel pinions, electric double bell alarm. Dial 2 inches, height 5 inches, base 6½ inches. 1 day, time.
(C. Parker Clock Co.)

**674**

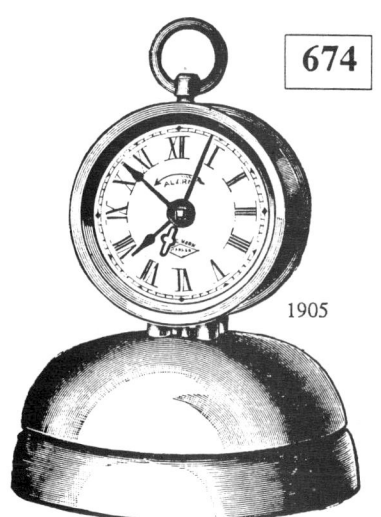

1905

**No. 601. $2.50**
Nickel, cut steel pinions.
1 day, time, alarm.
(C. Parker Clock Co.)

**675**

1905

**NO. 300. NICKEL OR GILT.**

1-Day, half-hour strike......................$4 75
Height, 5¼ inches; width, 6½ inches. Striking parts separate from time parts, so one can be repaired without interference with the other.

191

# THE PARKER CLOCK CO.

**Double Bell Rotary Alarm.**

No.  98.

**HEAVY BRASS, NICKEL PLATED CASE, DUST PROOF, RECESSED BACK.**

Dial, 6 inches. Height, 6½ inches.
Alarm will ring continuously 3¾ minutes, and is equipped with alarm shut off latch. The movement has three plates, quick-beat watch escapement.
Main springs can be removed or replaced without disturbing escapement or any part of the trains.
Has cut steel pinions throughout.
Heavy brass, nickel-plated, dust-proof case from which movement can be drawn by simply pressing push buttons at sides of case. A very desirable clock for the kitchen.
Price each ........................ $3 14

**Double Bell Rotary Alarm.**

No.  98.

**THIS CUT SHOWS NEW BACK AND KEYS OF No. 98, WITH DOUBLE BELL.**

Cut One-Third Size.

This clock has deep recessed back, and double bells, between which the hammer revolves, striking each alternately with electrical effect for 3½ minutes. Keys and hand sets stand out and can easily be taken hold of. No sore fingers from setting hands on our clock.
Imprints on back of case, giving names of arbors, and arrows indicating direction same should be turned.

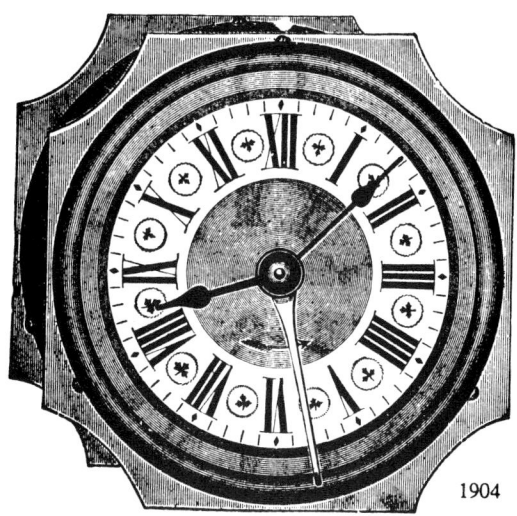

**Rotary Hammer Alarm.**

No.  104

**One Day Time Alarm, Heavy Gold Plated Case, Decorated Enamel Dials, Elegant Assorted Designs.**

Cut half size.

Dial 2⅝ inches. Height 3⅝ inches.

This clock has a very fine movement with reliable timekeeping qualities.
Case of French design, heavily gold plated and lacquered. Dials decorated enamel, assorted colors and design. Alarm easily set from front of dial. A nice ornament for a lady's boudoir.

Price, each ........................ $3.60

**Rotary Hammer Alarm.**

No.  102.

**HEAVY BRASS, NICKEL-PLATED CASE, DUST PROOF, WITH RECESSED BACK AND LARGE BELL.**

Dial, 4½ inches. Height, 5½ inches.
Heavy brass nickel-plated, dust-proof case, with recessed back and large bell. Long and loud ringing alarm which can be shut off by simply pressing push button of alarm, shut off latch at top of case.
Movement has quick beat, watch escapement, cut steel pinions throughout; main springs placed outside the plates, which allows removing or replacing without disturbing the escapement or trains. Can be drawn from case by pressing push buttons at side of case. Has all the best qualities of the Parker Clock.
**No. 102** replaces No. 97 which is discontinued.
Price, each .................... $1 80

**Rotary Hammer Alarm.**

No.  102.

**THIS CUT REPRESENTS THE GENERAL APPEARANCE OF NEW BACK AND KEYS OF PARKER ROTARY ALARM.**

We do away with the old style bell on top of clock, with hole in top of case through which dust easily penetrates. We place our bell on back of clock and use a hammer which revolves and strikes the bell with full force of alarm springs. Our clocks are easily wound or set to position. The keys stand out so that they are easily taken hold of, occasioning no sore fingers from winding or setting alarm.
Imprints on back of case giving names of arbors, and arrows indicating directions, same should be turned.

**Rotary Hammer Alarm.**

No. PARKER 100.

**HEAVY BRASS, NICKEL-PLATED CASE. DUST PROOF.**

Dial, 3 inches. Height, 4 inches.
The above is the smallest complete alarm clock we now make, and has movement of exactly same quality and construction as our larger models, embracing all the best features. Can be easily removed from case by pressing buttons at side.
Price, each ........................ $1 80
**No. 41.** Common one day time, same model as No. 100, but without alarm, the best small clock made.
Price, each ........................ $1 50

# THE PARKER CLOCK CO.

**682**

No. 40. TIME.
Diameter, 2¼ inches.

1904

No. 40N. Nickel Plate, $1.70    No. 40G. Gold Plate, $1.70
[EFFUSION]    [EGLANTINE]

**683**

1907

No. 1.
Two-third size,
GILT OR NICKEL.

**684**

1910

### Parker Rotary Hammer Alarm.

No.  109

**Dust-Proof, Cushion Beat, Long Alarm, Visible Balance.**

**Micrometer and Back Regulator.**

Dial 4 inches.    Height 4½ inches.
Rotary hammer and cut steel pinions. Glass hard and mirror polished pallets. Hard metal train. Thin and elastic mainspring. Quick beat escapement. Finely adjusted setting and winding mechanism. Mainspring can be removed and replaced without disturbing the train, dial or hands. Dealers can afford to guarantee these clocks on smaller margin of profit than warranted by any other make.

Price, each .................. $1.66
In case lots of 50 clocks, each.......... 1.60

**685**

1911

No. 150
## The Parker Princess Alarm

Has the following exclusive "PARKER" features.

Double Roller Ratchet Tooth Escapement. Solid Steel Highly Polished Pallets. Cut Steel Pinions and Cut Hard Brass Wheels. Oil Reservoirs on all Pivot Bearings. Easy Winding Keys and Easy Turning Sets. Light, Non-breakable Main Springs.

Also has attachment so alarm can be used either as an Intermittent or Long Alarm.

**686**

1910

### Rotary Hammer Alarm.

No.  103.

**ONE DAY TIME ALARM.**
**HEAVY BRASS NICKEL PLATED CASE.**

Dial 3½ inches.    Height 5 inches.
One day time alarm.
The construction is entirely new. It has front set, no glass or sash, indestructible, gilt or silver plated metal dial.
Heavy brass nickel plated dust-proof case.
The movement embraces all the new and best features found only in the construction of the Parker clocks.
Price, each ................... $1.60

**687**

1911

### THE TORNADO INTERMITTENT ALARM CLOCK.

THE BEST AND MOST ATTRACTIVE ALARM CLOCK MADE.

¶ This clock has a 4¼ inch dial; height 5½ inches, and is fitted with a 4-inch bell metal gong on back.

¶ It rings like a fire alarm every few seconds for fifteen minutes. The alarm can be stopped at pleasure by means of a patent switch.

## Price, each - - $1.90

# RUSSELL & JONES CLOCK COMPANY
## Pittsfield, Massachussetts

With the failure of the Terry Clock Company at Pittsfield, Mass. in 1888, the Pittsfield businessmen who were large investors, particularly brothers Hezekiah S. and Solomon N. Russell and Edward D. G. Jones, took over the operation. These men and their associates had supplied funds to buy the bankrupt Terry firm at Waterbury, Conn. in 1880 and move it to Massachusetts in the summer of that year. They built a new factory for the operation in 1883.

By January of 1889, the firm's name had been changed to Russell & Jones Clock Company and a trade catalog was issued with few changes, if any, from the old firm's line. In 1890, the new firm issued a new catalog with a substantial number of new and unusual models. As far as can be determined, the firm discontinued manufacture and was disbanded in 1893.

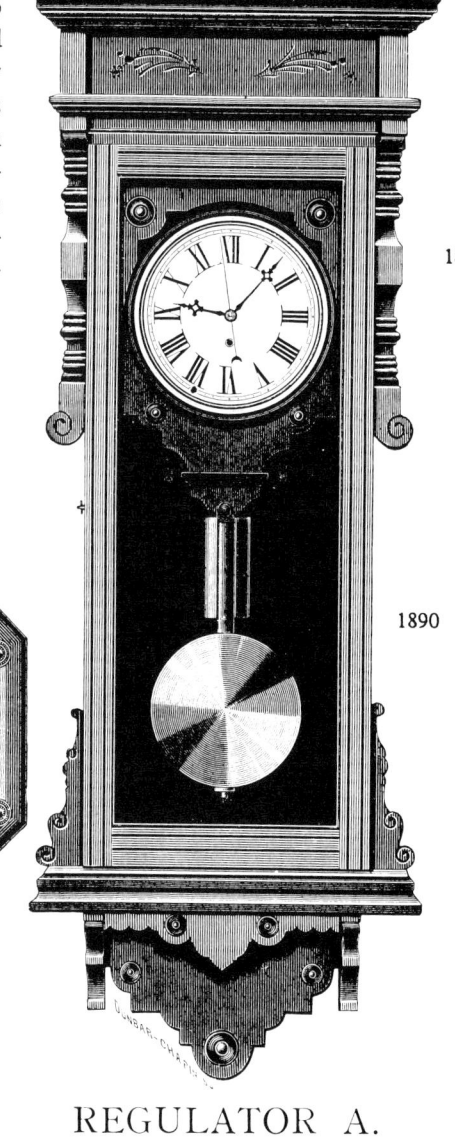

**688**

1890

**8, 10 & 12 INCH YESSO.**
POLISHED VENEERED.
BRASS TRIMMINGS
8 Day, Spring, Time.
8 Day, Spring, Strike.
8 Day, Spring, Strike, Calendar.

**689**

1890

## REGULATOR A.
10-inch Porcelain Dial.

**WALNUT AND ASH.**
**GLASS SIDES.   POLISHED MOVEMENT,**
**WITH RETAINING POWER.**
**BRASS WEIGHT.   SWEEP SECOND.**

Height 62 Inches.

Eight Day — Time.

**690**

1890

## REGULATOR B.
12-inch Porcelain Dial.

**WALNUT, OAK AND MAHOGANY.**
**GLASS SIDES.**
**FINE POLISHED MOVEMENT,**
**WITH RETAINING POWER.**
**BEVELED FRONT PLATE GLASS.**
**SWEEP SECOND.**
**BRASS WEIGHT.**

Eight Day Time.   78 Inches.

# RUSSELL & JONES

**691**

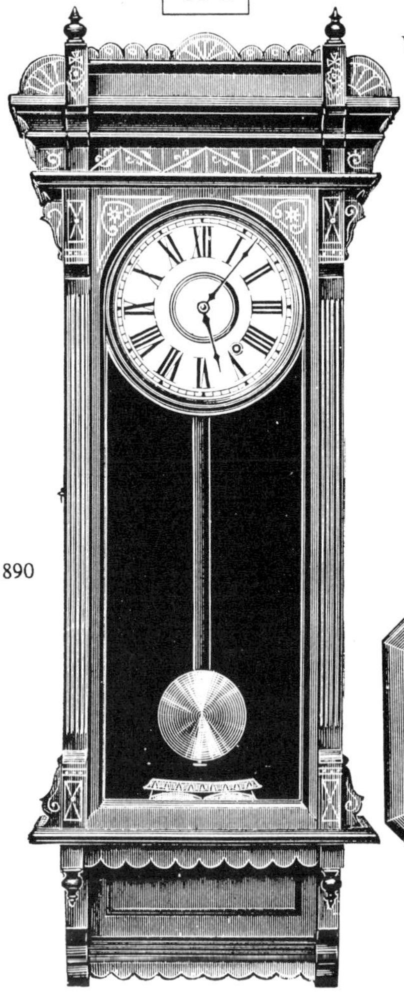

1890

## GARLAND.

WALNUT, OAK AND CHERRY.
8-inch Dial. Height 37 Inches.
8 Day, Spring, Time.
8 Day, Spring, Strike.
8 Day, Spring, Time, Calendar.
8 Day, Spring, Strike, Calendar.

**692**

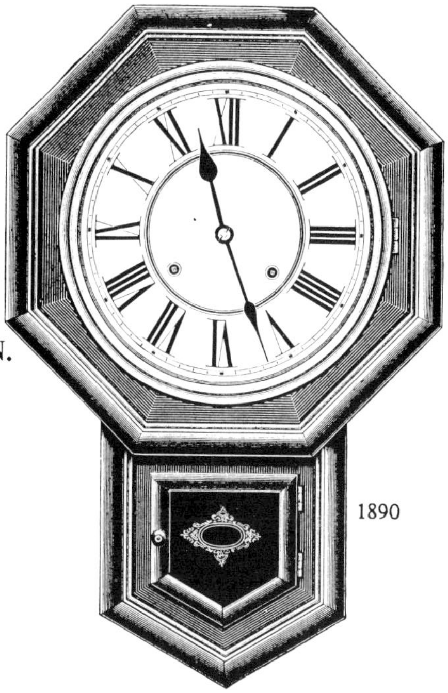

1890

## 8, 10 & 12 INCH DROP OCTAGON.

POLISHED VENEERED. GILT.
8 Day, Spring, Time.
8 Day, Spring, Strike.
8 Day, Spring, Strike, Calendar.

**693**

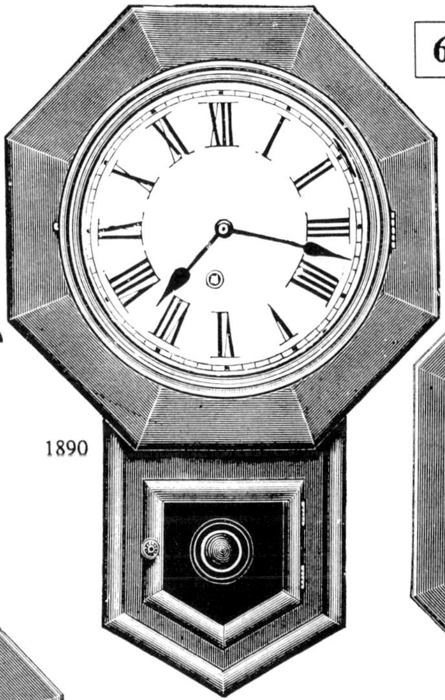

1890

## 8, 10 & 12 INCH DROP OCTAGON, R. C.

WALNUT, OAK AND VENEERED.
8 Day, Spring, Time.
8 Day, Spring, Strike, Alarm.

**694**

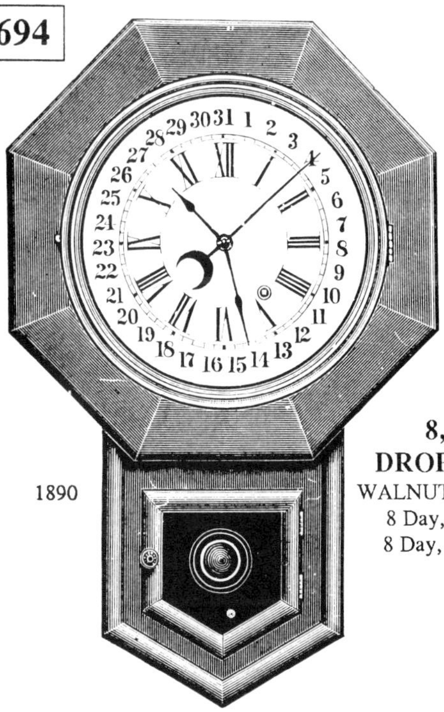

1890

## 8, 10 & 12 INCH DROP OCTAGON, R. C.

WALNUT, OAK AND VENEERED.
8 Day, Spring, Time, Calendar.
8 Day, Spring, Strike, Calendar.

**695**

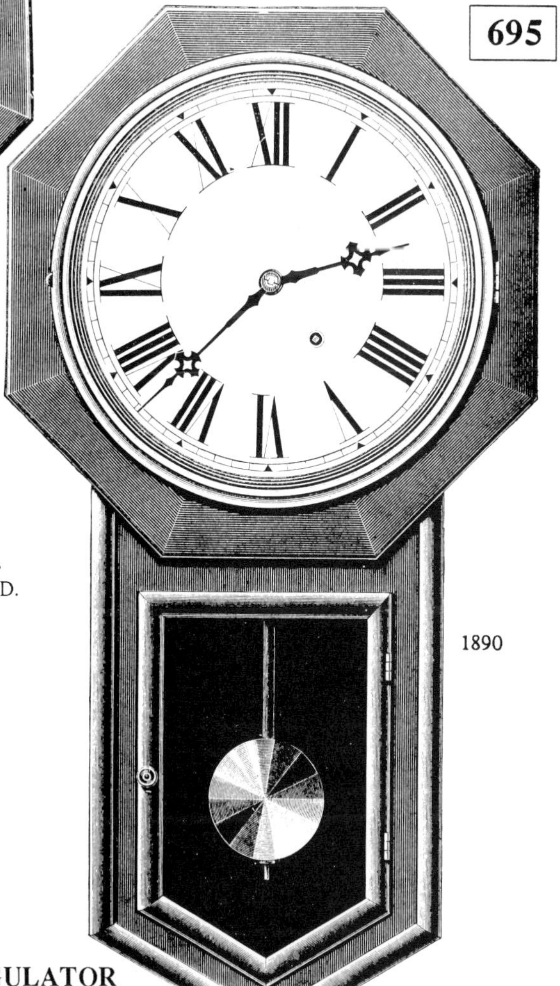

1890

## DROP REGULATOR

POLISHED VENEERED
12-inch Dial. Height 31 Inches.
8 Day Time - 8 Day Strike
8 Day Strike Calendar

# RUSSELL & JONES

1890

**BERKSHIRE DROP.**
THERMOMETER &
LEVEL. 6-INCH DIAL.
Base 15 1/4 Inches. Height 28 Inches.
8 Day, Strike.
8 Day, Strike, Alarm.
8 Day, Strike, Gong.
8 Day, Strike, Gong, Alarm.

1890

**BERKSHIRE DROP.**
6-INCH DIAL.
Base 15 1/4 Inches. Height 28 Inches.
8 Day, Strike.
8 Day, Strike, Alarm.
8 Day, Strike, Gong.
8 Day, Strike, Gong, Alarm.

*To receive top prices, clocks must be in ORIGINAL MINT condition and in good running order - A replaced or damaged dial, missing parts, will reduce prices accordingly.*

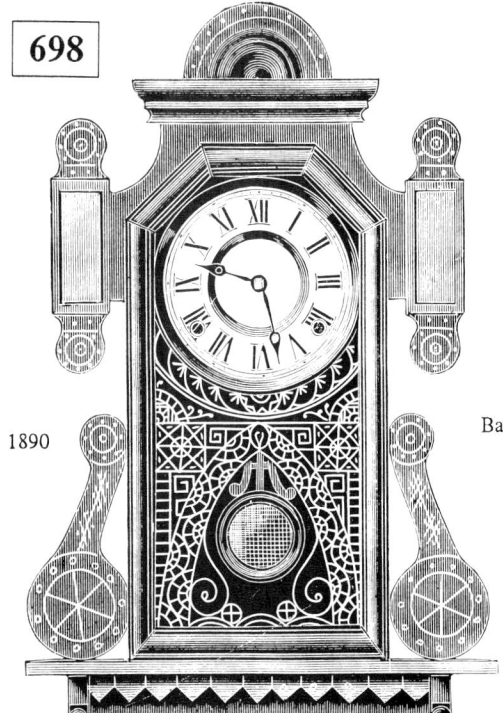

1890

1890

**LOWELL DROP.**
6-INCH DIAL.
Base 15 Inches. Height 26 Inches.
8 Day, Strike.
8 Day, Strike, Alarm.
8 Day, Strike, Gong, Alarm.

**LOWELL DROP.**

**HIGHLAND DROP.**
6-INCH DIAL.
Base 14 1/2 Inches. Height 26 Inches.
8 Day, Strike.
8 Day, Strike, Alarm.
8 Day, Strike, Gong.
8 Day, Strike, Gong, Alarm.

# RUSSELL & JONES

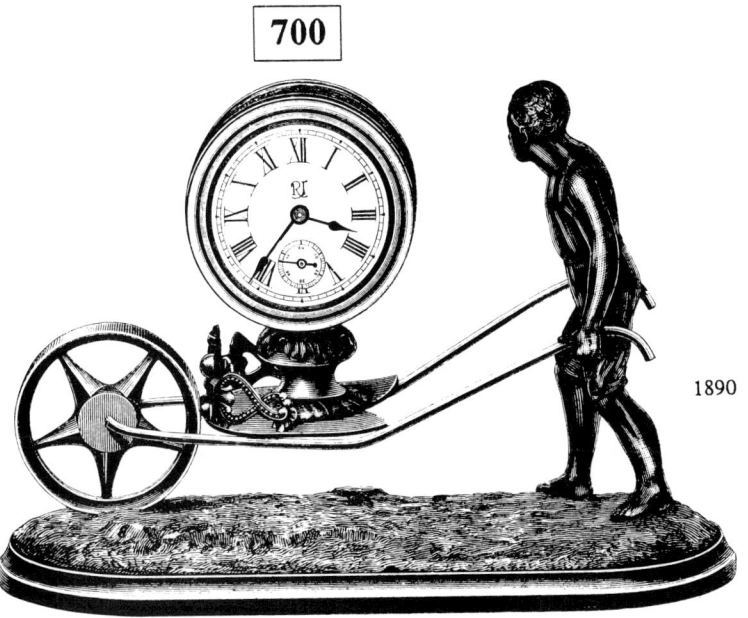

**700**

### PLANTATION.
Base 7¾ Inches.   Height 5 Inches.

**701**

### EIFFEL TOWER.
Base 3¾ Inches.
Height 11½ Inches.

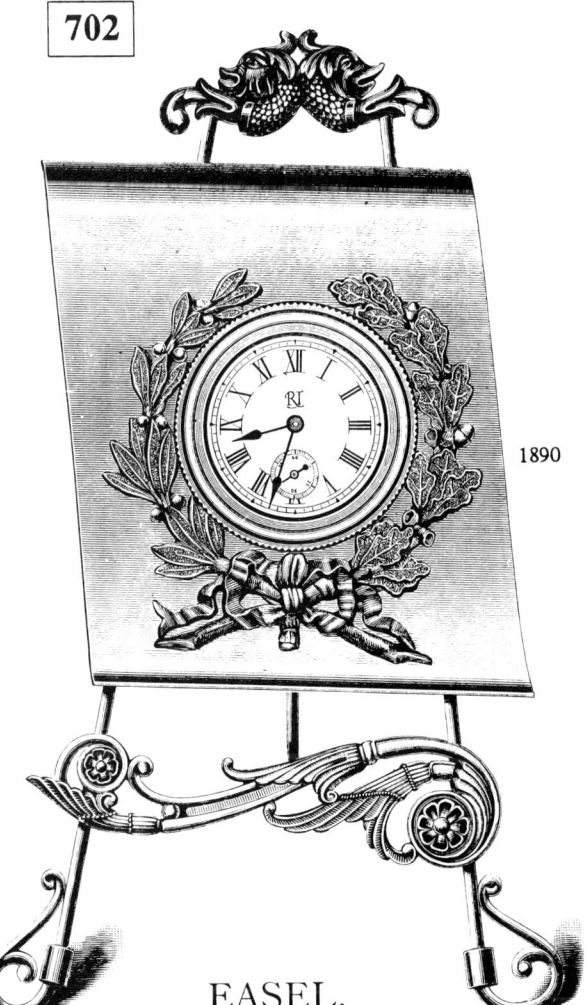

**702**

### EASEL.
Base 6 Inches.   Height 10 Inches.

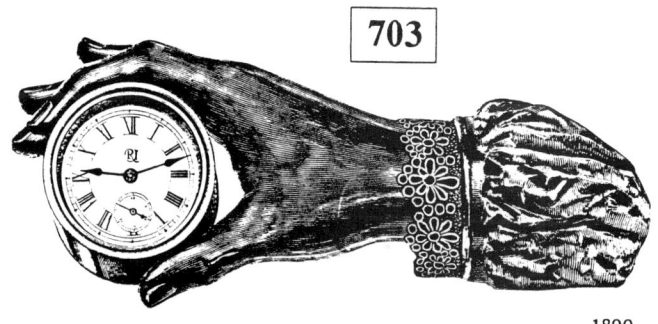

**703**

### HAND.
Base 7 Inches.   Height 3¼ Inches.

# SETH THOMAS CLOCK COMPANY
## Plymouth Hollow, Connecticut
## (After 1865) Thomaston, Connecticut

The Seth Thomas Clock Company was organized as a joint stock corporation on May 3, 1853 to succeed the earlier clockmaking operation of the founder. Seth Thomas (1785-1859) had been manufacturing clocks at the site since 1814.

After Thomas' death in 1859, his son Aaron became President and began to add new products to a conservative line. About 1862, the firm purchased the patent rights of Wait T. Huntington and Hervey Platts of Ithaca, New York and added three models to their line that year. The earliest of the clocks indicate only three patent dates on the dials, September 19, 1854, November 17, 1857 and January 31, 1860. The fourth and final patent of March 1, 1862 is carried on most of their calendar clocks manufactured until 1875 or 1876. On February 15, 1876 Randall T. Andrews, Jr., a Thomas relative and workman in the factory, received a patent on an improved mechanism. This was put into production and utilized on all later perpetual calendar clocks until the last model was dropped in 1917.

The Seth Thomas Clock Company was very prosperous into the 20th century and were considered the "Tiffanys" of Connecticut clock manufacture, even by their competitors. Between 1865 and 1879 they operated a subsidiary firm known as Seth Thomas' Sons & Company that manufactured a higher grade 15 day mantel clock movement and during that period were major supporters of a New York sales outlet known as the American Clock Company. They also became a major manufacturer of tower and street clocks after 1872 and in between 1915 were manufacturers of jewelled watches.

On January 1, 1931, the firm became a subsidiary of General Time Instruments Corporation and soon passed from family control. The firm's decline was gradual over the next 50 years and culminated in the firm's removal from Connecticut to Norcross, Georgia about 1975. It has been reported in 1988 that the firm is all but dissolved.

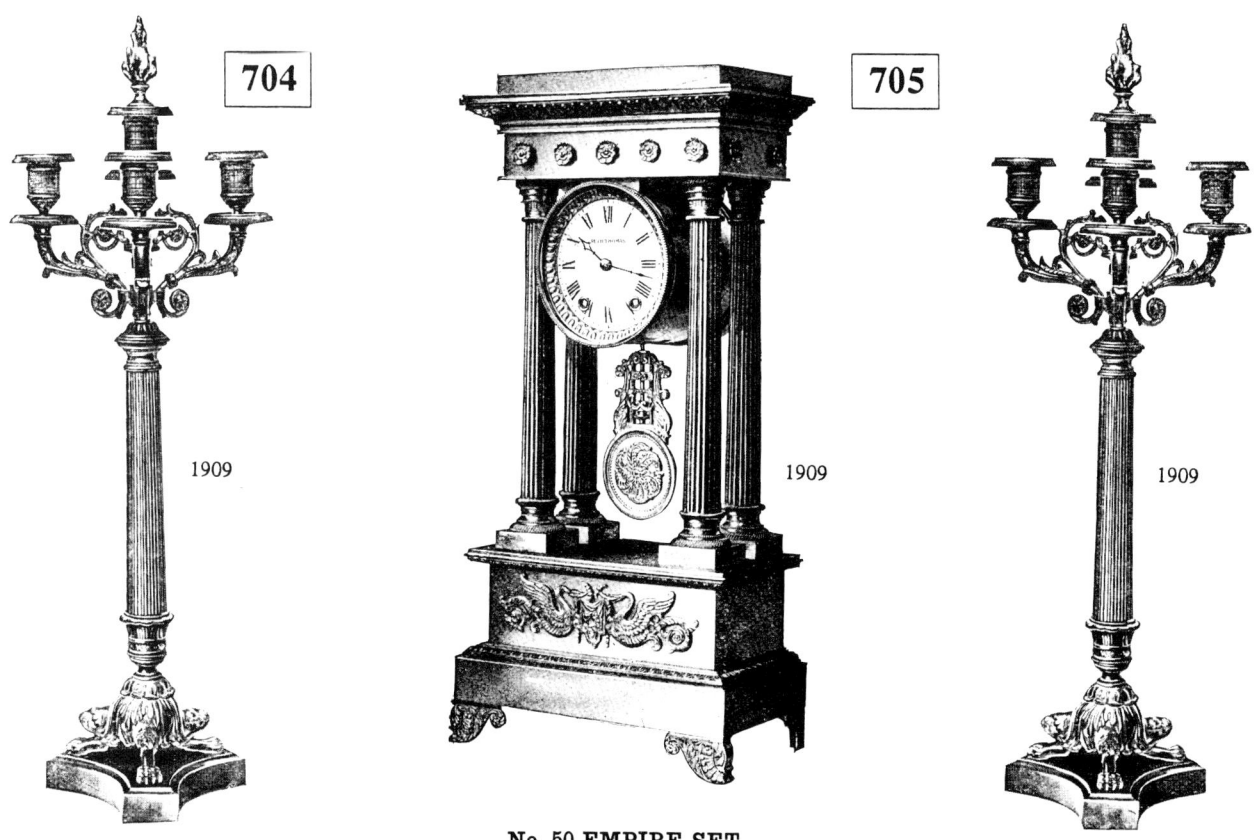

**No. 50 EMPIRE SET**

Rich Gold Finish and Highly Burnished. Fifteen Day, Fine Polished Movement, Half Hour Strike, Cathedral Bell. 3 inch Porcelain Dial. Height of Clock, 18½ inches; Base, 9 inches. Height of Candelabra, 21½ inches; Base, 7 inches.

List Price of Clock, $135.00    List Price of Set, $200.00    List Price of Candelabra, per pair, $65.00

# SETH THOMAS

**706**

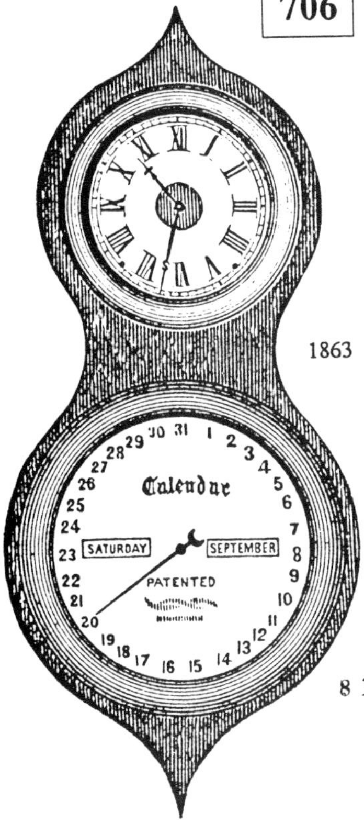

*For comprehensive listing of Calendar Clocks, see Tran Duy Ly "Calendar Clocks - A Guide to Identification & Prices."*

1863

**OFFICE CALENDAR.—No. 3.**

8 Day—Spring—Time—6 inch Time Dial.

Hight, 24 inches.

**OFFICE CALENDAR No. 3.**

**707**

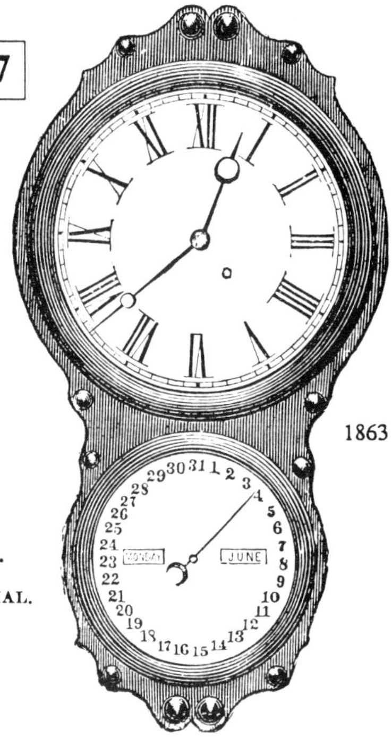

1863

**OFFICE CALENDAR.—No. 4.**

8 Day—Spring—Time—12 inch Dial.

Hight, 28 inches.

Seth Thomas
**OFFICE CALENDAR. No. 4.**

**708**

*The year entered beside each clock is the year of the catalog from which the illustration was taken. Some clocks may have been produced a few years earlier.*

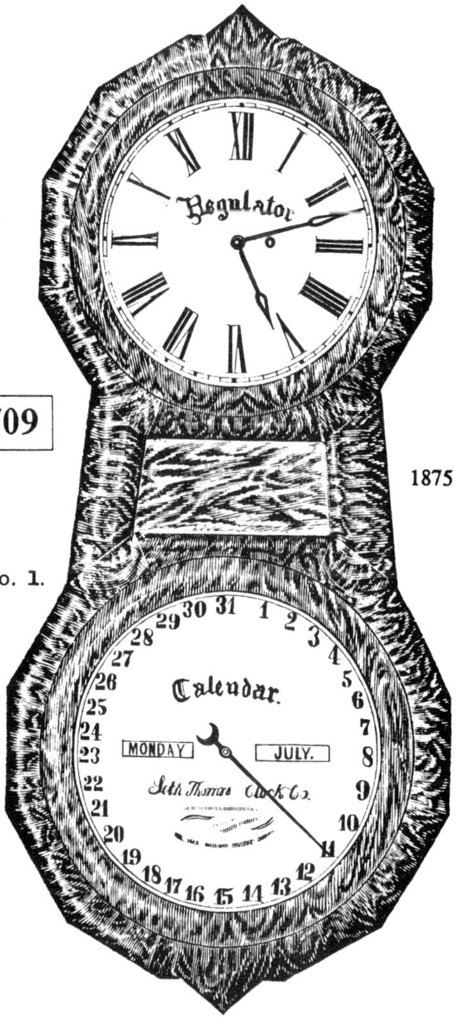

1875

Seth Thomas
**OFFICE CALENDAR. No. 1.**

Weight.

12 inch time dial.

Height, 40 inches.

EIGHT DAY, TIME.

**OFFICE CALENDAR—No. 2.**

8 Day.    Time.    Weight.    14 Inch Dial.

Hight 42 1-2 inches.

**OFFICE CALENDAR—No. 2.**

**709**

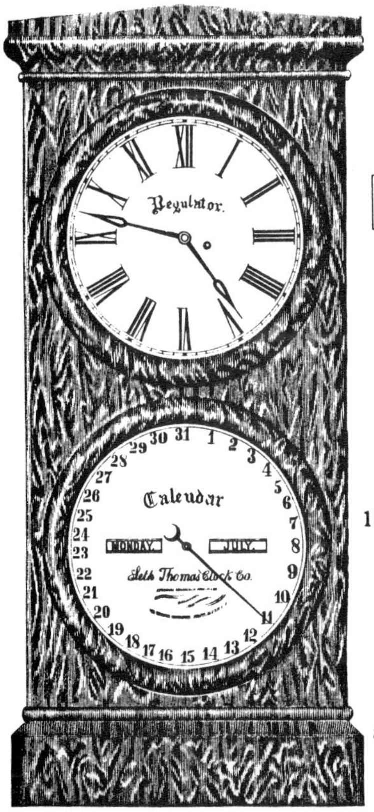

1875

**OFFICE CALENDAR.—No. 1.**

# SETH THOMAS

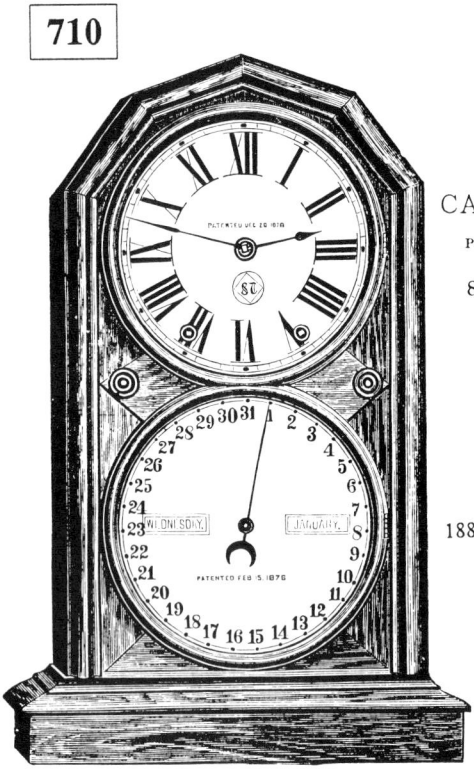

PARLOR CALENDAR No. 5.

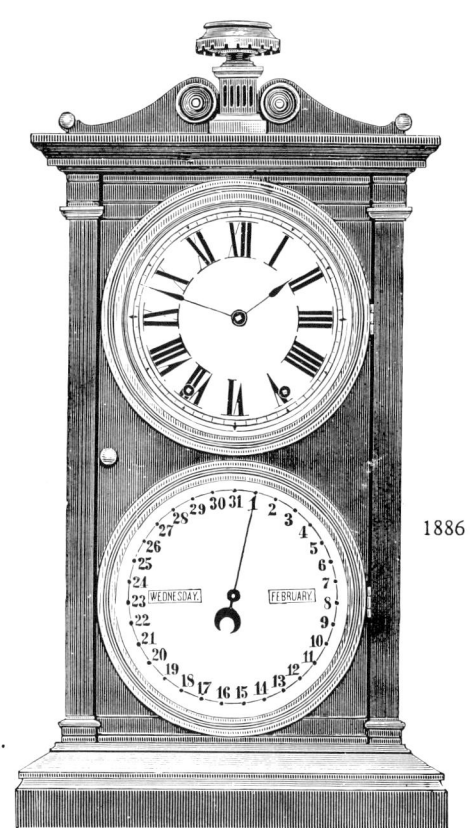

### PARLOR CALENDAR No. 5.
POLISHED WALNUT VENEER.

8 Day, Spring, Strike.

8 inch Dials.

Hight, 20 inches.

PERPETUAL CALENDAR.

1886

### PARLOR CALENDAR No. 6.
WALNUT, OAK OR OLD OAK.

8 Day, Spring, Strike, Cup Bell.

8 inch Dials.

Hight, 27 inches.

PERPETUAL CALENDAR.

1886

PARLOR CALENDAR No. 6

*The year entered beside each clock is the year of the catalog from which the illustration was taken. Some clocks may have been produced a few years earlier.*

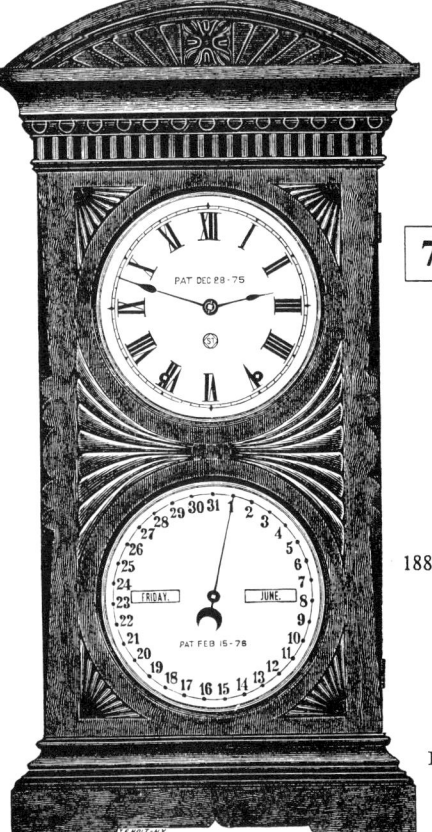

PARLOR CALENDAR No. 7.

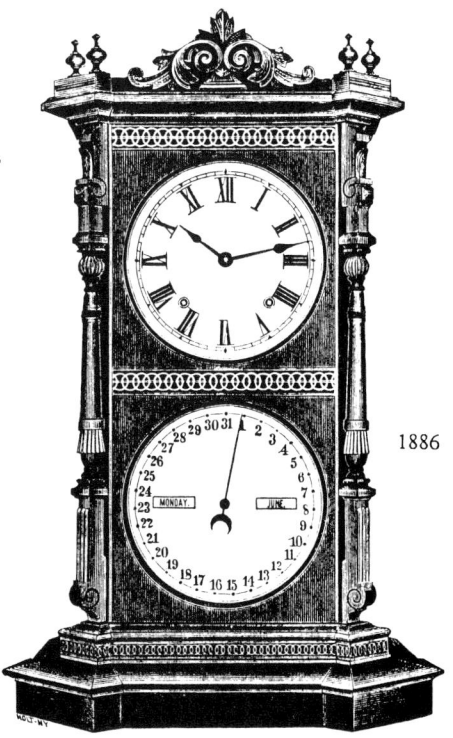

### PARLOR CALENDAR No. 8.
WALNUT CASE.

MAPLE TRIMMINGS.

8 Day, Spring, Strike.

Cathedral Bell.

8 inch Dials.

Hight, 27½ inches.

PERPETUAL CALENDAR.

1885

1886

### PARLOR CALENDAR No. 7.
Hand Carved.   Walnut Case.

Height, 29 inches.    8-inch Dials.

8 Day, Spring, Strike....$20 25
Perpetual Calendar.

PARLOR CALENDAR No. 8.

# SETH THOMAS

*For comprehensive listing of Calendar Clocks, see Tran Duy Ly "Calendar Clocks - A Guide to Identification & Prices."*

### 714

### 715

### OFFICE CALENDAR No. 6.

10 inch Calendar Dial.

12 inch Time Dial.

8 Day, Spring, Time.

8 Day, Spring, Strike.

Hight, 32 inches.

PERPETUAL CALENDAR.

OFFICE CALENDAR No. 6.

### 716

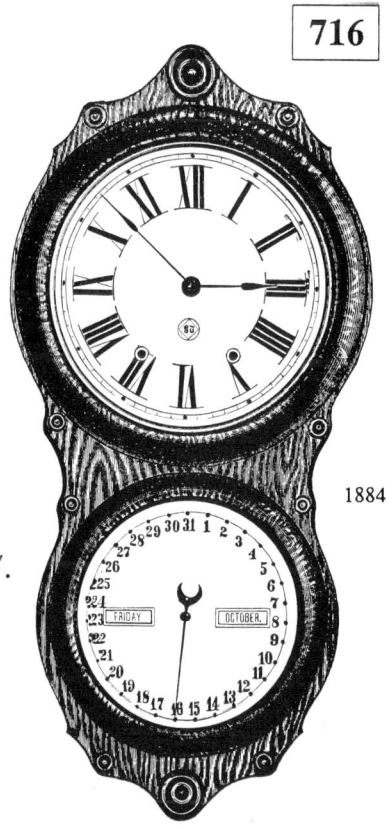

### OFFICE CALENDAR No. 5.

Walnut case, with polished French walnut trimmings.

8 Day, Weight, Time, 14 inch Dials.

Hight, 50 inches.

PERPETUAL CALENDAR.

### OFFICE CALENDAR No. 7.

10 inch Time Dial.

8 inch Calendar Dial.

8 Day, Spring, Time.

8 Day, Spring, Strike.

Hight, 26½ inches.

PERPETUAL CALENDAR.

OFFICE CALENDAR No. 7.

# SETH THOMAS

*For comprehensive listing of Seth Thomas Clocks, see Tran Duy Ly "Seth Thomas Clocks & Movements - A Guide to Identification & Prices."*

*For comprehensive listing of Calendar Clocks, see Tran Duy Ly "Calendar Clocks - A Guide to Identification & Prices."*

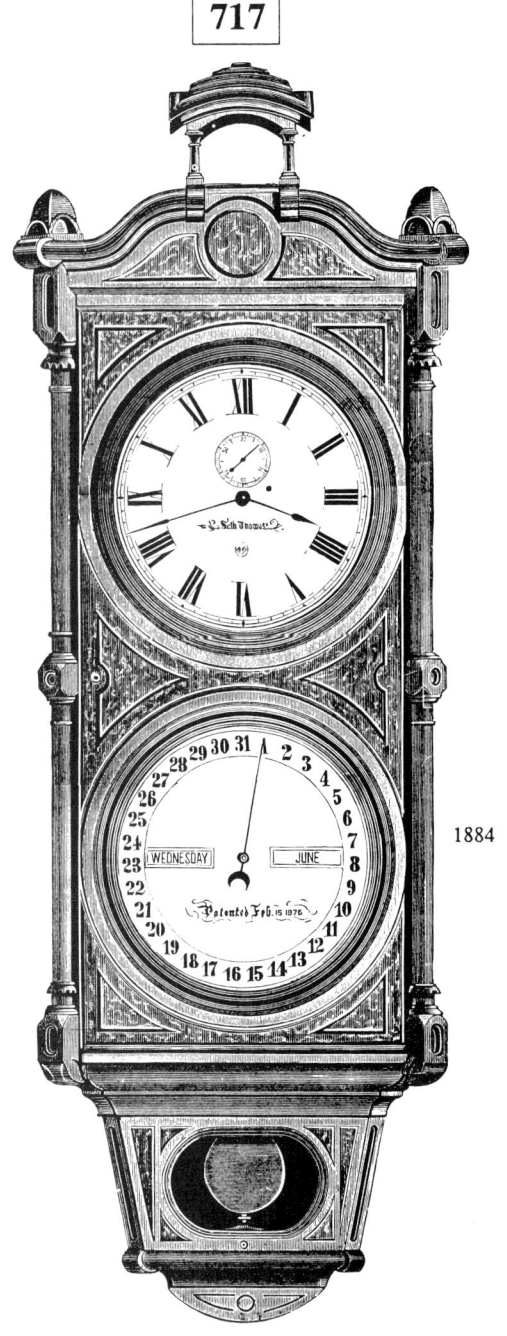

**717**

1884

### OFFICE CALENDAR No. 8.

8 Day, Weight, Time. Beats Seconds.

Hight, 66 inches. 14 inch Dials.

PERPETUAL CALENDAR.

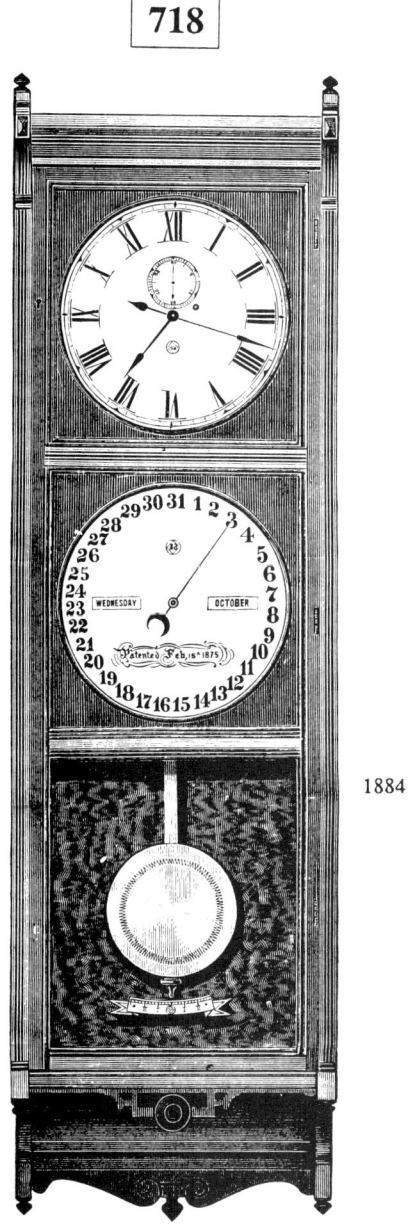

**718**

1884

### OFFICE CALENDAR No. 9.

MADE IN WALNUT AND OAK.

8 Day, Weight, Time. Beats Seconds.

Hight, 68 inches. 14 inch Dials.

PERPETUAL CALENDAR.

203

# SETH THOMAS

*The year entered beside each clock is the year of the catalog from which the illustration was taken. Some clocks may have been produced a few years earlier.*

*For comprehensive listing of Calendar Clocks, see Tran Duy Ly "Calendar Clocks - A Guide to Identification & Prices."*

**OFFICE CALENDAR No. 11.**

WALNUT, MAHOGANY, OAK OR OLD OAK.

8 Day, Weight, Time. Beats Seconds.

14 inch Dials. Hight, 68½ inches.

PERPETUAL CALENDAR.

**OFFICE CALENDAR No. 10.**

MADE IN WALNUT, OAK AND CHERRY.

8 Day, Weight, Time.
10 inch Dials.
Hight, 49 inches.

PERPETUAL CALENDAR.

**OFFICE CALENDAR No. 10.**

WALNUT, CHERRY, OAK OR OLD OAK.

8 Day, Weight, Time.
Hight, 49 inches.    10 inch Dials.

PERPETUAL CALENDAR.

Price, $40.00.

# SETH THOMAS

*For comprehensive listing of Calendar Clocks, see Tran Duy Ly "Calendar Clocks - A Guide to Identification & Prices."*

*To receive top prices, clocks must be in ORIGINAL MINT condition and in good running order - A replaced or damaged dial, missing parts, will reduce prices accordingly.*

**723**

1911

### FLAKE

Eight-Day, Pendulum. 12-inch Dial. Wood Case, White Adamantine Finish. Can be cleaned with a damp cloth any number of times without injury to the finish. Diameter, 16 inches.

| | |
|---|---|
| Time .................. List | $10.00 |
| Strike ................. List | 12.00 |
| Time and Calendar .... List | 11.00 |
| Strike and Calendar ... List | 13.00 |

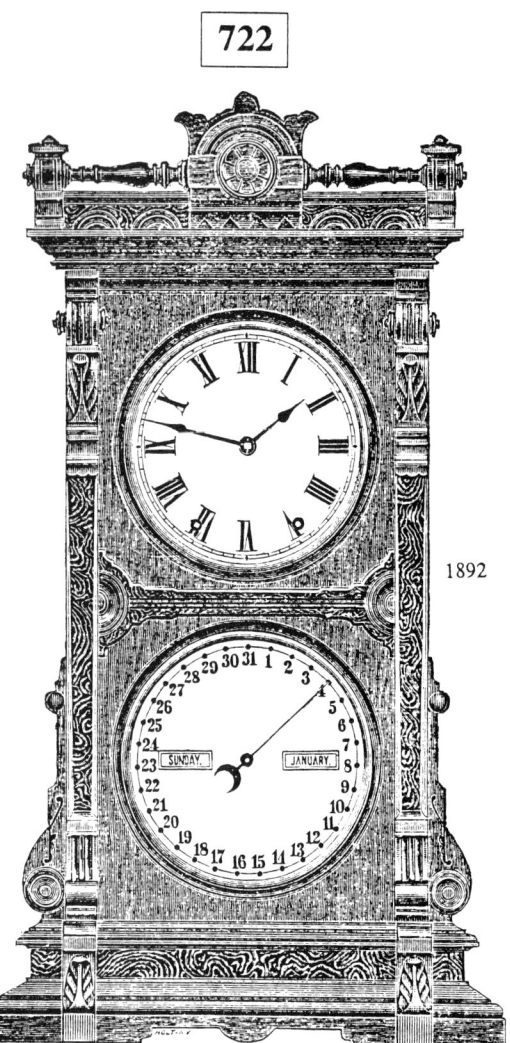

**722**

1892

### PARLOR CALENDAR No. 11.

OLD OAK.

8 Day, Strike.
Cathedral Bell.
8 inch Dials.
Hight, 30 inches.

PERPETUAL CALENDAR.

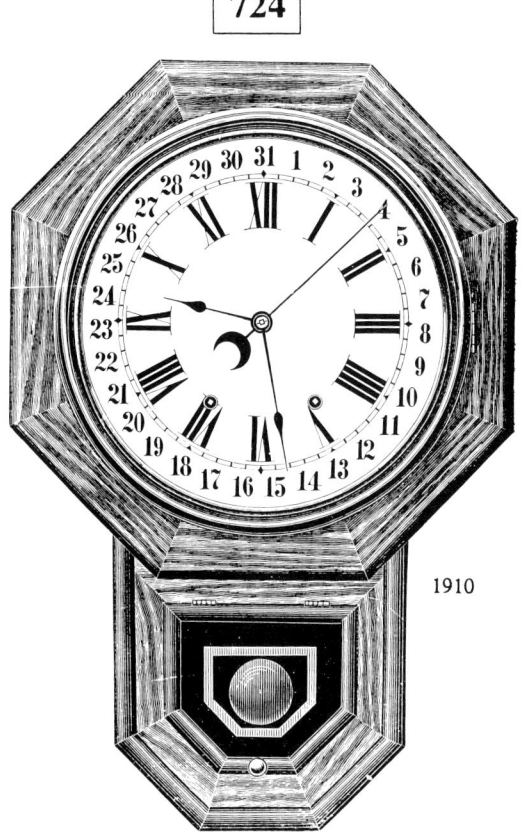

**724**

1910

### DROP OCTAGON CALENDAR

Mahogany Veneer and Old Oak Polished. Eight Day, Spring. 12 inch Dial. Height, 23½ inches.

| | |
|---|---|
| Time .................. List, | $8.30 |
| Strike ................. List, | 9.90 |

# SETH THOMAS

**725**

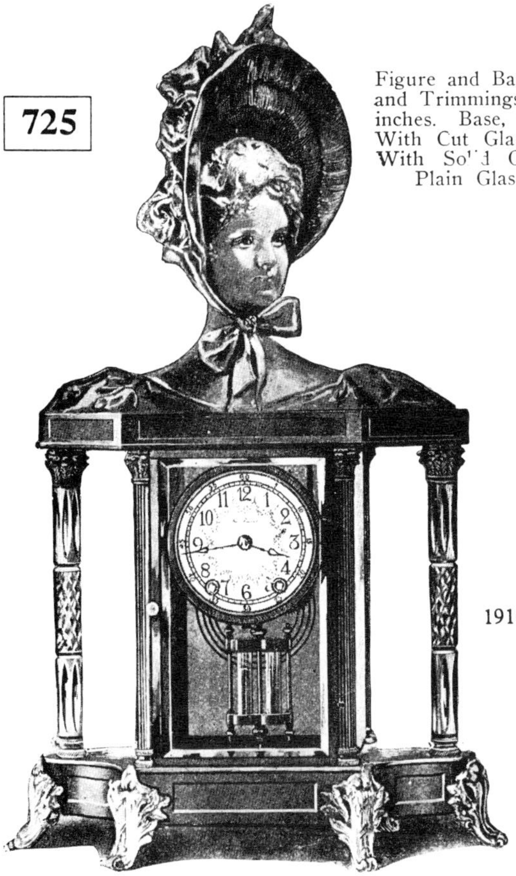

1910

**EMPIRE No. 31**

**EMPIRE No. 31**

Figure and Base Bronze, Barbedienne Finish, Body and Trimmings Rich Gold, Burnished. Height, 25 inches. Base, 14½ inches.
With Cut Glass Columns..............List, $60.00
With So'·d Gold Flower Decorations on Plain Glass Columns...............List, 62.00

**726**

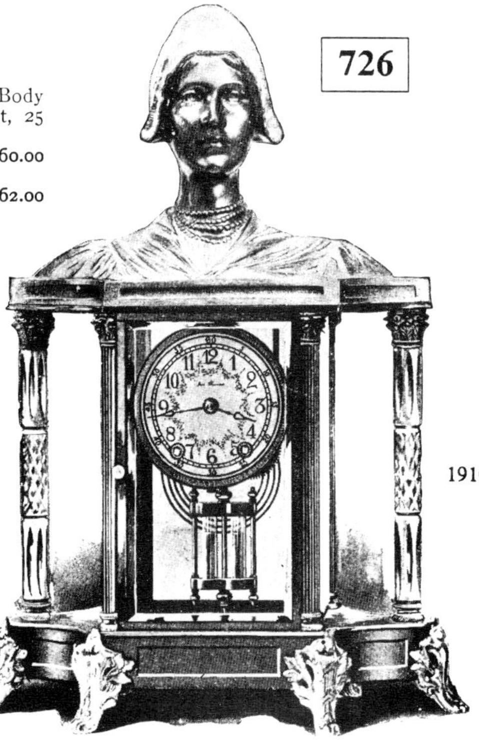

1910

**EMPIRE No. 32**

Figure Bronze Art Nouveau, Base Barbedienne Finish, Body and Trimmings Rich Gold, Burnished. Height, 20 inches. Base, 14½ inches.
With Cut Glass Columns...............List, $60.00
With Solid Gold Flower Decorations on Plain Glass Columns..............List, 62.00

**728**

**727**

1910

1910

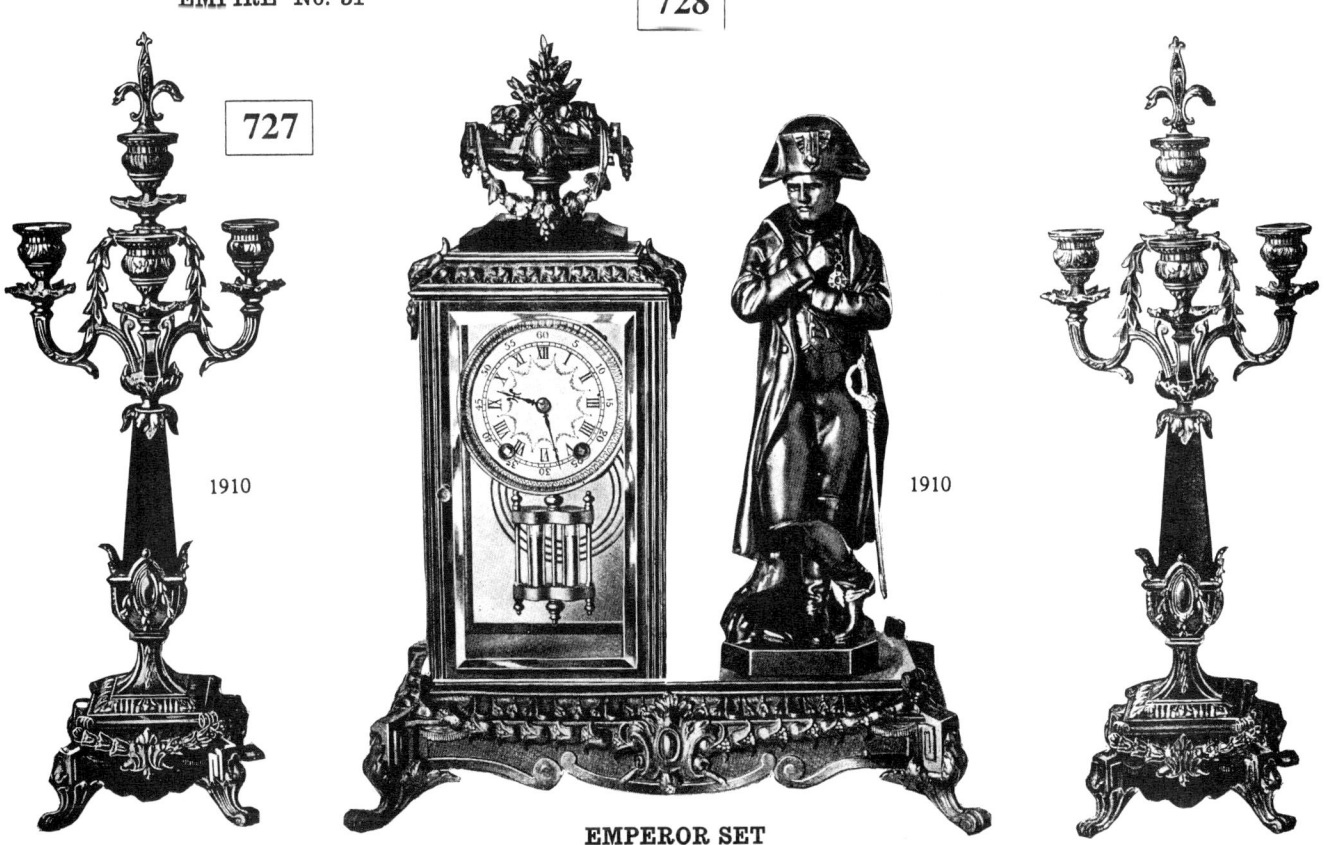

**EMPEROR SET**

Bronze Art Nouveau Finish. Fifteen Day, Fine Polished Movement, Half Hour Strike, Cathedral Bell. 3 inch Decorated Porcelain Dial. Height of Clock, 16¾ inches; Base, 14½ inches. Height of Candelabra, 17 inches; Base, 5¼ inches.
List Price of Clock, $70.00 List Price of Set, $100.00 List Price of Candelabra, per pair, $30.00

# SETH THOMAS

**729**

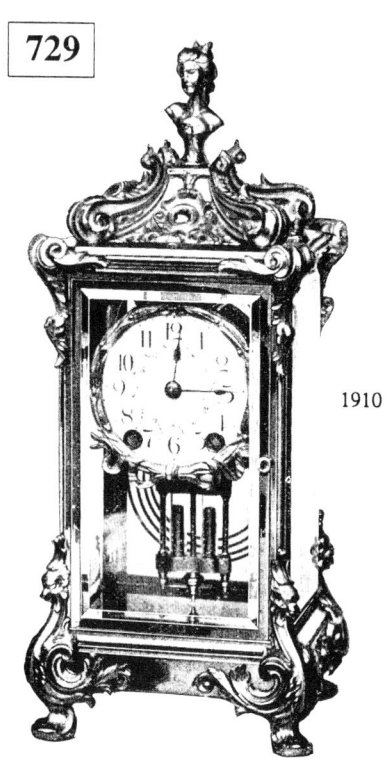

1910

**EMPIRE No. 41**
Rich, Gold Finish, Decorated Dial.
Height, 13½ inches. Base, 6 inches.
List, $47.50

**730**

1910

**CUPID EMPIRE**
Real Bronze Finish, except Base plate and the four uprights of case holding glass, which are gold plated. Fifteen Day, Fine Movement. Cathedral Bell. 3 inch Decorated Porcelain Dial.
Height, 14 inches. Base, 14½ inches.
List, $70.00

**731**

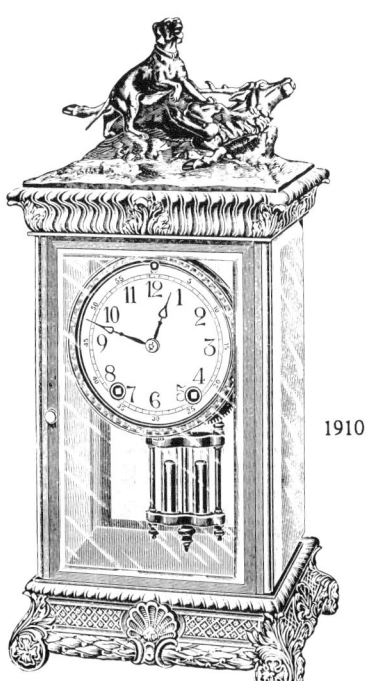

1910

**EMPIRE No. 14**
Rich Gold Finish.
Height, 16 inches. Base, 8 inches.
List, $43.00

**732**

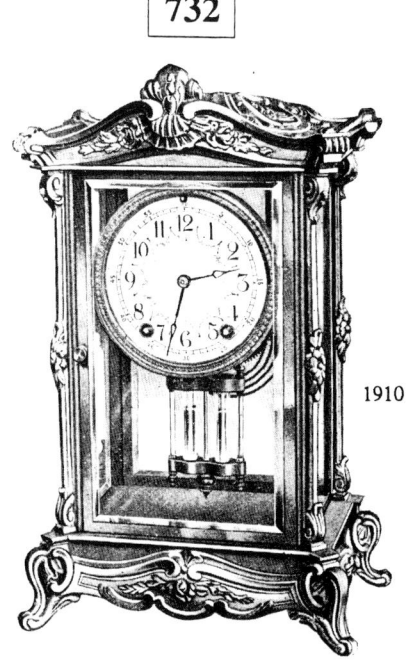

1910

**EMPIRE No. 9**
Rich Gold Finish, Decorated Dial.
Height, 13 inches. Base, 8 inches.
List, $38.00

**733**

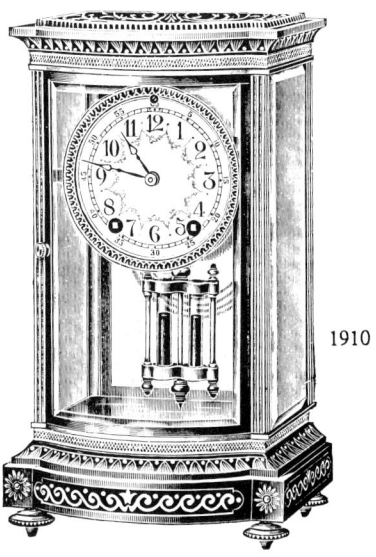

1910

**EMPIRE No. 8**
Top and Base made in Rich Gold, and Bronze Finishes. Convex Beveled Glass Front, Decorated Dial. Height, 12½ inches. Base, 7 inches.
List, $40.80

# SETH THOMAS
## METAL CASES, GOLD PLATED AND LACQUERED

Beveled Plate Glass Front, Sides and Back. Convex Fronts, 3½ inch Cream Porcelain Dials. Fifteen Day, High Grade Movement

Convex Fronts, 4 inch Cream Porcelain Dials. Eight Day, Fine Movements,

### 734
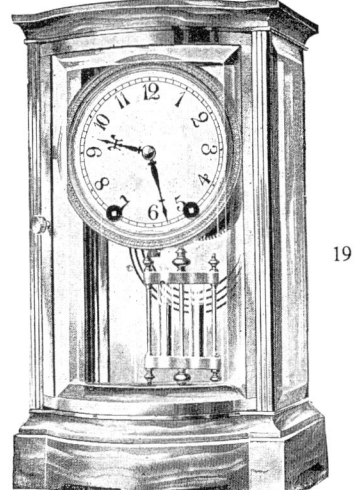

**EMPIRE No. 148**

Onyx Top and Base, Polished Gold Body, Convex Front.
Height, 9 inches.
List, $52.00

### 735
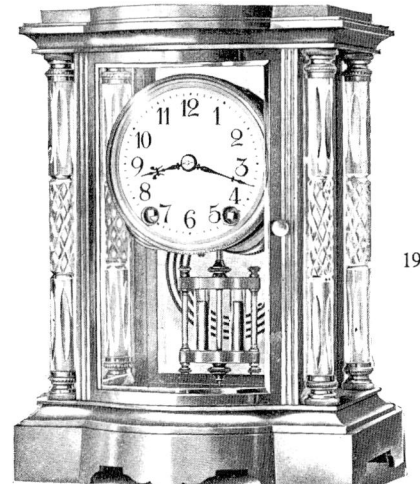

**EMPIRE No. 49**

Rich Gold, Convex Front.
Height, 9½ inches. Base, 7 inches.
With Metal Columns......List, $54.00
With Cut Glass Columns...List, 61.50

### 736

**EMPIRE No. 48**

Polished Gold, Convex Front.
Height, 8¾ inches.
List, $43.00

**EMPIRE No. 48 EXTRA**

Rich Gold, Convex Front.
Height, 8¾ inches.
List, $47.50

### 737

**EMPIRE No. 65**

Polished Gold, Convex Front.
Height, 11 inches. Base, 9½ inches.
With Metal Columns......List, $65.00
With Cut Glass Columns....List, 81.50
With Gold Decorated Cols.List, 86.50

**EMPIRE No. 65 EXTRA**

Rich Gold Case and Cols..List, 71.50

### 738
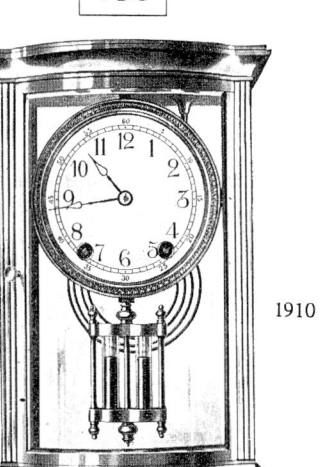

**EMPIRE No. 60**

Polished Gold, Convex Front.
Height, 11 inches.
List, $45.50

**EMPIRE No. 60 EXTRA**

Rich Gold, Convex Front.
Height, 11 inches.
List, $50.00

### 739

**EMPIRE No. 160**

Onyx Top and Base, Polished Gold Body, Convex Front.
Height, 11 inches.
List, $56.00

# SETH THOMAS

**740**

**Empire No. 23**
BRAZILIAN ONYX TOP AND BASE.
15 Day, Half Hour Strike.
Fine Polished Movement.
Hight, 12¾ inches.  Base, 8½ inches.
Price, $54.00

**741**

**EMPIRE No. 29**
Bronze Top and Base. Rich Gold Body. Fifteen Day, Fine Movement. 3½ inch Decorated Dial. Height, 17 inches. Base, 8½ inches.

List, $54.50

**742**

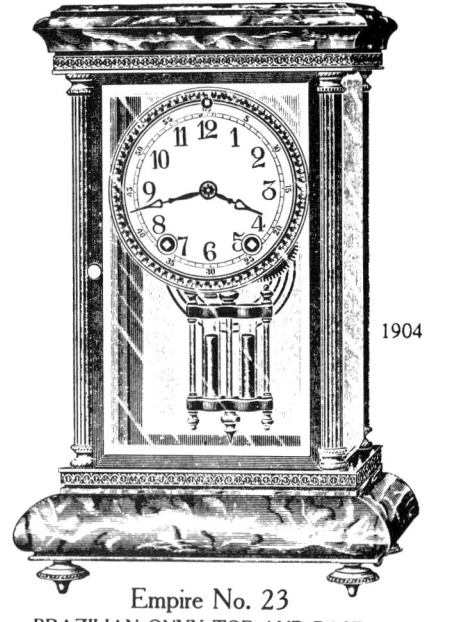

**EMPIRE No. 26**
Rich Gold Finish.
Height, 19½ inches.  Base, 10 inches.
Price of Clock only..........List, $54.50
Price of Clock with Lion.....List, 58.50

**743**

**EMPIRE No. 25**
Eight Day. Syrian Bronze Finish.
Height, 18½ inches.  Base, 10 inches.
List, $54.50

**744**

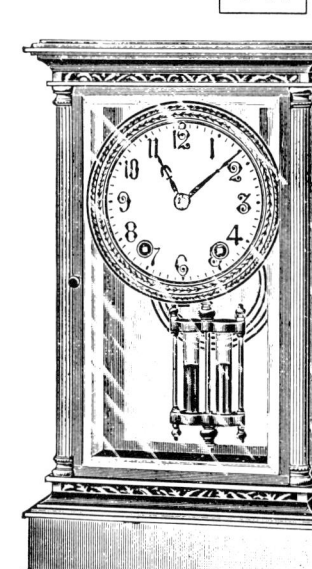

**EMPIRE No. 22**
Rich Gold Finish.
Height, 13½ inches.  Base, 7¾ inches.
List, $56.00

**745**

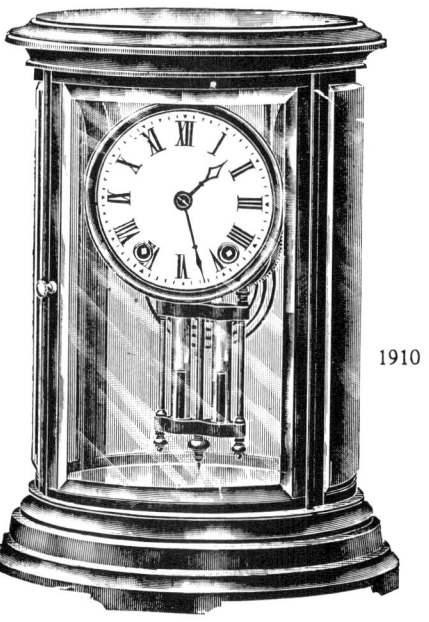

**ORCHID No. 6**
Polished Gold Finish.
Height, 11 inches.  Base, 8½ inches.
Oval Base and Top, fitted with Convex Beveled Glass on all sides. A plain design, but made of highest grade material and of best workmanship.
List, $55.00

# SETH THOMAS

**EMPIRE No. 7 WITH BUST**
Rich Gold Base, Top and Bust.
Polished Gold Upright.
Height, 17 inches.  Base, 8½ inches.
List, $33.00

**EMPIRE No. 7 WITH URN**
Rich Gold Base, Top and Urn.
Polished Gold Uprights.
Height, 15½ inches.  Base, 8½ inches.
List, $32.00

**EMPIRE No. 47**
Rich Gold Base and Top, Polished Gold Uprights.
Height, 10 inches.  Base, 6 inches.
List, $37.00

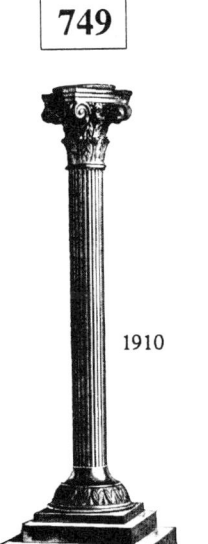

**No. 65 EMPIRE AND LION SET**

Convex Front Case.  Columns and Caps and Lion, Rich Gold, Highly Burnished.  Base of Candlesticks and Base and Top of Clock Polished Gold.
Also made in Verde Antique, Base, Top and Lion, with Columns Rich Gold and balance of parts Polished Gold.
Eight Day, Fine Movement, Half Hour Strike, Cathedral Bell.  4 inch Decorated Porcelain Dial.
Height of Clock, 14 inches; Base, 9½ inches.  Height of Candlesticks, 11½ inches; Base, 3¾ inches.

List Price of Clock, $68.50     List Price of Set, $87.50     List Price of Candlesticks, per pair, $19.00

**EMPIRE No. 16**
Rich Gold Ornaments and Polished Gold Uprights.
Height, 16 inches.  Base, 8 inches.
List, $43.00

# SETH THOMAS

**752**

*For comprehensive listing of Seth Thomas Clocks, see Tran Duy Ly "Seth Thomas Clocks & Movements - A Guide to Identification & Prices."*

1910

**EMPIRE No. 20**
Rich Gold Finish.
A Fine Pattern and Extra Fine Finish.
Height, 14 inches. Base, 8 inches.
List, $61.00

**753**

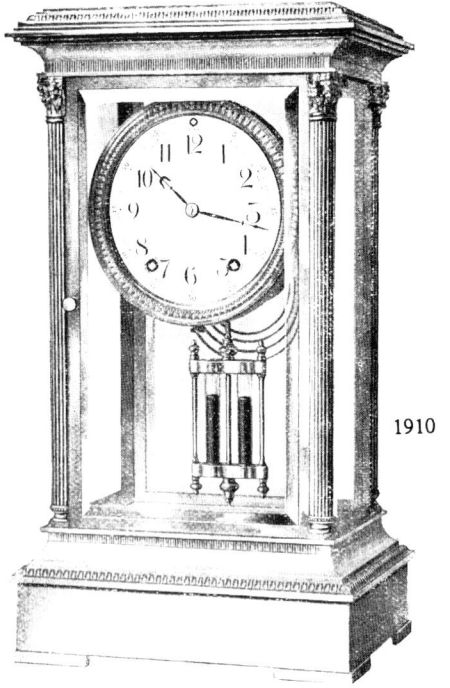

1910

Empire No. 10
Hight, 15¼ inches. Base, 8 inches.
Price, $35.00

**754**

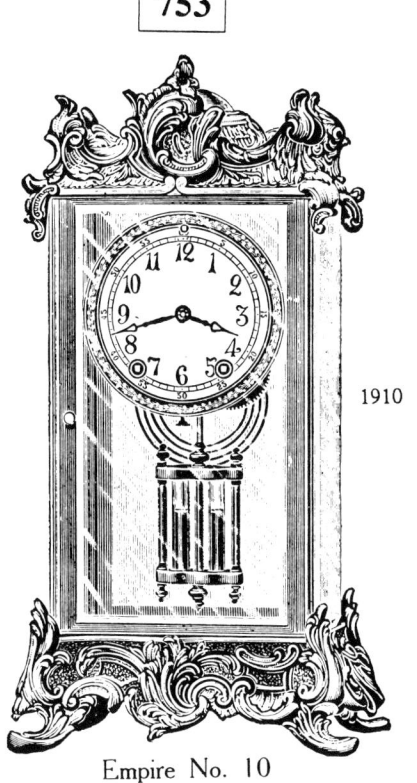

1910

**EMPIRE No. 11**
Rich Gold Finish.
Height, 14 inches. Base, 8 inches.
List, $31.50

**755**

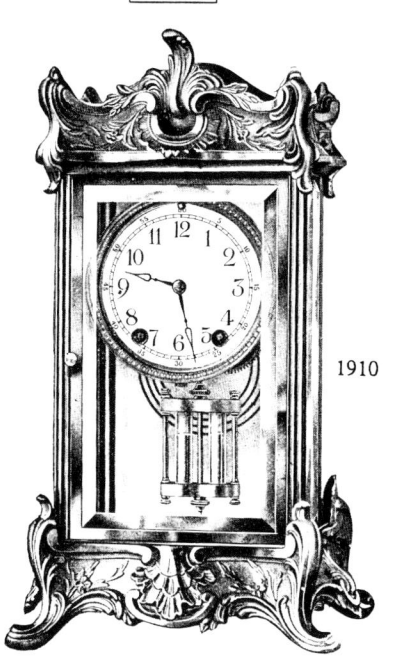

1910

**EMPIRE No. 15**
Rich Gold Finish.
Height, 15½ inches. Base, 8 inches.
List, $31.50

**756**

1910

**EMPIRE No. 12**
Rich Gold Finish.
Height, 14 inches. Base, 8 inches.
List, $31.50

**757**

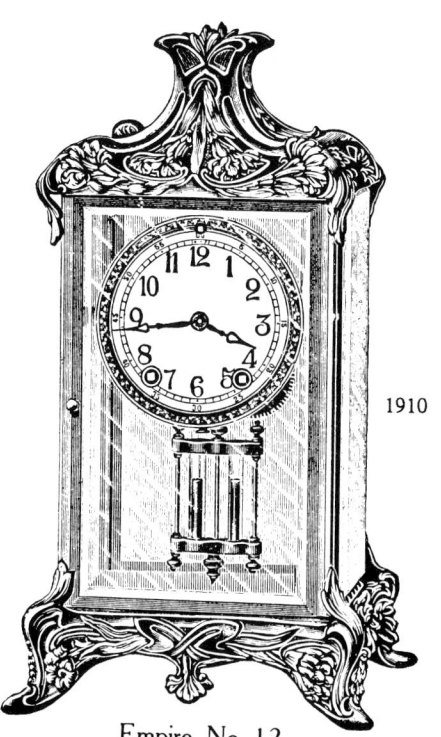

1910

Empire No 12
Hight, 14 inches. Base, 8 inches.
Price, $35.00

# SETH THOMAS

*For comprehensive listing of Seth Thomas Clocks, see Tran Duy Ly "Seth Thomas Clocks & Movements - A Guide to Identification & Prices."*

**758**

**759**

**760**

**761**

1884

1879

1864

1864

REGULATOR No. 1.
8 Day, Weight, Time.
Hight, 34 inches.
12 inch Dial.

Movement 4¼ x 5¼ inches, cut pinions of solid steel, Graham pallets, brass covered zinc ball, and wood rod. Eighty beats to the minute. Movement has retaining power.

REGULATOR No. 1.
8 Day, Weight, Time.
Hight, 34 inches.
12 inch Dial.

REGULATOR No. 1.
8 Day, Weight, Time.
Hight, 34 inches.
12 inch Dial.

REGULATOR No. 2.
8 Day, Weight, Time.
Hight, 34 inches.   12 inch Dial.

# SETH THOMAS

The year entered beside each clock is the year of the catalog from which the illustration was taken. Some clocks may have been produced a few years earlier.

For comprehensive listing of Seth Thomas Clocks, see Tran Duy Ly "Seth Thomas Clocks & Movements - A Guide to Identification & Prices."

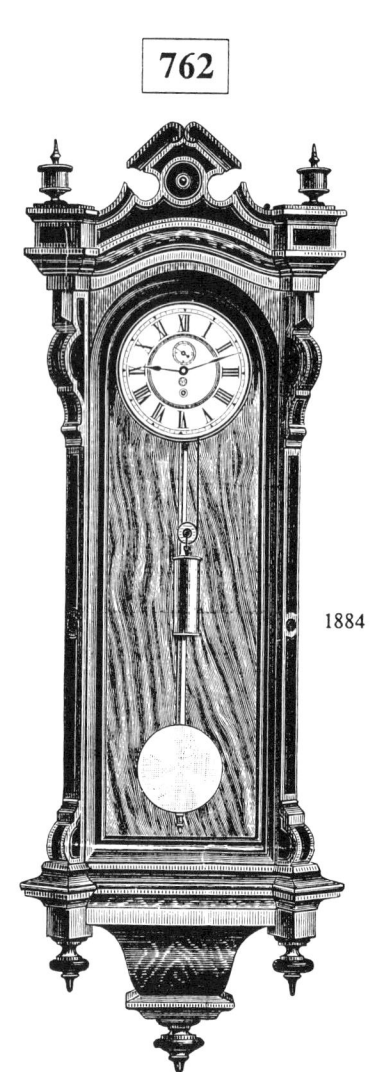

1884

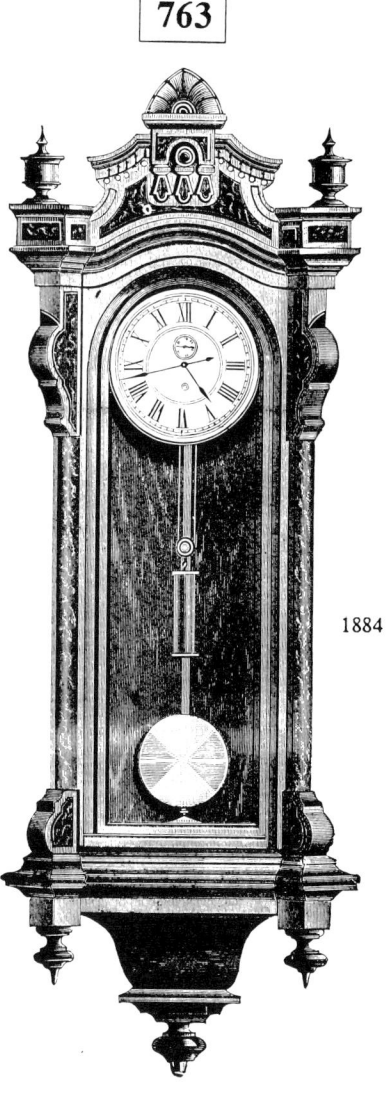

1884

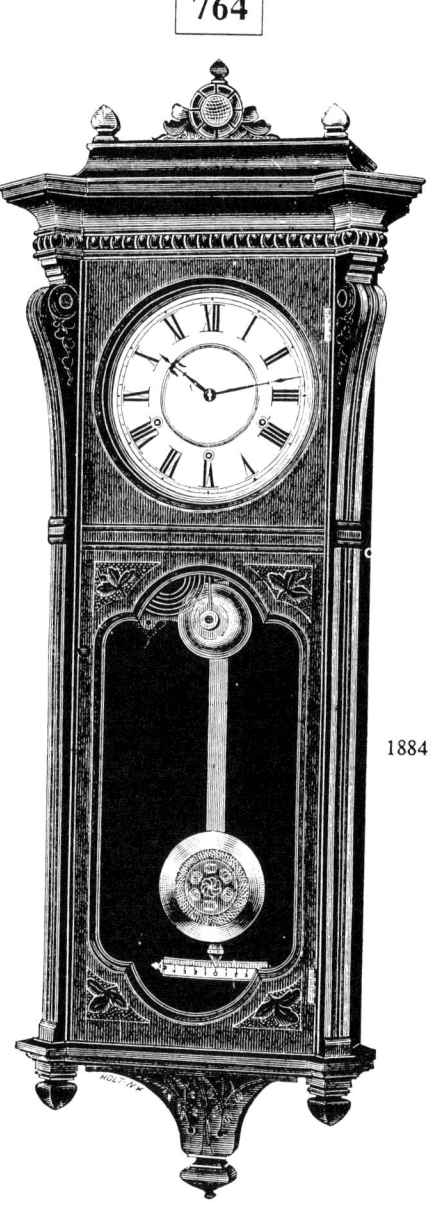

1884

### REGULATOR No. 4.
WALNUT CASE.

Hight, 47 inches.

8 Day, Weight, Time.

7 inch Porcelain Dial.

Movement 3¼ x 4½ inches, finished same as movement in Regulator No. 3, and hung on brass bracket. It has a fine enamel dial with sunk centre attached to movement, and mountings are gilded. Movement has retaining power.

### REGULATOR No. 5.

8 Day, Weight, Time.

Hight, 50 inches.

7½ inch Porcelain Dial.

Movement same as Regulator No. 4.

### MARCY
QUARTER STRIKE.

8 Day, Spring.

Dial, 8½ inches.

Hight, 46 inches.

Striking quarter hours on two Cup Bells and hours on a Cathedral Bell.

MADE IN WALNUT, OAK AND CHERRY.

# SETH THOMAS

*For comprehensive listing of Seth Thomas Clocks, see Tran Duy Ly "Seth Thomas Clocks & Movements - A Guide to Identification & Prices."*

### 765

1884

**QUEEN ANNE No. 1.**
Walnut Case. Oil Finish. Height, 36 inches.
8½-inch Dial.

No. 6052. Eight Day, Spring, Time........$13 50
  "    "    "    "    Strike...... 15 00
Also in Ebony finish, with a very superior movement and a Cathedral Bell, $21 00.

### 766

1884

**QUEEN ANNE.**
Hight, 36 inches.
8½ inch Dial.
Walnut Case.
8 Day, Spring, Time,
8 Day, Spring, Strike,
Made with Cathedral Bell and a very superior Movement,
in Ebony, Oak,
Mahogany and Cherry.

### 767

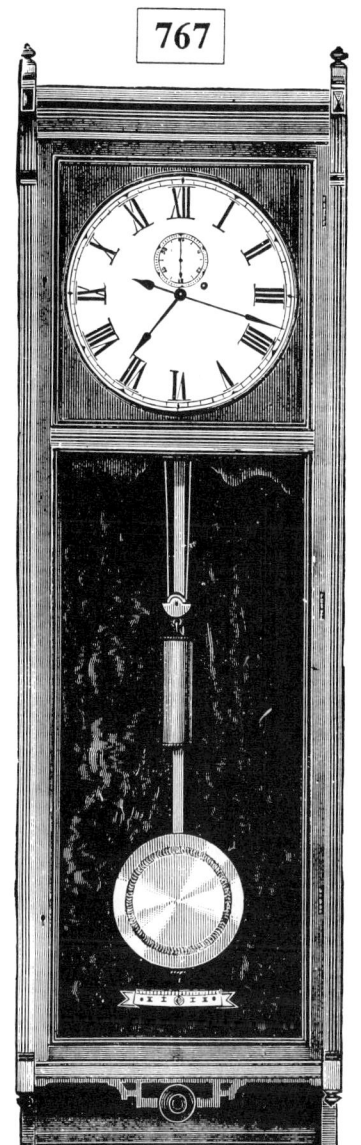

1884

**REGULATOR No. 17.**

MADE IN WALNUT AND OAK.

Made to order to indicate Double Time, same as Regulator No. 6, Double Time

8 Day, Weight, Time. Beats Seconds. Hight, 68 inches. 14 inch Dial.

MOVEMENT SAME AS IN No. 16 REGULATOR,
BUT INDEPENDENT SECONDS.

1910

**REGULATOR No. 17**

Cherry or Old Oak. Eight Day, Weight, Time. Beats Seconds. 14 inch Dial. Height, 68 inches. Large cut steel pinion movement, polished plates, wood rod, brass-covered zinc ball, Graham escapement, and maintaining power.

List, **$85.00**

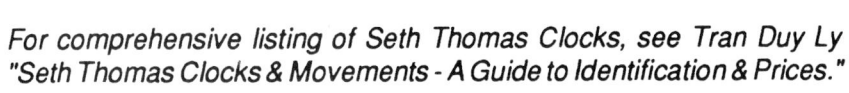

*For comprehensive listing of Seth Thomas Clocks, see Tran Duy Ly "Seth Thomas Clocks & Movements - A Guide to Identification & Prices."*

# SETH THOMAS

*The year entered beside each clock is the year of the catalog from which the illustration was taken. Some clocks may have been produced a few years earlier.*

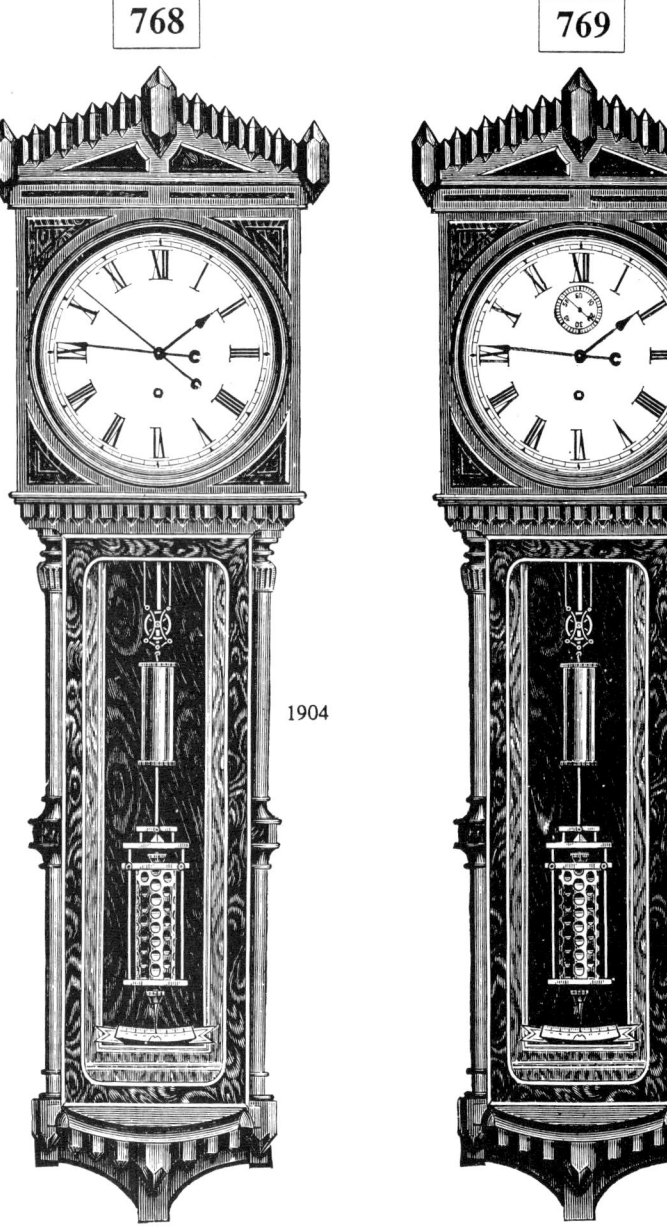

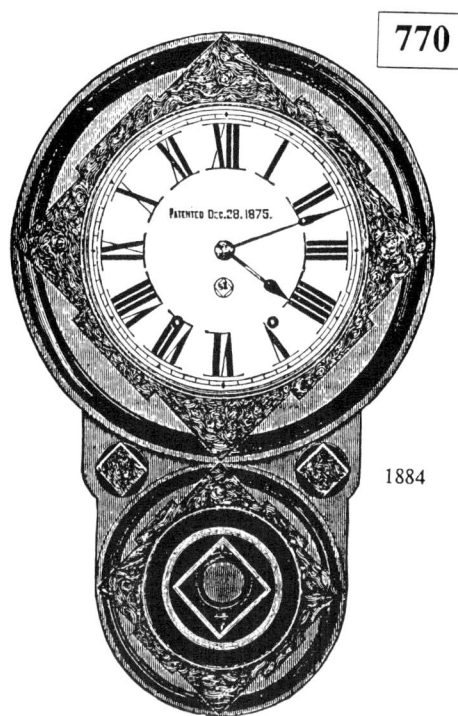

SIGNET.
8 Day, Spring, Time.
8 Day, Spring, Strike.
Hight, 23 inches.
10 inch Dial.

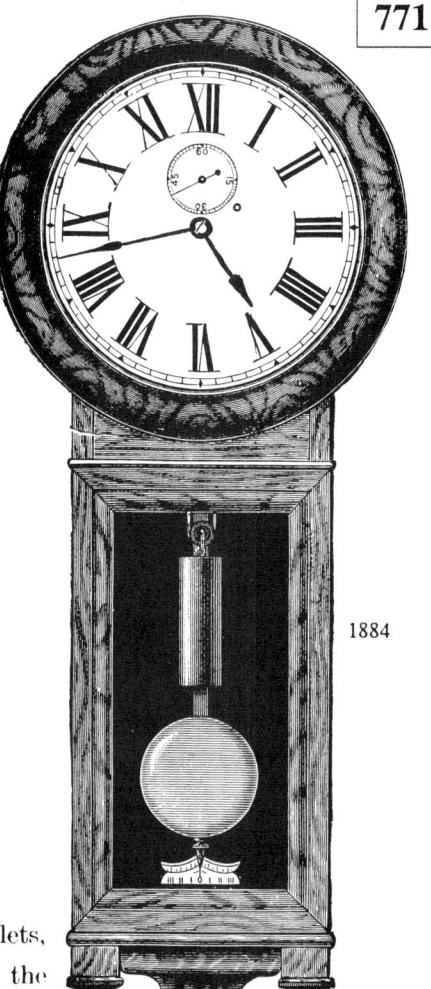

**REGULATOR No. 10**
Sweep Seconds.

**REGULATOR No. 12**
Independent Seconds.

8 Day, Weight, Time.
Hight, 72 inches. 14 inch Dial.
Price, $325.00          Price, $325.00

### REGULATOR No. 2.
8 Day, Weight, Time.
Hight, 34 inches. 12 inch Dial.

Movement 2⅜ x 4¼ inches, lantern pinions, Graham pallets, brass covered zinc ball, and wood rod. Eighty beats to the minute. Movement has retaining power.

# SETH THOMAS

*For comprehensive listing of Seth Thomas Clocks, see Tran Duy Ly "Seth Thomas Clocks & Movements - A Guide to Identification & Prices."*

**772**

**773**

**774**

### LUNAR
Mahogany or Old Oak. Eight Day, Strike, Cathedral Bell. 12 inch Dial. Height, 41 inches. Dial shows the Moon's Phases. Clock is wound by pulling center chains.
List, $38.00

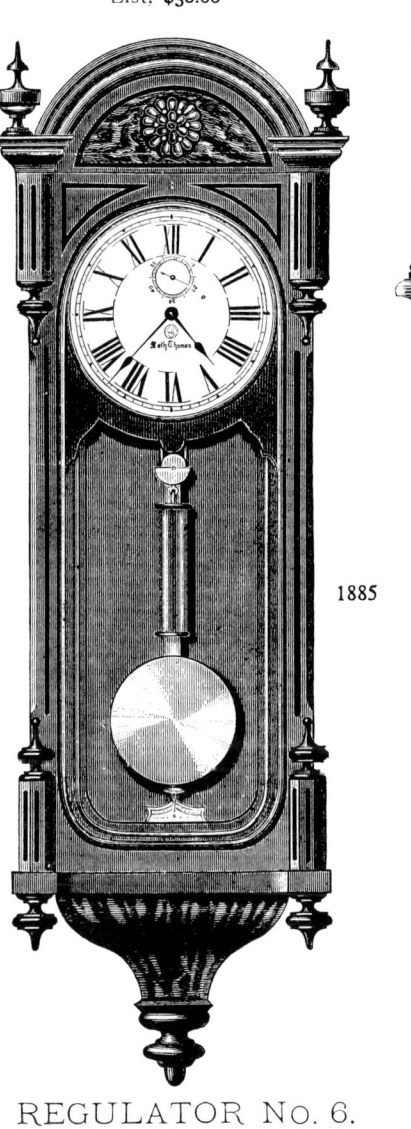

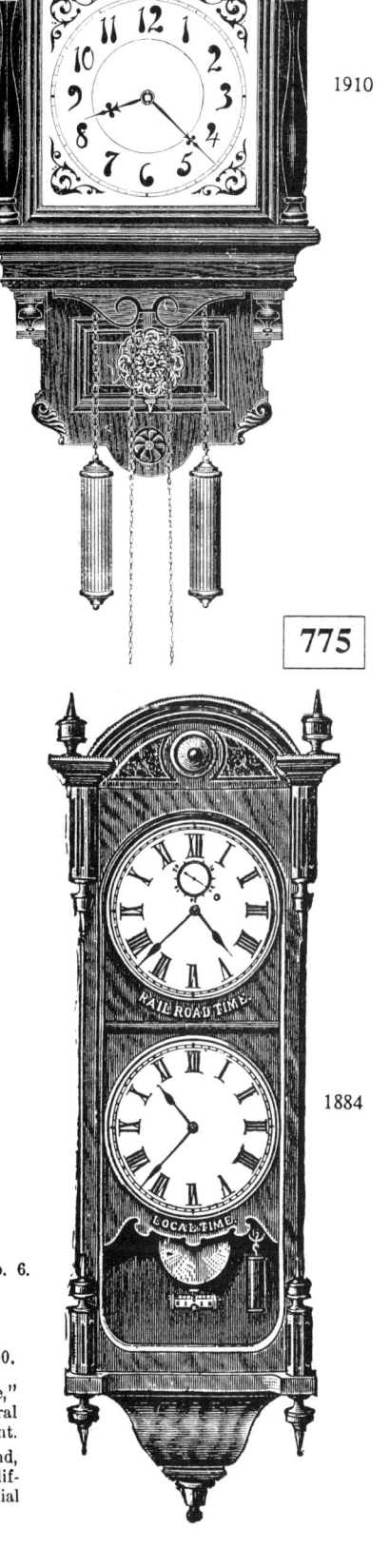

1910

1885

1910

**775**

1884

### REGULATOR No. 6.
WALNUT CASE.
8 Day, Weight, Time.
Hight, 49 inches.   10 inch Dial.
Movement same as Regulator No. 2.

### JUPITER
Mahogany and Old Oak. Eight Day, Strike, Cathedral Bell. 12 inch Dial. Height, 59 inches. Dial shows the Moon's Phases. Clock is wound by pulling the center chains.
List, $60.00

### SETH THOMAS REGULATOR No. 6.
Double Time.   Walnut Case.
Height, 49 inches.   10-inch Dials.

No. 6077. 8 Day, Time, Weight, $35 00.

Serves to indicate "Railroad Time," "Local Time," "Eastern Time," "Central Time," or any other similar requirement.

Dials are run by one movement, and, being once set, will maintain relative difference. Moving hands on upper dial will change lower dial to same extent.

# SETH THOMAS

Except for those fabulous clocks for which you wouldn't mind cutting a hole in the ceiling, Grandfather or Hanging Clocks under eight feet are the most desirable and practical. Some manufacturers produced certain of these clocks in two styles: one to stand on the floor, and the other to hang on a wall. Of the clocks designed with two styles, the hanging style is considered to be the most valuable.

The year entered beside each clock is the year of the catalog from which the illustration was taken. Some clocks may have been produced a few years earlier.

776

1884

### FLORA.

MADE IN WALNUT, EBONY, OAK,
MAHOGANY AND CHERRY.
CASE HAND CARVED.

8 Day, Weight, Strike.
Cathedral Bell.

8 inch Dial.
Brass Weights. Wooden Rod.

Hight, 38 inches.

NEW AND SUPERIOR MOVEMENT.

777

1884

### SUEZ.

MADE IN WALNUT, EBONY, OAK AND CHERRY.

8 Day, Spring, Strike, Cathedral Bell.
Hight, 44 inches.

A Superior Movement. Wooden Rod.
8½ inch Dial.

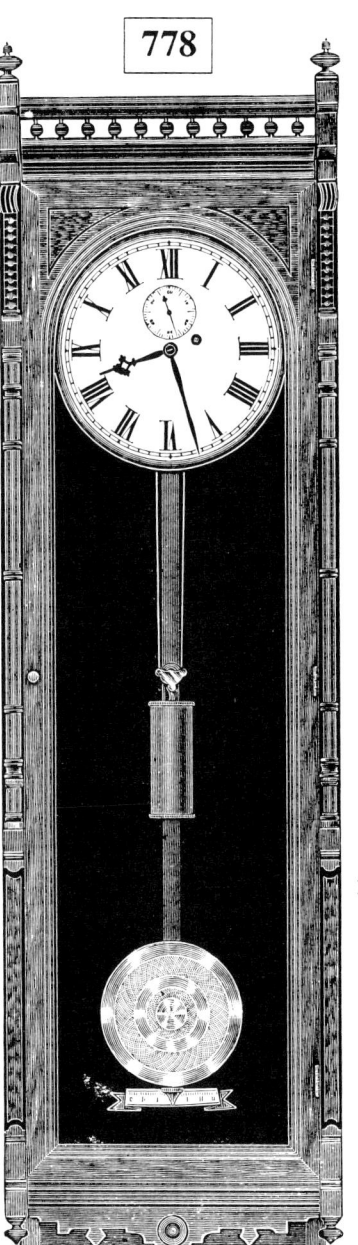

778

1910

### REGULATOR No. 32

Made in Old Oak only. Eight Day, Weight. Time. Beats Seconds. 12 inch Dial. Height, 68 inches. Large cut steel pinion movement, with polished plates, wood rod, brass-covered zinc ball. Graham dead-beat escapement, and maintaining power.

List, $100.00

# SETH THOMAS

For comprehensive listing of Seth Thomas Clocks, see Tran Duy Ly
"Seth Thomas Clocks & Movements - A Guide to Identification & Prices."

### 779

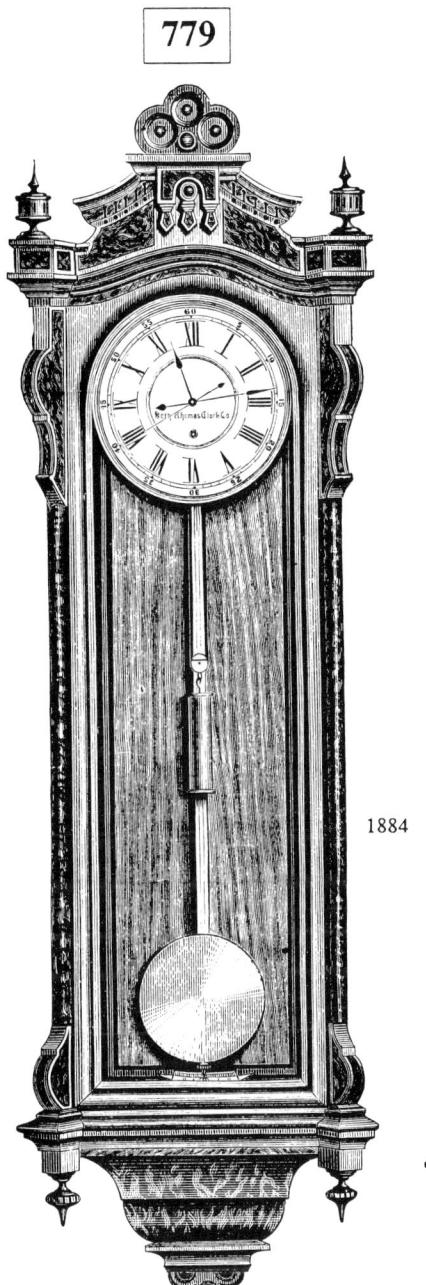

1884

**REGULATOR No. 16.**

MOVEMENT HAS CUT PINIONS
OF SOLID STEEL,

8 Day, Weight, Time.

PALLETS, BRASS COVERED
ZINC BALL. WOOD ROD.
POLISHED PLATES, GRAHAM

Hight, 75 inches.   12 inch Dial.

Beats Seconds.

### 780

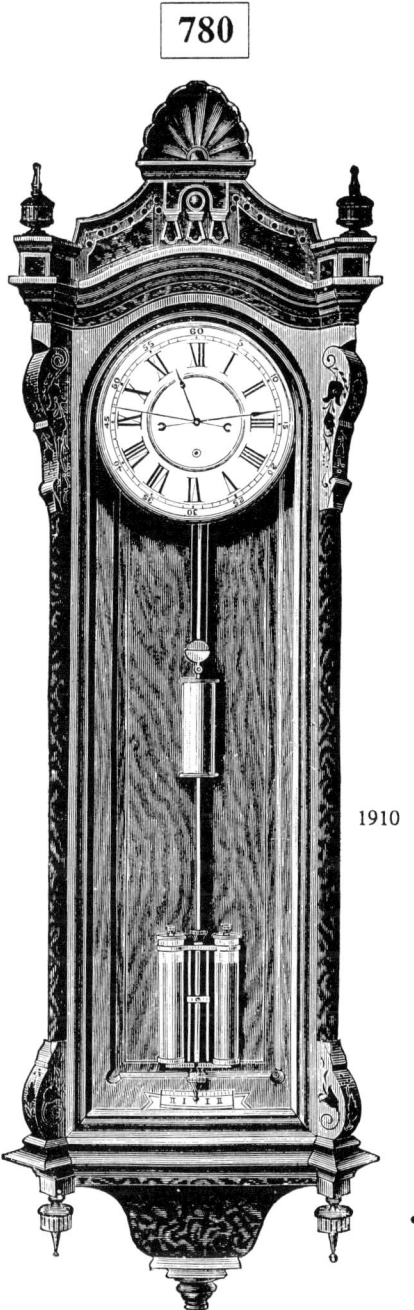

1910

**REGULATOR No. 19**

(Sweep Seconds)

Mahogany and Old Oak. Eight Day, Weight, Time. Beats Seconds. 12 inch Dial. Height, 75 inches. Large cut steel pinion movement, with polished plates, wood rod, Mercury pendulum, Graham dead-beat escapement, maintaining power. An accurate timepiece for Railroads, Public Buildings, etc.

List, $225.00

### 781

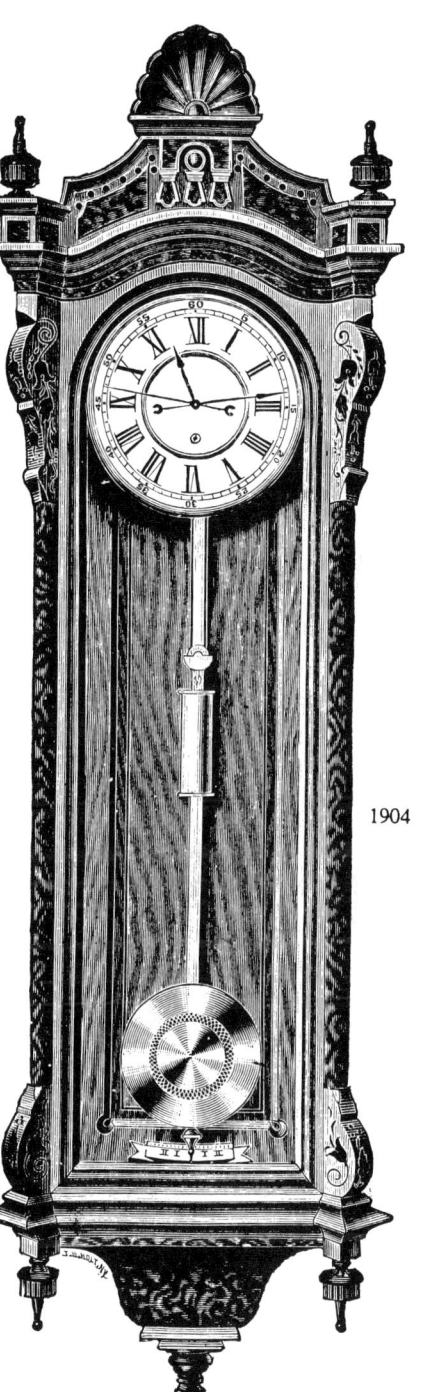

1904

**REGULATOR No. 16**

WALNUT. OAK OR OLD OAK.

8 Day, Weight, Time. Beats Seconds.

Hight, 75 inches.   12 inch Dial.

Movement has cut pinions of solid steel, polished plates, Graham pallets, brass covered zinc ball, wood rod.

Price, $125.00

# SETH THOMAS

*To receive top prices, clocks must be in ORIGINAL MINT condition and in good running order - A replaced or damaged dial, missing parts, will reduce prices accordingly.*

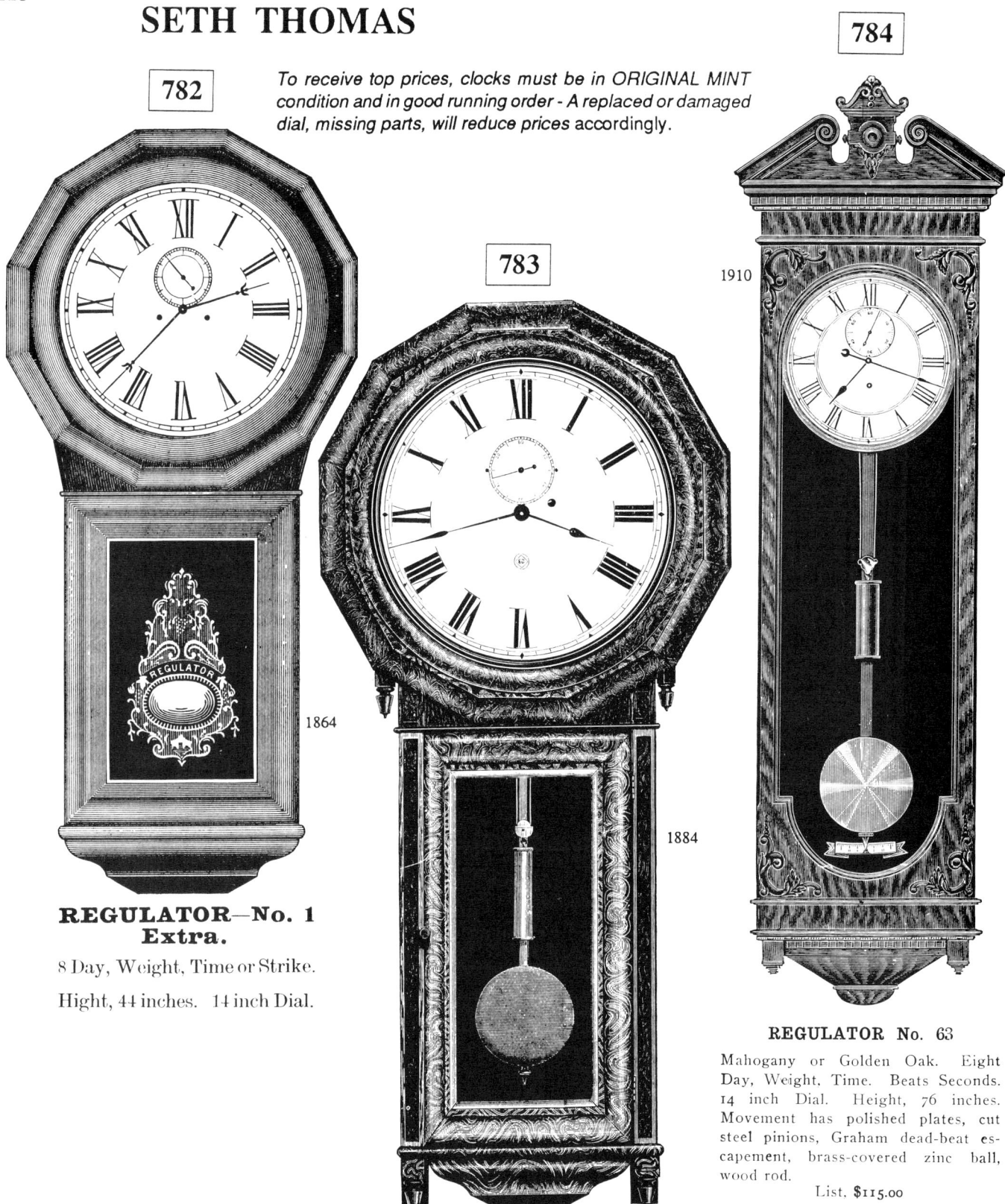

**REGULATOR—No. 1 Extra.**

8 Day, Weight, Time or Strike.

Hight, 44 inches. 14 inch Dial.

**REGULATOR No. 63**

Mahogany or Golden Oak. Eight Day, Weight, Time. Beats Seconds. 14 inch Dial. Height, 76 inches. Movement has polished plates, cut steel pinions, Graham dead-beat escapement, brass-covered zinc ball, wood rod.

List. $115.00

REGULATOR No. 3.
ROSEWOOD OR WALNUT VENEER, POLISHED.

8 Day, Weight, Time or Strike. Hight, 44 inches. 14 inch Dial.

This movement 4¼ x 5¼ inches, polished, cut pinions of solid steel, Graham pallets, brass covered zinc ball, wood rod. Seventy-two beats to the minute.

Strike movement has lantern pinions, plates not polished; seventy-six beats to the minute. Both movements have retaining power.

219

# SETH THOMAS

*For comprehensive listing of Seth Thomas Clocks, see Tran Duy Ly "Seth Thomas Clocks & Movements - A Guide to Identification & Prices."*

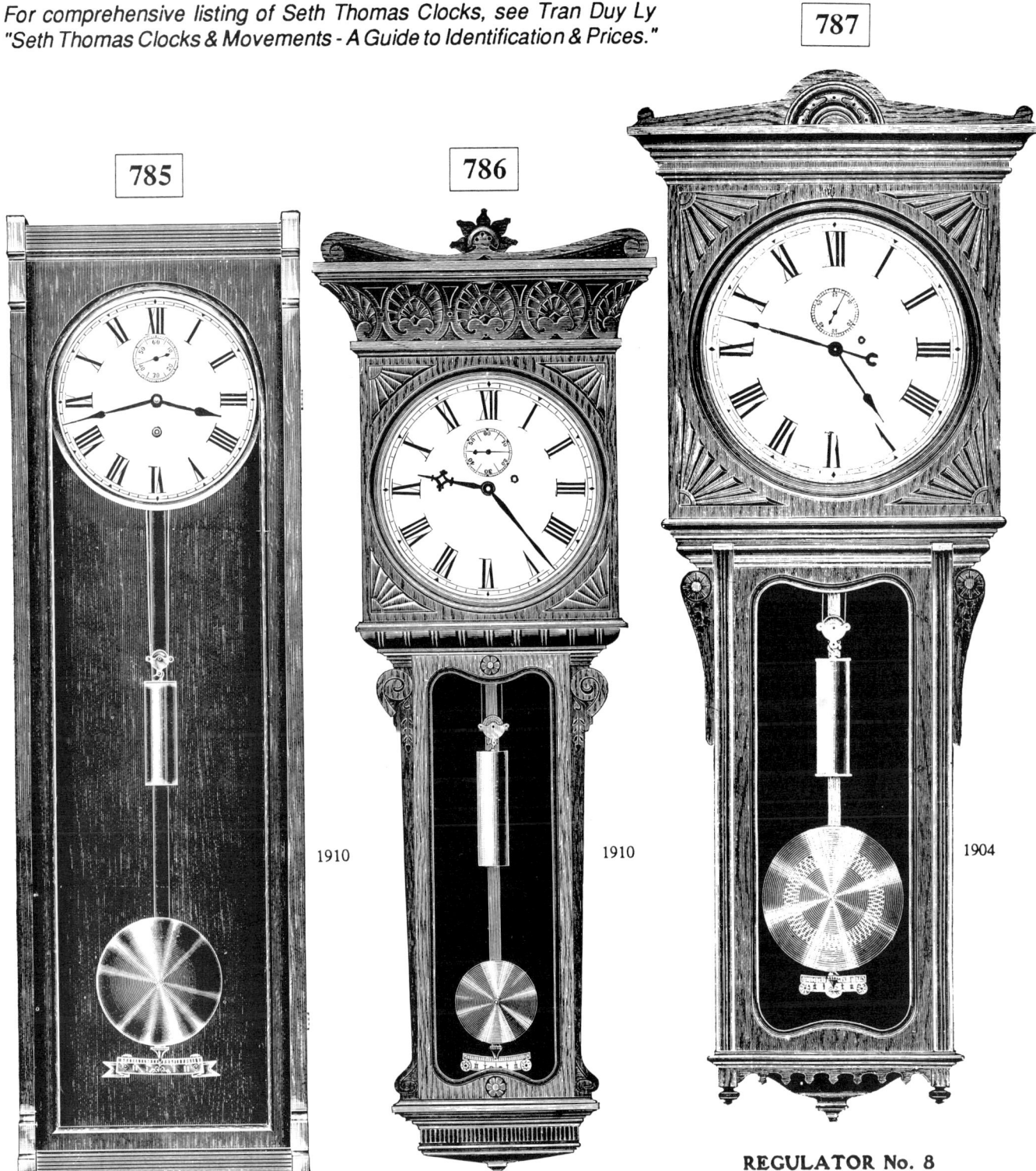

**REGULATOR No. 11**

Mahogany and Golden Oak. Eight Day, Weight, Time. Beats Seconds. 12 inch Dial. Height, 56 inches. Large, cut steel pinion movement, with polished plates. Wood rod, brass-covered zinc ball. Graham dead-beat escapement, and maintaining power.

List, $75.00

**REGULATOR No. 7**

Cherry or Walnut Hand Carved Cabinet Case. 8 Day, Weight, Time. 12 inch Dial. Height, 45 inches. Cut steel pinion movement, with polished plates. Wood rod, brass-covered zinc ball. Graham dead-beat escapement, maintaining power. Seventy-two beats to the minute.

List, $60.00

**REGULATOR No. 8**

MADE IN WALNUT ONLY.

8 Day, Weight, Time. Hight, 56 inches. 14 inch Dial. Polished movement; cut steel pinions, Graham pallets, wood rod, brass covered ball; retaining power; 68 beats to the minute.

Price, $50.00

# SETH THOMAS

### 14 INCH LOBBY
Mahogany and Old Oak. 14 inch Dial. Height, 30½ inches. Width, 20 inches.
List, $27.00

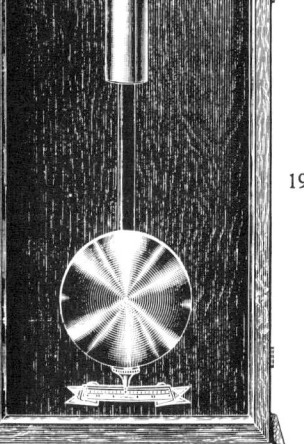

### REGULATOR No. 20
Cherry, Mahogany Finish and Old Oak. Eight Day, Weight, Time. Beats Seconds. 14 inch Dial. Height, 62 inches. Large movement, wood rod, brass-covered zinc ball, Graham dead-beat escapement, and maintaining power.

The round top and two side-pieces can be removed if a plain top case is desired.

List, $55.00

### OFFICE No. 6

### REGULATOR No. 31
Mahogany and Old Oak. Eight Day. Weight. Time. Beats Seconds. 18 inch Dial. Height, 68 inches. Roman or Arabic. Large cut steel pinion movement, with polished plates, wood rod, brass-covered zinc ball, Graham dead-beat escapement, and maintaining power. Suitable for a very large room.

List, $100.00

### OFFICE No. 6
Mahogany Finish and Old Oak. Eight Day, Spring. Time only. 12 inch Dial. Height, 36 inches.
Time .............. List, $8.00
Top can be taken off if a plainer case is preferred.

# SETH THOMAS

**792**

1910

### GALLERY
Mahogany, Old Oak and Gold Gilt. 18 inch Dial. Back measurement, 25½ inches. 24 inch Dial. Back measurement, 32 inches.

| | |
|---|---:|
| 18 inch Mahogany or Old Oak | $35.00 |
| 18 inch Gilt | 54.00 |
| 24 inch Mahogany or Old Oak | 50.00 |
| 24 inch Gilt | 65.00 |

**793**

1910

### REGULATOR No. 25
Flemish Oak. Eight Day, Weight Time. 12 inch Dial. Height, 32 inches. Movement same as in Regulator No. 2. The design and finish of this case is suitable to go with any Mission furniture in Flemish, Fumed or Weathered Oak.

List, $27.00

**794**

1910

### OFFICE No. 5
Made in Old Oak only.
Eight Day, Pendulum. 12 inch Dial.
Height, 23¼ inches. Width, 19¼ inches.

| | |
|---|---:|
| Time | List, $10.80 |
| Strike | List, 12.40 |

**795**

1910

### LITCHFIELD

### LITCHFIELD
Fitted with a Thirty Day Time, Large Double Spring Movement, with Graham Dead-beat Escapement. 12 inch Dial. Height, 31 inches. Mahogany Finished Case. Gold Leaf Border on Upper and Lower Glasses.

List, $21.50

*For comprehensive listing of Seth Thomas Clocks, see Tran Duy Ly "Seth Thomas Clocks & Movements - A Guide to Identification & Prices."*

# SETH THOMAS

**796**

1910

### DOUBLE DIAL CLOCK

Hand Carved Cabinet Case. Made in Mahogany and Golden Oak. Two 18 inch Dials, placed back to back, so that time can be read from two sides. Case arranged to hang, to project from wall, or stand. When ordering, specify which way Clock is to be used. Height, 29 inches. Base, 43 inches.
List, **$170.00**

**797**

1910

### REGULATOR No. 30

Cherry or Old Oak. Eight Day, Weight, Time. 12 inch Dial. Height, 49 inches. Movement same as in Regulator No. 2.
List. **$47.50**

**798**

1910

### OFFICE No. 1.

ROSEWOOD.

Spring.

12 inch dial. Height, 25 inches.

EIGHT DAY, TIME. EIGHT DAY, STRIKE.

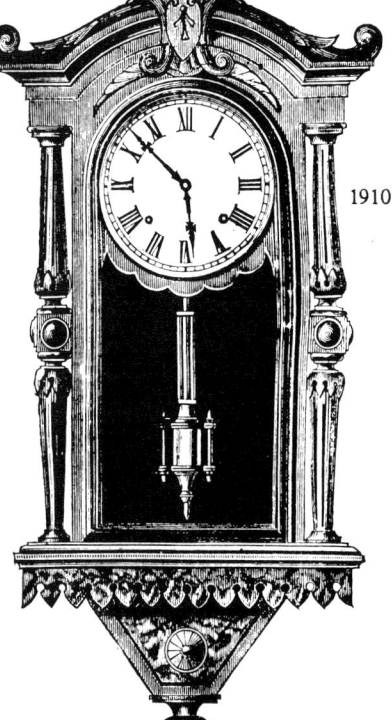

**799**

1910

### REGULATOR No. 30

### REGULATOR No. 30

WALNUT.

Spring.

Height, 37 inches.

EIGHT DAY, TIME OR STRIKE.

Seth Thomas movement.

# SETH THOMAS

### 800

**REGULATOR No. 60**

GOLDEN OAK OR MAHOGANY.

8 Day, Weight, Time.
Beats Seconds.

Hight, 58½ inches.  14 inch Dial.

Movement has cut pinions of solid steel, finished plates, Graham escapement, brass covered zinc ball, wood rod.

Price, $87.50

### 801

**REGULATOR No. 62**

MAHOGANY OR GOLDEN OAK.

8 Day, Weight, Time.  Beats Seconds.
Hight, 60 inches.
14 inch Painted Steel Dial.
Large lantern pinion movement, Graham pallets, retaining power, brass covered zinc ball, wood rod, brass weight.

Price, $56.00

### 802

1904

### 801

1910

**BRIGHTON**

Made in Mahogany only.  Eight Day.
Spring, Strike.  12 inch Dial.
Height, 22¼ inches.
List, $9.00

### 803

**RIO**

Mahogany Veneer, Polished.  Eight Day,
Spring, Strike.  12 inch Dial.
Height, 25¼ inches.
List, $10.50

# SETH THOMAS

*For comprehensive listing of Seth Thomas Clocks, see Tran Duy Ly "Seth Thomas Clocks & Movements- A Guide to Identification & Prices."*

### GLOBE
Mahogany and Old Oak Veneer, Polished. Eight Day, Spring, Wire Bell. 12 inch Dial. Height, 31 inches.
Time ............List, $11.00
Strike ..........List, 12.60

GLOBE

### REGULATOR No. 18
Mahogany and Old Oak Veneer, Polished. Eight Day, Weight, Time. 14 inch Dial. Height, 54 inches. Beats Seconds. Movement has polished plates and is secured to an iron bracket, and pendulum is suspended from same. Wood rod, brass-covered zinc ball, Graham dead-beat escapement, maintaining power.
List, $40.00

### WORLD
Mahogany and Old Oak Veneer. Fifteen Day, Time. Has a large double spring movement with Graham dead-beat escapement. An excellent timepiece.
Eight Day, Spring, Strike, Cathedral Bell. Height, 32 inches. 12 inch Dial.
Time ..........List, $14.00
Strike ..........List, 16.00

### 8 INCH DROP OCTAGON
Mahogany Finish and Fumed Oak. Eight Day Spring. Height, 17 inches.
Time ...................List, $5.40
Strike ..................List, 7.00

# SETH THOMAS

**808**

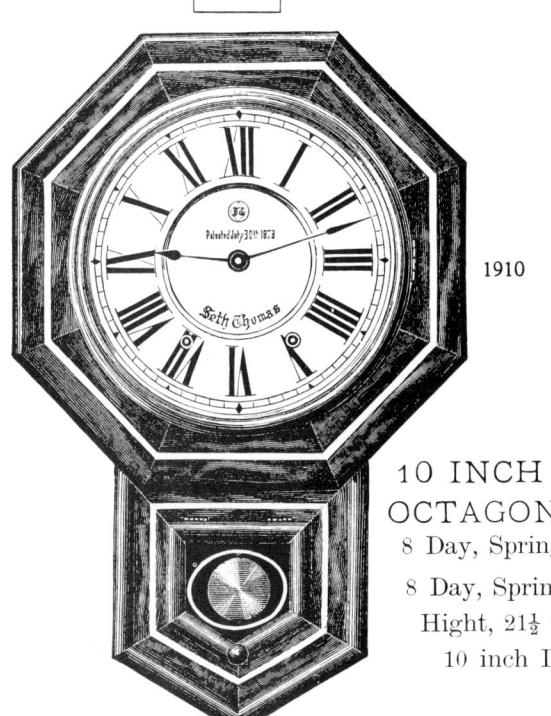

1910

**10 INCH DROP OCTAGON, GILT.**
8 Day, Spring, Strike.
8 Day, Spring, Time.
Hight, 21½ inches.
10 inch Dial.

**809**

1910

**OFFICE No. 3**
Mahogany Veneer, Polished. Eight Day, Spring, Strike. 10 inch Dial.
Height, 21½ inches.
List, $9.50

**810**

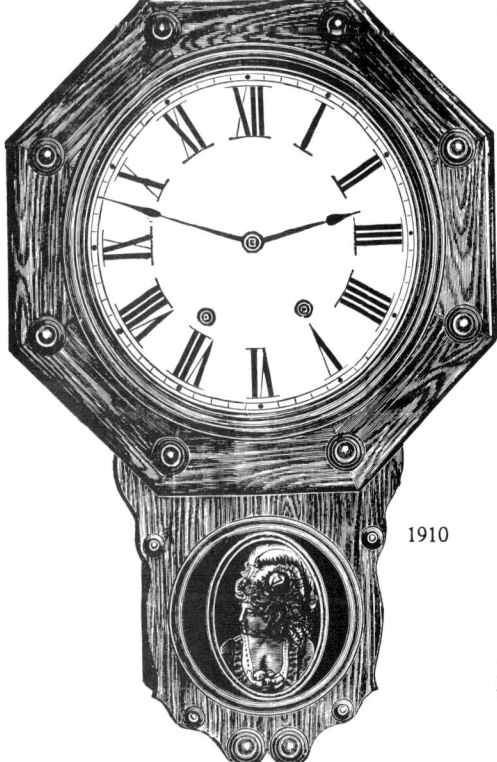

1910

*For comprehensive listing of Seth Thomas Clocks, see Tran Duy Ly "Seth Thomas Clocks & Movements - A Guide to Identification & Prices."*

**811**

1910

**OFFICE No. 2**
Mahogany Veneer, Polished. Eight Day, Spring. 12 inch Dial. Height, 26 inches.
Time .................... List, $8.00
Strike ................... List, 9.60

**12 INCH DROP OCTAGON**
Mahogany Veneer and Old Oak, Polished. Eight Day, Spring.
Height, 23½ inches.
Time .................... List, $7.80
Strike ................... List, 9.40

# SETH THOMAS

### ARCADE

Fitted with a Thirty Day Time. Large Double Spring Movement, with Graham Dead-beat Escapement. 18 inch Dial. Mahogany Finish and Old Oak Cases. Front measure over all, 23½ inches. Back is an oblong box arrangement for movement, and measures 19 inches high, 10 inches wide, and 4 inches deep. The entire depth from front to back is 5½ inches.

List, $21.50

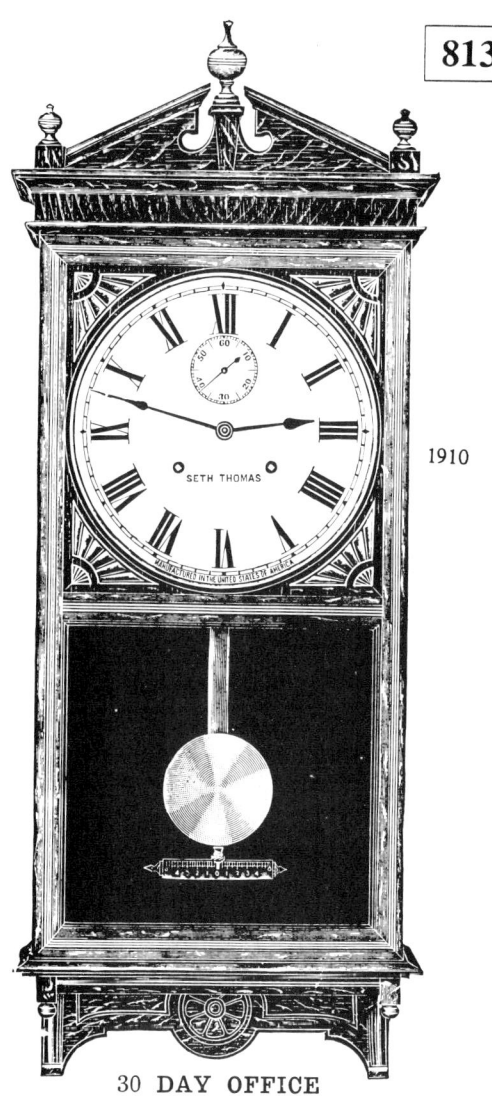

### 30 DAY OFFICE

Large Double Spring Time Movement, with Graham Dead-beat Escapement. 12 inch Dial. Height, 42 inches. Fine Hand Carved Cabinet, Old Oak Case.

List, $21.50

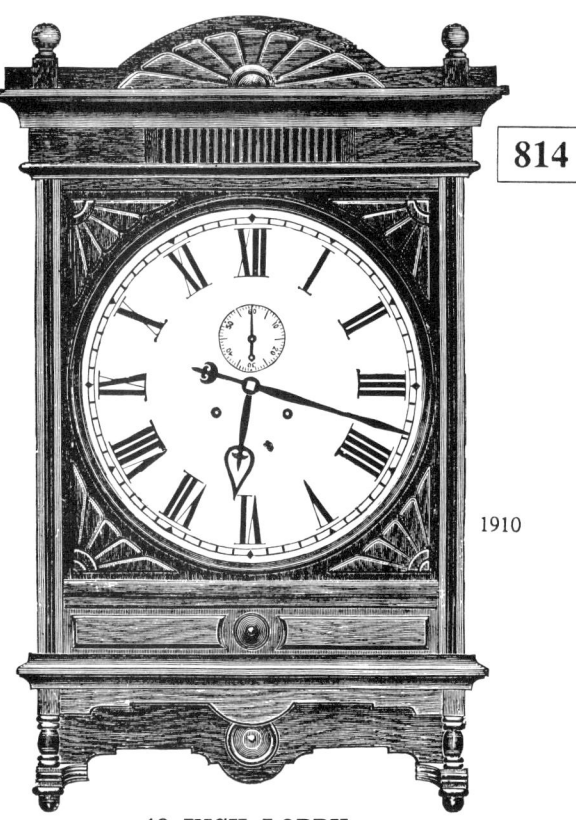

### 18 INCH LOBBY

Mahogany and Old Oak. 18 inch Dial. Height, 38 inches. Width, 25 inches.

List, $35.00

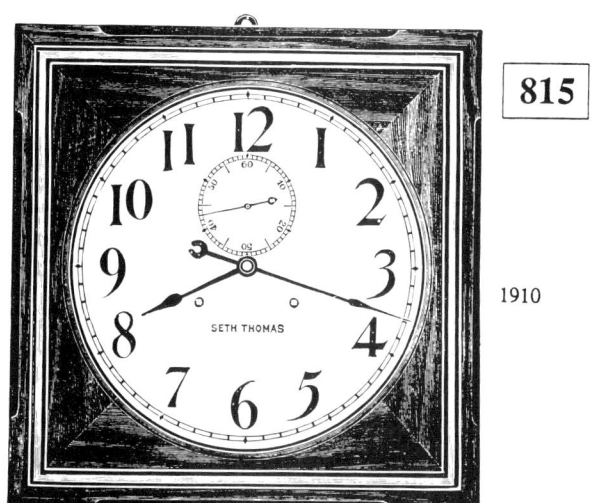

### HUDSON

Fitted with a Thirty Day Time. Large Double Spring Movement, with Graham Dead-Beat Escapement. 14 inch Dial. Case, 18 inches square. Mahogany inlaid.

List, $21.50

# SETH THOMAS

*The year entered beside each clock is the year of the catalog from which the illustration was taken. Some clocks may have been produced a few years earlier.*

**816**

1910

**GOTHIC No. 0**
Real Bronze.
Fifteen Day Strike. 3¼ inch Dial.
Height, 9 inches. List, $35.00
Made also in brass finish.
Same price.

Fifteen Day, High Grade Movement, with Round, Polished Plates, Cut Steel Pinions and Springs in Barrels. Hour and Half Hour Strike on Cathedral Bell.

**817**

1910

**GOTHIC No. 00**
Real Bronze.
2¾ inch Dial. Height, 5½ inches.
This size only has 7 jewel lever, time movement and etched dial, described on page 22.
List, $35.00
Made also in brass, brush finish.
Same price.

**818**

1910

**GOTHIC No. 1**
Real Bronze.
Fifteen Day Strike. 5 inch Dial.
Height, 11¾ inches.
List, $40.00

**819**

1910

**GOTHIC No. 2**
Real Bronze.
Fifteen Day Strike. 6 inch Dial.
Height, 14 inches.
List, $50.00

**820**

1910

**GOTHIC No. 3**
Real Bronze.
Fifteen Day Strike. 8 inch Dial.
Height, 18 inches. Depth, 8½ inches.
List, $70.00

# SETH THOMAS

*For comprehensive listing of Seth Thomas Clocks, see Tran Duy Ly "Seth Thomas Clocks & Movements - A Guide to Identification & Prices."*

Fifteen Day, High Grade Movement with Round, Polished Plate, Cut Steel Pinions and Springs in Barrels. Hour and Half Hour Strike on Cathedral Bell.

1910

1910

1910

**DORIC No. 0**
Real Bronze.
Fifteen Day Strike. 3¾ inch Dial.
Height, 8 inches.
List, $35.00
Made also in brass, brush finish.
Same price.

**DORIC No. 00**
Real Bronze.
Eight Day Time in this case. 2¾ inch Dial. Height, 5 inches.
This size only, has 7 jewel, lever, time movement and etched dial described on page 22.
List, $35.00
Made also in brass, brush finish. Same price.

**DORIC No. 1**
Real Bronze.
Fifteen Day Strike. 5 inch Dial.
Height, 10¾ inches.
List, $40.00

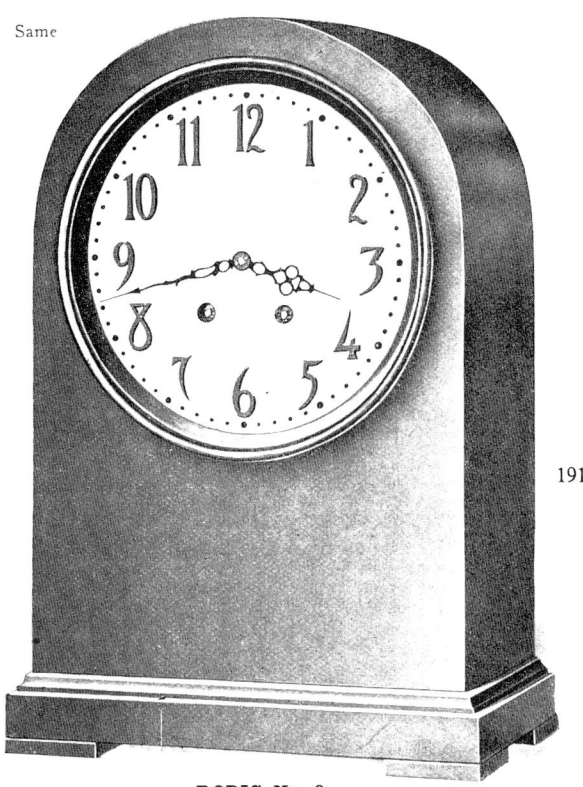

1910

1910

**DORIC No. 2**
Real Bronze.
Fifteen Day Strike. 6 inch Dial.
Height, 13 inches.
List, $50.00

**DORIC No. 3**
Real Bronze.
Fifteen Day Strike. 8 inch Dial.
Height, 16 inches. Depth, 8½ inches.
List, $70.00

# SETH THOMAS

Eight Day, Hour and Half Hour Strike, Porcelain Dials.

**826 — DUKE**
Bronze and Verde Finishes. 4½ inch Dial. Five Gold Plated Top Ornaments. Height, 13 inches.
List, $18.00

**827 — LA REINE**
Bronze Finish. 3½ inch Decorated Dial. French Sash and Beveled Glass. Height, 13½ inches.
List, $19.00

**828 — REX**
Brass Antique and Bronze Finishes. 4½ inch Dial. Height, 12½ inches.
List, $16.00

**829 — DUCHESS WITH LION**
Bronze and Verde Finishes. 4½ inch Dial. Height, 13¾ inches.
List, $18.00

**830 — DUCHESS**
Bronze Finish. 4½ inch Dial. Height, 9¾ inches.
List, $15.00

# SETH THOMAS

**831** — These three clocks are 8 Day, Half Hour Strike, Cathedral Bell, 4½ inch, Porcelain Dials French Sash and Beveled Glass.

**VISTA**
Height, 12¾ inches. Base, 8 inches.
Rich Gold, List, $22.70
Special Syrian Finish, List, $19.00

**832 — LA FLEUR**
Height, 14 inches. Base, 7½ inches.
Rich Gold, List, $23.70
Special Syrian Finish, List, $19.00

**833 — LA NORMA**
Height, 12 inches. Base, 7 inches.
Rich Gold, List, $22.70
Special Syrian Finish, List, $18.50

**834 — MEDITATION**
Bronze with Gold Panel. 8 Day. Movt. as in Lily. 4 inch Gold Dial with Porcelain Numerals. No Sash or Glass. Winding arbors piped to prevent dust getting in movement. Height, 11½ inches. Base, 8½ inches. List, $10.50.

These two patterns are also made in Rich Gold with Porcelain Dials, French Sash and Beveled Glass.
List, $12.50

**835 — LILY**
Bronze with Gold Panel. 8 Day, Strike Hours on Cathedral Bell and Half Hours on Cup Bell. 4 inch Dial. Height, 11½ inches. Base, 8½ inches.
List, $10.50

# SETH THOMAS

**COLUMN WEIGHT.**

ROSE OR WALNUT VENEER POLISHED.

WITH SHELL OR GILT COLUMNS.

1 Day, Weight, Strike.

Hight, 25 inches.

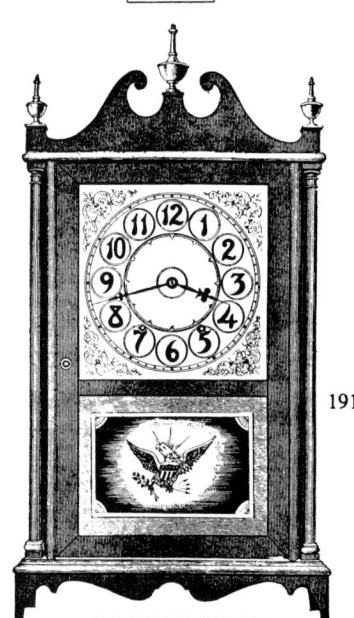

**CONTINENTAL**

Mahogany Case, Brass or Wood Tips. Eight Day, Spring, Half Hour Strike, Cathedral Bell. 10 inch Dial. Height, 32 inches. Base, 17½ inches.

List, $30.70

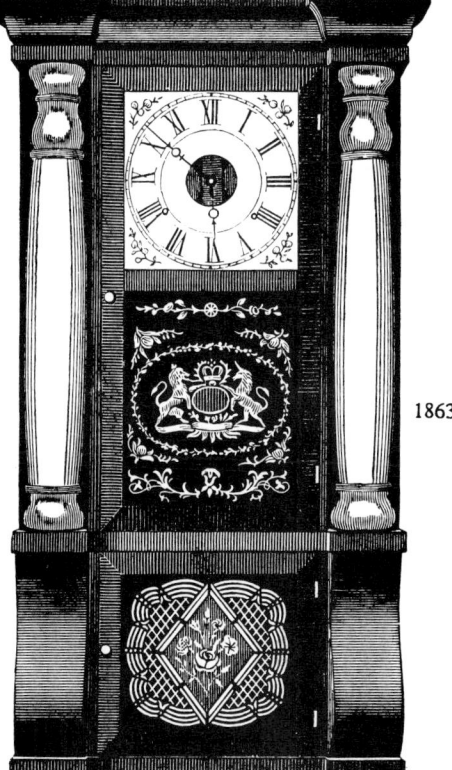

**8 DAY.**

ROSEWOOD—GILT OR GILT CAP—STRIKE—WEIGHT.

Height of Clock 32 inches.

**YORKTOWN**

Mahogany Case, Hand Carved Columns. Feet and Ornament. Eight Day, Spring. Strike, Cathedral Bell, 10 inch Dial. Height, 36 inches. Base, 18 inches.

List, $32.70

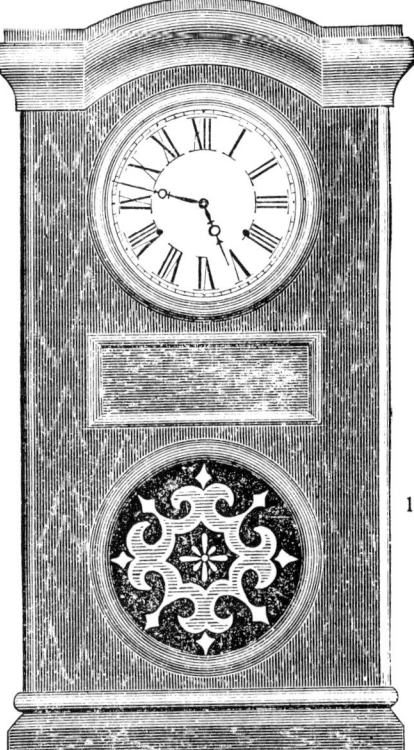

**8 DAY PARLOR—No. 1.**

STRIKE—WEIGHT.

Hight, 32 inches.

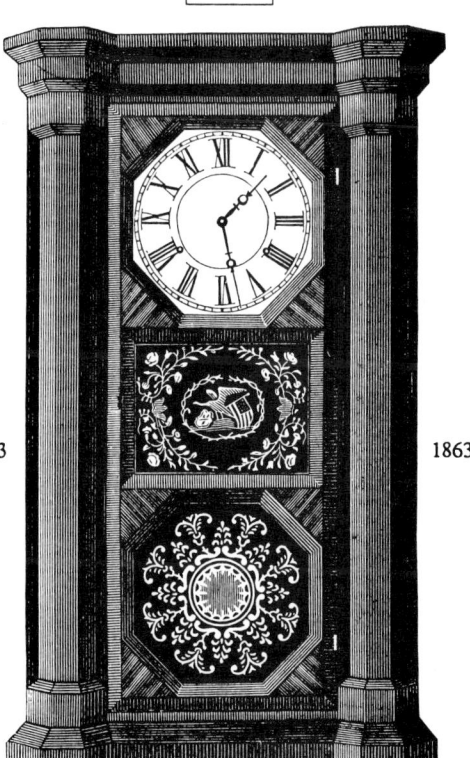

**8 DAY.**

PARLOR—STRIKE—WEIGHT.

Height of Clock 30¼ inches.

# SETH THOMAS

**842**

The year entered beside each clock is the year of the catalog from which the illustration was taken. Some clocks may have been produced a few years earlier.

**843**

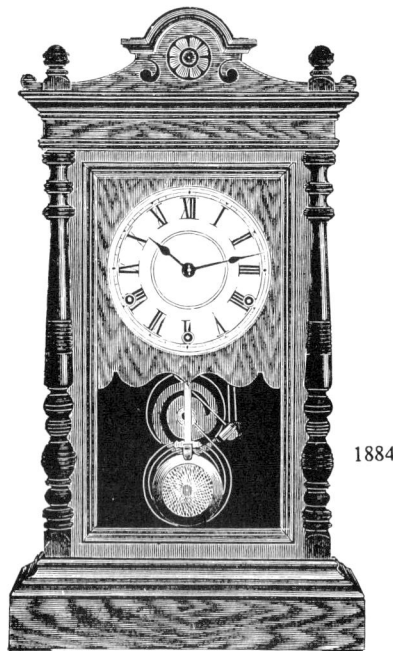

ATLAS.

QUARTER STRIKE.

MADE IN WALNUT, OAK AND CHERRY.

Striking quarter hours on two

Cup Bells and hours on

8 Day, Spring.

Cathedral Bell.

6 inch Dial.

Hight, 22½ inches.

1884

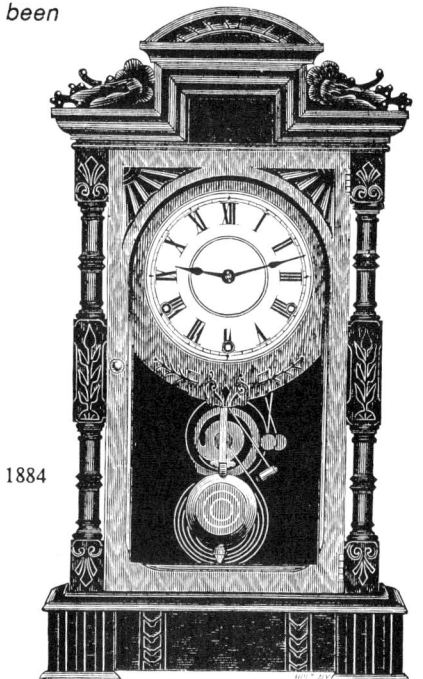

1884

HECLA.

For comprehensive listing of Seth Thomas Clocks, see Tran Duy Ly "Seth Thomas Clocks & Movements - A Guide to Identification & Prices."

**844**

**845**

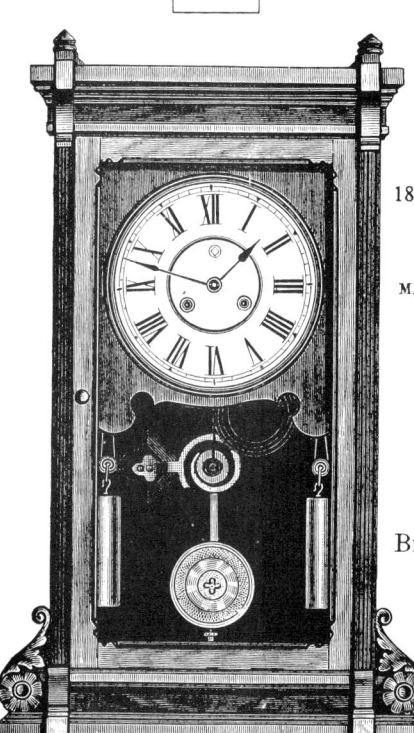

LINCOLN.

Hight, 27 inches.

1884

MADE IN WALNUT, EBONY, OAK AND MAHOGANY.

ENTIRELY NEW AND SUPERIOR MOVEMENT.

8 Day, Weight, Strike.

Wooden Rod.

Cathedral Bell.

8 inch Dial.

Brass Weights and Ball Nickel Plated.

NOT MADE WITH ALARM.

1884

GARFIELD.

Hight, 29 inches.

# SETH THOMAS

### 846
### DALLAS.
METAL ORNAMENTS AT BASE.
MADE IN WALNUT, OAK AND CHERRY.
New Movement Springs in Barrel.
Striking half hours on
Cup Bell and hours on
Cathedral Bell.
8 Day, Spring, Strike.
4½ inch Porcelain Dial.
Hight, 13 inches.
NOT MADE WITH ALARM.

### 847
### BEE.
CABINET CASE, MARQUETERIE PANEL.
MADE IN WALNUT, EBONY, OAK AND MAHOGANY.
Strikes hours and half hours
on a Cathedral Bell.
Has fine polished 15 day Movement.
Hight, 14 inches.

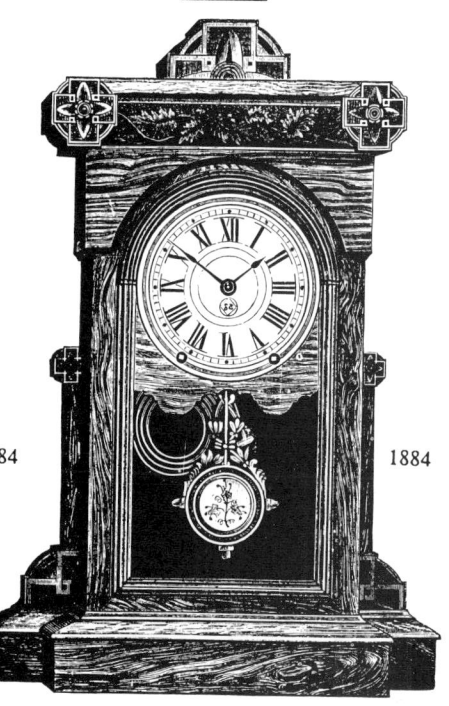

### 848
### PITTSBURGH.
WALNUT VENEER POLISHED.
Cathedral Bell.
8 Day, Spring, Strike.
6 inch Dial.
Hight, 23 inches.
NOT MADE WITH ALARM.

### 849
### STERLING.
MADE IN WALNUT, EBONY,
OAK, MAHOGANY AND OLIVE.
Hight, 12 inches.

### 850
### NORMANDY.
MADE IN WALNUT AND ASH.
Hight, 15 inches.
CABINET CASE.
Has fine polished 15 day Movement.
Strikes hours and half hours on a Cathedral Bell.

### 851
### MINSTER
MADE IN WALNUT, EBONY,
OAK, MAHOGANY AND OLIVE.
Hight, 12 inches.

# SETH THOMAS

**852**

*The year entered beside each clock is the year of the catalog from which the illustration was taken. Some clocks may have been produced a few years earlier.*

1884

### BUFFALO.
WALNUT CASE.
BRASS PILLARS.

8 Day, Spring, Strike
Cathedral Bell.
6 inch Dial.
Hight, 20½ inches.
WITH AND WITHOUT ALARM.

**853**

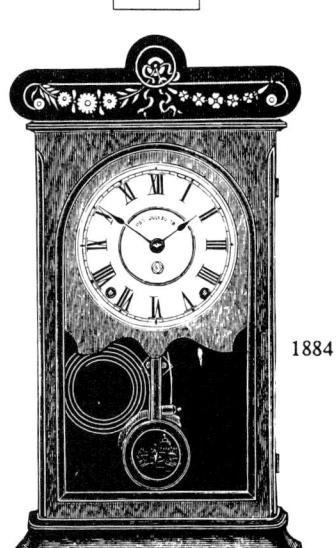

1884

### OREGON.
COCOBOLA FINISH. MARQUETERIE TOP.

8 Day, Spring, Strike.
Cathedral Bell.
6 inch Dial.
Hight, 19½ inches.
NOT MADE WITH ALARM.

**854**

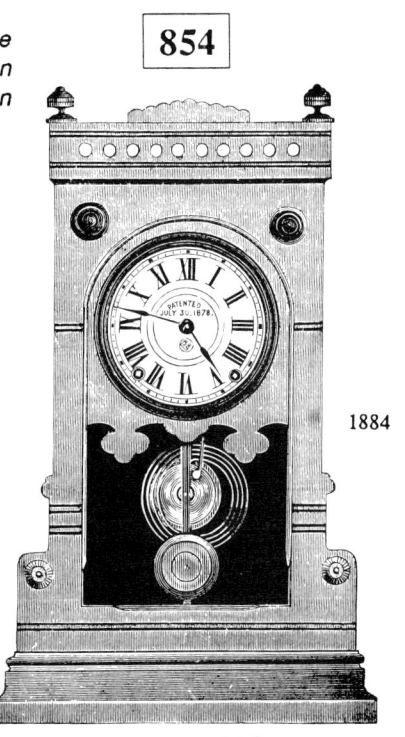

1884

### BOSTON.
WALNUT CASE.

8 Day, Spring, Strike.
Cathedral Bell.
6 inch Dial.
Hight, 19 inches.
WITH AND WITHOUT ALARM.

**855**

1884

### ERIE.
WALNUT CASE.

**856**

1884

8 Day, Spring, Strike.
Cathedral Bell.
6 inch Dial.
Hight, 19 inches.
WITH AND WITHOUT ALARM.

### OMAHA.
MADE IN WALNUT, EBONY, OAK AND CHERRY.

# SETH THOMAS

*Most white metal and brass ornamentals and clock cases are gilded or silver plated. Great care must be taken when cleaning these surfaces. The thin layer of gilding or silver plating will be easily removed by cleaning solutions and strong rubbing. If you believe cleaning is necessary, hire the services of a professional and avoid the expense of re-gilding or re-plating.*

**857** — 1910

**WAGNER**
Bronze Top and Base, Rich Gold Pillars and Center.
Height, 8 inches.
List, $5.50

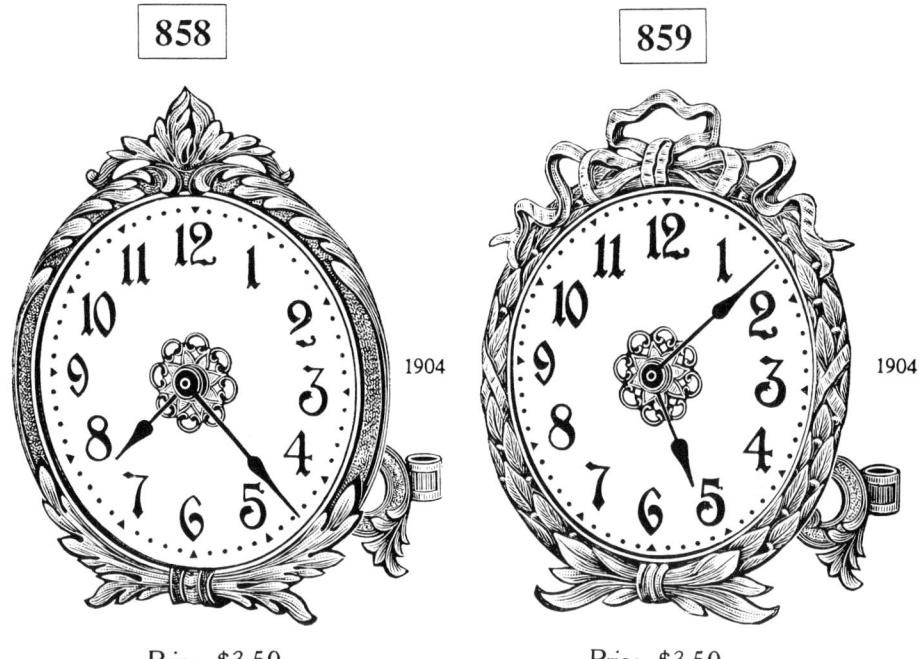

**858** — 1904

Price, $3.50

**NIGHT CLOCK "A"**

**859** — 1904

Price, $3.50

**NIGHT CLOCK "B"**

GILDED CAST METAL FRAME WITH EASEL BACK TO STAND,
OR CAN BE ATTACHED TO GAS BURNER.

5 inch Opal Glass Dial.

Hight, 7¼ inches.

**860** — 1910

**SHAKESPEARE**
Bronze Top and Base, Rich Gold Pillars and Center.
Height, 8 inches.
List, $5.50

**861** — 1910

**SCHILLER**
Bronze Top and Base, Rich Gold Pillars and Center. Height, 8 inches.
List, $5.50

**862** — 1910

**MOZART**
Bronze Top and Base, Rich Gold Pillars and Center.
Height, 8 inches.
List, $5.50

**863** — 1910

**GOETHE**
Bronzed Top and Base, Rich Gold Pillars and Center.
Height, 8 inches.
List, $5.50

# SETH THOMAS

*For comprehensive listing of Seth Thomas Clocks, see Tran Duy Ly "Seth Thomas Clocks & Movements - A Guide to Identification & Prices."*

### 864

1910

**COLIN**
Rich Gold and Bronze Art Nouveau
Height, 6 inches.
List, $4.30

### 865

1910

**PADDOCK**
Rich Gold and Bronze Art Nouveau.
Height, 7½ inches.
List, $5.00

### 866

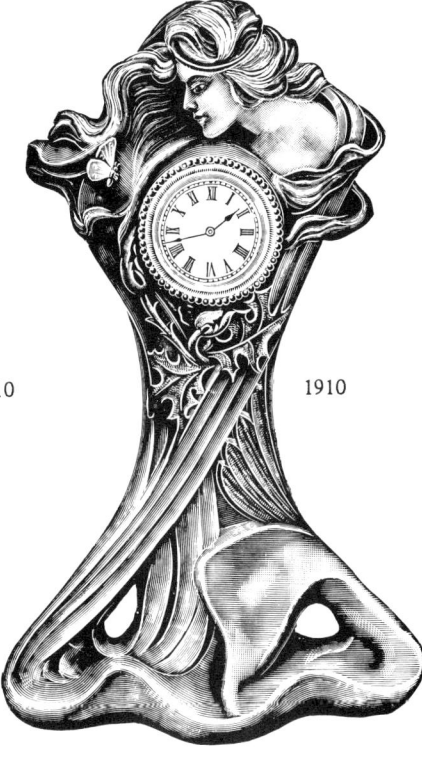

1910

**JOE**
Rich Gold, also Art Nouveau Bronze.
Height, 12½ inches.
List, $9.30

### 867

1910

**PETITE**
Actual Size. Metal Case.
Gold Plated. Also Bronze Finish. One Day, Lever Time.
Height, 3½ inches. List, $2.00

### 868

1910

**BONA**
Rich Gold Finish.
Height, 5¾ inches.
List, $3.70

### 869

1910

**MODE ALARM**
Embossed Metal Case. Made in two finishes, Gun Metal with Polished Brass Flowers, or Rich Gold Plated and Lacquered. One Day, Lever, Time Alarm. 3 inch Dial. Height, 6½ inches.
Gun Metal, List, $2.50
Gold Plated, List, $3.20

# SETH THOMAS

*Except for those fabulous clocks for which you wouldn't mind cutting a hole in the ceiling, Grandfather or Hanging Clocks under eight feet are the most desirable and practical. Some manufacturers produced certain of these clocks in two styles: one to stand on the floor, and the other to hang on a wall. Of the clocks designed with two styles, the hanging style is considered to be the most valuable.*

*The year entered beside each clock is the year of the catalog from which the illustration was taken. Some clocks may have been produced a few years earlier.*

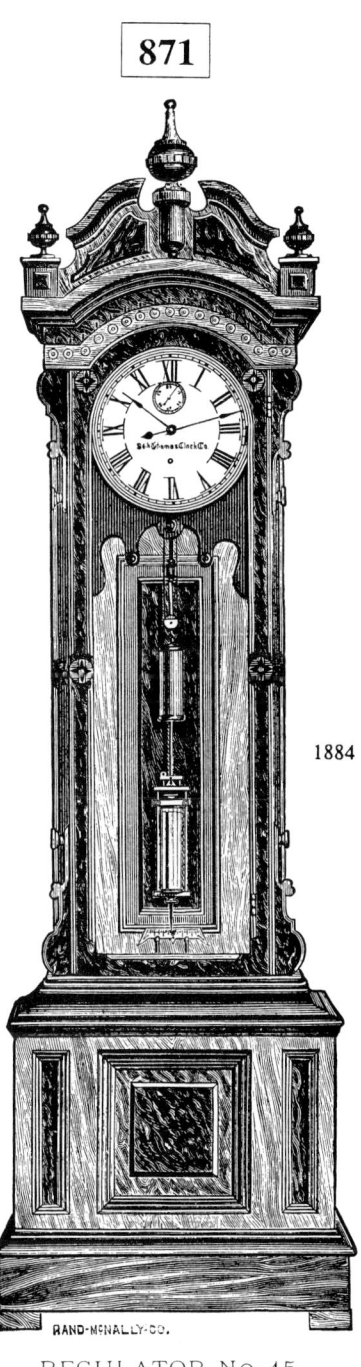

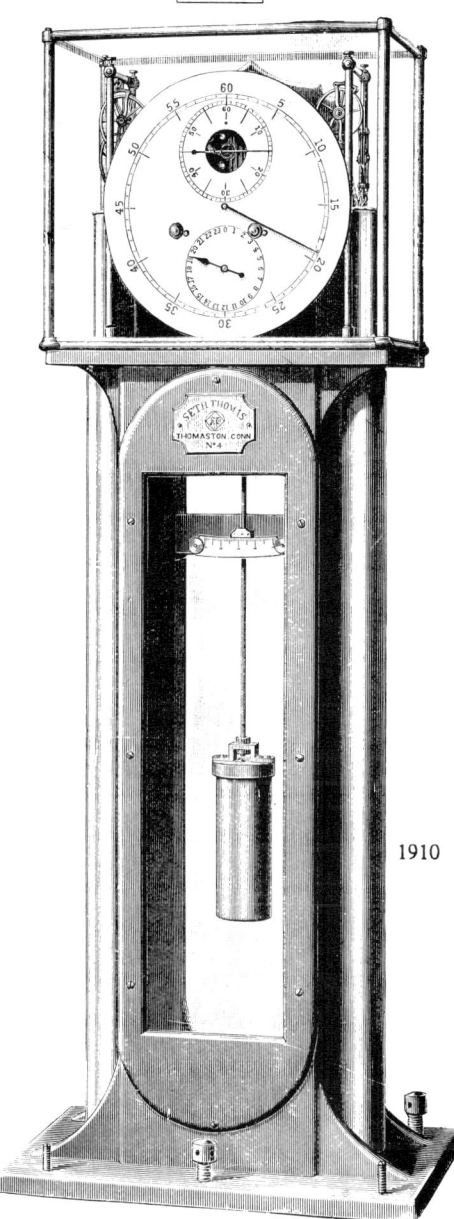

870 — 1884
REGULATOR No. 14.
Sweep Second
MERCURY PENDULUM.

871 — 1884
REGULATOR No. 15.
BEATS SECONDS.
MERCURY PENDULUM.

8 Day, Weight, Time.
Hight, 100 inches. 14 inch Dial.

872 — 1910
**PRECISION CLOCK**
With Dead-beat Escapement, Thirty-two Day. With Gravity Escapement, Eight Day. 14 inch Dial. Height, 62 inches. With dead-beat escapement.
List, $1080.00

With Gravity Escapement.
List, $1350.00

# SETH THOMAS

For comprehensive listing of Seth Thomas Clocks, see Tran Duy Ly "Seth Thomas Clocks & Movements - A Guide to Identification & Prices."

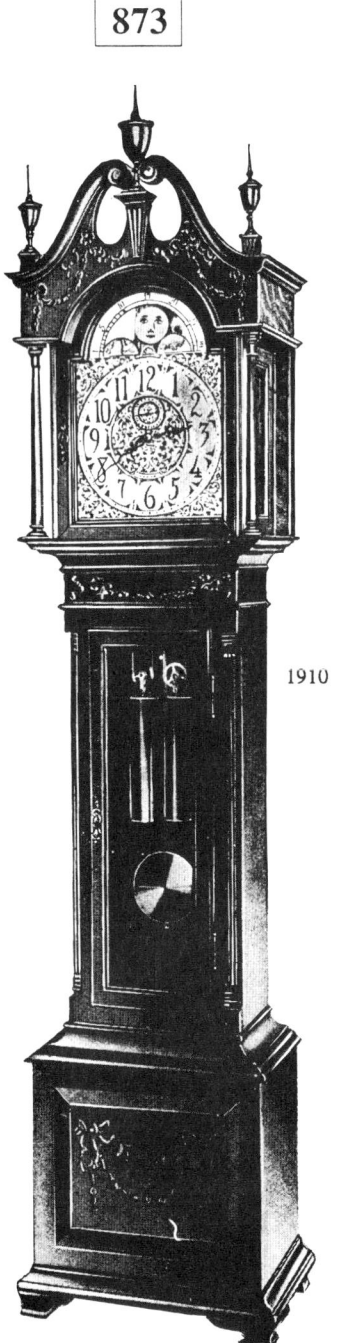

**873** — 1910

**HALL CLOCK No. 28**
Strike and Moon. Made in Mahogany or Golden Oak. 98 inches High, 23 inches Wide, 14 inches Deep. Eight Day, Weight, striking hours and half hours on Cathedral Bell. Shows Moon changes. Fitted with Movement No. 72 B. Dial, Pierced Gilt.
List, $275.00

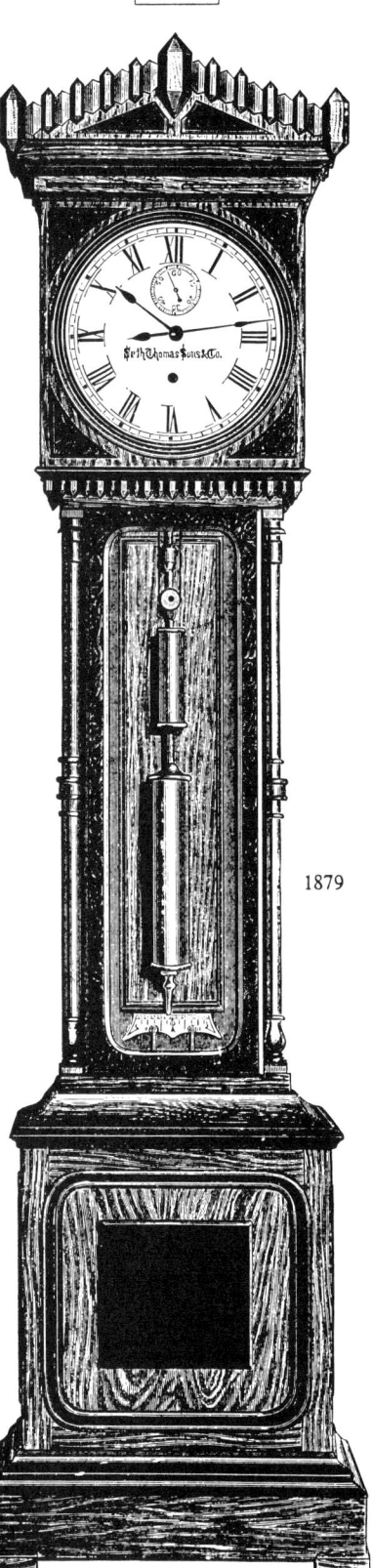

**874** — 1879

**REGULATOR NO. 13**
Independent Second.

**875** — 1879

**REGULATOR NO. 11**
Sweep Second.

8 Day, Weight, Time.
Hight, 96 inches.

# THE TERRY CLOCK COMPANY

Waterbury, Conn. & Pittsfield, Mass.

Silas Burnham Terry (1807-1876) was trained by his famous clockmaking father, Eli Terry. Commencing his own business in 1831 at Terrysville, CT, young S. B. Terry manufactured weight-driven shelf and wall clocks and later spring-driven clocks under his own name until 1852, though not without financial difficulties, particularly in the latter 1840's.

In June of 1852, Terry formed a partnership with a nephew and another relative, known as S. B. Terry & Company, no doubt because of his need for operating capital. This firm operated only about a year when a joint stock corporation called the Terryville Manufacturing Company was formed, particularly to manufacture a torsion escapement clock for which Terry had received patents in 1852 and 1853. A new factory was built about a mile south at Pequabuck, Conn. and S. B. Terry became the manager of the operation until the autumn of 1854 when he sold out his interest and resumed business in his old shop.

On January 1, 1859 Silas B. Terry was bankrupt and his clock shop and other assets were sold. He subsequently moved to Winsted, Conn. for about two years as manager of the clock movement department of W. L. Gilbert & Company. He thereafter went to Waterbury, Conn. and held a similar position for the Waterbury Clock Company.

In 1867, Silas B. Terry and his four sons formed the Terry Clock Company at Waterbury, Conn. renting a factory building from the American Flask & Cap Company. Concerned about Terry's poor financial record in the past, the firm was incorporated in 1868.

Though the Terry Clock Company produced some wooden cased clocks, the majority of their production prior to 1880 was for clocks with cases that were primarily cast iron, painted black and had various degrees of hand striping and decoration. Not only did Terry receive some patents for movement escapements and cast iron case fronts, but also for finishing reels which the firm was manufacturing in the 1870's.

The Terry Clock Company's business was moderately successful, but the firm never had adequate operating capital and lived a curious existence of annually borrowing from one creditor to pay the past one. After Silas B. Terry died of heart disease in 1876, his sons ran the operation until 1880, but the firm went bankrupt in May of that year.

Subsequently the operation was purchased by a group of investors from Pittsfield, Massachusetts and the operation moved there and set up a steam flour mill on the second floor. In 1883, a new three story building was built for them on the Housatonic River. The Terry Clock Company name was retained and three of the Terry brothers went to Pittsfield and ran the operation. In 1888, the firm failed and was taken over by its creditors who changed the name to Russell & Jones Clock Company and operated about four years.

# TERRY

**876** — 1875

### ONE DAY MANTEL, SPRING.
BRONZE.
With Terry's Patent Pendulum.

2½ Inch Dial.   7¼ Inches High.

| | | |
|---|---|---|
| 110. | One Day Time, . . . | $4.75 |
| 111. | " " " Alarm, . | 5.25 |

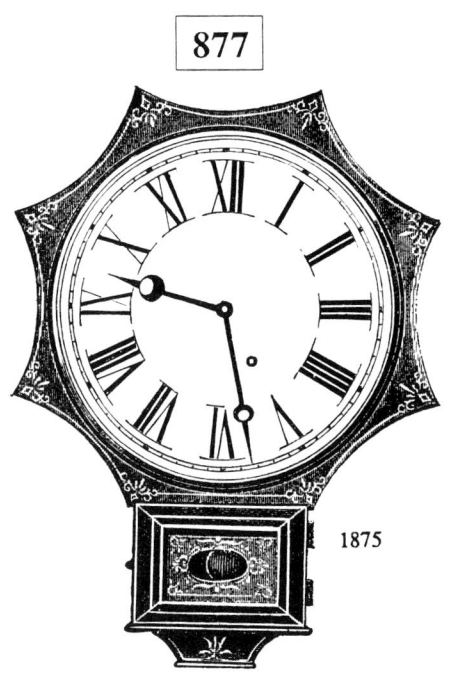

**877** — 1875

### EIGHT DAY OCTAGON TOP DROP, SPRING.
21 Inches High.   12 Inch Dial.
Pendulum beats just half seconds.

| | |
|---|---|
| 8 Day Time, Black and Gilt, . | $6.50 |
| " Strike, " " " . | 7.75 |

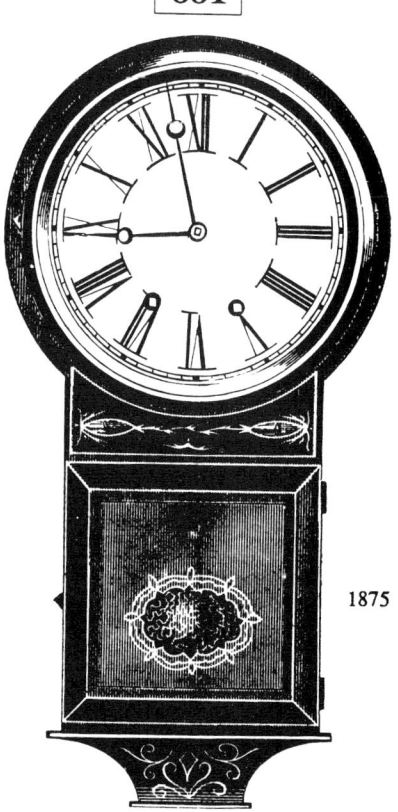

**878** — 1875

### ONE DAY CHAPEL, SPRING.
BRONZE.
With Terry's Patent Pendulum.

9¼ Inches High.

3½ Inch Dial.

| | |
|---|---|
| One Day Time, . . . | $6.50 |
| " " " Alarm, . | 7.00 |

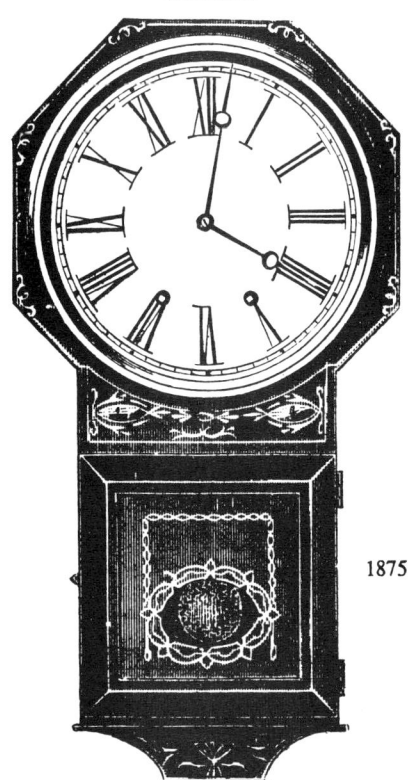

**879** — 1875

### EIGHT DAY OCTAGON TOP DROP, SPRING.
19 Inches High.   8 Inch Dial.
Pendulum beats just half seconds.

| | |
|---|---|
| Eight Day Time, Black and Gilt, | $6.00 |
| " Strike, Black and Gilt, . | 7.25 |

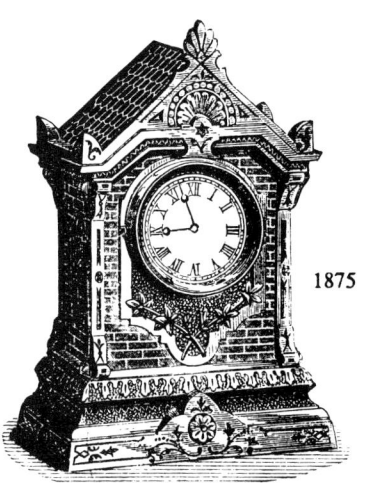

**880** — 1875

### EIGHT DAY CHAPEL, SPRING.
BRONZE.
With Terry's Patent Pendulum.

4 Inch Dial.

13 Inches High.

Eight Day Strike, . $13.25

**881** — 1875

### EIGHT DAY ROUND TOP DROP, SPRING.
19 Inches High.   8 Inch Dial.
Pendulum beats just half seconds.

| | |
|---|---|
| Eight Day Time, Black and Gilt, . | $6.00 |
| " Strike, Black and Gilt, . | 7.25 |

# TERRY

**882**

### EIGHT DAY SPRING.

4 Inch Dial.  9 Inches High.

The above cut represents Eight Day Strike, No. 31. Other varieties of the same case are as follows, viz. :

### EIGHT DAY TIME.

| NO. | | EACH. |
|---|---|---|
| 27. | Plain Black, | $3.75 |
| 28. | Black and Gilt, | 4.00 |

### EIGHT DAY STRIKE.

| 30. | Plain Black, | 5.25 |
|---|---|---|
| 31. | Black and Gilt, | 5.50 |

### EIGHT DAY TIME ALARM.

| 33. | Plain Black, | 4.25 |
|---|---|---|
| 34. | Black and Gilt, | 4.50 |

### EIGHT DAY STRIKE ALARM.

| 36. | Plain Black, | 5.75 |
|---|---|---|
| 37. | Black and Gilt, | 6.00 |

1875

**883**

### EIGHT DAY ROSE COLUMN.

22½ Inches High.  15 Inches Wide.

8 Inch Dial.

Eight Day Rose Column,  $8.00

**885**

1875

### EIGHT DAY MANTEL, SPRING.

Either Round or Octagon Top.

19 Inches High.  8 Inch Dial.

Pendulum beats just half seconds.

| 8 Day Round Top Time, Black and Gilt, | $6.25 |
|---|---|
| "    "    Strike,  "    " | 7.50 |
| "  Octagon Top Time,  "    " | 6.25 |
| "    "    Strike,  "    " | 7.50 |

The cut represents Eight Day Strike, No. 47. Other varieties of the same case, are as follows, viz:

13 Inches High.  5 Inch Dial.

### ONE DAY.

| NO. | | EACH. |
|---|---|---|
| 43. | One Day Strike, Black and Gilt, | $5.00 |
| 44. | "   "   Alarm, Black and Gilt, | 5.50 |

### EIGHT DAY.

| 45. | Eight Day Time, Black and Gilt, | 4.75 |
|---|---|---|
| 46. | "   "   Alarm, Black and Gilt, | 5.25 |
| 47. | Eight Day Strike, Black and Gilt, | 6.25 |
| 48. | "   "   Alarm, Black and Gilt, | 6.75 |

**884**

1875

### ONE AND EIGHT DAY GOTHIC SPRING.

# TIFFANY ELECTRIC MANUFACTURING COMPANY
Buffalo, N. Y.

On March 8, 1904, a patent was granted to George S. Tiffany of Buffalo, N.Y. for an electrically wound clock. The earlier clocks are generally under a glass dome and larger with a pendulum having two round ball weights.

By about 1918, some seven models were offered, all with later style pendulums with the two weights having domed tops and flat bottoms. At that time three models were offered under glass domes, one 12 1/2" tall at $31.50 and two smaller ones 9 3/4" tall at $23.00 and $19.00. Four models with square "crystal regulator" cases were offered, all 9 1/2" tall, three of which were priced at $26.00 and one at $23.00. Lower priced models had polished brass cases instead of gold or bronze plated ones.

Though fairly expensive, the "Tiffany Never-Winds" enjoyed some success as they are relatively common compared to other pre-alternating current electrics. Clocks using similar mechanisms are found with the names Cloister Manufacturing Company, Buffalo, N.Y., the National Magnetic Clock Company, New York, N.Y. and the Niagara Clock Company.

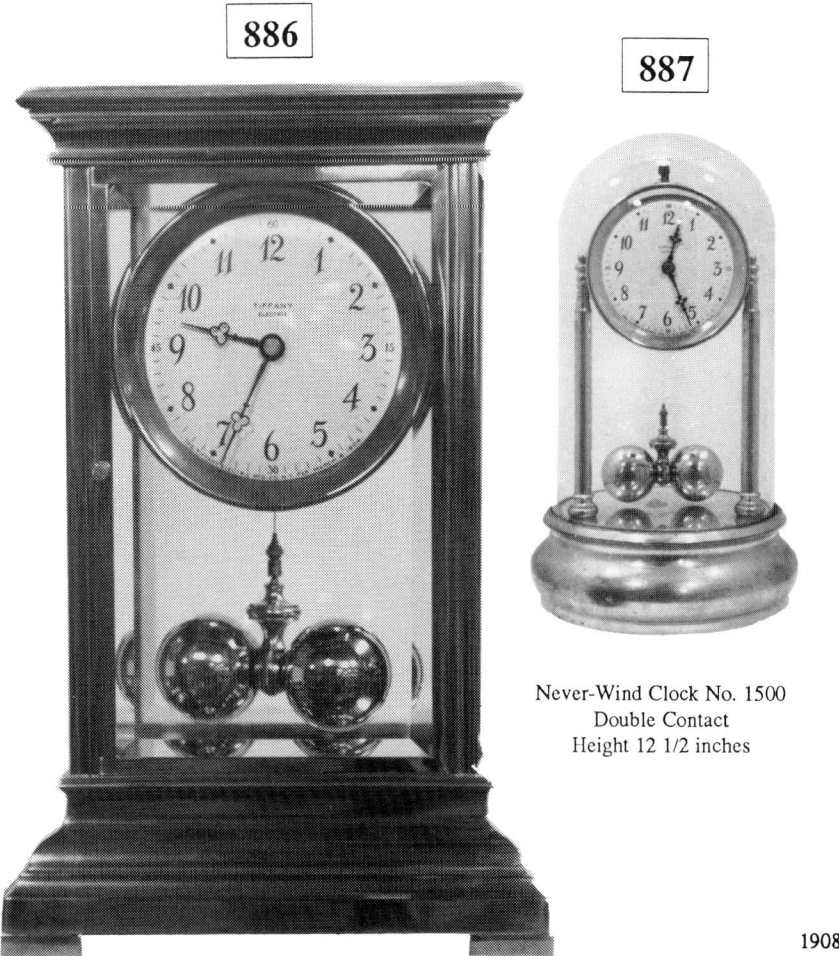

Never-Wind Clock No. 1000
Double Contact
Height 12 inches

Never-Wind Clock No. 1500
Double Contact
Height 12 1/2 inches

1908

NEVER-WIND CLOCK

THE CARE-FREE CLOCK

Tireless time keeping without tiresome winding. This sensational clock is never wound. Without outside aid it runs from year's end to year's end as faithfully as the sun rises. The motive power is furnished by a powerful little electric dry-cell battery hidden in the base. Each battery is guaranteed for one year and usually lasts longer. It is a standard size and can be replenished anywhere. Money could not make a better clock. And it is an ornament of great beauty. The frame is plated with real gold—will never tarnish. Never-Wind is 11 inches high, 7 inches across the base; 4½ inch porcelain dial. You slip in a new battery as easily as you do in a pocket flashlight. It is a clock of marvelous time-keeping accuracy.
No. 2000  Never-Wind Clock, each......$42.00

## WALTHAM CLOCK COMPANY
Waltham, Mass.

During January of 1897, the Waltham Clock Company was formed at Waltham, MA. for the manufacture of high grade tall clock movements and to compete with some of the English and German firms which were importing tall clock movements and clocks in increasing numbers. This firm succeeded the Waltham Electric Clock Company which was dissolved and the manufacture of electric clocks discontinued.

Tall clock offerings increased for the firm during the first decade of the 20th century, dozens of case styles were being offered with several optional movements. By 1915, the best quarter-striking movement with three different chime selections played on nine tabular "bells," a moon dial and solid mahogany colonial style case could be purchased for $560.00 retail.

In 1913, the Waltham Clock Company was purchased by the Waltham Watch Company which was wanting to diversify its watchmaking business. The Waltham Clock Company name was continued until 1923 when, during a reorganization of the parent firm, the name Waltham Watch and Clock Company was adopted. This name lasted only about two years when the name Waltham Watch Company was adopted for all operations.

Production of tall clocks declined during the 1920's though more mantel and wall clocks were added to the line. Production of pendulum clocks ceased about the time of the Great Depression, though electric clocks and speedometers were manufactured until about 1940.

889

1905

Pattern No. 82
9 feet high
30 inches wide
20 inches deep

244

# WALTHAM

*The year entered beside each clock is the year of the catalog from which the illustration was taken. Some clocks may have been produced a few years earlier.*

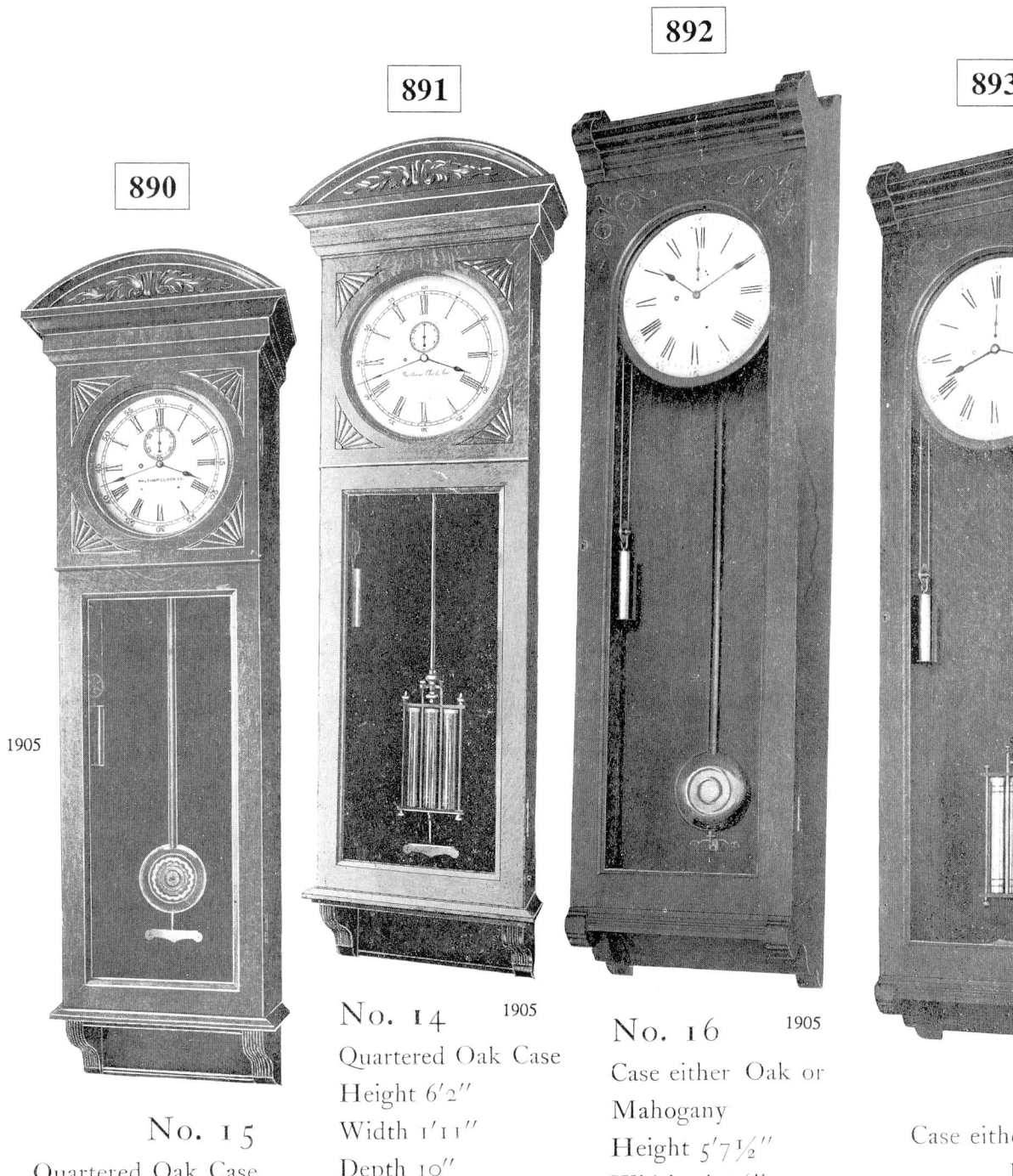

1905

890

No. 15
Quartered Oak Case
Height 6'2"
Width 1'10"
Depth 10"

891

No. 14   1905
Quartered Oak Case
Height 6'2"
Width 1'11"
Depth 10"

892

No. 16   1905
Case either Oak or
Mahogany
Height 5'7½"
Width 1'5½"
Depth 10½"

893

1905

No. 13
Case either Oak or
Mahogany
Height 5'10"
Width 1'6½"
Depth 10½"

# WALTHAM

Only clocks in ORIGINAL MINT condition and in good running order will command top prices. A replaced or damaged dial, missing parts, tablets which are either cracked, flacking, redone will **greatly** reduce the price.

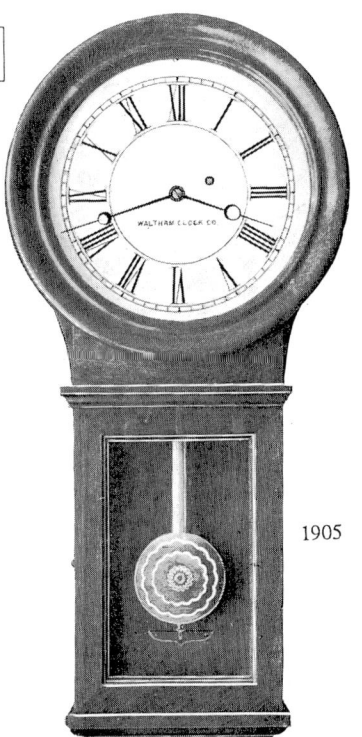

**895**

1905

### No. 34
Height 3'
Width 1'4"
Depth 5½"

**894**

1905

### No. 33
Case either Oak or Mahogany
Height 4'7"
Width 1'10½"
Depth 9"

**897**

1930

**WALTHAM CHRONOMETER**
8-Day, Lever Escapement, Winding Indicator Movement Adjusted to Changes in Temperature and Isochronism, enclosed in a Dust and Weather Proof Case Suspended on Gimbals in Solid Mahogany Box
8525 — 15 Jewel . . . . . . Price $131.70

**896**

## WALTHAM
### Curtis
### Girandole

Only four clocks of this beautiful banjo design were made by Lemuel Curtis, of Concord Massachusetts, during the period 1810 to 1818. The clock at the left is now classified as a museum piece with a reported value of $7000. Waltham has reproduced this famous clock so carefully that none of its original color, style, or charm is lost. The time-keeping accuracy of this reproduction is guaranteed by Waltham's dead-beat escapement pendulum movement.

#1510

Actual Height 49 Inches
$350.00

# WALTHAM

**898**

*Only clocks in ORIGINAL MINT condition and in good running order will command top prices. A replaced or damaged dial, missing parts, tablets which are either cracked, flacking, redone will **greatly** reduce the price.*

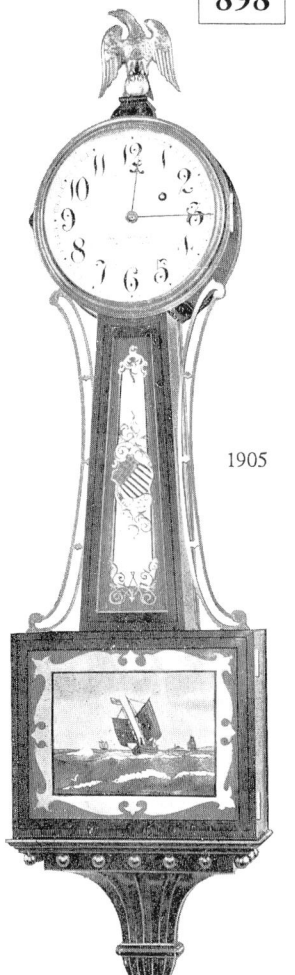

1905

No. 31
Colonial
Height 3′ 4½″
Width 10½″
Depth 4″

**899**

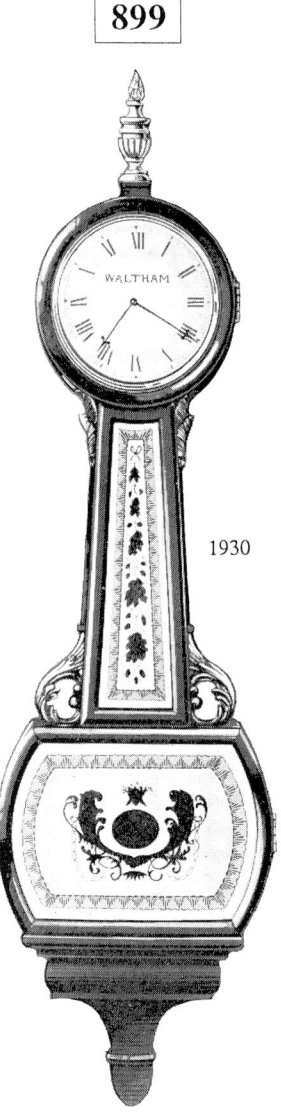

1930

Mahogany
42 inches high, 10½ inches wide,
4 inches deep
1505 . **Price $114.50**

**900**

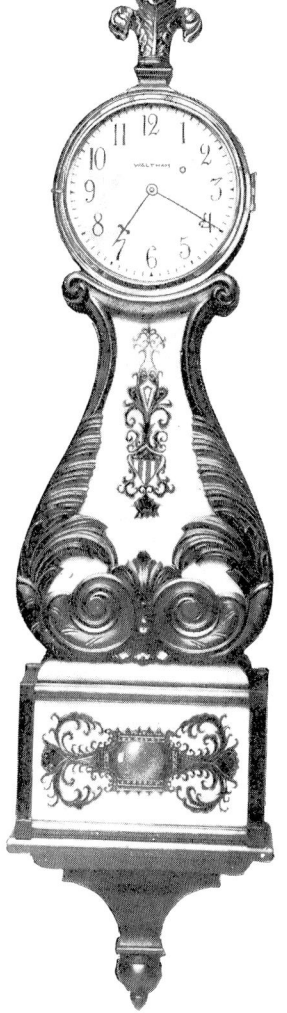

1930

Lyre
Mahogany only
Hand-Carved Case
42 inches high, 10½ inches wide,
4 inches deep
1548—Mahogany, **Price $171.80**

**901**

## WALTHAM
### Abbot Lyre

The original of this remarkable Waltham copy of a banjo lyre clock is reported to be 104 years old, and is now in the possession of a well-known Massachusetts family. The original dial bears the inscription **S. Abbot, Boston**, and no more than a dozen copies were ever offered by the maker. Here, again, Waltham has reproduced this famous clock with absolute fidelity. The accuracy of this beautiful clock is guaranteed by Waltham's finest pendulum movement.

#1515

Actual Height 43 Inches
$225.00

**902**

1930

This happy adaptation of a quaint Colonial Clock is an exclusive Waltham design. The gilded eagle, hand-painted glasses, and carved base are reminiscent of the early American period. Especially pleasing is the brilliant contrast of brass ornaments and gilt rope against the rich lustre of solid mahogany. Furnished with ivory dial with black figures. 14½ inches high, 5¼ inches wide

1470 — 7 Jewel . . . . . **Price $57.30**
1471 — 15 Jewel . . . . . **Price $64.20**

# WALTHAM

247

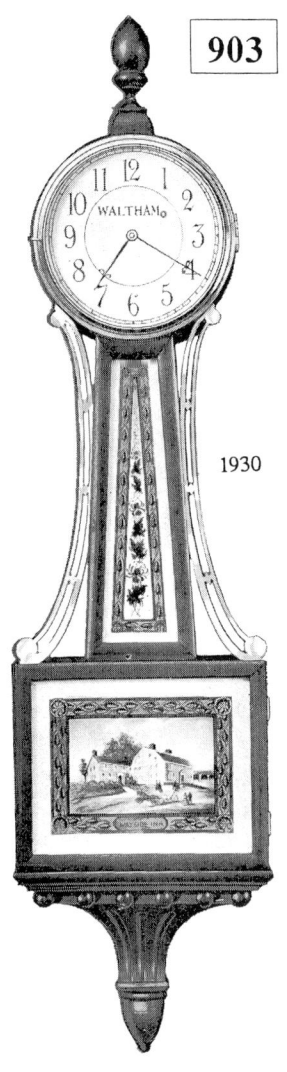

**903**

1930

Mahogany or Walnut
42 inches high, 10½ inches wide,
4 inches deep
1500 — Mahogany **Price $85.90**
1501 — Walnut . **Price $85.90**

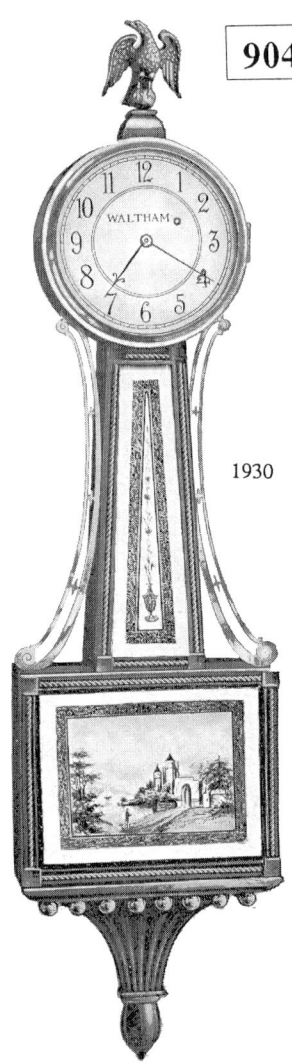

**904**

1930

Mahogany or Walnut. Gilt Rope
Panels
42 inches high, 10½ inches wide,
4 inches deep
1543 — Mahogany **Price $97.30**
1544 — Walnut . **Price $97.30**

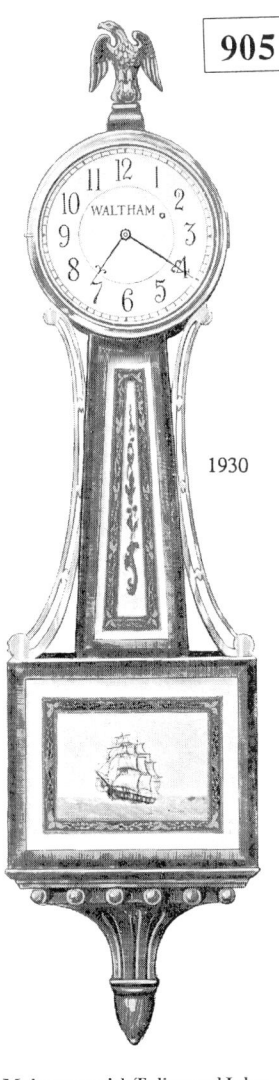

**905**

1930

Mahogany with Tulipwood Inlay
or Walnut with Satinwood
Inlay
42 inches high, 10½ inches wide,
4 inches deep
1546 — Mahogany **Price $85.90**
1547 — Walnut . **Price $85.90**

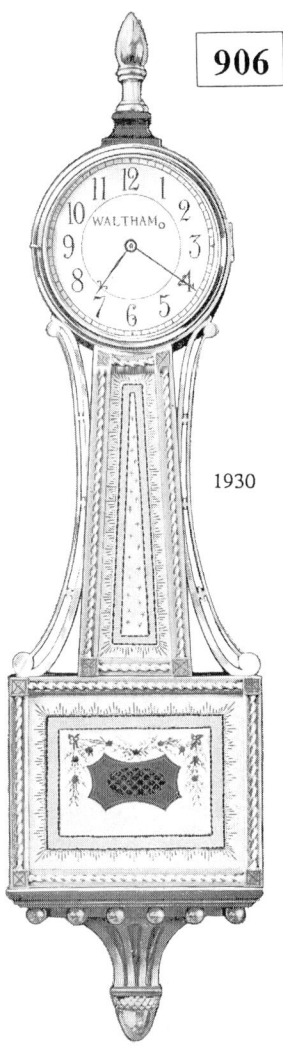

**906**

1930

Mahogany with 23K Gold Leaf
Front
42 inches high, 10½ inches wide,
4 inches deep
1525 . . **Price $143.20**

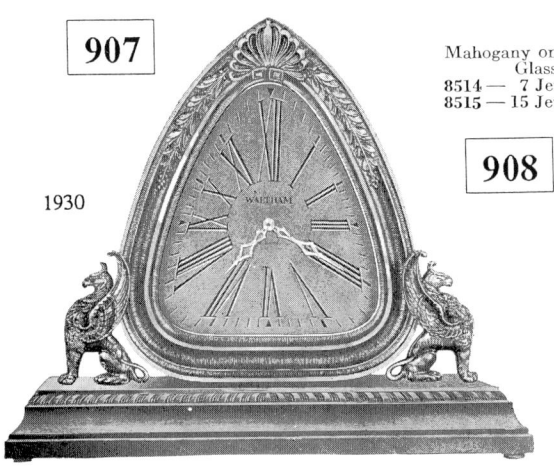

**907**

1930

**SOLID BRONZE**
Natural or Verde Finish. 11¾ inches high, 13½ inches
wide, 1½ inches deep
8516 — 15 Jewel, Natural Finish . **Price $85.90**
8517 — 15 Jewel, Verde Finish . **Price $85.90**

Mahogany only. Brass Trimmings, Hand-Painted Panel
Glasses. 12 inches high, 5¾ inches wide
8514 — 7 Jewel . . . . . **Price $57.30**
8515 — 15 Jewel . . . . . **Price $64.20**

**908**

1930

**909**

1930

**SOLID BRONZE**
Natural or Verde Finish. 11 inches high, 11½ inches
wide, 1⅝ inches deep
8518 — 15 Jewel, Natural Finish . **Price $85.90**
8519 — 15 Jewel, Verde Finish . **Price $85.90**

# WALTHAM

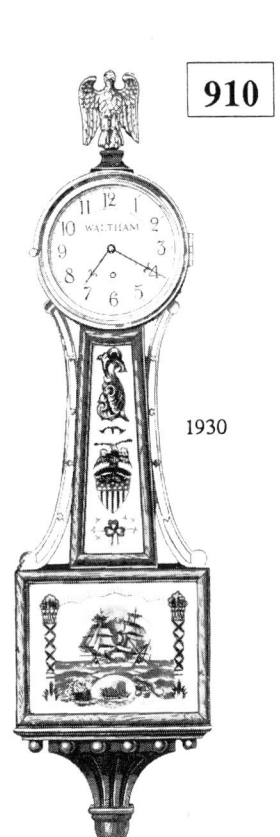

910

1930

Mahogany or Walnut
Special Movement with Lever Escapement
30 inches high, 7¾ inches wide,
3 inches deep
1570 — Mahogany **Price $74.50**
1571 — Walnut . **Price $74.50**

## MOVEMENTS

The large clocks have 8-day, weight-driven pendulum movements with heavy brass plates.

## ALL WALTHAM WILLARD CLOCKS HAVE HAND-PAINTED GLASSES WITH A CHOICE OF THE FOLLOWING SUBJECTS:

MT. VERNON — WASHINGTON
CONSTITUTION AND GUERRIERE
WALTHAM DESIGN     LONE SHIP
OLD ENGLISH CASTLE     BOSTON STATE HOUSE
MONTICELLO — JEFFERSON     PERRY'S VICTORY
WAYSIDE INN     OLD IRONSIDES
INDEPENDENCE HALL — LIBERTY BELL

| | |
|---|---|
| Luminous Dots and Hands, small models, extra | $4.50 |
| Luminous Dots and Hands, large models, extra | $6.50 |
| Dial with Luminous Figures and Hands on half-size models only, extra | $9.00 |
| 15 Jewel Movement, small models, extra | $6.99 |
| Gold Leaf Base, half size, extra | $12.50 |
| Gold Leaf Base, full size, extra, (Model 1525 excepted) | $19.00 |

### ORNAMENTS
Choice of brass eagle or carved acorn top ornament, solid brass side rails and hand-painted glasses.

### DIALS
Ivory with Arabic or Roman figures. Colored Circle Dials to match colored cases.

911

1930

Mahogany or Walnut
21 inches high, 5¼ inches wide,
2 inches deep
1556 — Mahogany **Price $45.80**
1557 — Walnut . **Price $45.80**

912

1930

Mahogany or Walnut. Gilt Rope
21 inches high, 5¼ inches wide,
2 inches deep
1553 — Mahogany **Price $45.80**
1554 — Walnut . **Price $45.80**

913

## WALTHAM CHRONOMETER

1930

8-Day, Lever Escapement, Winding Indicator Movement Adjusted to Changes in Temperature and Isochronism, enclosed in a Dust and Weather Proof Solid Mahogany Box.

#8725 — 15 Jewels ...................... **$60.00**

# WALTHAM

*Except for those fabulous clocks for which you wouldn't mind cutting a hole in the ceiling, Grandfather or Hanging Clocks under eight feet are the most desirable and practical. Some manufacturers produced certain of these clocks in two styles: one to stand on the floor, and the other to hang on a wall. Of the clocks designed with two styles, the hanging style is considered to be the most valuable.*

## 914

1905

No. 11
Mahogany
Height 7'10"
Width 2'5"
Depth 1'8"

## 915

1905

No. 12
Mahogany
Height 6'6"
Width 2'7"
Depth 11"

## 916

1905

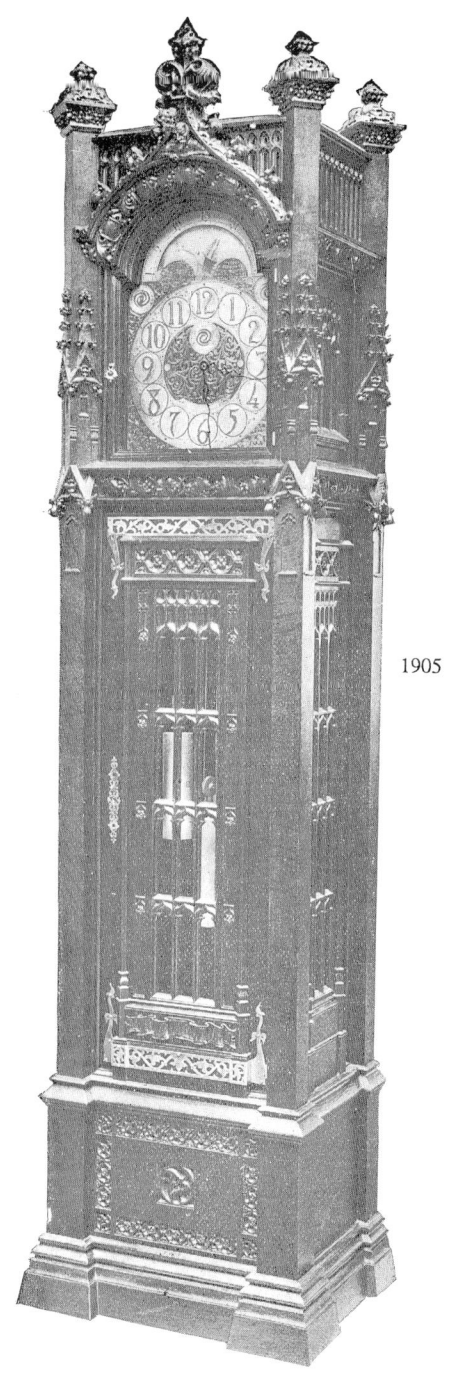

Pattern No. 81
8 feet 4 inches high
26½ inches wide
20 inches deep

# WALTHAM

*The year entered beside each clock is the year of the catalog from which the illustration was taken. Some clocks may have been produced a few years earlier.*

### 917

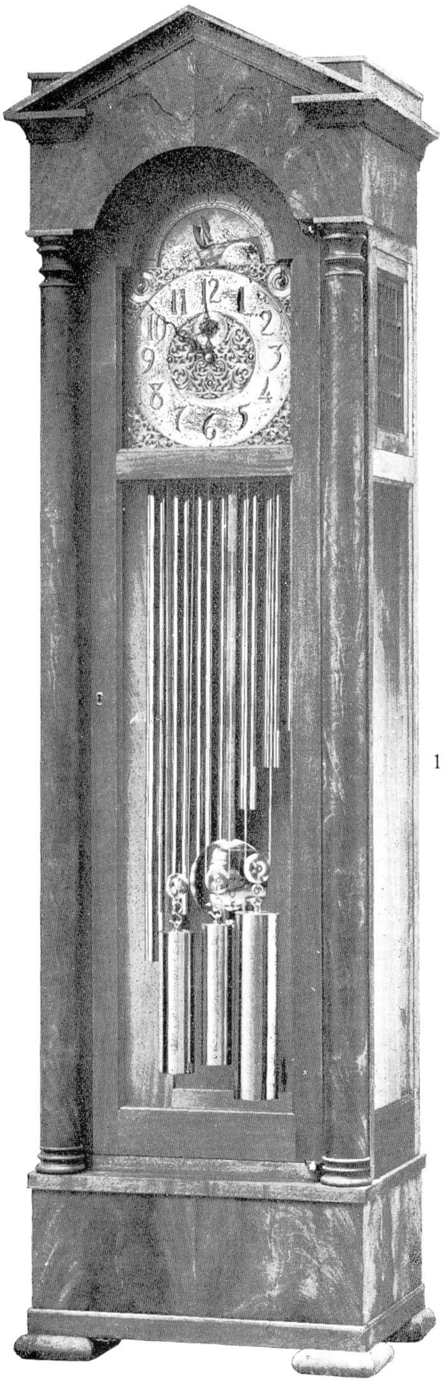

1905

Pattern No. 65
- 7 feet 10 inches high
- 27 inches wide
- 19 inches deep

### 918

1905

Pattern No. 2
- 8 feet high
- 22¼ inches wide
- 14¾ inches deep

### 919

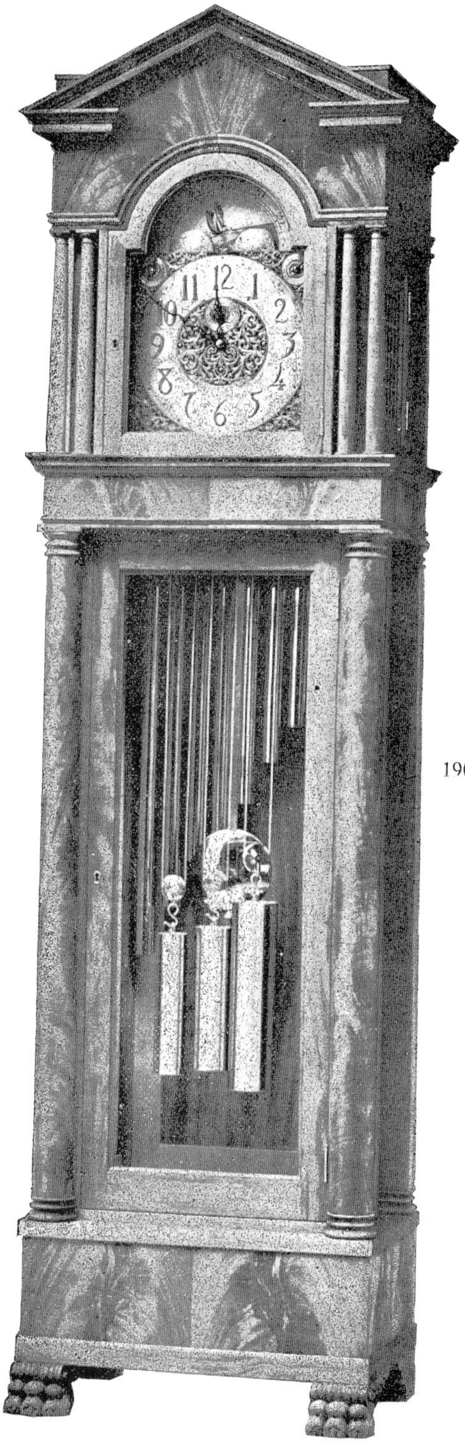

1905

Pattern No. 67
- 8 feet 2 inches high
- 27 inches wide
- 18 inches deep

# WALTHAM

*To receive top prices, clocks must be in ORIGINAL MINT condition and in good running order - A replaced or damaged dial, missing parts, will reduce prices accordingly.*

## 920

## 921

## 922

1905

Pattern No. 64
8 feet 6 inches high
29 inches wide
20 inches deep

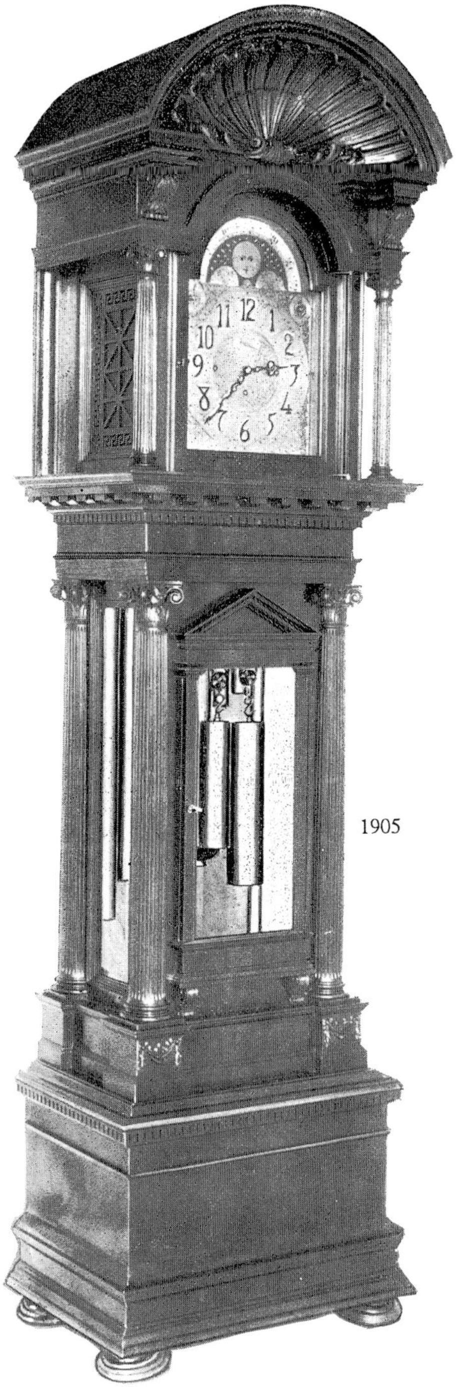

Pattern No. 63
8 feet 9 inches high
30 inches wide
20 inches deep

Pattern No. 62
8 feet high
24 inches wide
16 inches deep

# WALTHAM

*Except for those fabulous clocks for which you wouldn't mind cutting a hole in the ceiling, Grandfather or Hanging Clocks under eight feet are the most desirable and practical. Some manufacturers produced certain of these clocks in two styles: one to stand on the floor, and the other to hang on a wall. Of the clocks designed with two styles, the hanging style is considered to be the most valuable.*

| 923 | 924 | 925 |
|---|---|---|
| Pattern No. 14 | Pattern No. 71 | Pattern No. 69 |
| 8 feet 7 inches high | 8 feet 8 inches high | 8 feet high |
| 29 inches wide | 30 inches wide | 28 inches wide |
| 18 inches deep | 20 inches deep | 18 inches deep |

1905

# WALTHAM

Pattern No. 60
   8 feet 6 inches high
   27 inches wide
   19 inches deep

Pattern No. 40
   8 feet 7 inches high
   29 inches wide
   18 inches deep

Pattern No. 35
   8 feet 6 inches high
   28 inches wide
   19½ inches deep

# WALTHAM

### 929
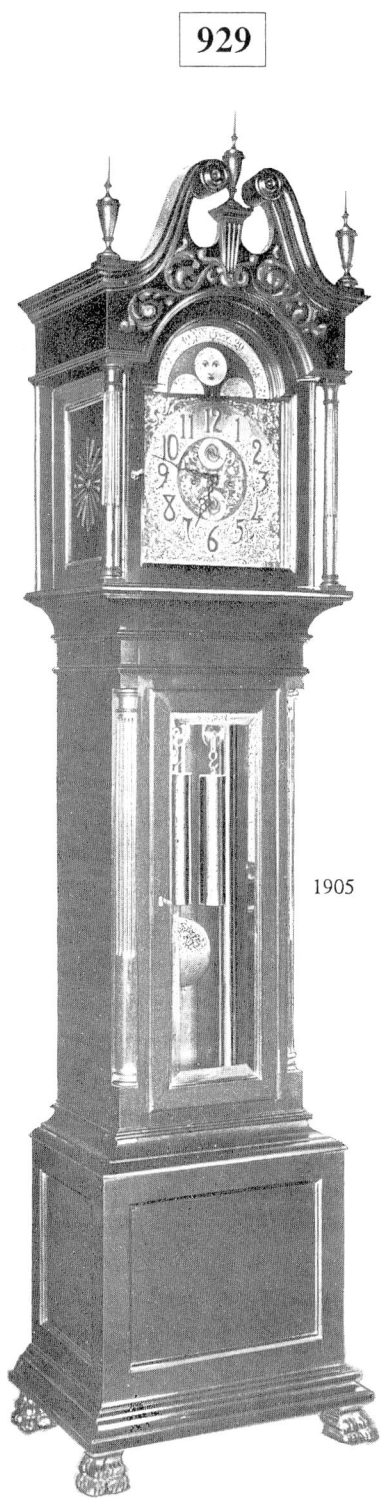

Pattern No. 7
8 feet 4 inches high
26 inches wide
15 inches deep

### 930
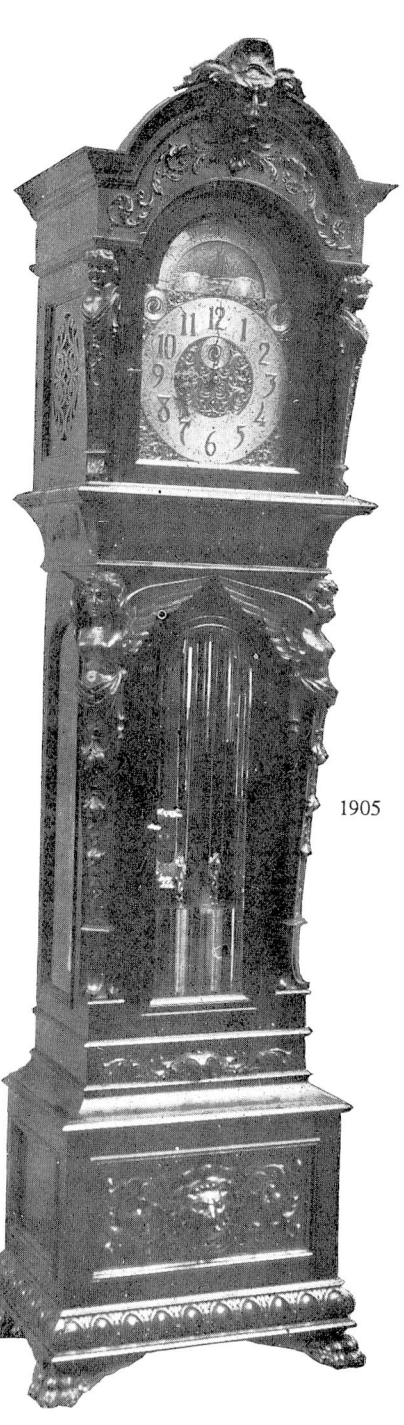

Pattern No. 68
8 feet 5 inches high
27 inches wide
18 inches deep

### 931
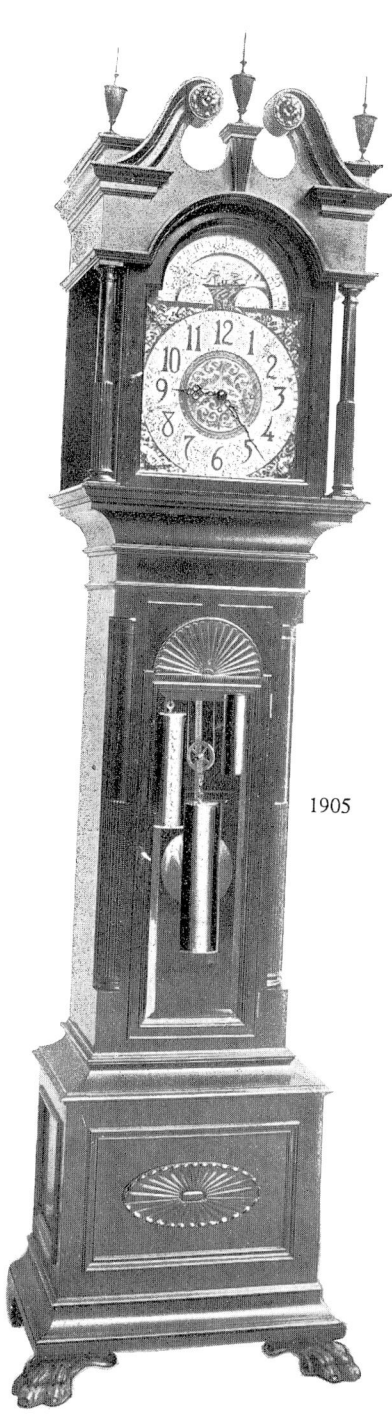

Pattern No. 6
8 feet 6 inches high
25 inches wide
15 inches deep

# WALTHAM

*To receive top prices, movements must be complete and original, and they must be in good running or repairable condition. Damaged, worn-out or missing parts will reduce prices accordingly.*

### 932

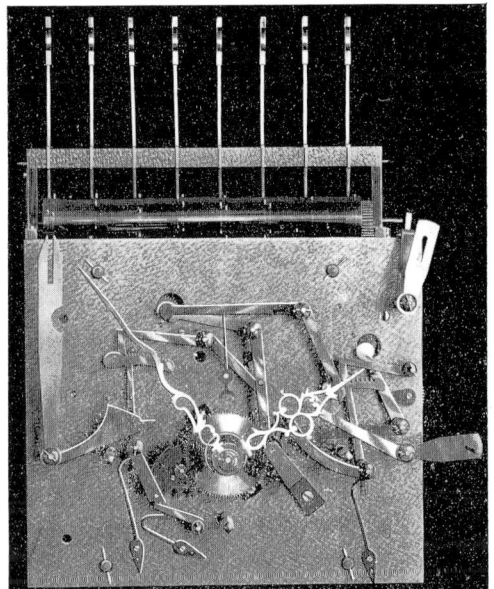

1905

Nine Tube Movement

### 933

1905

Nine Tube Movement, in case

### 934

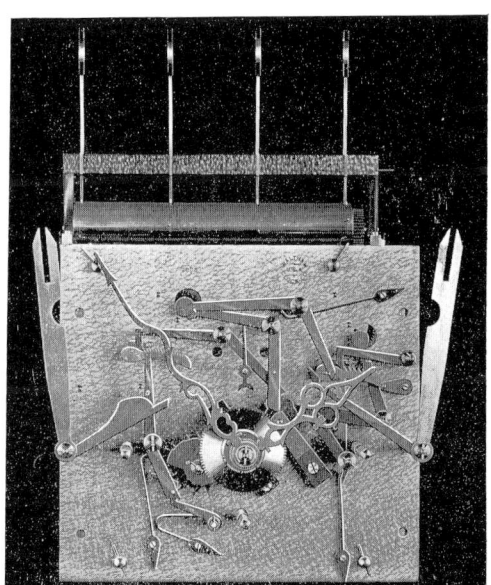

1905

Five Tube Movement

### 935

1905

Chime Dial No. 1

# WALTHAM

### 936

1905

Chime Dial No. 2

### 937

1905

Half-hour Strike Dial, Second Grade

### 938

1905

Half-Hour Strike Dial, Painted

### 939

1905

Half-hour Strike Dial, First Grade

# WALTHAM

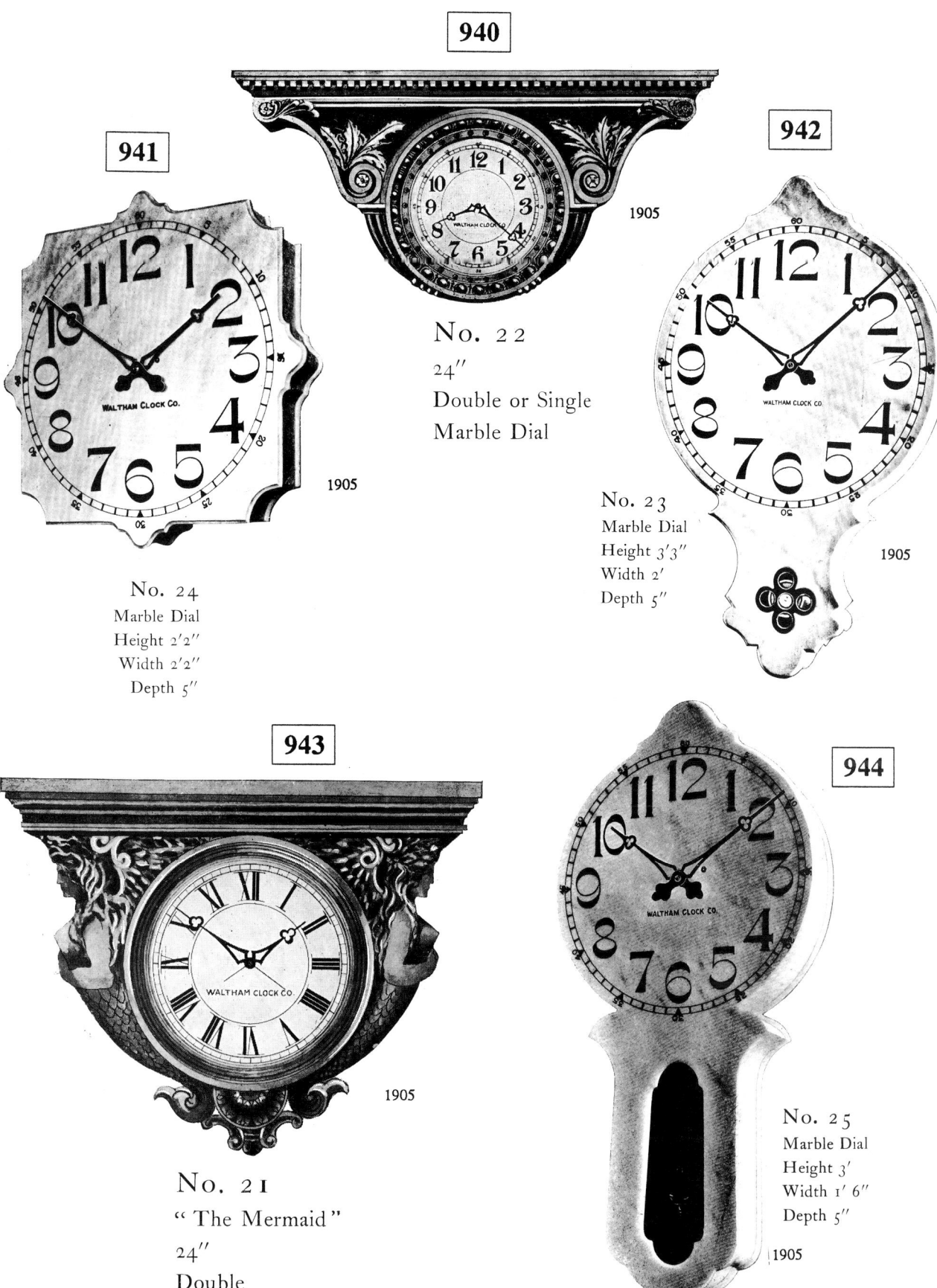

**940** — No. 22, 24″ Double or Single Marble Dial, 1905

**941** — No. 24, Marble Dial, Height 2′2″, Width 2′2″, Depth 5″, 1905

**942** — No. 23, Marble Dial, Height 3′3″, Width 2′, Depth 5″, 1905

**943** — No. 21, "The Mermaid", 24″ Double Marble Dial, 1905

**944** — No. 25, Marble Dial, Height 3′, Width 1′ 6″, Depth 5″, 1905

# WARREN TELECHRON COMPANY
## Ashland, Mass.

Henry Ellis Warren received the first of his many patents on an electrical clock on July 13, 1909. Warren's first experimental electric clock, a family banjo clock originally made about 1830 and converted to electric drive prior to 1909, is now part of the permanent collection of the American Clock & Watch Museum at Bristol, CT.

In 1912, the Warren Clock Company was founded at Ashland, MA and in 1915, commercial production of battery clocks commenced. Warren's greatest achievement was the development of alternating current electric clocks in 1916, which opened the door for inexpensive domestic electric clocks.

Warren's alternating current motor was well protected by patents, giving fits to other clock manufacturers who wanted to enter this promising market in the latter 1920's. Though competitors eventually introduced domestic electrics, their earliest attempts were often inferior and troublesome, causing them endless grief for several years.

In 1926, Warren's firm became known as Warren Telechron Company, and in 1946 the firm's name was changed to Telechron, Inc. In 1951, the firm was merged with General Electric which had owned an interest in the firm for some years. By 1955, General Electric's trade catalogs called some clock models "General Electric Telechron" and the trademark "Telechron" has been used on some General Electric models through the present day.

945

TYPE "B" MASTER CLOCK
34 1/2 inches high
16 1/2 inches wide

946

Fig. 6. Table Clock.
12" high, 8" wide

947

1919

TYPE "A" MASTER CLOCK
56 1/2 inches high
17 1/4 inches wide

## The WATERBURY CLOCK COMPANY
Waterbury, Connecticut

Formed March 5, 1857, as a joint stock corporation, the Waterbury Clock Company from its inception was an operation designed to be a major user of brass produced by the parent firm, the Benedict & Burnham Manufacturing Company. Utilizing the best talent available to them, they hired veteran clockmaker Chauncey Jerome to set up the new firm's casemaking shop and his brother Noble Jerome, a well-known clock mechanic, to set up the movement manufacturing operation.

Though this firm became a major clock producer, after 1890 they became a major manufacturer of non-jewelled pocket watches, supplying R. H. Ingersoll & Brother, a major mail order firm. Large-scale production and profitability were enjoyed for more than two decades with this association. Major factory expansions between 1900 and 1915 made this the largest clock manufacturing facility in America.

In 1922, the Waterbury Clock Company purchased the Ingersoll operation whose business had begun to sour after 1910 and had gone bankrupt two years previous because of poor management. Waterbury's operation began to decline and was particularly hard hit by the Great Depression.

By 1932, their huge factory complex was little used. They barely avoided bankruptcy, but the firm was reorganized as the Ingersoll-Waterbury Company with investors raising half a million dollars in new capital. During this period the popular "Mickey Mouse" character watch was made and electric clocks were added to the line.

After America entered World War II, the Ingersoll-Waterbury Company switched almost 100% to manufacturing war products. In 1942, the operation was purchased by a group of Norwegian investors and a new factory was built at Middlebury, CT. In 1944, the firm became known as United States Time Corporation and introduced the popular "Timex" watch shortly after the war. In November, 1969, U. S. Time was succeeded by Timex Corporation which continues business at Middlebury, CT.

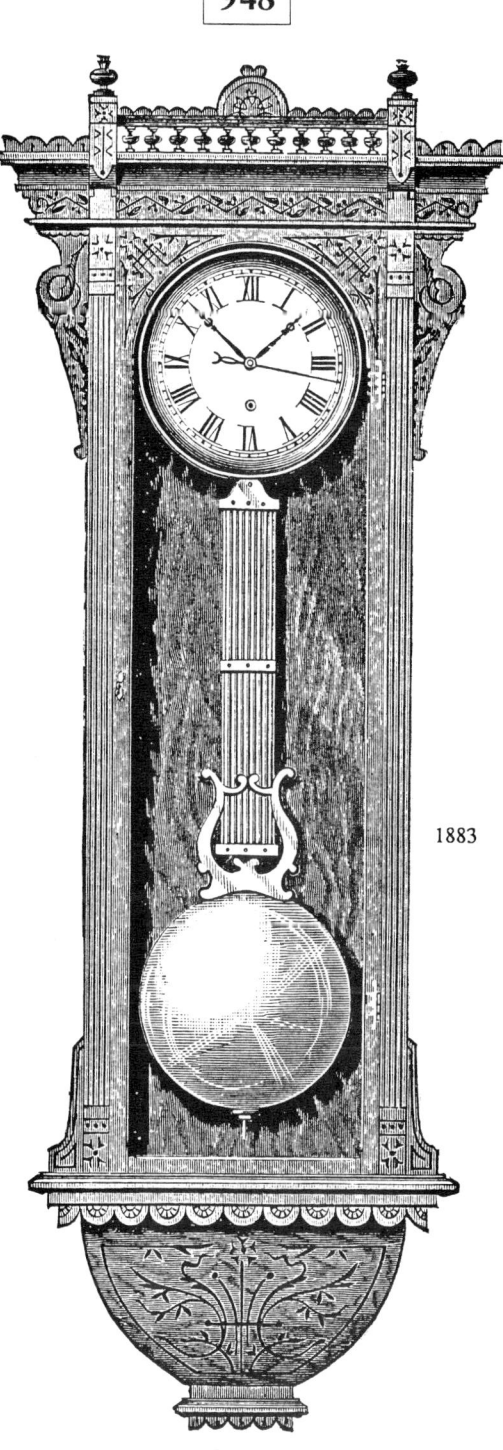

948

1883

REGULATOR, No. 10.
WITH SWEEP SECOND.
PIN ESCAPEMENT,

Weight—Time.

12 inch Dial.

Eight Day,

Height, 85 inches.

# WATERBURY

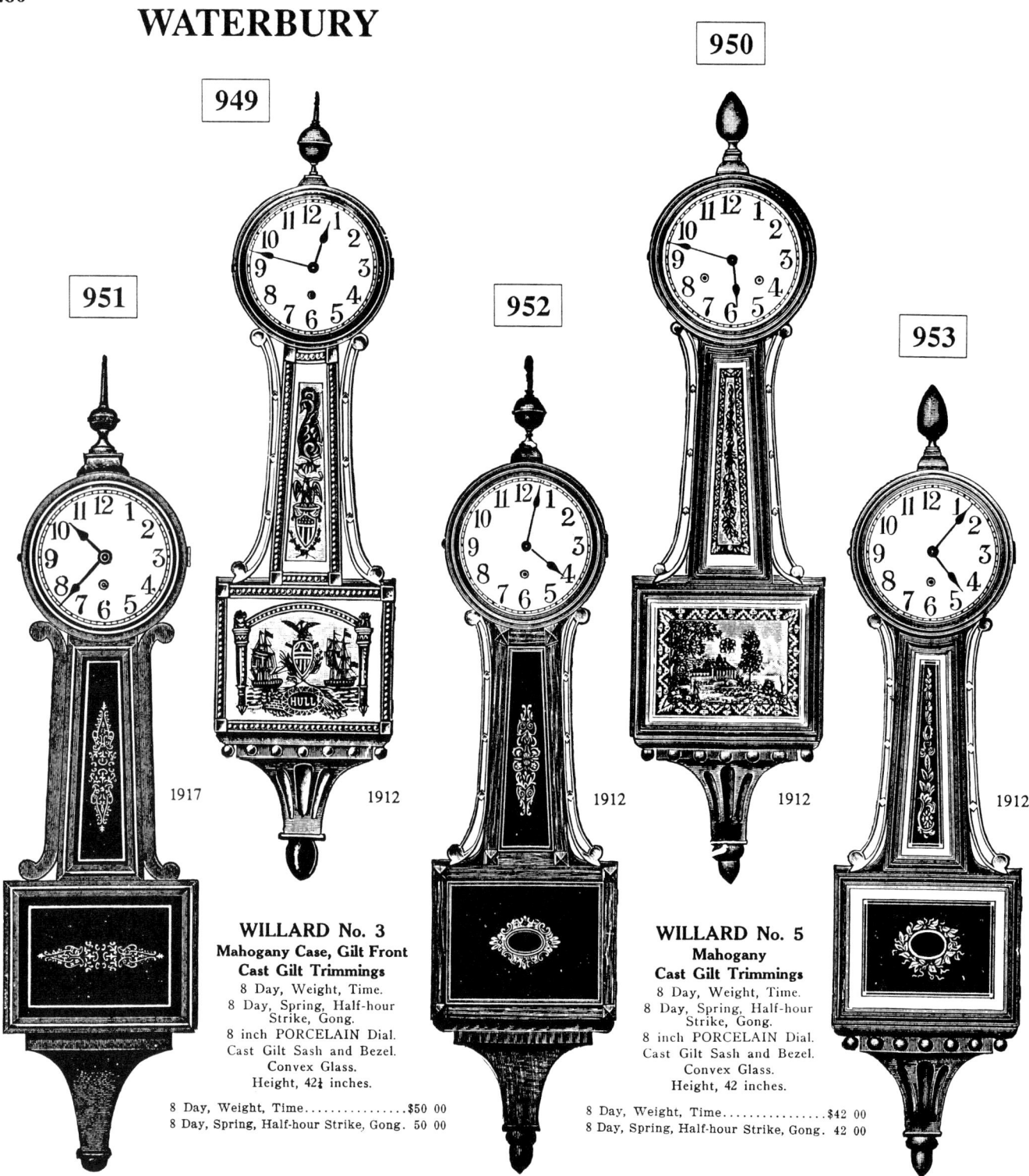

**WILLARD No. 3**
Mahogany Case, Gilt Front
Cast Gilt Trimmings
8 Day, Weight, Time.
8 Day, Spring, Half-hour Strike, Gong.
8 inch PORCELAIN Dial.
Cast Gilt Sash and Bezel.
Convex Glass.
Height, 42¼ inches.

8 Day, Weight, Time.............$50 00
8 Day, Spring, Half-hour Strike, Gong. 50 00

**WILLARD No. 5**
Mahogany
Cast Gilt Trimmings
8 Day, Weight, Time.
8 Day, Spring, Half-hour Strike, Gong.
8 inch PORCELAIN Dial.
Cast Gilt Sash and Bezel.
Convex Glass.
Height, 42 inches.

8 Day, Weight, Time.............$42 00
8 Day, Spring, Half-hour Strike, Gong. 42 00

**WILLARD No. 11**
Birch Mahogany
or Mission Oak
8 Day, Spring, Time.
8 inch PORCELAIN Dial.
Convex Glass.
Height, 42¼ inches.

**WILLARD No. 1**
Flemish Oak
Cast Gilt Trimmings
8 Day, Weight, Time.
8 Day, Spring, Half-hour Strike, Gong.
8 inch PORCELAIN Dial.
Cast Gilt Sash and Bezel.
Convex Glass.
Height, 43 inches.

8 Day, Weight, Time.............$25 00
8 Day, Spring, Half-hour Strike, Gong. 25 00

**WILLARD No. 2**
Mahogany
Cast Gilt Trimmings
8 Day, Weight, Time.
8 Day, Spring, Half-hour Strike, Gong.
8 inch PORCELAIN Dial.
Cast Gilt Sash and Bezel.
Convex Glass.
Height, 42 inches.

8 Day, Weight, Time.............$40 00
8 Day, Spring, Half-hour Strike, Gong. 40 00

# WATERBURY

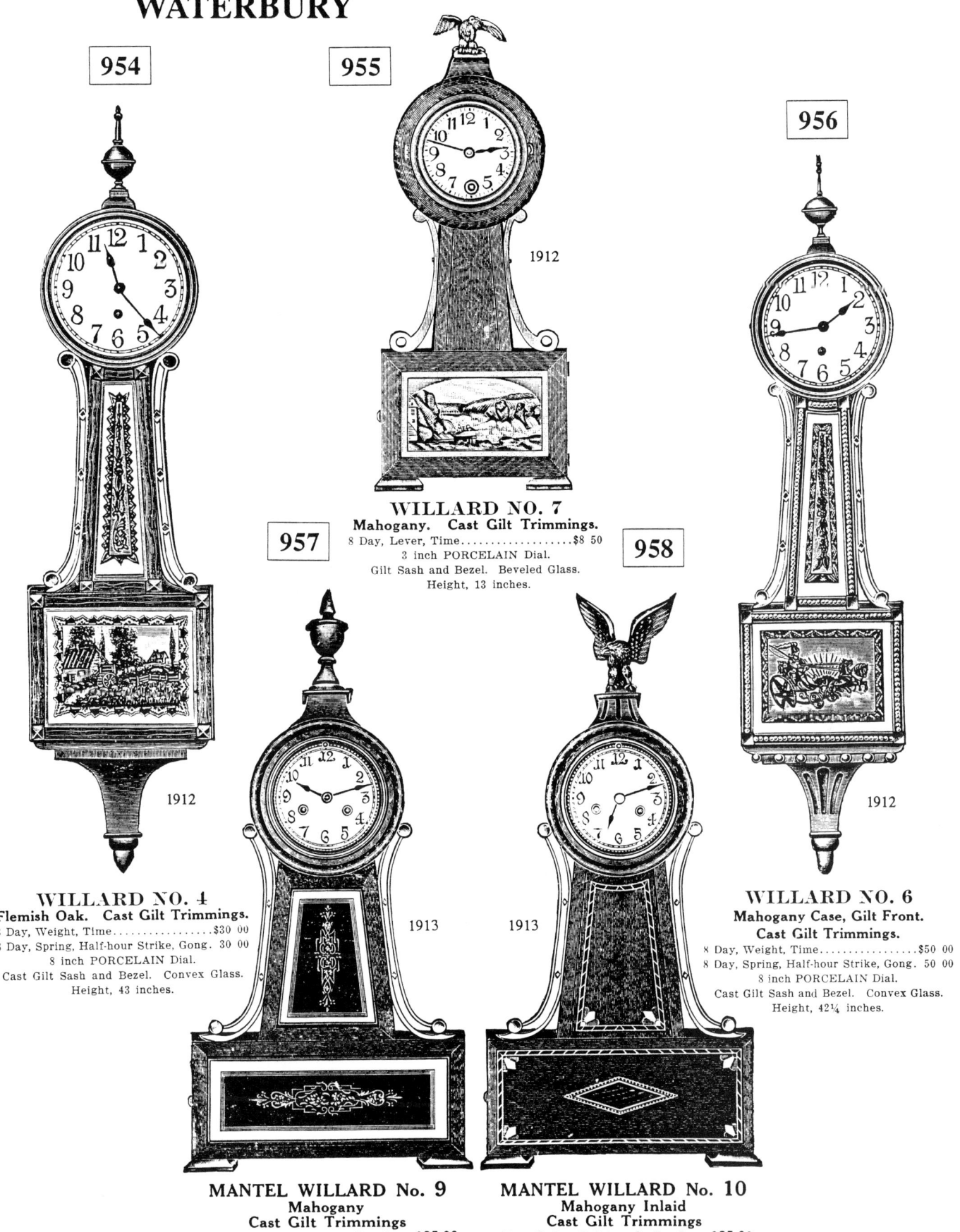

**954** — **WILLARD NO. 4**
Flemish Oak. Cast Gilt Trimmings.
8 Day, Weight, Time.............$30 00
8 Day, Spring, Half-hour Strike, Gong. 30 00
8 inch PORCELAIN Dial.
Cast Gilt Sash and Bezel. Convex Glass.
Height, 43 inches.

**955** — **WILLARD NO. 7**
Mahogany. Cast Gilt Trimmings.
8 Day, Lever, Time...............$8 50
3 inch PORCELAIN Dial.
Gilt Sash and Bezel. Beveled Glass.
Height, 13 inches.

**956** — **WILLARD NO. 6**
Mahogany Case, Gilt Front.
Cast Gilt Trimmings.
8 Day, Weight, Time.............$50 00
8 Day, Spring, Half-hour Strike, Gong. 50 00
8 inch PORCELAIN Dial.
Cast Gilt Sash and Bezel. Convex Glass.
Height, 42¼ inches.

**957** — **MANTEL WILLARD No. 9**
Mahogany
Cast Gilt Trimmings
8 Day, Spring, Half-hour Strike, Gong **$25.00**
4½ inch PORCELAIN Dial.
Cast Gilt Sash and Bezel.
Convex Beveled Glass.
Height, 22¼ inches.
Width, 10¾ inches.

**958** — **MANTEL WILLARD No. 10**
Mahogany Inlaid
Cast Gilt Trimmings
8 Day, Spring, Half-hour Strike, Gong **$25.00**
4½ inch PORCELAIN Dial.
Cast Gilt Sash and Bezel.
Convex Beveled Glass.
Height, 22½ inches.
Width, 10¾ inches.

# WATERBURY

### TOPSEY.
ONE-DAY. WINKER. IRON.
Height, 16½ Inches.

### SAMBO.
ONE-DAY. WINKER. IRON.
Height, 16 Inches.

### CONTINENTAL.
ONE-DAY. WINKER. IRON.
Height, 16 Inches.

### ORGAN GRINDER.
ONE-DAY. WINKER. IRON.
Height, 17½ Inches.

# WATERBURY

*For comprehensive listing of Calendar Clocks, see Tran Duy Ly "Calendar Clocks - A Guide to Identification & Prices."*

**963**

### CALENDAR No. 22.

OAK. BRASS RAILING.

CABINET FINISH. GLASS SIDES.

8 Day, Weight, Time.
8 Day, Weight, Half-hour Strike, Gong.
Brass Weight.

Beats Seconds, Dead-beat Escapement, Retaining Power

Solid Polished Movement Frames.
Dials 10 inches.   Height 76¼ inches.

8 Day, Weight, Time.......................$85 00
8 Day, Weight, Half-hour Strike, Gong......105 00

**964**

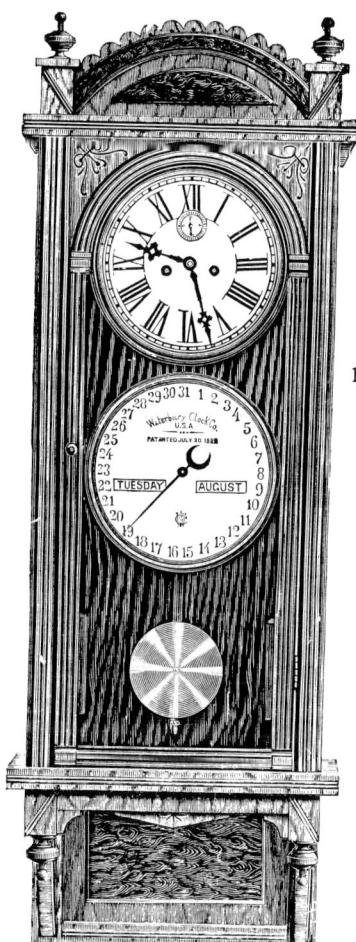

1893

### CALENDAR No. 26.

OAK, WALNUT OR MAHOGANY.

CABINET FINISH.

8 Day, Weight, Time.
Brass Weights.
Dead-beat Escapement, Retaining Power.
Solid Movement Frames.
Dials 9 inches.   Height 46 inches.

8 Day, Weight, Time............$28 00

### CALENDAR No. 44.

WALNUT.

8 Day, Half-hour Strike, Gong.
Also with Alarm.
Dials 6 inches.   Height 24 inches.

**965**

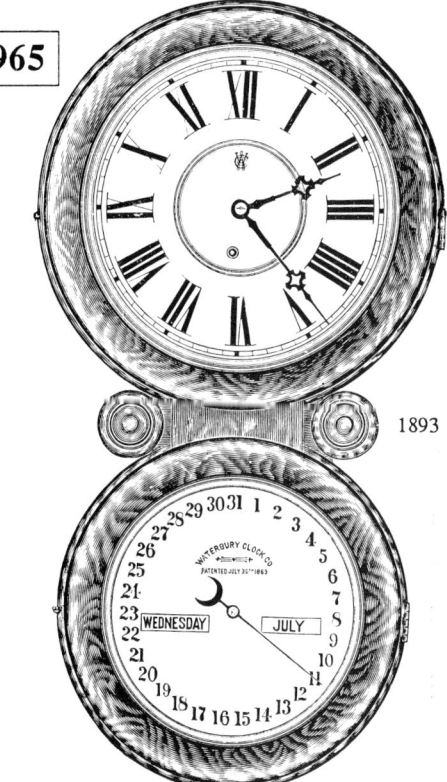

1893

### CALENDAR No. 34

Height, 29½ inches.

OAK OR ROSEWOOD

8 Day, Time................................$13 00
8 Day, Half-hour Strike, Gong............. 15 00
Time Dial, 12 inches.   Calendar Dial, 10 inches.

**966**

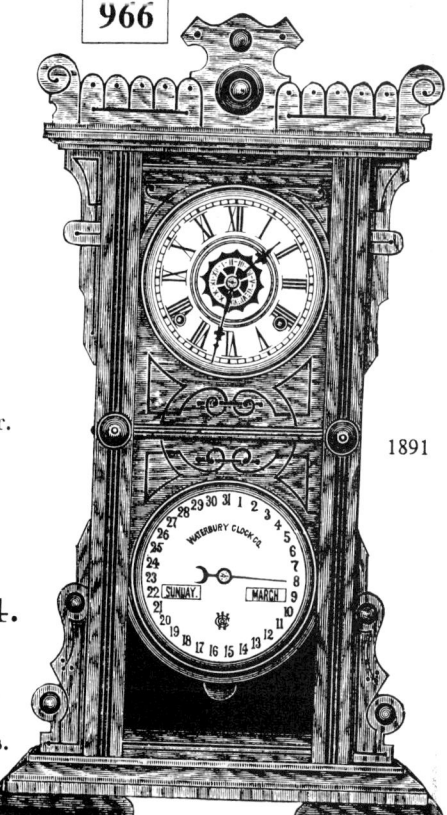

1891

**CALENDAR NO. 44**

# WATERBURY

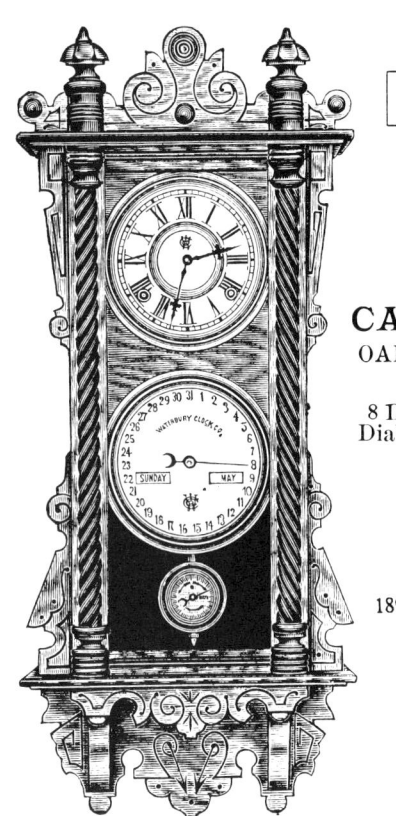

CALENDAR No. 36.

### 967

**CALENDAR No. 36.**
OAK, WALNUT OR CHERRY.

8 Day, Time.
8 Day, Half-hour Strike, Gong.
Dials 6 inches.   Height 28 inches.

### 968

**CALENDAR No. 38.**
WALNUT.

8 Day, Time.
8 Day, Half-hour Strike, Gong.
Dials 6 inches.   Height 28 inches.

CALENDAR No. 38.

*The year entered beside each clock is the year of the catalog from which the illustration was taken. Some clocks may have been produced in an earlier year.*

CALENDAR No. 40.

### 969

**CALENDAR No. 40.**
OAK, WALNUT OR CHERRY.

8 Day, Half-hour Strike, Gong.
Dials 6 inches.   Height 24 inches.

### 970

**CALENDAR No. 42.**
WALNUT.

8 Day, Half-hour Strike, Gong.
Also with Alarm.
Dials 6 inches.   Height 24 inches.

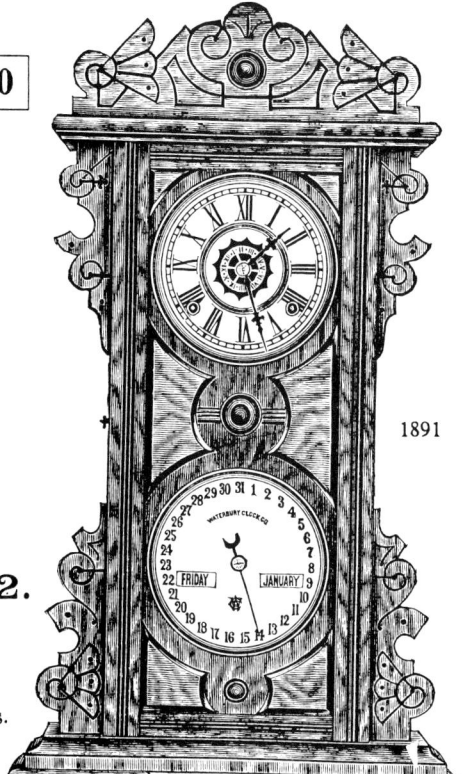

CALENDAR No. 42.

# WATERBURY

*For comprehensive listing of Calendar Clocks, see Tran Duy Ly "Calendar Clocks - A Guide to Identification & Prices."*

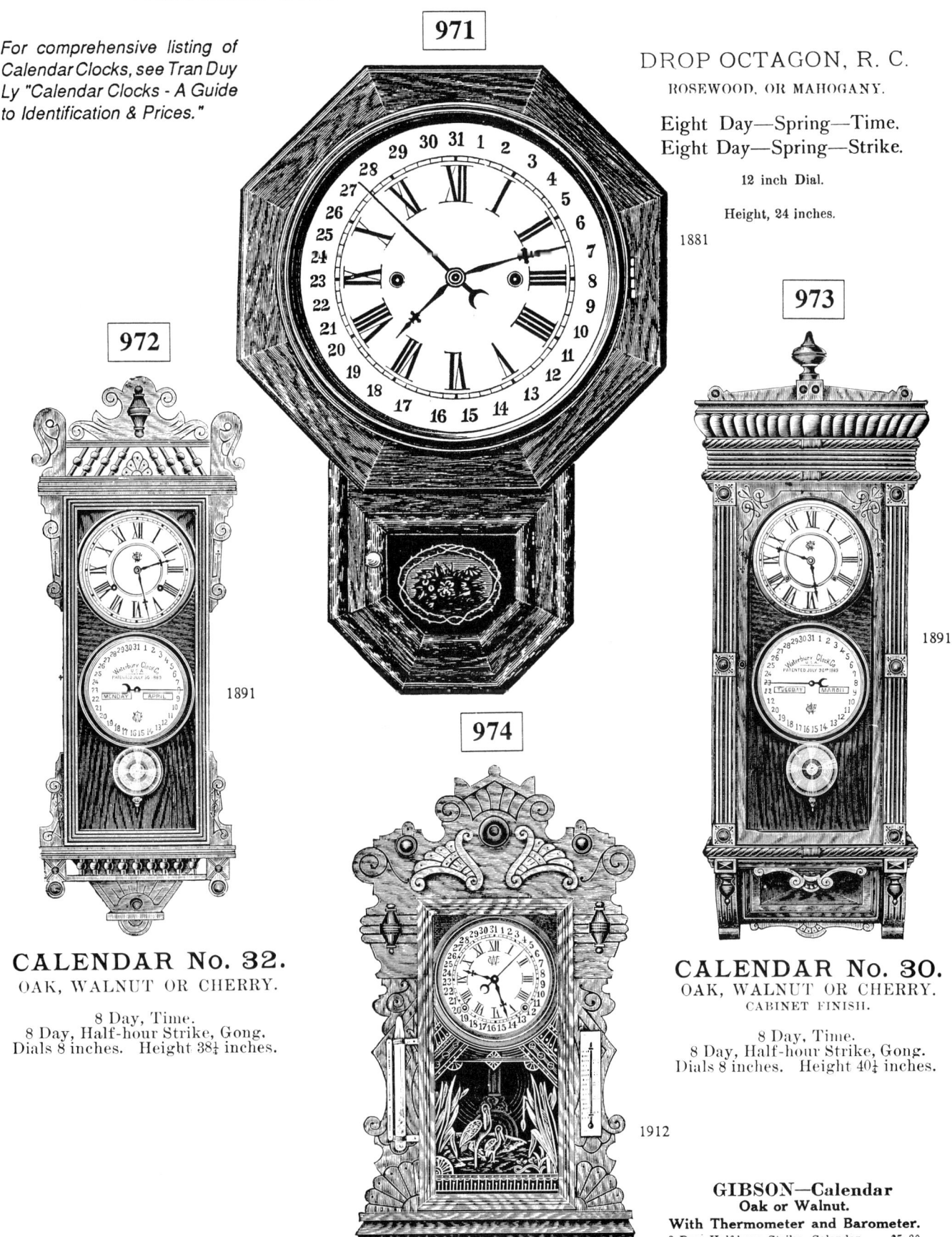

**971**

DROP OCTAGON, R. C.

ROSEWOOD, OR MAHOGANY.

Eight Day—Spring—Time.
Eight Day—Spring—Strike.

12 inch Dial.

Height, 24 inches.

1881

**972**

**CALENDAR No. 32.**
OAK, WALNUT OR CHERRY.

8 Day, Time.
8 Day, Half-hour Strike, Gong.
Dials 8 inches.   Height 38¼ inches.

**973**

**CALENDAR No. 30.**
OAK, WALNUT OR CHERRY.
CABINET FINISH.

8 Day, Time.
8 Day, Half-hour Strike, Gong.
Dials 8 inches.   Height 40¼ inches.

**974**

GIBSON

**GIBSON—Calendar**
Oak or Walnut.
With Thermometer and Barometer.

8 Day, Half-hour Strike, Calendar.....$5 80
8 Day, Half-hour Strike, Gong, Calendar  6 30
Cannot be fitted with Alarm.
6 inch Rococo Dial. Height, 24 inches.

# WATERBURY

The year entered beside each clock is the year of the catalog from which the illustration was taken. Some clocks may have been produced in an earlier year.

**975**

**GLENWOOD, CALENDAR.**

Eight Day—Spring—Time.
Eight Day—Spring—Strike.

12 inch Dial.

Height, 25½ inches.

**BUFFALO.** WALNUT.

8 Day, Half-hour Strike, Calendar.
Can not be fitted with Alarm.
8 Day, Half-hour Strike.
8 Day, Half-hour Strike, Gong.
Also with Alarm.
Dial 8 inches.   Height 26⅞ inches.

1881

**977**

1881

**REGENT—CALENDAR**
OAK OR POLISHED VENEERED
Eight Day—Spring—Time.
Eight Day—Spring—Strike.

12 inch Dial.
Height, 32 inches.

For comprehensive listing of Calendar Clocks, see Tran Duy Ly "Calendar Clocks - A Guide to Identification & Prices."

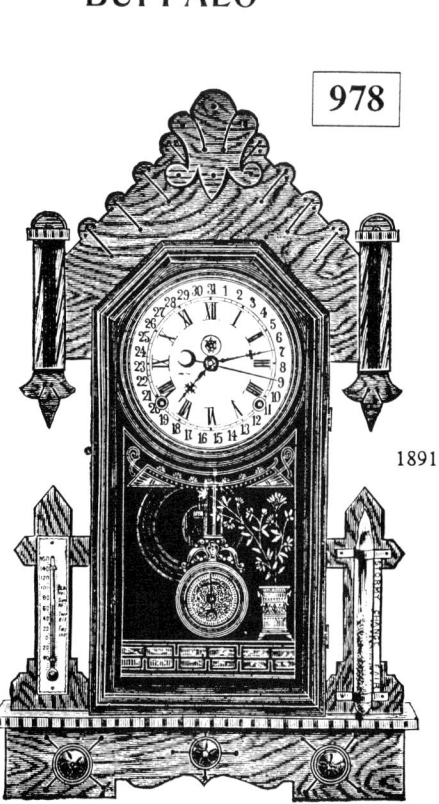

**976**

1891

**BUFFALO**

**978**

1891

**COMBINE, CALENDAR.**
WALNUT.

8 Day, Half-hour Strike, Calendar.
Can not be fitted with Alarm.
Dial 6 inches.   Height 22 inches.

# WATERBURY

### SPRINGFIELD
**Oak.**

8 Day, Spring, Time..................$9 00
8 Day, Spring, Half-hour Strike, Gong.10 50
8 Day, Spring, Time, Calendar......... 9 50
8 Day, Spring, Half-hour Strike, Gong,
   Calendar .........................11 00
8 inch Dial.
Height, 39⅜ inches.  Width, 16½ inches.

979

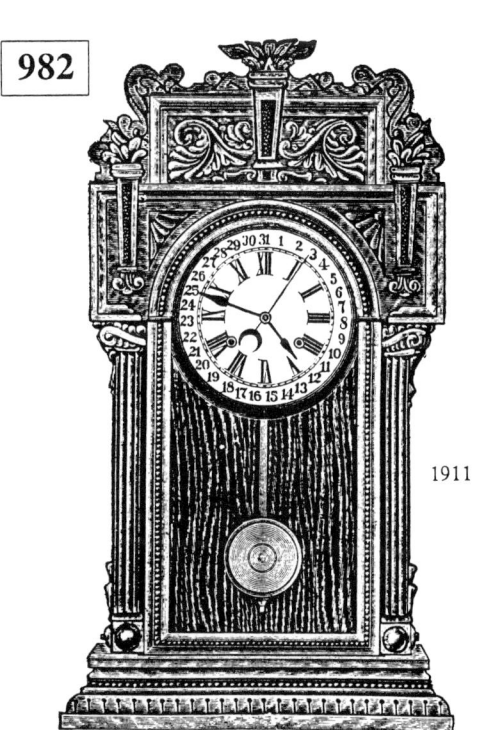
1912

980

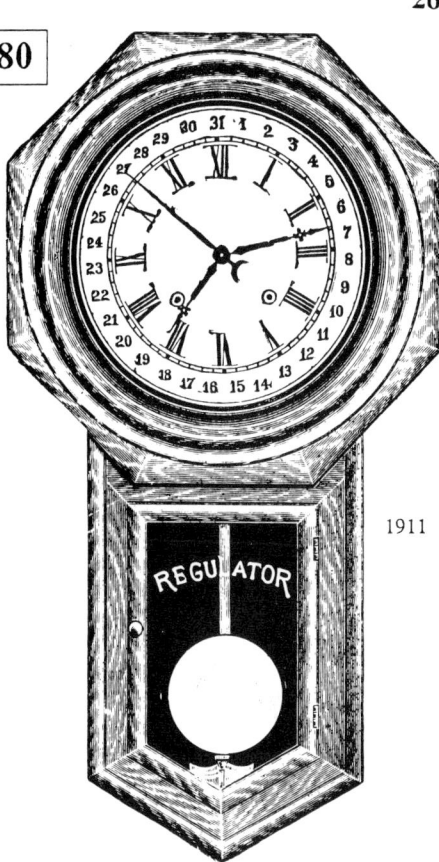
1911

### CONSORT—Oak or Walnut.

8 Day, Time ........................$9 00
8 Day, Half-hour Strike..............10 00
8 Day, Half-hour Strike, Gong........10 50
8 Day, Time, Calendar................ 9 50
8 Day, Half-hour Strike, Calendar....10 50
8 Day, Half-hour Strike, Gong, Calendar.11 00
12 inch Dial.  Height, 32 inches.

981

1912

### WALTON
**Oak.**

8 Day, Spring, Time.................$11 00
8 Day, Spring, Half-hour Strike, Gong. 12 50
8 Day, Spring, Time, Calendar....... 11 50
8 Day, Spring, Half-hour Strike, Gong,
   Calendar ......................... 13 00
10 inch Dial.
Height, 41 inches.  Width, 15¼ inches.

982

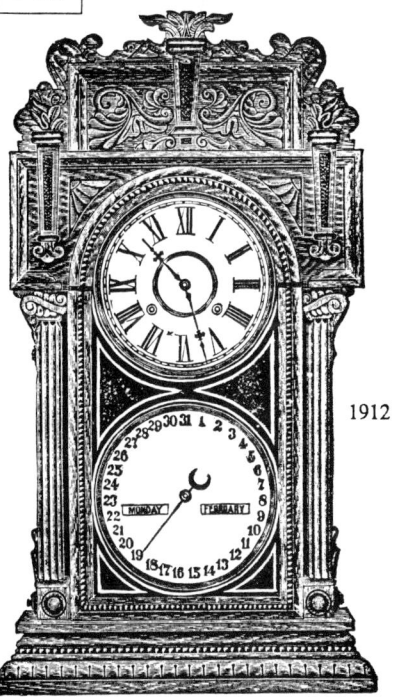
1911

### NIAGARA
**Walnut.**

8 Day, Half-hour Strike..............$7 50
8 Day, Half-hour Strike, Calendar..... 8 00
8 inch Rococo Dial.  Height, 28¼ inches.

983

1912

### CALENDAR NO. 43
**Oak or Walnut.**

8 Day, Half-hour Strike, Gong.......$14 40
8 inch Dials.  Height, 28¼ inches.

# WATERBURY

*For comprehensive listing of Calendar Clocks, see Tran Duy Ly "Calendar Clocks - A Guide to Identification & Prices."*

**984** ADMIRAL—Oak.

| | |
|---|---|
| 8 Day, Time | $7 50 |
| 8 Day, Half-hour Strike | 8 50 |
| 8 Day, Half-hour Strike, Gong | 9 00 |
| 8 Day, Time, Calendar | 8 00 |
| 8 Day, Half-hour Strike, Calendar | 9 00 |
| 8 Day, Half-hour Strike, Gong, Calendar | 9 50 |
| 30 Day, Time, with Second Hand | 9 50 |

12 inch Dial. Height, 32 inches.

**985** ENGLISH DROP No. 2
VENEERED, INLAID

Dial 12 inches. Height 27¾ inches.
8 Day, Time, Calendar.
8 Day, Half-hour Strike, Calendar.

**986** ENGLISH DROP No. 3
VENEERED, INLAID

Dial 12 inches. Height 27¾ inches.
8 Day, Time, Calendar.
8 Day, Half-hour Strike, Calendar.
Can not be fitted with Alarm.

**987** ENGLISH DROP No. 1
VENEERED

8 Day, Time, Calendar.
8 Day, Half-hour Strike, Calendar.
Dial 12 inches. Height 24⅜ inches.

# WATERBURY

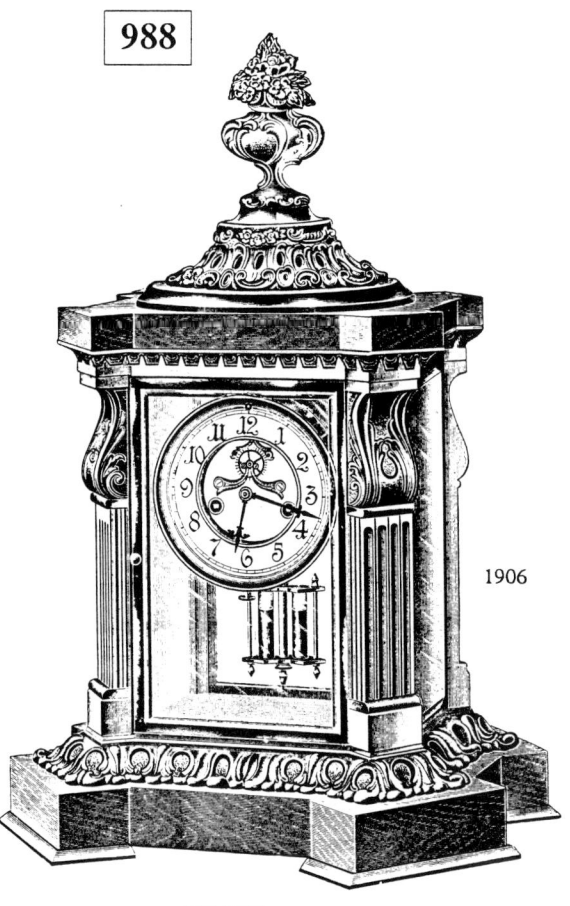

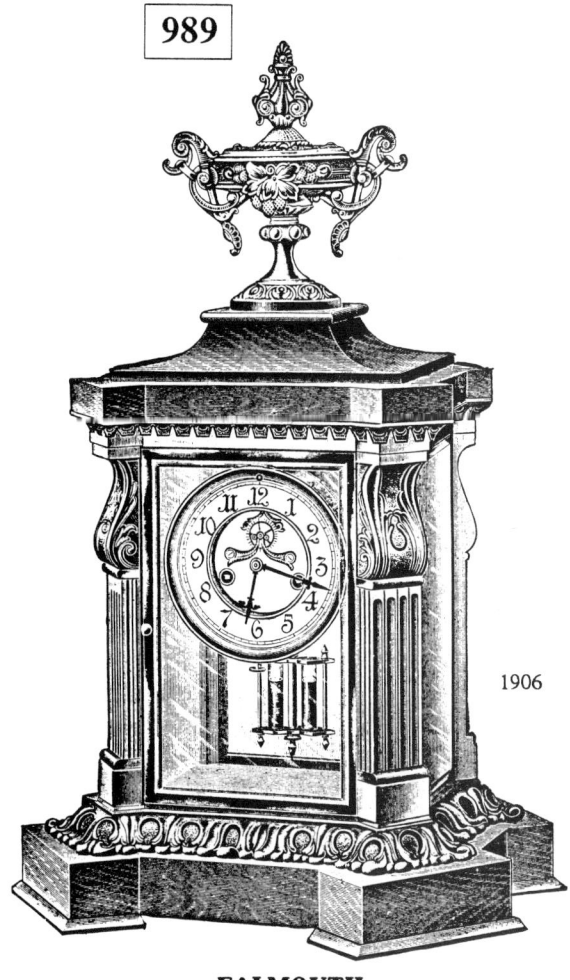

### PARIS
**Rich Gold Plated.**

Polished Mahogany Base and Top.
8 Day, Half-hour Strike, Gong..........$60 00
4½ inch IVORY Dial. Ivory Center. Visible Escapement.
Cast Gilt Bezel. Beveled Glass Front, Sides and Back.
Height, 19 inches. Width, 13¼ inches.

### FALMOUTH
**Rich Gold Plated.**

Polished Mahogany Base and Top.
8 Day, Half-hour Strike, Gong ...... $60 00
4½ inch IVORY Dial. Ivory Center. Visible Escapement.
Cast Gilt Bezel. Beveled Glass Front, Sides and Back.
Height, 21¼ inches. Width, 13¼ inches.

*For comprehensive listing of Waterbury Clocks, See Tran Duy Ly "Waterbury Clocks - History, Identification & Price Guide."*

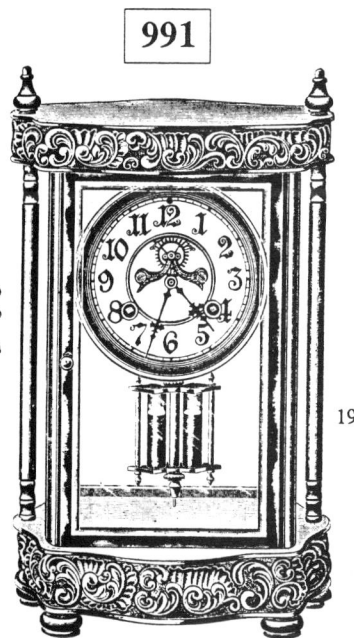

### CALAIS
**Rich Gold Plated.**

8 Day, Half-hour Strike, Gong......... $45 00
4¼ inch IVORY Dial. Ivory Center. Visible Escapement.
Cast Gilt Bezel. Beveled Glass Front, Sides and Back.
Height, 11⅞ inches. Width 7 inches.

### FLANDERS
**Rich Gold Plated.**

Polished Mahogany Base and Top.
8 Day, Half-hour Strike, Gong ......... $34 00
4¼ inch IVORY Dial. Ivory Center. Visible Escapement.
Cast Gilt Bezel. Beveled Glass Front, Sides and Back.
Height, 12⅛ inches. Width, 6⅝ inches.

# WATERBURY

**992**

*Only clocks in ORIGINAL MINT condition and in good running order will command top prices. A replaced or damaged dial, missing parts, dull finish, worn-off gold gild, re-painted or chipped beveled glass will reduce prices accordingly.*

**993**

### CHALONS
**Rich Gold Plated or Syrian Bronze.**
8 Day, Half-hour Strike, Gong ....... $34 00
4¼ inch IVORY Dial. Ivory Center. Visible Escapement.
Cast Gilt Bezel. Beveled Glass Front, Sides and Back.
Height, 14 inches. Width, 9 inches.

### ALBURTIS
**Rich Gold Plated.**
8 Day, Half-hour Strike, Gong ....... $40 00
4½ inch IVORY Dial. Ivory Center.
Visible Escapement. Cast Gilt Bezel.
Beveled Glass Front, Sides and Back.
Height, 16 inches. Width, 10 inches.

1906

**ALBURTIS**

**CHALONS**

**994**

*Most white metal and brass ornamentals and clock cases are gilded or silver plated. Great care must be taken when cleaning these surfaces. The thin layer of gilding or silver plating will be easily removed by cleaning solutions and strong rubbing. If you believe cleaning is necessary, hire the services of a professional and avoid the expense of re-gilding or re-plating.*

**995**

### AVIGNON
**Rich Gold Plated or Golden Bronze**
8 Day, Half-hour Strike, Gong.. $40 00
4½ inch IVORY Dial. Ivory Center.
Visible Escapement. Cast Gilt Bezel.
Beveled Glass Front, Sides and Back.
Height, 17½ inches. Width 8⅝ inches.

### BORDEAUX
**Rich Gold Plated.**
8 Day, Half-hour Strike, Gong.. $40 00
4½ inch IVORY Dial. Ivory Center.
Visible Escapement. Cast Gilt Bezel.
Beveled Glass Front, Sides and Back.
Height, 18¼ inches. Width, 8⅝ inches.

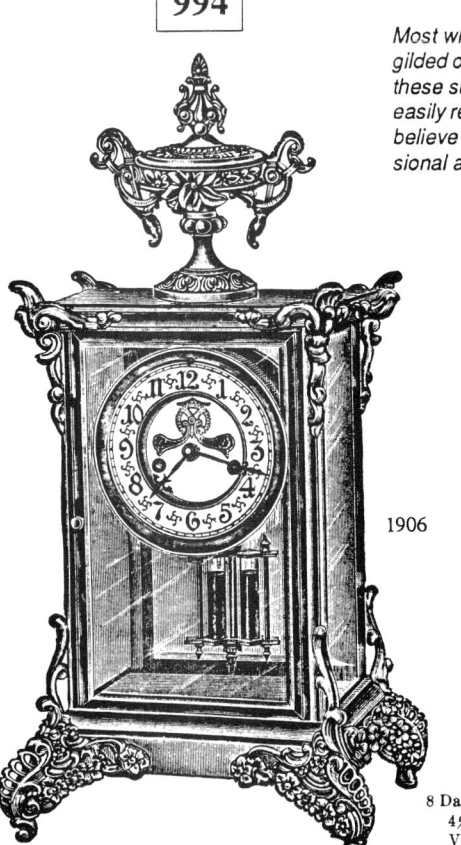

1906

**BORDEAUX**

1906

**AVIGNON**

# WATERBURY

1906

### NEVERS
**Rich Gold Plated or Golden Bronze.**
8 Day, Half-hour Strike, Gong ........ $35 00
3½ inch IVORY Dial. Ivory Center. Visible Escapement.
Cast Gilt Bezel. Beveled Glass Front, Sides and Back.
Height, 13⅝ inches. Width, 9 inches.

1906

### VERSAILLES
**Rich Gold Plated.**
8 Day, Lever, Half-hour Strike, Gong .. $22 00
2½ inch IVORY Dial. Ivory Center.
Beveled Glass Front and Sides.
Height, 10 inches. Width, 6¾ inches.

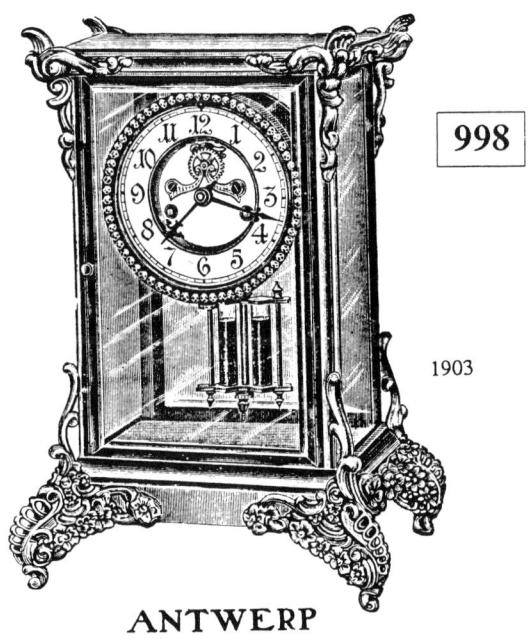

1903

### ANTWERP
**Rich Gold Plated.**
8 Day, Half-hour Strike, Gong
   (without Jeweled Sash).......... $43 00
8 Day, Half-hour Strike, Gong
   (Jeweled Sash) ................. 46 00
4½ inch IVORY Dial. Ivory Center. Visible Escapement.
Cast Gilt Bezel. Beveled Glass Front, Sides and Back.
Height, 12¼ inches. Width, 8⅝ inches.
For Dial without Jeweled Sash see illustration of Alburtis.

1903

### SAVOY
**Rich Gold Plated.**
8 Day, Half-hour Strike, Gong ........ $40 00
8 Day, Half-hour Strike, Gong
   (Jeweled Sash) ................. 43 00
3½ inch IVORY Dial. Ivory Center.
Cast Gilt Bezel. Beveled Glass Front, Sides and Back.
Height, 10⅝ inches. Width, 7 inches.

# WATERBURY

### 1000

1905

**ORLEANS**
Rich Gold Plated.
8 Day, Half-hour Strike, Gong........ $42 00
3½ inch IVORY Dial. Ivory Center. Visible Escapement.
Cast Gilt Bezel. Beveled Glass Front, Sides and Back.
Height, 14¾ inches. Width, 7 inches.

### 1001

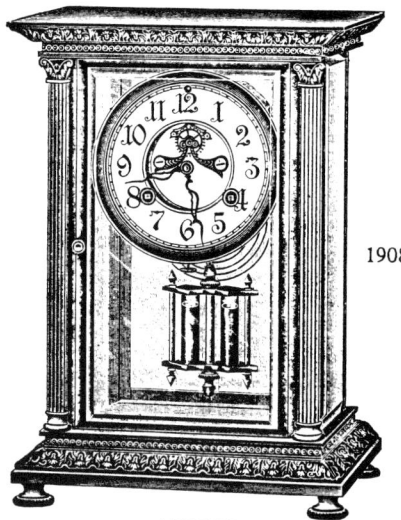

1908

**AUDE**
Rich Gold Plated.
8 Day, Half-hour Strike, Gong........ $35 00
4¼ inch IVORY Dial. Ivory Center. Visible Escapement.
Cast Gilt Bezel. Beveled Glass Front, Sides and Back.
Height, 10⅛ inches. Width, 7¾ inches.

### 1002

1905

**MOGUL**
Rich Gold Plated.
8 Day, Half-hour Strike, Gong........ $42 00
3½ inch IVORY Dial. Ivory Center. Visible Escapement.
Cast Gilt Bezel. Beveled Glass Front, Sides and Back.
Height, 15½ inches. Width, 7 inches.

### 1003

1908

**CANTAL**
Rich Gold Plated.
8 Day, Half-hour Strike, Gong........ $36 00
4¼ inch IVORY Dial. Ivory Center. Visible Escapement.
Cast Gilt Bezel. Beveled Glass Front, Sides and Back.
Height, 10⅜ inches. Width, 6⅝ inches.

### 1004

1905

**RENNES**
Rich Gold Plated.
8 Day, Half-hour Strike, Gong ....... $42 00
3½ inch IVORY Dial. Ivory Center. Visible Escapement.
Cast Gilt Bezel. Beveled Glass Front, Sides and Back.
Height, 13½ inches. Width, 8¼ inches.

### 1005

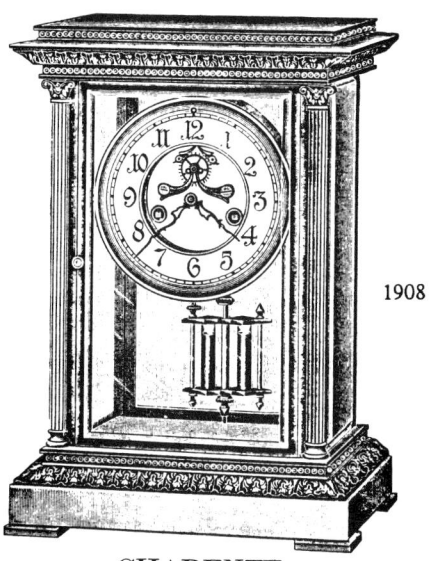

1908

**CHARENTE**
Rich Gold Plated.
8 Day, Half-hour Strike, Gong........ $38 00
4½ inch IVORY Dial. Ivory Center. Visible Escapement.
Cast Gilt Bezel. Beveled Glass Front, Sides and Back.
Height, 11⅜ inches. Width, 8⅛ inches.

# WATERBURY

### 1006

1905

**VERONA**

8 Day, Half-hour Strike, Gong. 3½ inch IVORY Dial. Ivory Center.
Visible Escapement. Cast Gilt Sash and Bezel. Beveled Glass.
Height, 13½ inches. Width, 12 inches.
Finished in Japanese Bronze............................................$18.00

### 1007

1905

**VALDORA**

8 Day, Half-hour Strike, Gong.
4½ inch IVORY Dial. Ivory Center. Visible Escapement.
Rococo Cast Gilt Sash and Bezel. Beveled Glass.
Height, 13¾ inches. Width, 19 inches.
Finished in Japanese Bronze............................................$18 50

### 1008

1905

**VALE**

8 Day, Half-hour Strike, Gong.
4½ inch IVORY Dial. Ivory Center. Visible Escapement.
Rococo Cast Gilt Sash and Bezel. Beveled Glass.
Height, 13½ inches. Width, 12¾ inches.
Finished in Japanese Bronze........................$15 50
Finished in Rich Gold................................ 18 75

### 1009

1908

**VALENCIA**

8 Day, Half-hour Strike, Gong.
3½ inch IVORY Dial. Ivory Center. Visible Escapement.
Cast Gilt Sash and Bezel. Beveled Glass.
Height, 12½ inches. Width, 12½ inches.
Finished in Japanese Bronze............................................$20 00
Finished in Syrian Bronze............................................ 23 00

# WATERBURY

*For comprehensive listing of Waterbury Clocks, See Tran Duy Ly "Waterbury Clocks - History, Identification & Price Guide."*

### 1010

1905

**VERNDALE**
8 Day, Half-hour Strike, Gong.
5½ inch IVORY Dial. Ivory Center.
Cast Gilt Sash and Bezel. Beveled Glass.
Height, 21 inches. Width, 20 inches.
Finished in Japanese Bronze .......................... $32 00

### 1011

1905

**VANDYKE**
8 Day, Half-hour Strike, Gong.
5½ inch IVORY Dial. Ivory Center. Visible Escapement.
Cast Gilt Sash and Bezel. Beveled Glass.
Height, 20¼ inches. Width, 20 inches.
Finished in Japanese Bronze .......................... $28 50

### 1012

1905

**VOTAW**
8 Day, Half-hour Strike, Gong.
5½ inch IVORY Dial. Ivory Center.
Cast Gilt Sash and Bezel. Beveled Glass.
Height, 21 inches. Width, 20 inches.
Finished in Japanese Bronze .......................... $30 00

### 1013

1905

**VIRGINS**
8 Day, Half-hour Strike, Gong.
5½ inch IVORY Dial. Ivory Center. Visible Escapement.
Cast Gilt Sash and Bezel. Beveled Glass.
Height, 20¼ inches. Width, 20 inches.
Finished in Japanese Bronze .......................... $30 50

# WATERBURY

*The year entered beside each clock is the year of the catalog from which the illustration was taken. Some clocks may have been produced a few years earlier.*

**1014**

**VICTOR** — 1905
8 Day, Half-hour Strike, Gong.
4½ inch IVORY Dial. Ivory Center. Visible Escapement.
Rococo Cast Gilt Sash and Bezel. Beveled Glass.
Height, 14 inches. Width, 19¾ inches.
Finished in Japanese Bronze ................................ $21 50

**1016**

**VASSAR** — 1905
8 Day, Half-hour Strike, Gong.
4¼ inch IVORY Dial. Ivory Center. Visible Escapement.
Rococo Cast Gilt Sash and Bezel. Beveled Glass.
Height, 16⅞ inches. Width, 16¼ inches.
Finished in Japanese Bronze ................................ $23 50
Finished in Syrian Bronze ................................... 26 75

**1015**

**VESPER** — 1905
8 Day, Half-hour Strike, Gong.
4½ inch IVORY Dial. Ivory Center.
Rococo Cast Gilt Sash and Bezel. Beveled Glass.
Height, 16⅞ inches. Width, 16¼ inches.
Finished in Japanese Bronze ................................ $23 00
Finished in Rich Gold ...................................... 26 25

**1017**

**VENANGO** — 1905
8 Day, Half-hour Strike, Gong.
4½ inch IVORY Dial. Ivory Center. Visible Escapement.
Rococo Cast Gilt Sash and Bezel. Beveled Glass.
Height, 15½ inches. Width, 19¾ inches.
Finished in Japanese Bronze ................................ $25 50

# WATERBURY

Except for those fabulous clocks for which you wouldn't mind cutting a hole in the ceiling, Grandfather or Hanging Clocks under eight feet are the most desirable and practical. Some manufacturers produced certain of these clocks in two styles—one to stand on the floor, and the other to hang on a wall. Of the clocks designed with two styles, the hanging style is considered to be the most valuable.

**1018** — 1891

**1019** — 1905

**1020** — 1881

**FREEPORT**
Oak or Walnut.
8 Day, Half-hour Strike, Gong...... $11 50
30 Day, Time, with Second Hand .... 11 50
10 inch Dial. Height, 45¼ inches;
Width, 18⅛ inches.

**REGULATOR No. 9.**

OAK. BRASS RAILING.

CABINET FINISH. GLASS SIDES.
8 Day, Weight, Time.
8 Day, Weight, Half-hour Strike, Gong.
Brass Weight.

Beats Seconds, Dead-beat Escapement, Retaining Power.
Solid Polished Movement Frames.
Dial 10 inches. Height 76¼ inches.

**WALNUT REGULATOR, No. 9.**
PIN ESCAPEMENT, WITH SWEEP SECOND.

Eight Day,   Weight—Time.

Height, 88 inches.

12 inch Dial.

# WATERBURY

*For comprehensive listing of Waterbury Clocks, See Tran Duy Ly "Waterbury Clocks - History, Identification & Price Guide."*

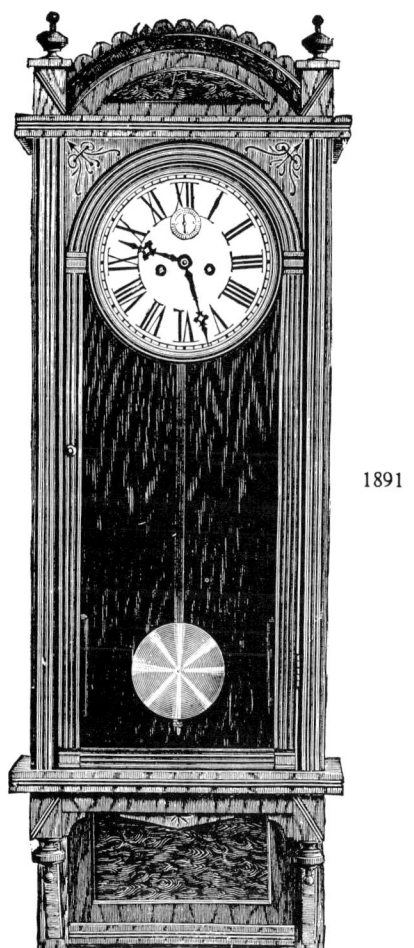
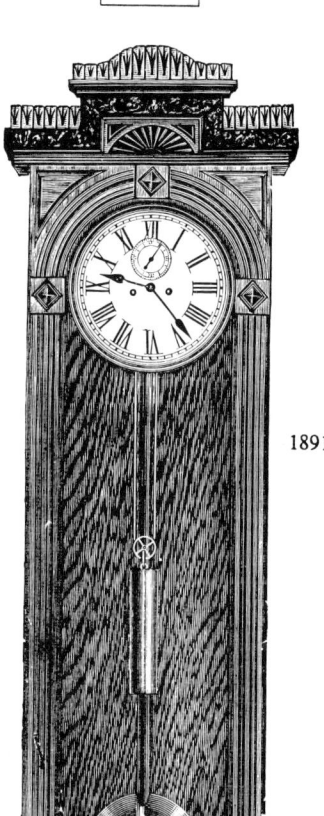
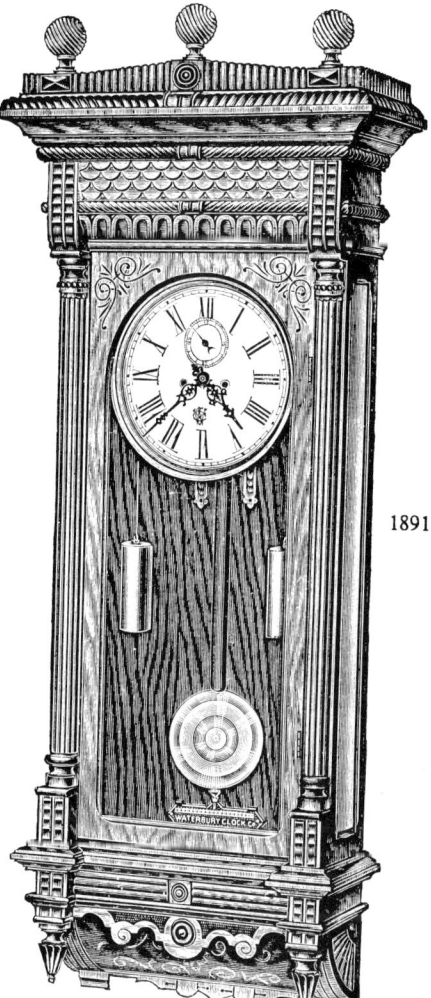

### REGULATOR No. 3.

OAK, WALNUT OR MAHOGANY.

CABINET FINISH.

8 Day, Weight, Time.
Brass Weights.
Dead-beat Escapement, Retaining Power.
Solid Movement Frames.
Dial 9 inches.  Height 46 inches.

### REGULATOR No. 5.

WALNUT.

CABINET FINISH.  GLASS SIDES.

8 Day, Weight, Time.
8 Day, Weight, Half-hour Strike, Gong.
Brass Weight.
Beats Seconds, Dead-beat Escapement, Retaining Power.
Solid Polished Movement Frames.
Dial 10 inches.  Height 69½ inches.

### REGULATOR No. 11.

OAK, WALNUT OR MAHOGANY.

CABINET FINISH.  GLASS SIDES.

8 Day, Weight, Time.
Brass Weights.
Dead-beat Escapement, Retaining Power.
Solid Polished Movement Frames.
Dial 10 inches.  Height 52¼ inches.

# WATERBURY

*Except for those fabulous clocks for which you wouldn't mind cutting a hole in the ceiling, Grandfather or Hanging Clocks under eight feet are the most desirable and practical. Some manufacturers produced certain of these clocks in two styles—one to stand on the floor, and the other to hang on a wall. Of the clocks designed with two styles, the hanging style is considered to be the most valuable.*

### 1024

1891

**REGULATOR No. 7.**

OAK, WALNUT OR CHERRY.

CABINET FINISH. GLASS SIDES.

8 Day, Weight, Time.
Brass Weight.
Finely finished Movement of best quality encased in Iron Box.
Dead-beat Pin Escapement, Sweep Second, Retaining Power.
Porcelain Dial 12 inches.
Height 82 inches.

### 1025
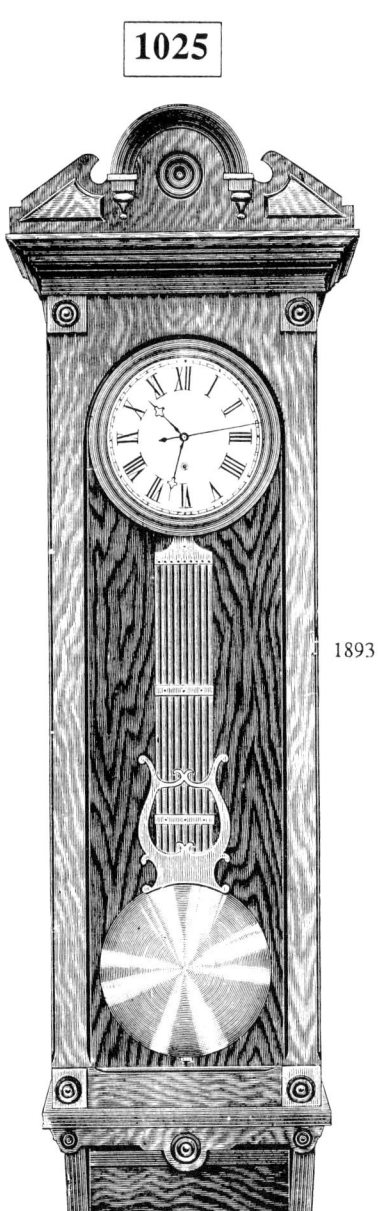
1893

**REGULATOR No. 51**

OAK, WALNUT OR CHERRY

8 Day, Weight, Time. No. 100 Movement.
Brass Weight.
Finely finished Movement of best quality encased in Iron Box.
Dead-beat Pin Escapement, Sweep Second, Retaining Power.
Porcelain Dial, 12 inches. Height, 83 inches.

With Gridiron Pendulum, oval Rods............$80 00
With Compensating Metal Pendulum............ 80 00
With Adjustable Compensating Pendulum....... 95 00
With Compensating Mercurial Pendulum.......112 00

### 1026
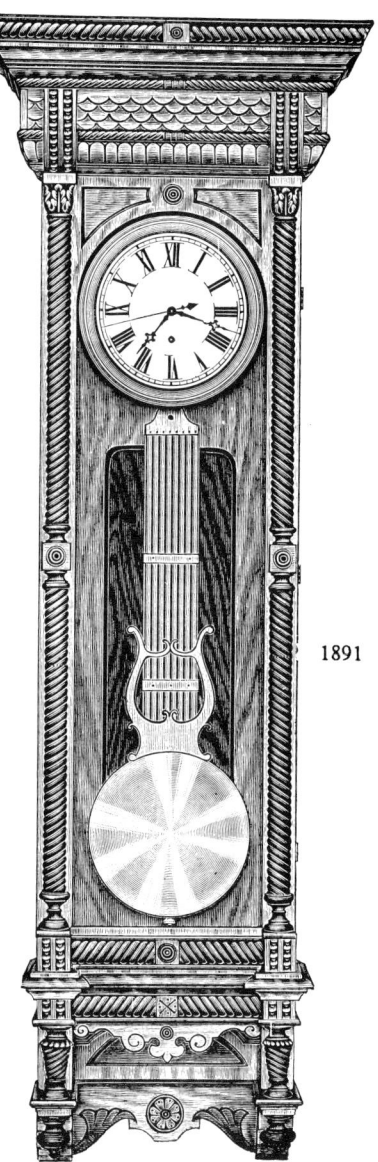
1891

**REGULATOR No. 12.**

OAK, WALNUT OR CHERRY.

CABINET FINISH. GLASS SIDES.

8 Day, Weight, Time.
Brass Weight.
Finely finished Movement of best quality encased in Iron Box.
Dead-beat Pin Escapement, Sweep Second, Retaining Power.
Porcelain Dial 12 inches.
Height 86 inches.

# WATERBURY

*The year entered beside each clock is the year of the catalog from which the illustration was taken. Some clocks may have been produced in an earlier year.*

### REGULATOR No. 17

OAK, WALNUT OR CHERRY

CABINET FINISH

Brass Weights.
Dead-beat Escapement, Retaining Power.
Solid Polished Movement Frames.
Silver Dial, 12 inches. Height, 56½ inches.

8 Day, Weight, Time................$45 00

### REGULATOR No. 21

OAK, WALNUT OR CHERRY

CABINET FINISH

Brass Weights.
Dead-beat Escapement, Retaining Power.
Solid Polished Movement Frames.
Dial, 12 inches. Height, 55½ inches.

8 Day, Weight, Time................$35 00

### REGULATOR NO. 54
Quartered Oak or Mahogany.

                        Oak    Mahogany
8 Day, Weight, Time........$32 00 $37 00
Brass Weights.
Dead-beat Escapement, Retaining Power.
Solid Polished Movement Frames.
12 inch Dial.
Height, 57 inches. Width, 21¼ inches.

# WATERBURY

*For comprehensive listing of Waterbury Clocks, See Tran Duy Ly "Waterbury Clocks - History, Identification & Price Guide."*

**1030**

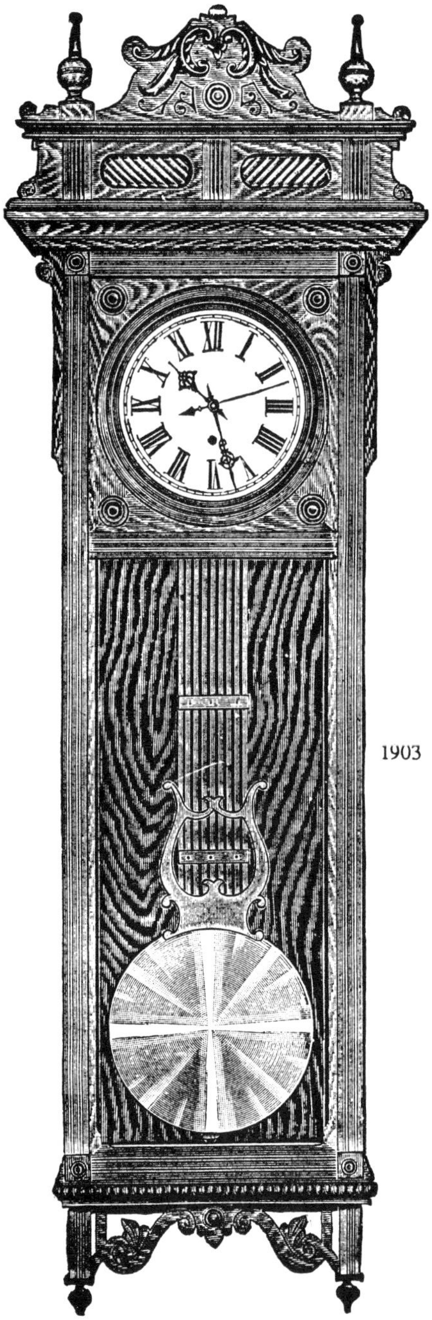

**1031**

**1032**

1903

### REGULATOR No. 60
Quartered Oak or Walnut.
Polished Finish.
8 Day, Weight, Time.
Brass Weight.
Dead-beat Escapement, Sweep Second, Retaining Power.
12 inch PORCELAIN Dial.
Height, 79½ inches. Width, 26 inches.
With Gridiron Pendulum, Oval Rods... $80 00
With Compensating Mercurial Pendulum 112 00
For Mercurial Pendulum see Regulator No. 70.

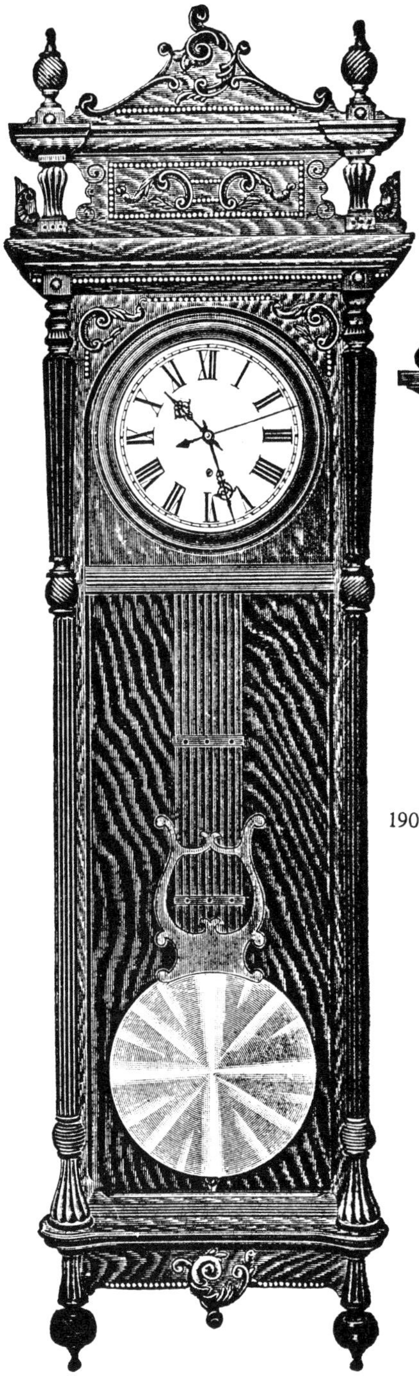

### REGULATOR No. 65
Quartered Oak, Walnut or Mahogany.
Polished Finish.
8 Day, Weight, Time.
Brass Weight.
Dead-beat Escapement, Sweep Second, Retaining Power.
12 inch PORCELAIN Dial.
Height, 81½ inches. Width, 24¼ inches.
With Gridiron Pendulum, Oval Rods... $90 00
With Compensating Mercurial Pendulum 122 00
For Mercurial Pendulum see Regulator No. 70.

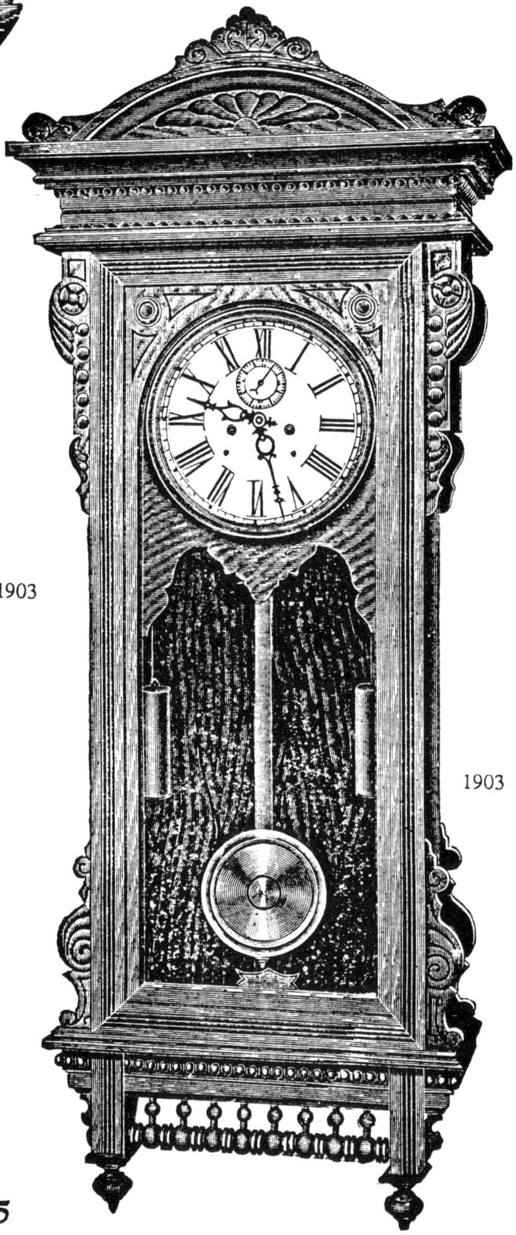

### REGULATOR No. 57
Quartered Oak or Walnut.
8 Day, Weight, Time.................. $23 00
Brass Weights.
Dead-beat Escapement, Retaining Power.
Solid Polished Movement Frames.
10 inch Dial.
Height, 49½ inches. Width, 19¼ inches.

# WATERBURY

*The year entered beside each clock is the year of the catalog from which the illustration was taken. Some clocks may have been produced in an earlier year.*

### REGULATOR NO. 20
**Oak.**

8 Day, Weight, Time........ $19 00
Dead-beat Escapement, Retaining Power.
Solid Movement Frames.
12 inch Dial. Height, 38 inches.

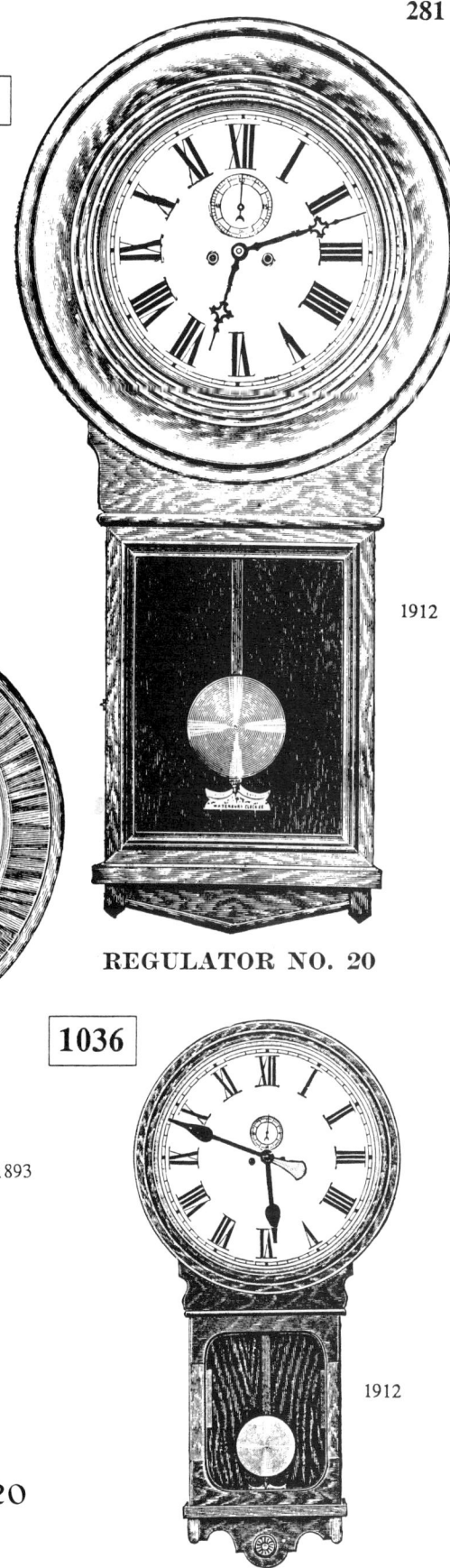

1912

### REGULATOR NO. 20

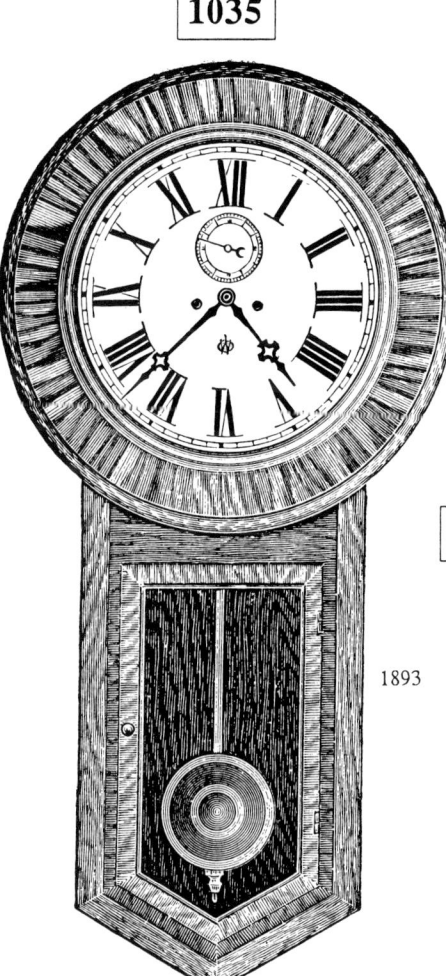

1903

1893

1912

### REGULATOR No. 66
**Quartered Oak or Walnut.**

8 Day, Weight, Time................ $52 00

8 Day, Weight, Time.
Brass Weight.
Beats Seconds, Dead-beat
Escapement, Retaining Power.
Solid Polished Movement Frames.
Dial 18 inches. Height 67½ inches.

### REGULATOR No. 20

**OAK OR ROSEWOOD**

Dead-beat Escapement, Retaining Power.
Solid Movement Frames.
Dial, 12 inches. Height, 34¾ inches.

8 Day, Weight, Time........... $17 50

### REGULATOR NO. 80
**Oak.**

8 Day, Weight, Time................$27 00
Dead-beat Escapement, Retaining Power.
Solid Movement Frames.
18 inch Dial.
Height, 41 inches.

# WATERBURY

For comprehensive listing of Waterbury Clocks, See Tran Duy Ly "Waterbury Clocks - History, Identification & Price Guide."

### 1039
### ADMIRAL
#### Oak.

| | |
|---|---|
| 8 Day, Time | $7 50 |
| 8 Day, Half-hour Strike | 8 50 |
| 8 Day, Half-hour Strike, Gong | 9 00 |
| 8 Day, Time, Calendar | 8 00 |
| 8 Day, Half-hour Strike, Calendar | 9 00 |
| 8 Day, Half-hour Strike, Gong, Calendar | 9 50 |
| 30 Day, Time, with Second Hand | 9 50 |

12 inch Dial. Height, 32 inches.

1912
ADMIRAL

### 1037

1891

### 24 INCH REGULATOR.
#### WALNUT.

CABINET FINISH. GLASS SIDES.

8 Day, Weight, Time.
Brass Weight.
Beats Seconds, Dead-beat
Escapement, Retaining Power.
Solid Polished Movement Frames.
Dial 24 inches. Height 70 inches.

### 1038

1908

#### YARMOUTH
#### Oak or Mahogany.

| | Oak | Mahogany |
|---|---|---|
| 8 Day, Spring, Time | $8 00 | $10 00 |
| 8 Day, Spring, Half-hour Strike, Gong | 9 50 | 11 50 |

6 inch Rococo Dial. Height, 30 inches.

### 1040

1912

#### REGULATOR NO. 81
#### Oak.

8 Day, Weight, Time ............. $35 00
Beats Seconds, Dead-beat Escapement.
Retaining Power.
Solid Movement Frames.
12 inch Dial.
Height, 57 inches.

REGULATOR NO. 81

# WATERBURY

283

**1041**

1891

**1042**

1917

**1043**

1912

### REGULATOR No. 4.

OAK, WALNUT OR MAHOGANY.

CABINET FINISH.

8 Day, Weight, Time.
Brass Weights.
Dead-beat Escapement, Retaining Power.
Solid Polished Movement Frames.
Dial 10 inches.   Height 51 inches.

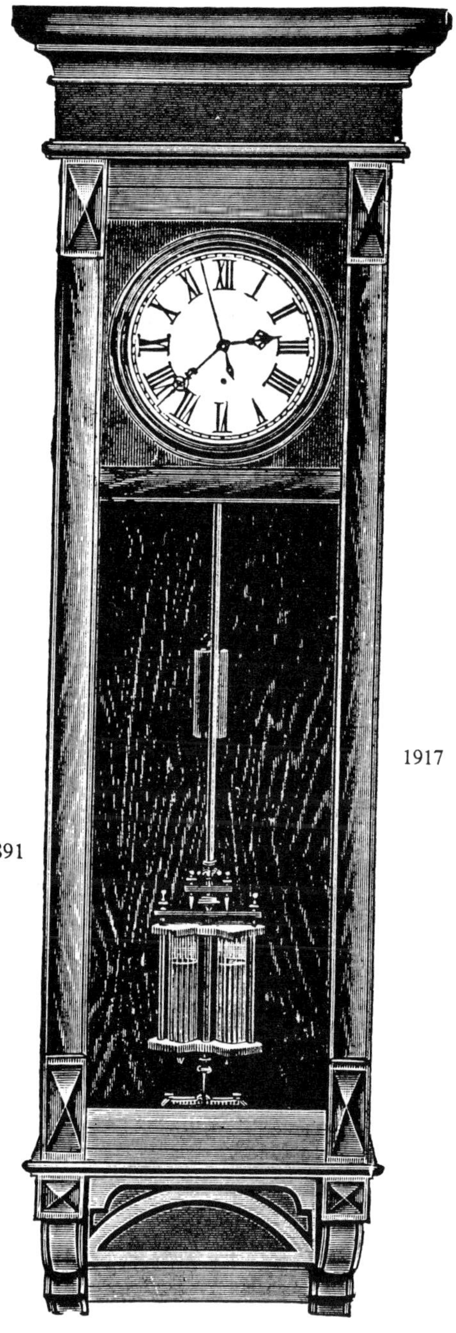

### REGULATOR No. 70
**Mahogany**

8 Day, Weight, Time.
Brass Weight.
Finely Finished Movement of Best Quality Encased in Iron Box.
Dead-beat Escapement, Sweep Second, Retaining Power.
12 inch PORCELAIN Dial.
Height, 82 inches.  Width, 26¼ inches.
With Gridiron Pendulum, Oval Rods.
With Compensating Mercurial Pendulum.
For Gridiron Pendulum See Regulator No. 60.

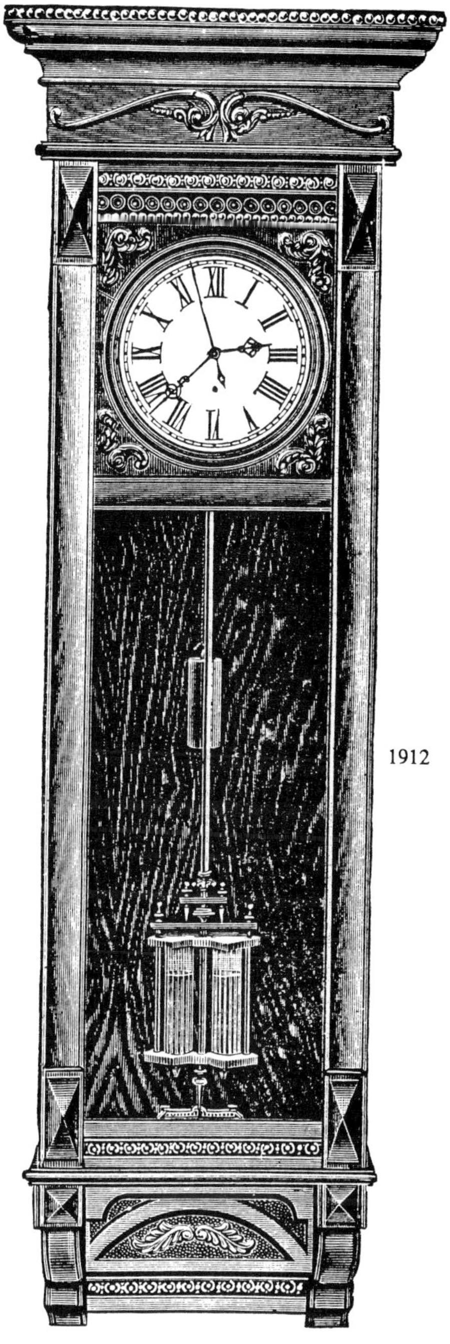

### REGULATOR NO. 70
**Quartered Oak or Mahogany.**

8 Day, Weight, Time.
Brass Weight.
Finely Finished Movement of Best Quality Encased in Iron Box.
Dead-beat Escapement, Sweep Second, Retaining Power.
12 inch PORCELAIN Dial.
Height, 82 inches.   Width, 26¼ inches.

| | Oak | Mahogany |
|---|---|---|
| With Gridiron Pendulum, Oval Rods | $95 00 | $105 00 |
| With Compensating Mercurial Pendulum | 133 50 | 143 50 |

For Gridiron Pendulum See Regulator No. 60.

# WATERBURY

### STUDY No. 4.
OAK.

CABINET FINISH. GLASS SIDES.

8 Day, Weight, Strike, Gong.
Silver Dial 8 inches.   Height 35¼ inches.

### STUDY No. 5.
OAK.

CABINET FINISH.

8 Day, Spring, Half-hour Strike, Gong.
Dial 10 inches.   Height 25⅝ inches.

### STUDY No. 2.
OAK.

CABINET FINISH. GLASS SIDES.

8 Day, Weight, Strike, Gong.
Silver Dial 8 inches.   Height 24¼ inches.

### STUDY No. 3. WEIGHT.
OAK.

CABINET FINISH. GLASS SIDES.

1 Day, Weight, Strike, Cast Bell.
Ord. Dial 8 inches.
8 Day, Weight, Strike, Gong.
Silver Dial 8 inches.   Height 22¾ inches.

### STUDY No. 3. SPRING.
OAK.

CABINET FINISH. GLASS SIDES.

8 Day, Spring, Time.
8 Day, Spring, Half-hour Strike, Gong.
Dial 8 inches.   Height 22¼ inches.

# WATERBURY

285

*For comprehensive listing of Waterbury Clocks, See Tran Duy Ly "Waterbury Clocks - History, Identification & Price Guide."*

## PENDANT.

OAK.
8 Day, Time.
8 Day, Half-hour Strike.
Also with Alarm.
Dial 10 inches.   Height 28¼ inches.

**1050**

**1049**

**1051**

1891

1893

1891

## PENDANT.

**1052**

1912

## AUGUSTA.

OAK. CABINET FINISH.

GLASS SIDES.   CAST BRASS ORNAMENTS.

8 Day, Weight, Strike, Gong.
Silver Dial 10 inches.   Height 50¾ inches.

## ATLANTA·

OAK. CABINET FINISH
GLASS SIDES

Silver Dial, 12 inches.   Height, 53 inches.

8 Day, Weight, Strike, Gong.......$40 00

## REGULATOR NO. 67
**Oak, Walnut or Mahogany.**

|  | Oak or Walnut | Mahogany |
|---|---|---|
| 8 Day, Weight, Time........ | $24 00 | $27 00 |

Dead-beat Escapement, Retaining Power.
Solid Polished Movement Frames.
12 inch Dial.
Height, 50 inches.   Width, 22 inches.

## REGULATOR NO. 67

# WATERBURY

**1053**

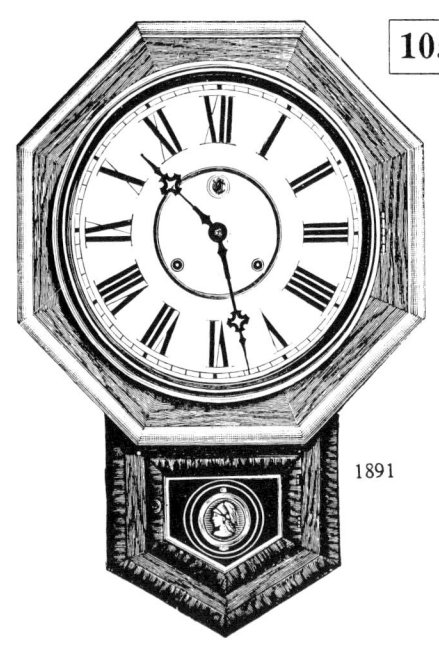

1891

*To receive top prices, clocks must be in ORIGINAL MINT condition and in good running order - A replaced or damaged dial, missing parts, will reduce prices accordingly.*

**1054**

### SIAM.
FULL GILT MOLDING.
POLISHED VENEERED,
8 Day, Time.
8 Day, Half-hour Strike.
Also with Alarm.
8 Day, Time, Calendar.
8 Day, Half-hour Strike, Calendar.
Dial 12 inches.  Height 23½ inches.

### DIGBY—Oak.
8 Day, Time ........................ $6 00
8 Day, Half-hour Strike............. 7 00
Alarm Additional ................... 50
8 Day, Time, Calendar............... 6 50
8 Day, Half-hour Strike, Calendar... 7 50
12 inch Dial.  Height, 27¼ inches.

1912

*The year entered beside each clock is the year of the catalog from which the illustration was taken. Some clocks may have been produced a few years earlier.*

**1057**

**1055**

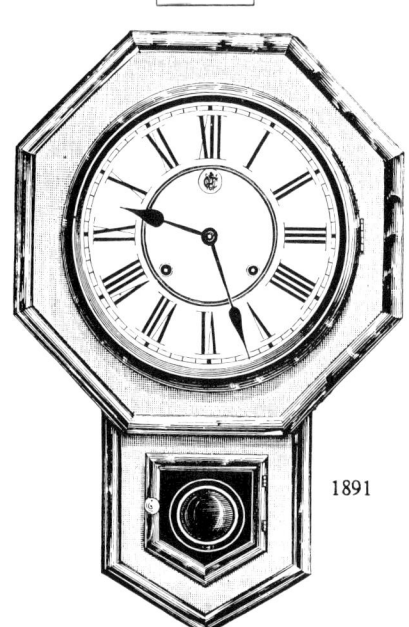

1891

**1056**

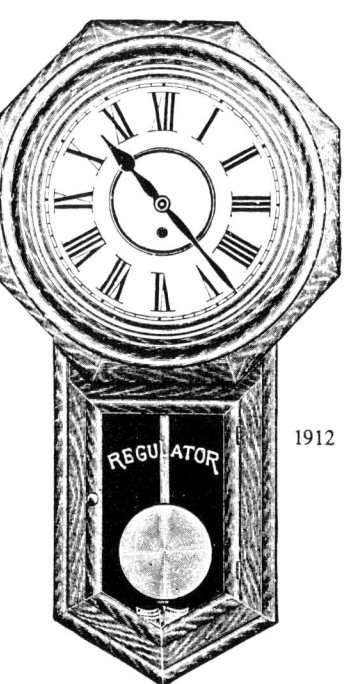

1912

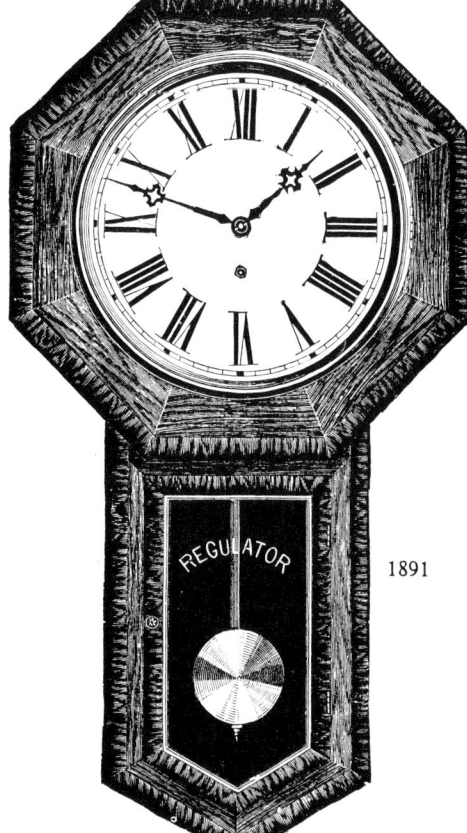

1891

### PEKING.
BRONZED, WITH GILT MOLDING.
8 Day, Time.
8 Day, Half-hour Strike.
Also with Alarm.
8 Day, Time, Calendar.
8 Day, Half-hour Strike, Calendar.
Dial 10 inches.  Height 21 inches.

### CONSORT—Oak or Walnut.
8 Day, Time ........................ $9 00
8 Day, Half-hour Strike.............10 00
8 Day, Half-hour Strike, Gong.......10 50
8 Day, Time, Calendar............... 9 50
8 Day, Half-hour Strike, Calendar...10 50
8 Day, Half-hour Strike, Gong, Calendar.11 00
12 inch Dial.  Height, 32 inches.

### REGENT.
OAK OR POLISHED VENEERED.
8 Day, Time.
8 Day, Half-hour Strike.
8 Day, Half-hour Strike, Gong.

Dial 12 inches.  Height 32 inches.

# WATERBURY

### 1058

### GALLERY
**Oak.**

8 Day, 12 inch, Time (Pendulum Movement) .................. $13 00
8 Day, 18 inch, Time (Pendulum Movement) .................. 20 00
30 Day, 24 inch, Time (Pendulum Movement) .................. 33 00
12, 18 and 24 inch Dial.
12 inch measures across the back 18¼ in.
18 inch measures across the back 25⅝ in.
24 inch measures across the back 34 in.

1912

### 1059

1912

### LYCEUM
**Oak.**

8 Day, Time (Pendulum Movement)...$7 50
12 inch Dial.
Height, 25 inches. Width, 17¾ inches.

### 1060

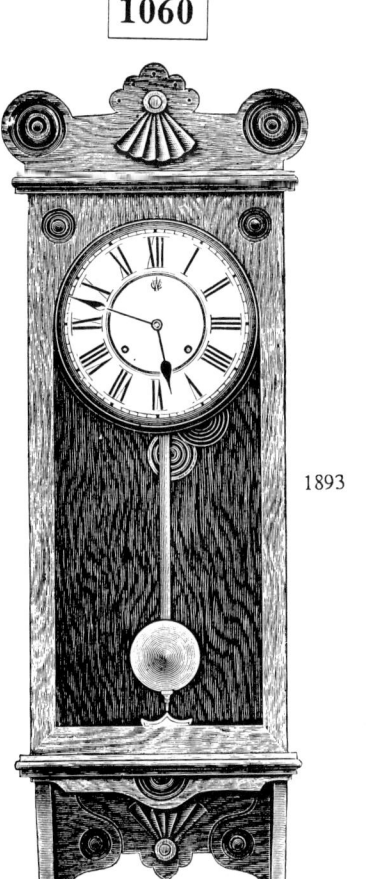

1893

### WALTHAM
OAK, WALNUT OR CHERRY

Dial, 10 inches. Height, 41 inches.

8 Day, Time.................$11 00
8 Day, Half-hour Strike, Gong.... 13 00

### 1061

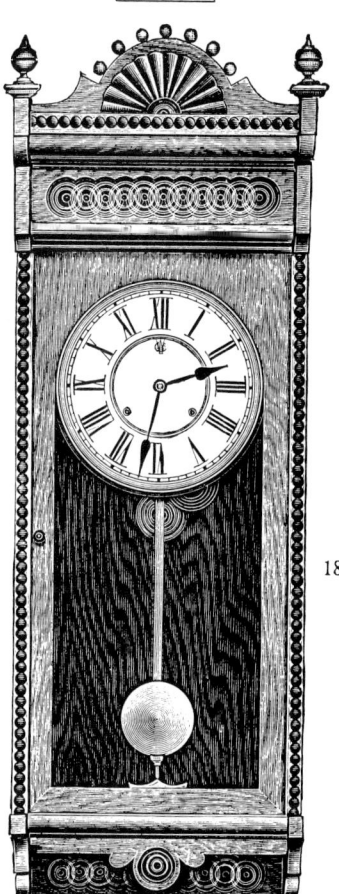

1893

### CHESHIRE
OAK, WALNUT OR CHERRY. CABINET FINISH

Dial, 10 inches. Height, 44¾ inches.

8 Day, Time.................$17 00
8 Day, Half-hour Strike, Gong........ 19 00

### 1062

1912

### LOBBY
**Oak.**

30 Day, Time (Pendulum Movement) .$35 00
18 inch Dial.
Height, 39 inches. Width, 28 inches.

# WATERBURY

**1063**

### VICKSBURG

8 Day, Half-hour Strike, Gong.
4¼ inch IVORY Dial. Ivory Center.
Rococo Cast Gilt Sash and Bezel. Beveled Glass.
Height, 12½ inches. Width, 8½ inches.
Finished in Japanese Bronze .......... $12 00
Finished in Rich Gold .................. 14 50
Finished in Syrian Bronze ............. 14 50

**1064**

### VALIANT

8 Day, Half-hour Strike, Gong.
5½ inch IVORY Dial. Ivory Center.
Cast Gilt Sash and Bezel. Beveled Glass.
Height, 18 inches. Width, 10⅞ inches.
Finished in Japanese Bronze .................. $18 00

**1065**

### VANCOUVER

8 Day, Half-hour Strike, Gong.
5½ inch IVORY Dial. Ivory Center.
Visible Escapement.
Cast Gilt Sash and Bezel. Beveled Glass.
Height, 17¼ inches. Width, 11¼ inches.
Finished in Japanese Bronze .................. $16 50

**1066**

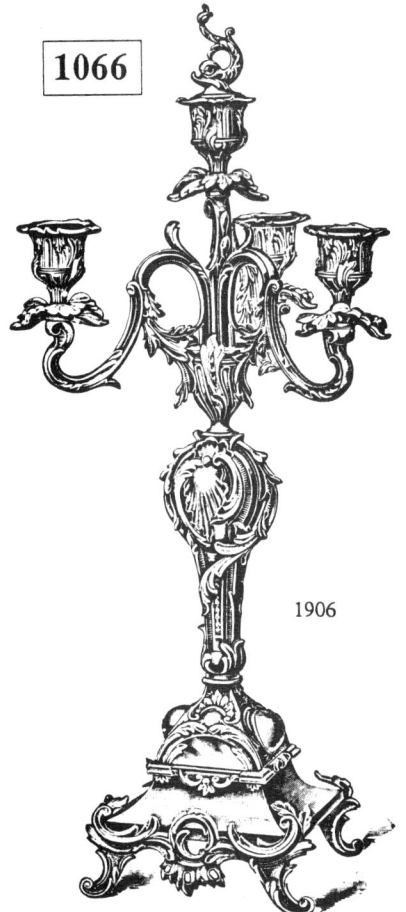

### CANDELABRA No. 11

**Rich Roman Gold Plated.**

Height, 18 inches.

Per Pair ........................ $20 00

**1067**

### CATAWBA

**Rich Roman Gold Plated.**

8 Day, Half-hour Strike, Gong ............ $30 00
4½ inch IVORY Dial. Ivory Center. Visible Escapement.
Rococo Cast Gilt Sash and Bezel. Beveled Glass.
Height, 17¼ inches. Width, 13¼ inches.

List Price of Set Complete (3 Pieces) .... $50 00

### CANDELABRA No. 11

**Rich Roman Gold Plated.**

Height, 18 inches.

Per Pair ........................ $20 00

# WATERBURY

**1068 — RAJAH**
**Rich Roman Gold Plated or Syrian Bronze.**
8 Day, Lever, Time.................................$11 00
2½ inch IVORY Dial.
Cast Gilt Sash and Bezel. Beveled Glass.
Height, 12⅜ inches. Width, 8½ inches.

**1069 — SARTORIS**
**Rich Roman Gold Plated.**
Decorated Porcelain Panel.
8 Day, Half-hour Strike, Gong..........$21 00
3½ inch IVORY Dial. Fancy Gilt Center.
Cast Gilt Sash and Bezel. Beveled Glass.
Height, 14 inches. Width, 7¼ inches.

**1070 — DURBIN**
**Rich Roman Gold Plated.**
8 Day, Lever, Time.......................................$9 00
2 inch IVORY Dial. Cast Gilt Sash and Bezel.
Beveled Glass.
Height, 11 inches. Width, 8 inches.

**1071 — CAVALIER**
**Rich Roman Gold Plated.**
8 Day, Half-hour Strike, Gong................$40 00
4½ inch IVORY Dial. Fancy Gilt Center. Visible Escapement.
Rococo Cast Gilt Sash and Bezel. Beveled Glass.
Height, 22¼ inches. Width, 17¼ inches.
List Price of Set Complete (3 Pieces)........$58 00

**1072 — DAUPHIN**
**Rich Roman Gold Plated.**
8 Day, Half-hour Strike, Gong..........$21 00
4¼ inch IVORY Dial. Fancy Gilt Center.
Rococo Cast Gilt Sash and Bezel. Beveled Glass.
Height, 18¼ inches. Width, 10⅝ inches.

# WATERBURY

*To receive top prices, clocks must be in ORIGINAL MINT condition and in good running order - A replaced or damaged dial, missing parts, will reduce prices accordingly.*

**1073**

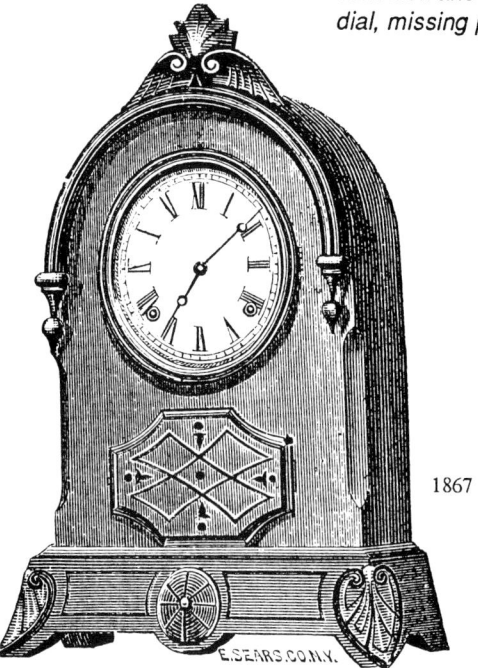

1867

### EXTRA CARVED WALNUT.
Eight Day Strike
Height, ..... 19 inches

**1074**

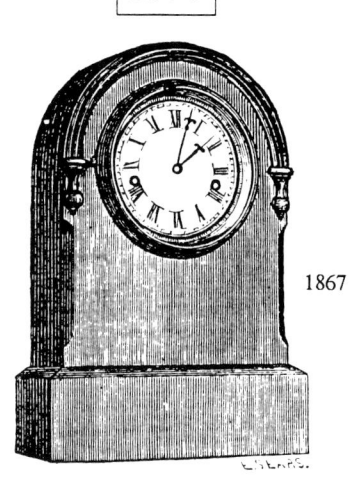

1867

### B. WALNUT ROUND TOP.
One-day. Time.
Height, ..... 10 inches

**1075**

1867

### BLACK WALNUT GOTHIC.
Eight Day Strike
Height, ..... 14 1/2 inches

**1076**

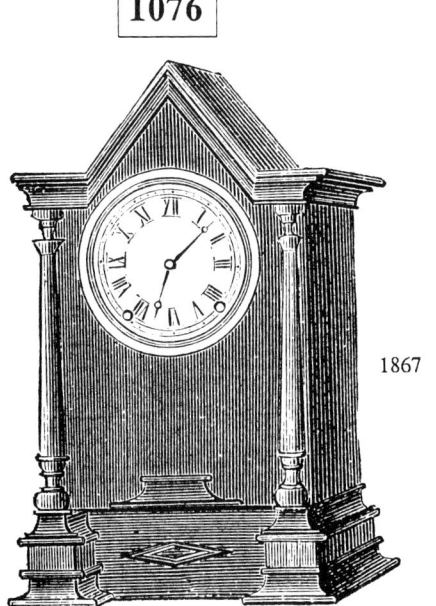

1867

### BLACK WALNUT EX. GOTHIC PILLAR.
Eight Day Strike
Height, ..... 15 inches

**1077**

1867

### BLACK WALNUT ARCH TOP
Eight Day Strike
Height, ..... 13 1/2 inches

**1078**

1867

### BLACK WALNUT ROUND GOTHIC.
One Day Strike
Height, ..... 13 inches

# WATERBURY

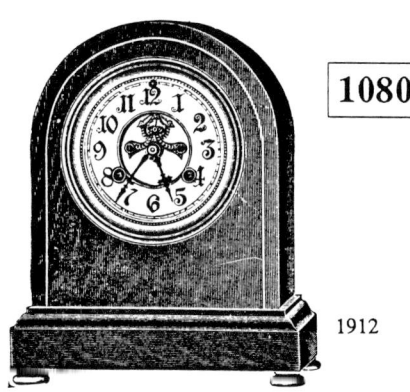

**1080**
**SOUTHAMPTON**
Mahogany Inlaid.
8 Day, Half-hour Strike, Gong........$14 00
4¼ inch PORCELAIN Dial.
Visible Escapement.
Cast Gilt Sash and Bezel.
Convex Beveled Glass.
Height, 9 inches. Width, 7⅜ inches.

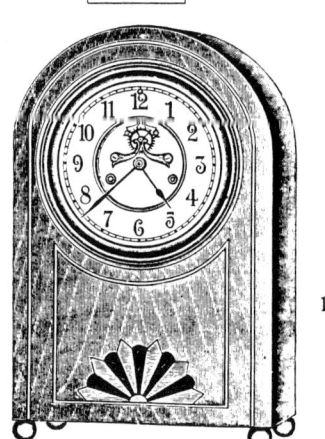

**1079**
**MONTGOMERY**
Mahogany Inlaid.
8 Day, Half-hour Strike, Gong........$15 00
5½ inch PORCELAIN Dial.
Visible Escapement.
Cast Gilt Sash and Bezel.
Convex Beveled Glass.
Height, 11⅝ inches. Width, 7⅜ inches.

**1081**
**BATH**
Mahogany Inlaid.
8 Day, Half-hour Strike, Gong........$18 00
5½ inch PORCELAIN Dial.
Visible Escapement.
Cast Gilt Sash and Bezel.
Convex Beveled Glass.
Height, 13½ inches. Width, 8⅝ inches.

**1082**
**LONDON**
Fumed Oak or Circassian Walnut.
8 Day, Half-hour Strike, Gong.
4¼ inch PORCELAIN Dial.
Cast Sash and Bezel.
Convex Beveled Glass.
Height, 9⅛ inches. Width, 8⅜ inches.
Fumed Oak ................$13 00
Circassian Walnut ............. 14 00

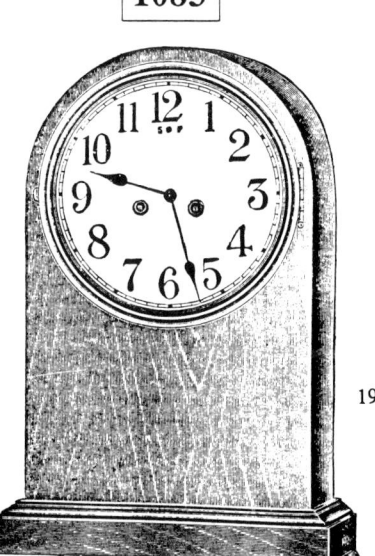

**1083**
**BIRMINGHAM**
Mahogany or Circassian Walnut.
8 Day, Half-hour Strike, Gong......$20 00
8 inch PORCELAIN Dial.
Cast Gilt Sash and Bezel. Convex Glass.
Height, 16⅜ inches. Width, 11 inches.

**1084**
**ELMONT**
Golden Oak or
Birch Mahogany Finish.
8 Day, Strike, Gong, Half-hour on Cup Bell.
5½ inch Dial. Convex Glass. Height, 11⅜ inches. Width, 8½ inches.
American White Dial, Arabic Figures, Plain Sash............$7 50

**1085**
**CHESTERTON**
Mahogany Inlaid.
8 Day, Half-hour Strike, Gong........$16 00
5½ inch IVORY Dial.
Visible Escapement.
Cast Gilt Sash and Bezel.
Convex Beveled Glass.
Height, 14¼ inches. Width, 9⅜ inches.

# WATERBURY

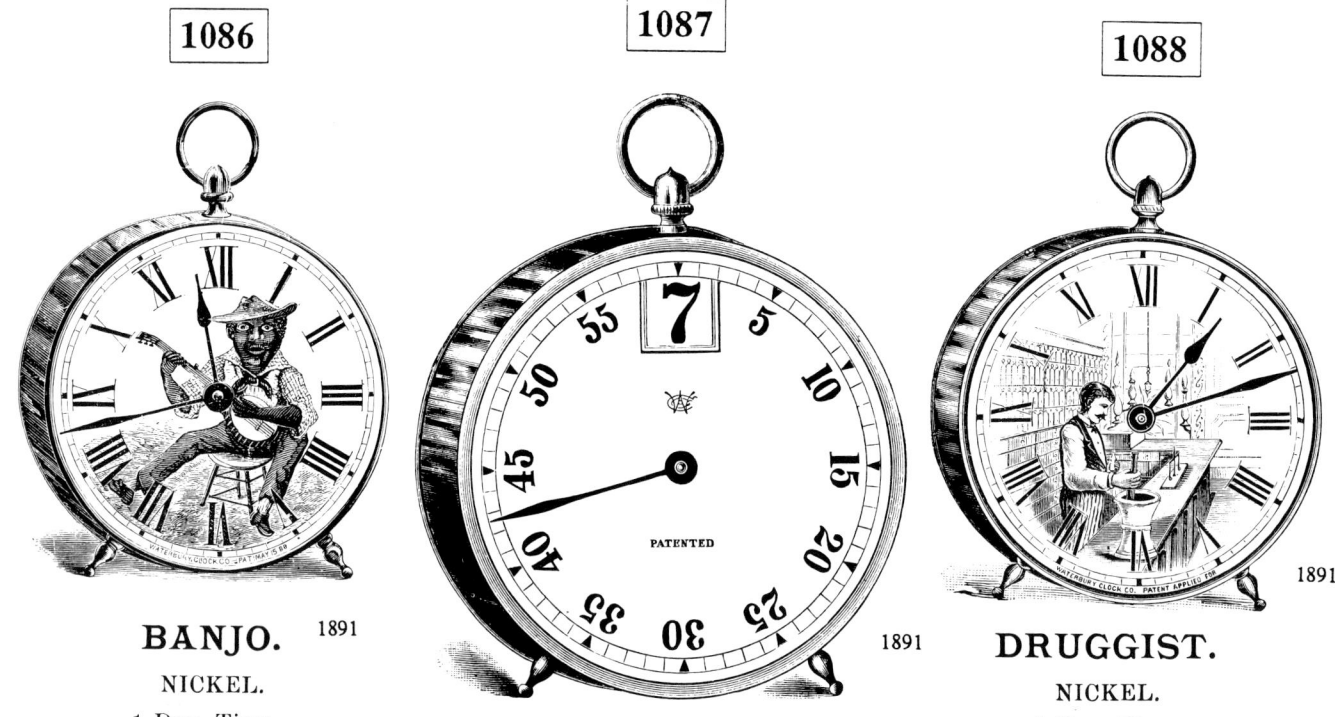

**1086** **BANJO.** 1891
NICKEL.
1 Day, Time.
1 Day, Time, Alarm. (Bell on Top.)
Dial 4½ inches. Illustration One-half Size.

**1087** **AUTOMATIC.**
NICKEL.
1 Day, Time.
Dial 4½ inches. Illustration One-half Size.
The figure showing the Hour changes Automatically.

**1088** **DRUGGIST.** 1891
NICKEL.
1 Day, Time.
1 Day, Time, Alarm. (Bell on Top.)
Dial 4½ inches. Illustration One-half Size.

**1089** **MIKADO.** 1891
NICKEL.
1 Day, Time.
1 Day, Time, Alarm. (Bell on Top.)
Dial 4½ inches. Illustration One-half Size.

**1090** **STEAMER.** 1891
NICKEL.
1 Day, Time.
Dial 5 inches. Illustration One-half Size.

**1091** **LAUNDRY.** 1891
NICKEL.
1 Day, Time.
1 Day, Time, Alarm. (Bell on Top.)
Dial 4½ inches. Illustration One-half Size.

**THESE CLOCKS SHOW MOTION WHILE RUNNING.**

# WATERBURY

Most white metal and brass ornamentals and clock cases are gilded or silver plated. Great care must be taken when cleaning these surfaces. The thin layer of gilding or silver plating will be easily removed by cleaning solutions and strong rubbing. If you believe cleaning is necessary, hire the services of a professional and avoid the expense of re-gilding or re-plating.

**1092** 1912

### SPIDER
**Rich Gold Plated.**
8 Day, Time, JEWELED MOVEMENT.......$6 75
2 inch IVORY Dial.
Beveled Glass. Height, 3½ inches.
Polished Movement, visible through a Glass Cylinder ⅛ inch thick.

**1093** 1912

### HORNET
**Rich Gold Plated.**
1 Day, Time.......................$3 50
2 inch IVORY Dial.
Beveled Glass. Height, 3½ inches.
Polished Movement, visible through a Glass Cylinder ⅛ inch thick.

**1094** 1904

### ROMP
**Rich Roman Gold Plated.**
1 Day, Time.......................$4 75
1½ inch IVORY Dial. Sunk Gilt Center.
Beveled Glass. Height, 6½ inches.

**1095** 1912

### FANG
**Rich Gold Plated.**
8 Day, Time.......................$4 00
2 inch IVORY Dial.
Beveled Glass. Height, 3½ inches.

**1096** 1891

### DWARF.
ONYX.
1 Day, Time.......................$3 50
1 Day, Time, Porcelain Dial............ 4 00
Dial, 2 inches. Beveled Glass.

**1097** 1912

### BEETLE
**Nickel.**
8 Day, Time.......................$3 00
2 inch Dial. Beveled Glass.
Height, 3 inches.

# WATERBURY

**STAND FOR BULB**
Gold Plated.

Stand (without Watch).................................$3 00
Stand (with Watch)...................................11 75
Height, 7⅝ inches. Width, 5 inches.

**BULB—A BALL WATCH**
Gilt.
PORCELAIN Dial.

1 Day, Time.............................................$8 75
Magnifying Lens Back and Front. Stem Wind and Pendant Set.
Height, 3⅜ inches. Width, 2 7/16 inches.
Back Lens beveled at bottom, permitting Watch to stand on Desk.
Bulb with Stand........................................$11 75

Pearl Clock as in use on Gas Burner or on Wall Bracket

**LILY**
Gilt
With Easel at Back.
1 Day, Time..........................$2.50
3½ inch PORCELAIN Dial.
Fancy Gilt Center. Gilt Bezel.
Height, 4¾ inches.

**PEARL**
Night Clock. Opal Dial, Rococo Gilt Bezel.
1 Day, Time (without Leather Case)..............$4 50
RED LEATHER CASE (without Clock)........... 3 50
5½ inch Dial. Height, 7 inches.
The swivel bracket which fits on the gas burner forms also a support for the Clock, so that it can be stood on the mantel; also has wall bracket.

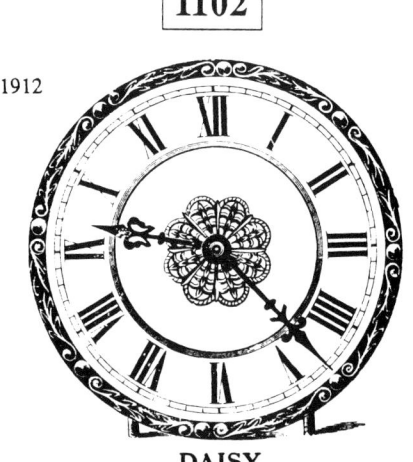

**DAISY**
Gilt
With Easel at Back.
1 Day, Time..........................$3.30
4½ inch PORCELAIN Dial.
Fancy Gilt Center. Beveled Glass.
Rococo Cast Gilt Bezel.
Height, 5 inches.

# WATERBURY

## Ship's Bell Clocks

Eight-Day Jeweled Movements.
Cast Brass Cases.
4½-inch Silvered Metal Dials.

**1103**

**SHIP'S BELL No. 17**
Width, 5⅝ inches; height, 8 inches.
Polished Brass .................. $60.00
Bronze Finish ................... 72.50
Raised Numerals or Radium Dial, $10.00 extra.

**1104**

**SHIP'S BELL No. 16**
Width, 12¾ inches; height, 8 inches.
Fitted with best grade Aneroid Barometer and Thermometer.
Polished Brass ............................................. $130.00
Bronze Finish .............................................. 150.00

**1105**

**SHIP'S BELL No. 18**
Height, 22⅜ inches; width, 7¾ inches; depth, 4 inches.
Polished Brass. Mahogany finish $68.00

# WATERBURY

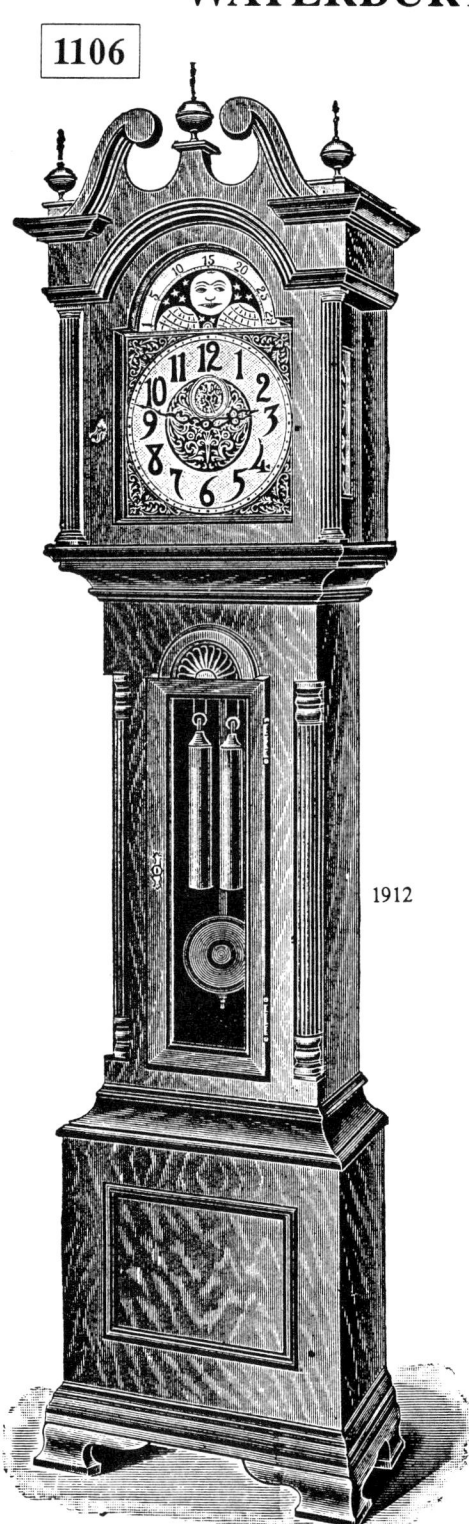
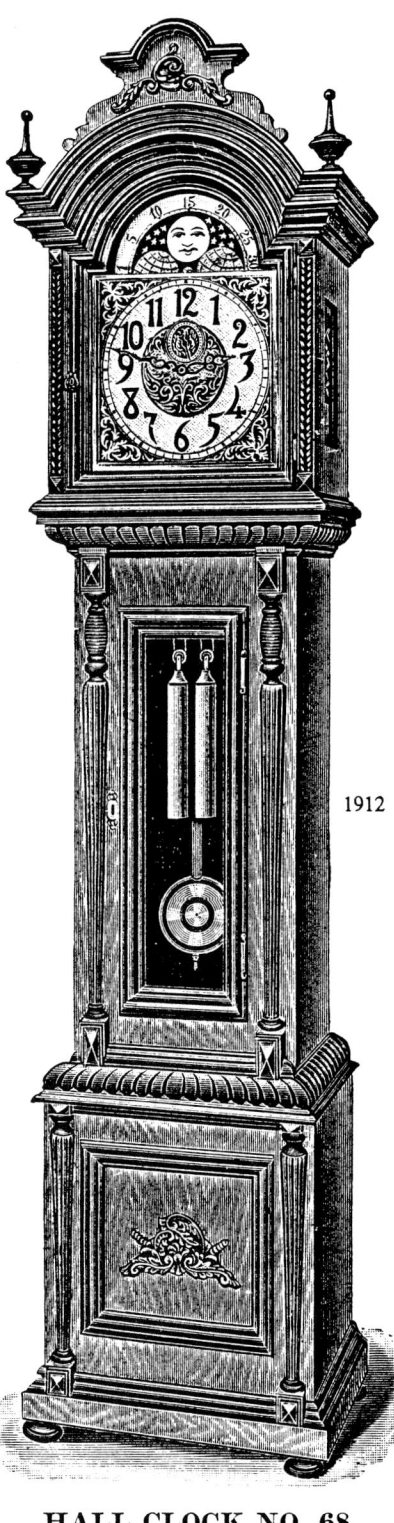
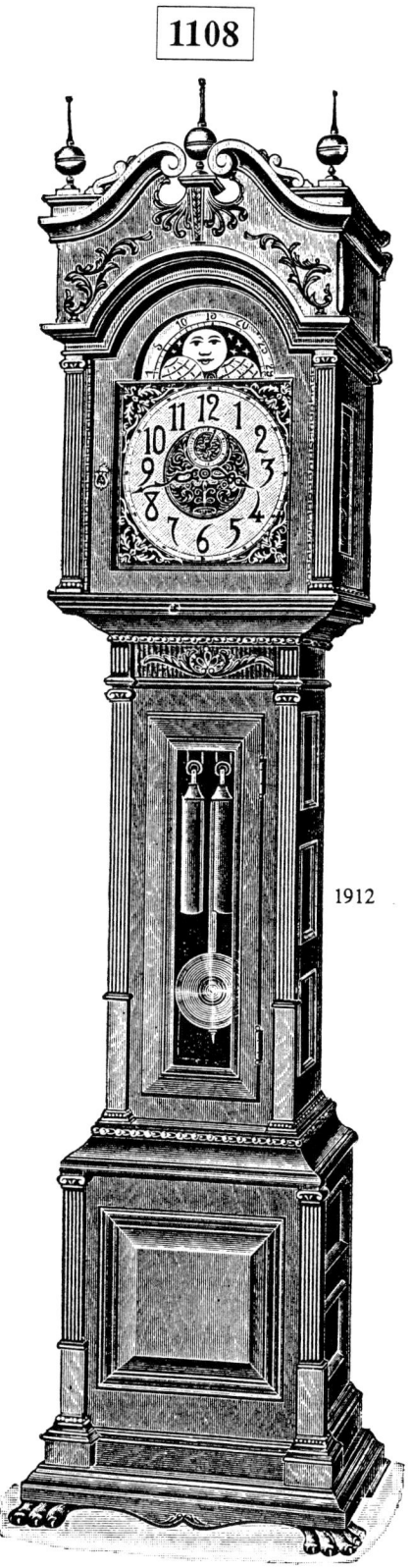

**HALL CLOCK NO. 58**
Mahogany. Gilt Ornaments, Bases and Caps.
or
Mahogany. Mahogany Ornaments, Bases & Caps.
Beveled Plate Glass Panel.
8 Day, Weight, Half-hour Strike,
TUBULAR GONG ............$150 00
Moon's Phases. Brass Weights.
Beats Seconds, Dead-beat Escapement,
Retaining Power.
Solid Polished Movement Frames.
12 inch GILT AND SILVER DIAL.
Height, 93¼ inches. Width, 24⅞ inches.

**HALL CLOCK NO. 68**
Mahogany.
Beveled Plate Glass Panel.
8 Day, Weight, Half-hour Strike,
TUBULAR GONG ............$150 00
Moon's Phases. Brass Weights.
Beats Seconds, Dead-beat Escapement,
Retaining Power.
Solid Polished Movement Frames.
12 inch GILT AND SILVER DIAL.
Height, 92¾ inches. Width, 22⅞ inches.

**HALL CLOCK NO. 72**
Quartered Oak.
Beveled Plate Glass Panel.
8 Day, Weight, Half-hour Strike,
TUBULAR GONG ............$170 00
Moon's Phases. Brass Weights.
Beats Seconds, Dead-beat Escapement,
Retaining Power.
Solid Polished Movement Frames.
12 inch GILT AND SILVER DIAL.
Height, 99 inches. Width, 25 inches.

# WATERBURY

*Except for those fabulous clocks for which you wouldn't mind cutting a hole in the ceiling, Grandfather or Hanging Clocks under eight feet are the most desirable and practical. Some manufacturers produced certain of these clocks in two styles—one to stand on the floor, and the other to hang on a wall. Of the clocks designed with two styles, the hanging style is considered to be the most valuable.*

### 1109

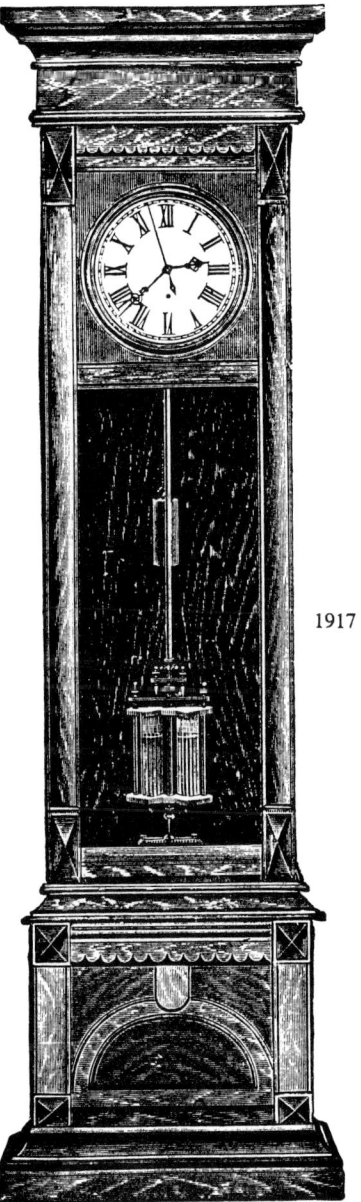

1917

**REGULATOR No. 71**
Quartered Oak or Mahogany
8 Day, Weight, Time.
Brass Weight.
Finely Finished Movement of Best Quality
Encased in Iron Box.
Dead-beat Escapement, Sweep Second,
Retaining Power.
12 inch PORCELAIN Dial.
Height, 96¼ inches. Width, 28 inches.
With Gridiron Pendulum, Oval Rods.
With Compensating Mercurial Pendulum.
For Gridiron Pendulum See Regulator
No. 61.

### 1110

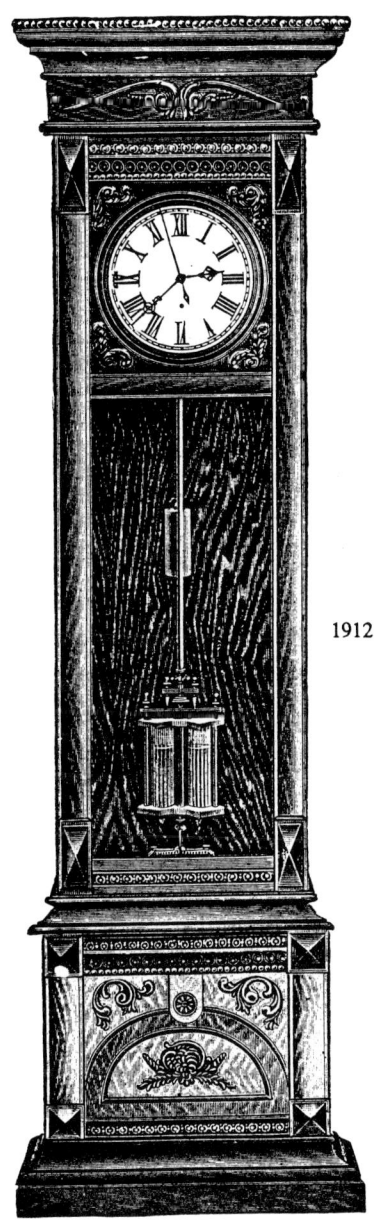

1912

**REGULATOR NO. 71**
Quartered Oak or Mahogany.
8 Day, Weight, Time.
Brass Weight.
Finely Finished Movement of Best Quality
Encased in Iron Box.
Dead-beat Escapement, Sweep Second,
Retaining Power.
12 inch PORCELAIN Dial.
Height, 96¼ inches. Width, 28 inches.
With Gridiron Pendulum,    Oak    Mahogany
   Oval Rods ............. $110 00   $120 00
With Compensating Mercurial
   Pendulum ............. 148 50    158 50
For Gridiron Pendulum See Regulator
No. 61.

### 1111

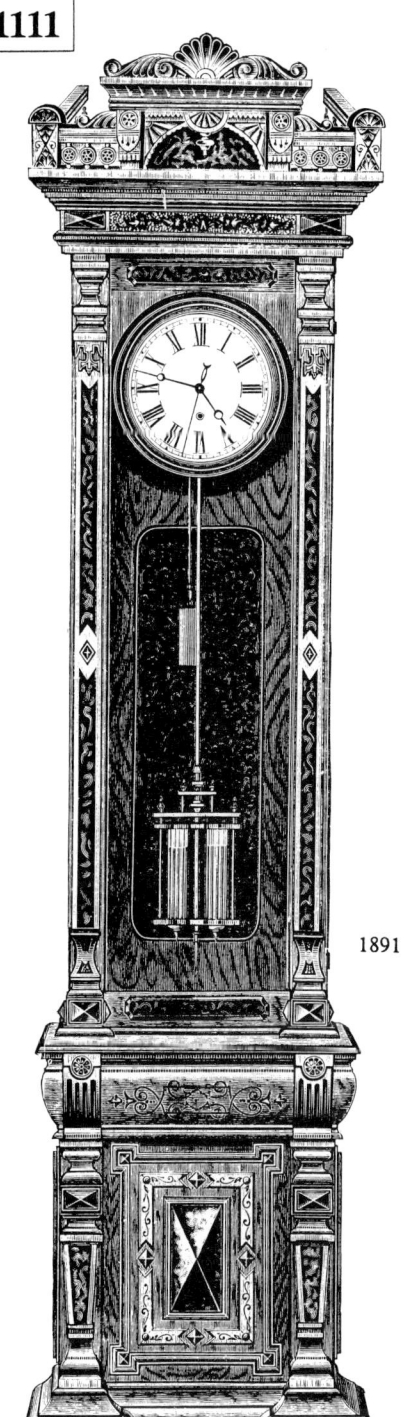

1891

**REGULATOR No. 16.**

OAK, WALNUT OR CHERRY.

CABINET FINISH.   GLASS SIDES.

Finely finished Movement of best
quality encased in Iron Box.

8 Day, Weight, Time.
Brass Weight.
Dead-beat Pin Escapement,
Sweep Second, Retaining Power.
Porcelain Dial 12 inches.
Height 108 inches.

# WATERBURY

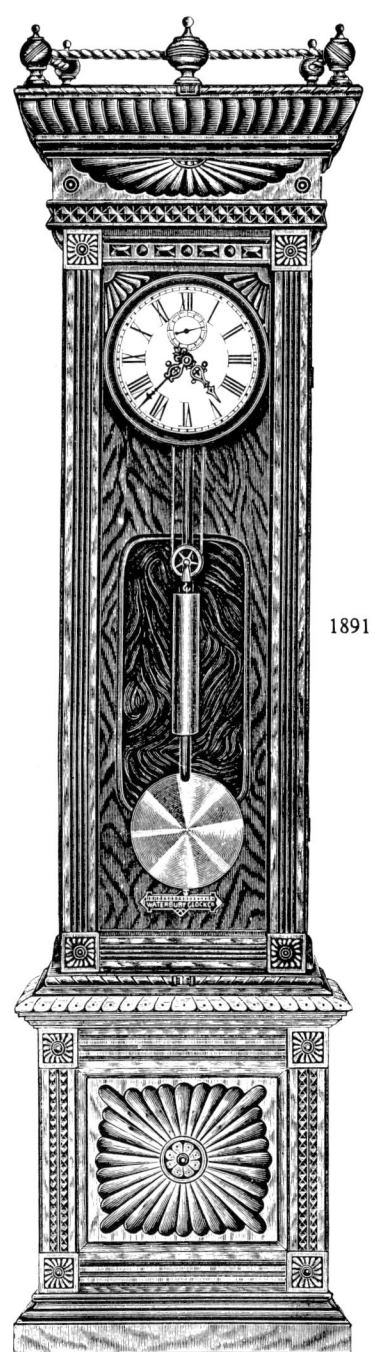

### REGULATOR No. 10.
OAK. BRASS RAILING.

CABINET FINISH. GLASS SIDES.

8 Day, Weight, Time.
8 Day, Weight, Half-hour Strike, Gong.
Brass Weight.
Beats Seconds, Dead-beat
Escapement, Retaining Power.
Solid Polished Movement Frames.
Dial 10 inches.   Height 91 inches.

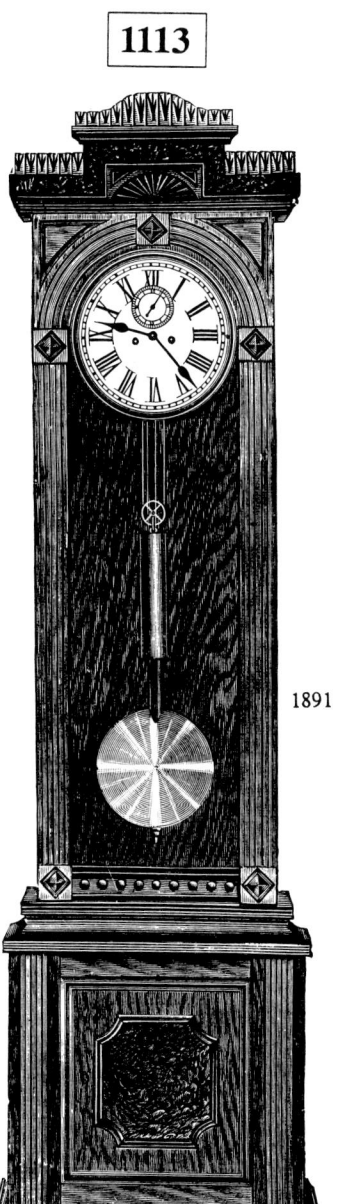

### REGULATOR No. 6.
WALNUT.

CABINET FINISH.  GLASS SIDES.

8 Day, Weight, Time.
8 Day, Weight, Half-hour Strike, Gong.
Brass Weight.
Beats Seconds, Dead-beat
Escapement, Retaining Power.
Solid Polished Movement Frames.
Dial 10 inches.   Height 87 inches.

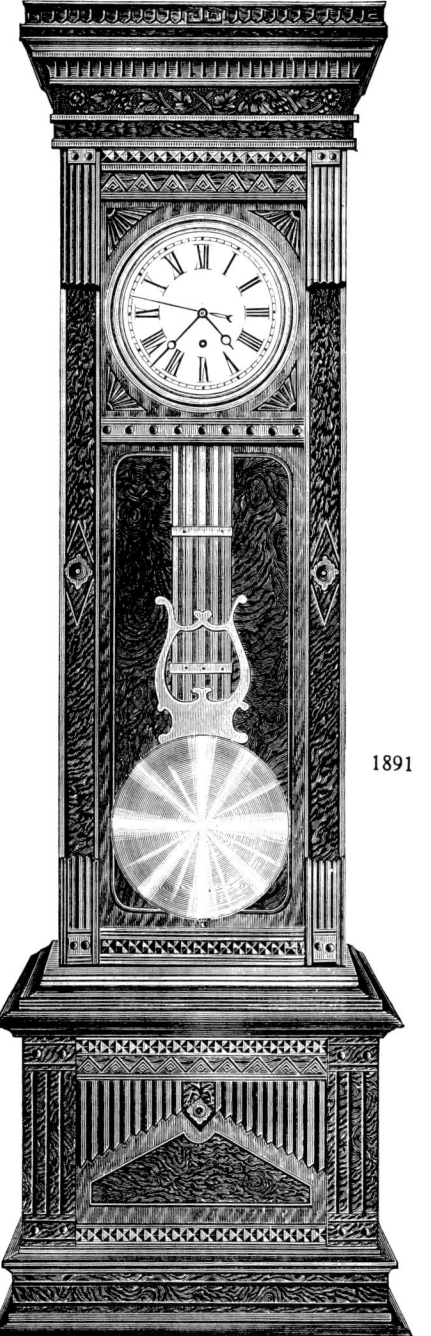

### REGULATOR No. 8.
OAK, WALNUT OR CHERRY.

CABINET FINISH.  GLASS SIDES.

8 Day, Weight, Time.
Brass Weight.
Finely finished Movement of best
quality encased in Iron Box.
Dead-beat Pin Escapement,
Sweep Second, Retaining Power.
Porcelain Dial 12 inches.
Height 102 inches.

# WATERBURY

*For comprehensive listing of Waterbury Clocks, See Tran Duy Ly "Waterbury Clocks - History, Identification & Price Guide."*

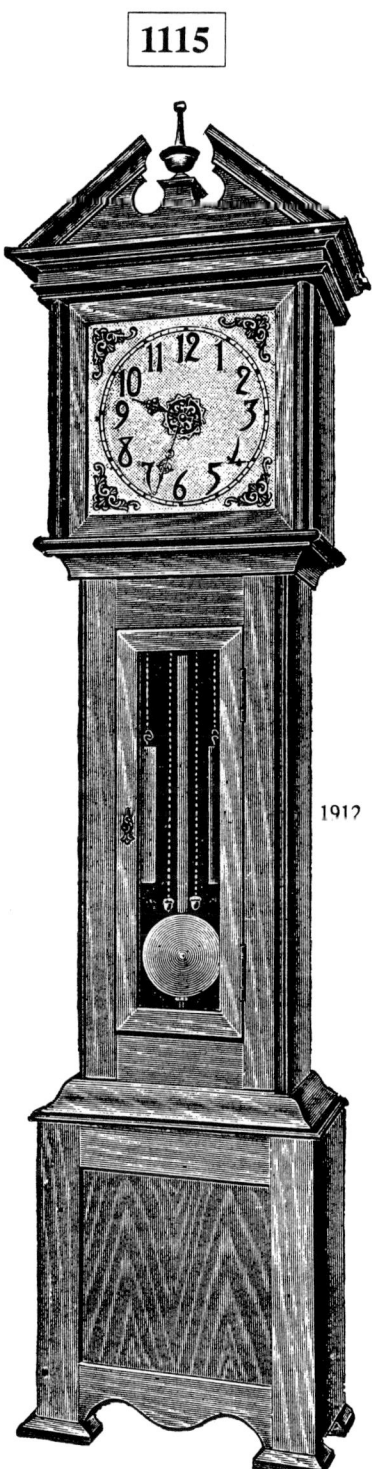

**1115**

**HALL CLOCK NO. 74**
**Oak or Mahogany Finish.**
8 Day, Chain Weight, Half-hour Strike,
Gong .............................. $68 00
Beats Seconds.
12 inch GILT AND SILVER DIAL.
Height, 88 inches.  Width, 22¾ inches.

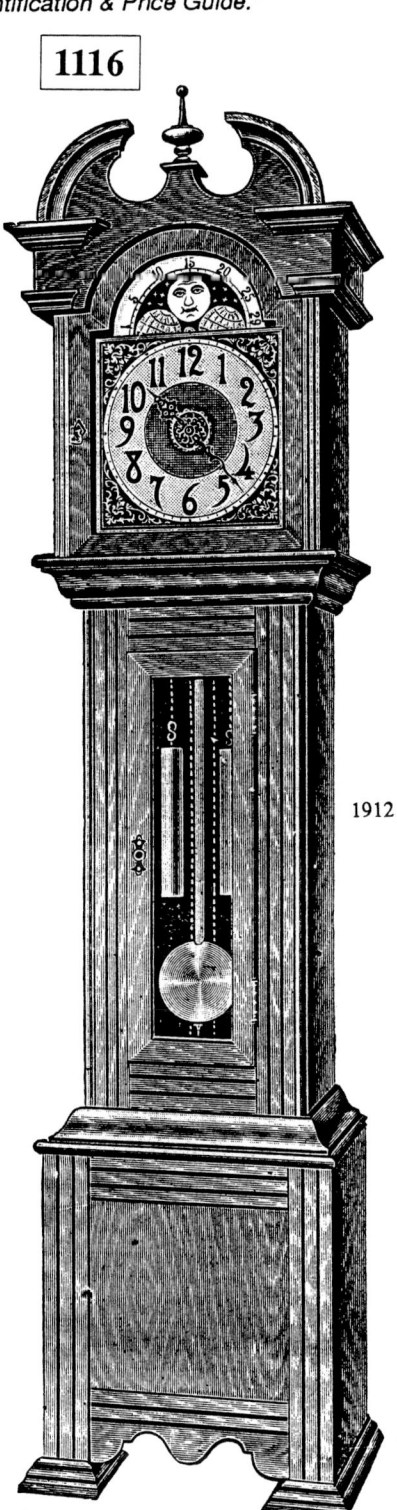

**1116**

**HALL CLOCK NO. 78**
**Oak or Mahogany Finish.**
8 Day, Chain Weight, Half-hour Strike,
Gong .............................. $86 00
Moon's Phases.
Beats Seconds.
12 inch GILT AND SILVER DIAL.
Height, 91½ inches.  Width, 23½ inches.

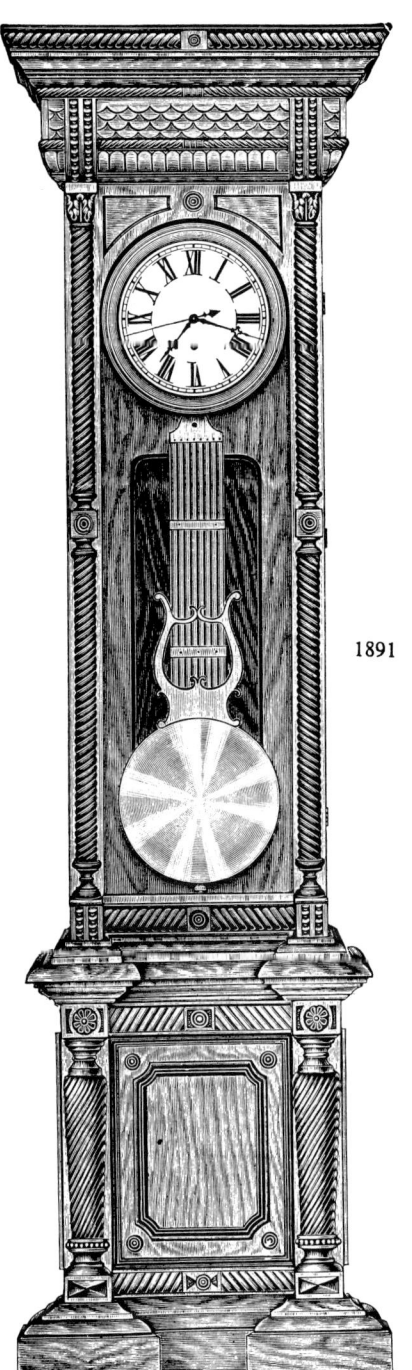

**1117**

**REGULATOR No. 13.**

OAK, WALNUT OR CHERRY.

CABINET FINISH.  GLASS SIDES.

Finely finished Movement of best
quality encased in Iron Box.

8 Day, Weight, Time.
Brass Weight.
Dead-beat Pin Escapement,
Sweep Second, Retaining Power.
Height 100½ inches.
Porcelain Dial 12 inches.

# WATERBURY

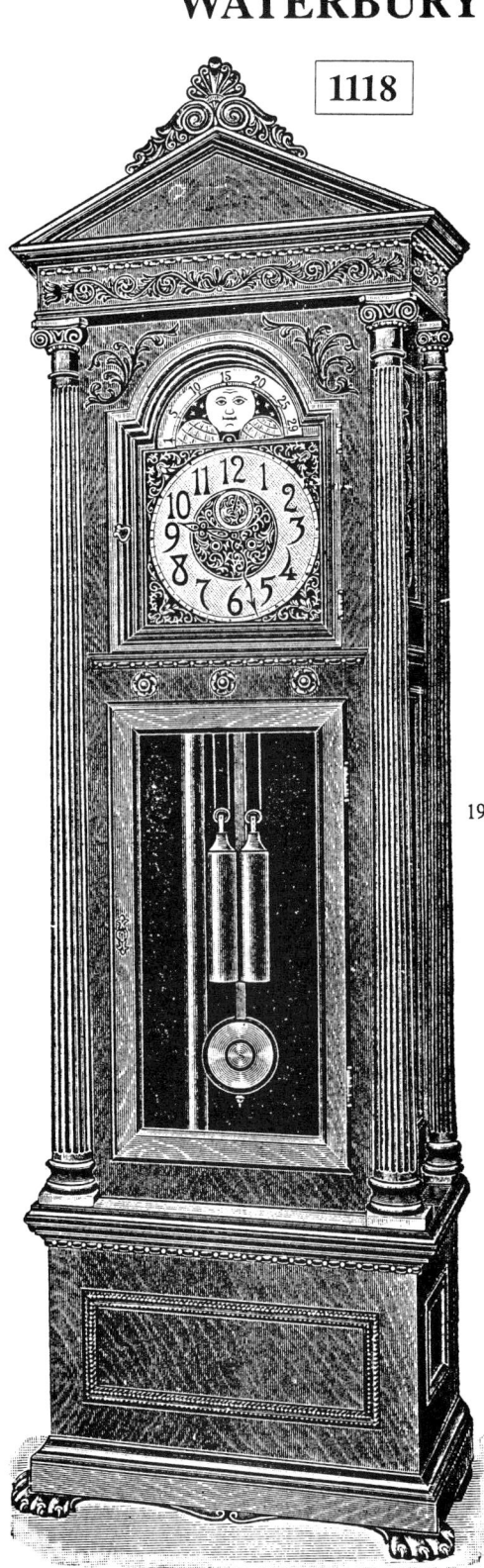

**1118** — 1912

**HALL CLOCK NO. 77**
**Mahogany.**
Beveled Plate Glass Panel.
8 Day, Weight, Half-hour Strike,
TUBULAR GONG ..............$210 00
Moon's Phases. Brass Weights.
Beats Seconds, Dead-beat Escapement,
Retaining Power.
Solid Polished Movement Frames.
12 inch GILT AND SILVER DIAL.
Height, 96 inches. Width, 29½ inches.

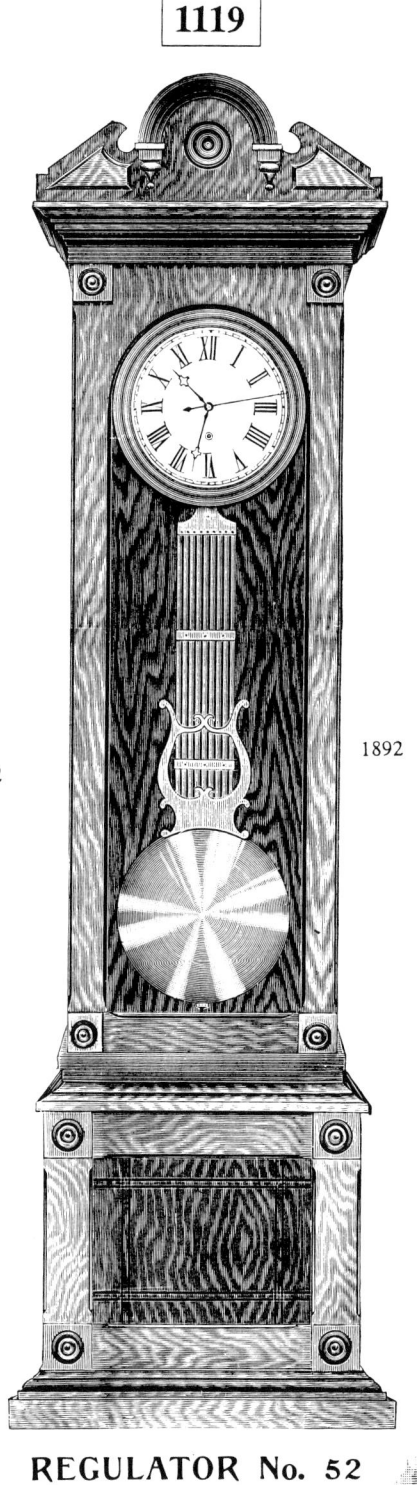

**1119** — 1892

**REGULATOR No. 52**
OAK, WALNUT OR CHERRY.
CABINET FINISH. GLASS SIDES.
Finely finished Movement of best
quality encased in Iron Box.
8 Day, Weight, Time.
Brass Weight.
Dead-beat Pin Escapement,
Sweep Second, Retaining Power.
Porcelain Dial 12 inches. Height 102 inches.

**1120** — 1912

**HALL CLOCK NO. 76**
**Mahogany.**
Beveled Plate Glass Panel.
8 Day, Weight, Half-hour Strike,
TUBULAR GONG ..............$200 00
Moon's Phases. Brass Weights.
Beats Seconds, Dead-beat Escapement,
Retaining Power.
Solid Polished Movement Frames.
12 inch GILT AND SILVER DIAL.
Height, 97 inches. Width, 27½ inches.

# WATERBURY

*The year entered beside each clock is the year of the catalog from which the illustration was taken. Some clocks may have been produced a few years earlier.*

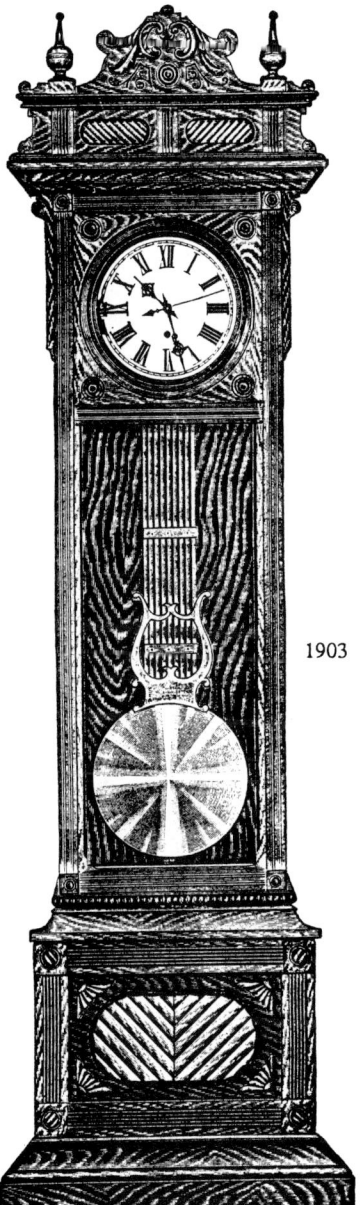

1121

1903

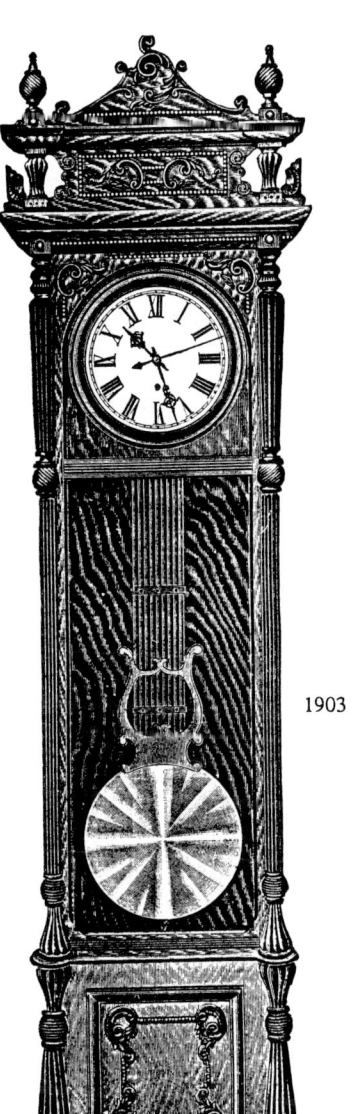

1122

1903

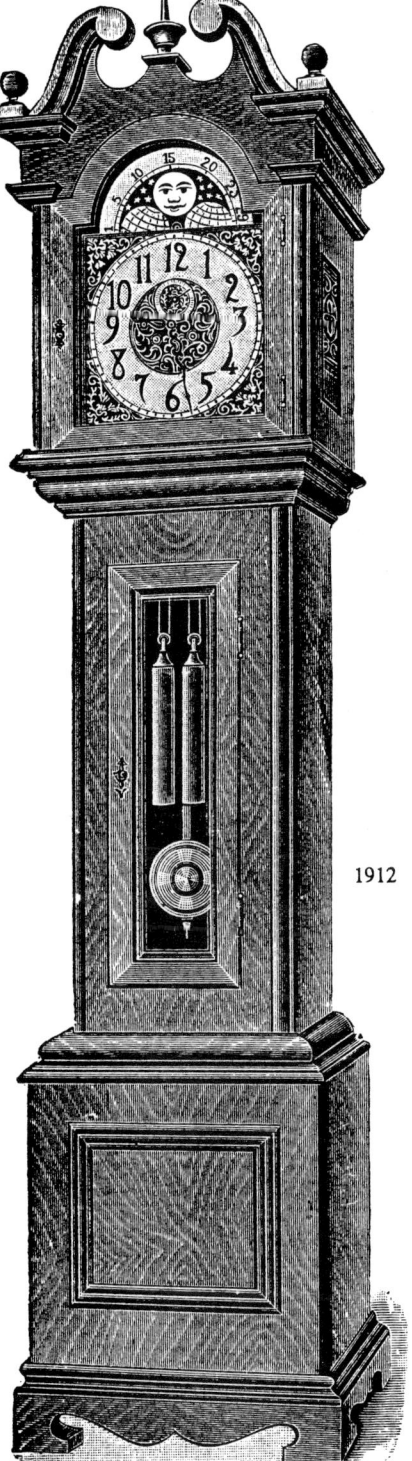

1123

1912

### REGULATOR No. 61
**Quartered Oak or Walnut.**
**Polished Finish.**
8 Day, Weight, Time.
Brass Weight.
Dead-beat Escapement, Sweep Second,
Retaining Power.
12 inch PORCELAIN Dial.
Height, 96¼ inches. Width, 28⅝ inches.
With Gridiron Pendulum, Oval Rods.... $90 00
With Compensating Mercurial Pendulum 122 00
For Mercurial Pendulum see Regulator No. 71.

### REGULATOR No. 69
**Quartered Oak, Walnut or Mahogany.**
**Polished Finish.**
8 Day, Weight, Time.
Brass Weight.
Dead-beat Escapement, Sweep Second,
Retaining Power.
12 inch PORCELAIN Dial.
Height, 96¼ inches. Width, 27 inches.
With Gridiron Pendulum, Oval Rods. .$100 00
With Compensating Mercurial Pendulum 132 00
For Mercurial Pendulum see Regulator No. 71.

### HALL CLOCK NO. 73
**Mahogany.**
Beveled Plate Glass Panel.
8 Day, Weight, Half-hour Strike,
TUBULAR GONG ................$140 00
Moon's Phases. Brass Weights.
Beats Seconds, Dead-beat Escapement,
Retaining Power.
Solid Polished Movement Frames.
12 inch GILT AND SILVER DIAL.
Height, 93 inches. Width, 23¾ inches.

# WATERBURY

*For comprehensive listing of Waterbury Clocks, See Tran Duy Ly "Waterbury Clocks - History, Identification & Price Guide."*

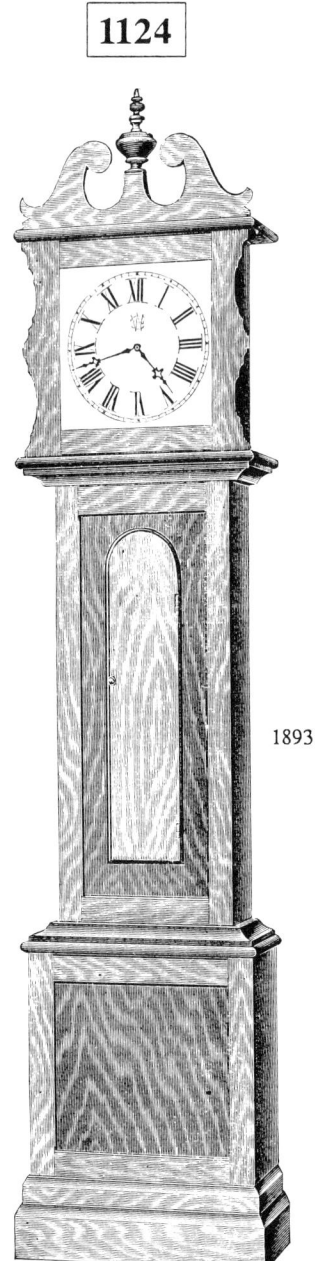

### HALL CLOCK No. 55

**OAK OR CHERRY**

**STAINED**

Dial, 12 inches. Height, 87¼ inches.

8 Day, Chain Weight, Strike, Gong..$40 00

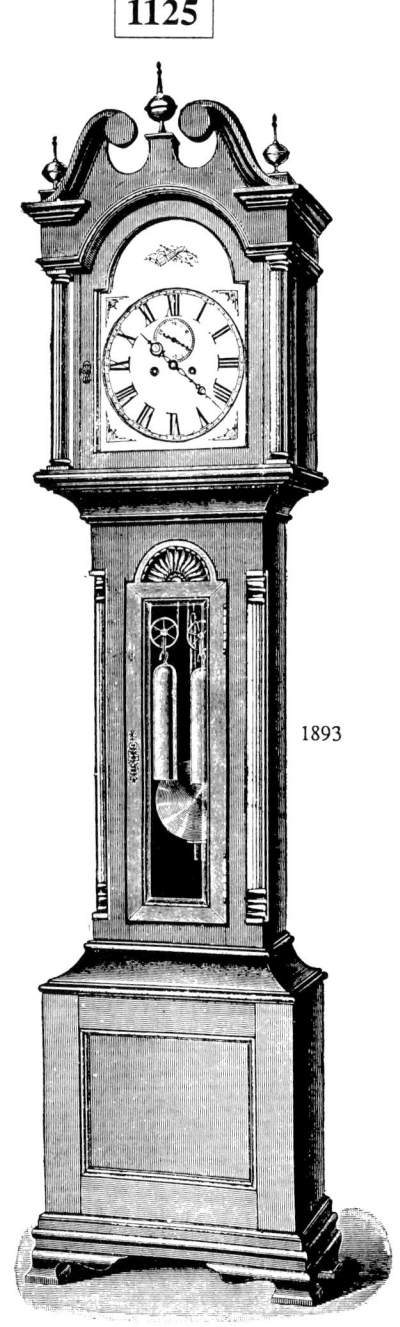

### HALL CLOCK No. 19

**OAK OR CHERRY**

CABINET FINISH

Beats Seconds, Dead-beat Escapement, Retaining Power.
Solid Polished Movement Frames.
Silver Dial, 12 inches. Height, 93¼ inches.

8 Day, Weight, Half-hour Strike, Gong.

Brass Weights, Beveled Plate Glass Panel... $165 00
Iron Weights, Wood Panel................ 150 00

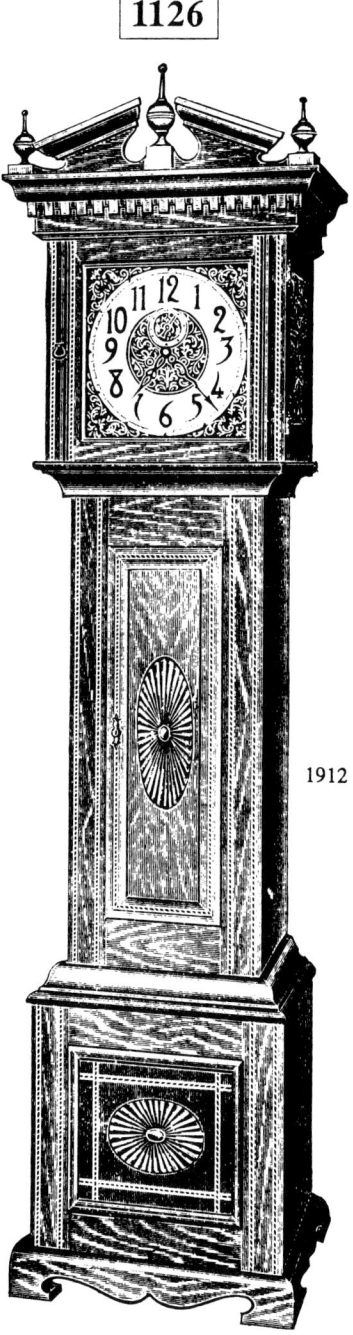

### HALL CLOCK NO. 79
**Mahogany Inlaid.**

8 Day, Chain Weight, Half-hour Strike, Gong......................$100 00
Beats Seconds.
12 inch GILT AND SILVER DIAL.
Height, 90 inches. Width, 23½ inches.

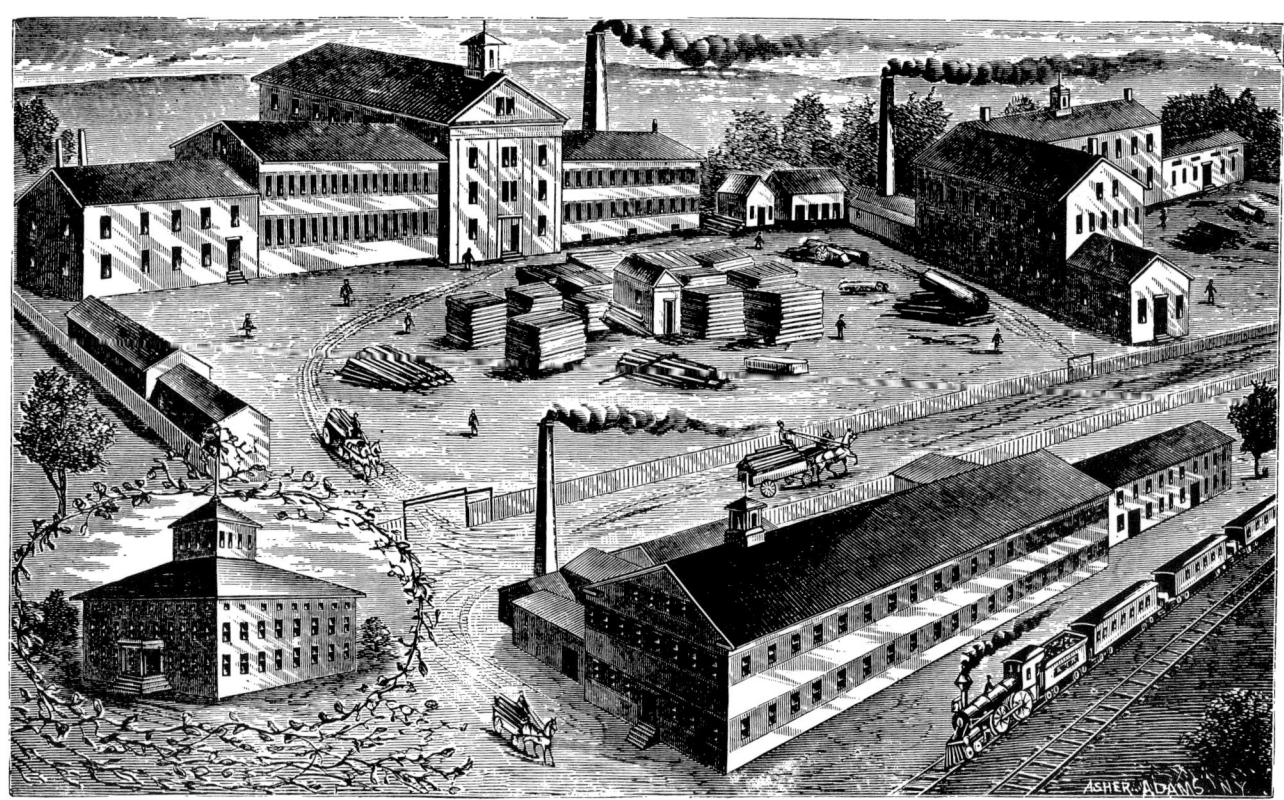

Partially Destroyed by Fire, March 17 and December 14, 1899.

# THE E. N. WELCH MFG. COMPANY
Forestville, Connecticut

The E. N. Welch Manufacturing Company was a joint stock corporation formed July 6, 1864 to succeed an older private firm making clocks under the name of E. N. Welch. Elisha N. Welch (1809-1887) had been making clocks at a factory site on East Main Street at Forestville, Conn. since taking over the bankrupt business of J. C. Brown about 1856.

A new movement shop was fitted up in 1869, adding to the two factories already in use by the firm. Between 1868 and 1884, a subsidiary firm called Welch, Spring & Company was formed to produce a more expensive line of clocks. This firm will be spoken of in more detail elsewhere. The Welch firm was well-known for its handsome rosewood cases, though in 1885, with changing styles in furniture, the surviving firm began to introduce new models with solid walnut cases and discontinued some of the older rosewood veneered cases.

After the death of Elisha Welch in 1887, the firm began to decline fast, selling off some of its assets and issuing new stock to raise capital. A new line of clocks was introduced for the year 1893, which appeared to be of much cheaper quality than their discounted line. In May of that year the factory was closed down and a receiver was appointed who spent nearly two years selling off stock and settling the debts of the firm. It was not for another year that the firm resumed production.

In 1899, two fires, one in March and a second in December reduced most of the Welch manufacturing complex to ashes. A new brick factory was built and occupied and by April, 1900, they could meet their liabilities with the interruption in their cash flow. Members of the wealthy Sessions family were busy at this time buying out former stockholders and eventually took control of the firm in 1902 and changed the name to the Sessions Clock Company on January 9, 1903.

# WELCH

*For comprehensive listing of Calendar Clocks, see Tran Duy Ly "Calendar Clocks - A Guide to Identification & Prices."*

**1127**

**AUBER, B. W.**
WEIGHT.
Lewis Calendar.

Height, 36½ inches.   10 inch Dials.
Height, 42 inches.    12 inch Dials.

Eight Day.   Strike.

**WAGNER, B. W.**

Lewis Calendar.
Spring
7 inch Dials.
Height, 32½ inches.
Eight Day.   Strike.

**1129**

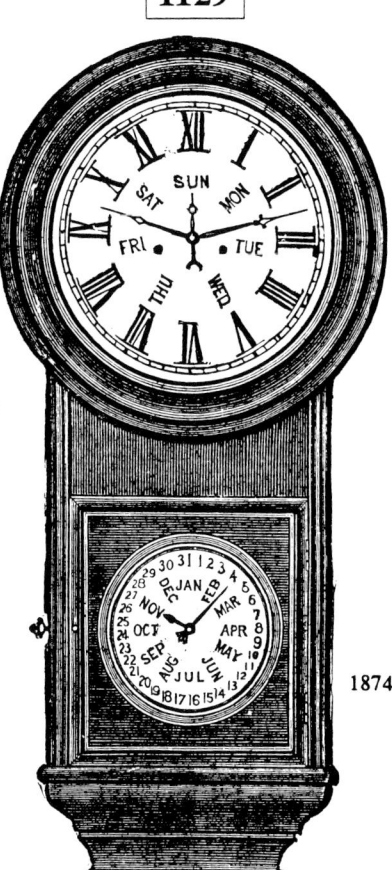

Welch, Spring & Co.
**REGULATOR CALENDAR. No. 2.**
Weight.
Time dial 12 inches.   Calendar dial, 8 inches.
Height, 34 inches.
EIGHT DAY, TIME.

**1128**

1880

**1130**

1880

**WAGNER, B. W.**
Lewis Calendar.
Eight Day.   Strike.
SPRING.
7 inch Dials.
Height, 26 inches.

# WELCH

*For comprehensive listing of Welch Clocks, See Tran Duy Ly "Welch Clocks - History, Identification & Price Guide."*

### 1131

**REGULATOR, G.**

WALNUT, ASH OR MAHOGANY.
HANGING.

8 Day, Weight, Time.

10 inch White or Black Dials

Height, 51½ inches.

### 1132

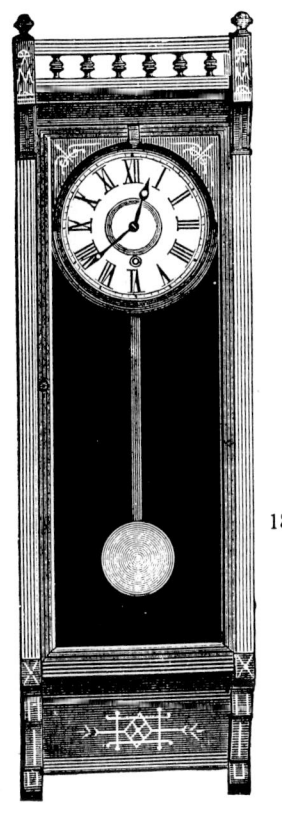

1875

**REGULATOR, H.**

WALNUT ASH OR MAHOGANY.
HANGING.

Height, 38 inches.   Dial, 8 inches.
8 Day.   Spring.   Time.
8 Day.   Spring.   Strike.
Cathedral Bell.
ALSO,
Time Calendar.
Strike Calendar.

*For comprehensive listing of Welch Clocks, see Tran Duy Ly "Welch Clocks - A Guide to Identification & Prices."*

### 1133

1875

**REGULATOR, (I) EYE.**

MAHOGANY, BLACK WALNUT OR ASH.
HANGING.

Fine Movement.
SWEEP SECOND.
8 Day, Weight, Time.
10 inch Dial, White or Black.
Height, 62 inches.   Width, 16¾ inches

# WELCH

*Except for those fabulous clocks for which you wouldn't mind cutting a hole in the ceiling, Grandfather or Hanging Clocks under eight feet are the most desirable and practical. Some manufacturers produced certain of these clocks in two styles—one to stand on the floor, and the other to hang on a wall. Of the clocks designed with two styles, the hanging style is considered to be the most valuable.*

**1134**

**No. 7 REGULATOR.**
SWEEP SECOND-HAND.
WEIGHT.

Height, 47 inches.   Dial, 8 inches.
8 Day.   Time.

**1135**

**SINICO.**
BLACK WALNUT.
Hanging.

8 Day Spring,
"Verdi" Movement.
Height, 45 inches.   Dial, 8 inches.
Cathedral Bell.
Half-hour Strike.
Turn back.

**1136**

**No. 8 REGULATOR.**
POLISHED BLACK WALNUT.
Extra Fine Finish.
Hanging.

Superior Fine Movement.
9 inch. Porcelain Dial.   Sweep Second Hand.
Pendulum beats Seconds.   Dead beat Escapement.
8 Day.   Weight.   Time.
Height 66 inches.   Width, 16 inches.

# WELCH

307

*For comprehensive listing of Welch Clocks, See Tran Duy Ly "Welch Clocks - History, Identification & Price Guide."*

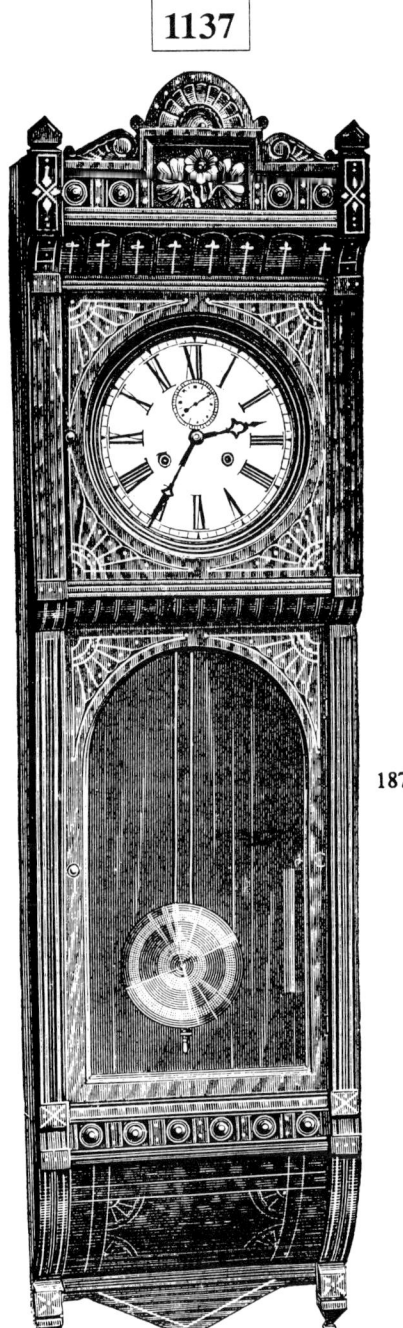

### 1137

**REGULATOR, E.**
WALNUT, ASH OR MAHOGANY.

8 Day, Weight, Time.
10 inch White or Black Dials.
Height, 56 inches.

1875

### 1138

**REGULATOR, (J) JAY.**
MAHOGANY, BLACK WALNUT OR ASH.

Fine Movement.
SWEEP SECOND.
8 Day, Weight, Time.
10 Inch Dial, White or Black.
Height, 62 Inches.   Width, 14 Inches.

1875

### 1139

**REGULATOR, F.**
WALNUT, ASH OR MAHOGANY.

8 Day, Weight, Time
10 inch White or Black Dials
Height, 56 inches.

1875

# WELCH

1875

PATTI No. 1—VERY FINE MOVEMENT.
Eight Day—Strike. $15 00.
Eight Day—Strike—Pat. Alarm. $16 00.
Eight Day Cathedral Bell. $15 60.
Dial 5 Inches. Height 18½ Inches.

1875

## PATTI, No. 2.

POLISHED ROSEWOOD.   GLASS SIDES.

Very Fine Movement.

Height, 10½ inches.   Dial, 3 inches.
8 Day.   Strike.   Cup Bell.

## SCALCHI.

POLISHED BLACK WALNUT.

With or without gilding.

With "Patti" Movement.
Height, 19½ inches.   Dial, 5 inches.
8 Day Cathedral Bell.

1875

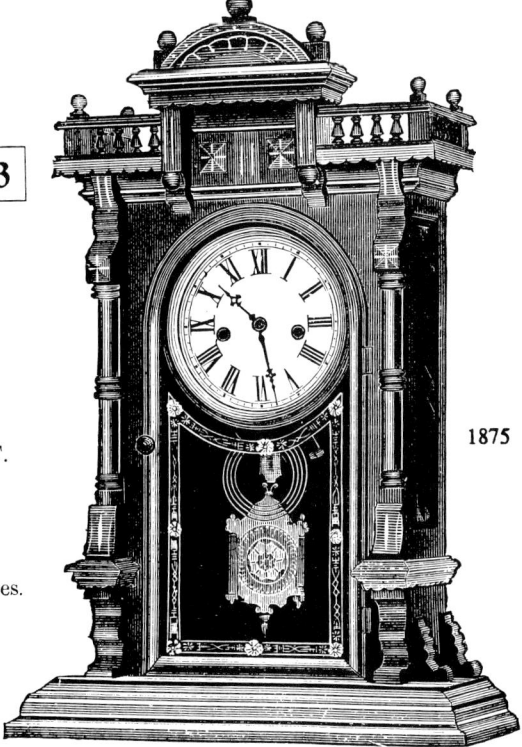

1875

## NILSSON.

POLISHED BLACK WALNUT.
Also MAHOGANY.
With "Patti" Movement.
Height, 22 inches.   Dial, 6 inches.
Eight Day—Strike—Cathedral Gong—Glass Side. $20 00.

# WELCH

*The year entered beside each clock is the year of the catalog from which the illustration was taken. Some clocks may have been produced a few years earlier.*

**1144**

1880

### CARY, V. P.
Height, 20 inches.   5 inch Dial.
SPRING.
Eight Day.   Strike.

**1145**

1875

### EVELINE.
MAHOGANY, WALNUT OR OAK.
With "Patti" Movement.
Height, 17½ inches.   Dial, 5 inches.
8 Day Cathedral Bell.

**1146**

1880

### GERSTER, V. P.
Eight Day.   Strike.
SPRING.
Height, 18½ inches.   5 inch Dial.

**1147**

1875

### PAREPA, V. P.
SPRING.
Height, 22 inches.   Dial, 6 inches.
8 Day.   Strike.

*The year entered beside each clock is the year of the catalog from which the illustration was taken. Some clocks may have been produced a few year earlier.*

**1148**

1889

### JUDIC.
POLISHED MAHOGANY.
With "Patti" Movement.
Height, 20 inches.   Dial, 5 inches.
8 Day.   Hour and Half-hour Strike.
Cathedral Bell.

# WELCH

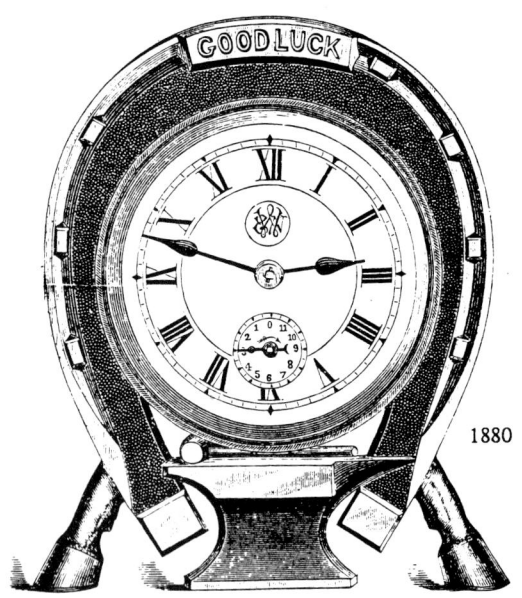

*Most white metal and brass ornamentals and clock cases are gilded or silver plated. Great care must be taken when cleaning these surfaces. The thin layer of gilding or silver plating will be easily removed by cleaning solutions and strong rubbing. If you believe cleaning is necessary, hire the services of a professional and avoid the expense of re-gilding or re-plating.*

### 1149 — GOOD LUCK.
Height, 6 inches.   3 inch Dial.
Gilt or Nickel.
One Day.   Lever.   Time.

### 1150 — BICYCLE.
ORNAMENTAL FINISH WITH NICKEL AND BRASS.
Height, 8 inches.   Dial, 4 inches.
1 Day.   Lever.   Time.   Alarm.

### 1151 — FIRE BUG.
Height, 8 inches.   3 inch Dial.
Nickel.
One Day.   Lever.   Time.   Alarm.
This Clock *lights* the Lamp AUTOMATICALLY at the hour for which the Alarm is set.

### 1152 — MIKADO.
GOLD AND SILVER FINISHED FRAME.   BEVEL PLATE MIRROR.
Height, 11 inches.   Width, 18½ inches.
Porcelain Dial, 2½ inches.   Jewel Movement.
Hour and Half-hour.   Repeating Strike on Cathedral Bell.
1 Day.   Lever.   Also without the Strike.

# WELCH

### 1153

### 1154

1884

### ARMOUR.
(Porcupine.)

**GOLD AND SILVER FINISH.**

One Day—Lever—Time—"Jewel" Movement—
Solid Steel Pinions. Porcelain Dial. $7 00.

1880

### L'IMPERIAL.

Height, 9¼ inches.　　3 inch Dial.

Eight Day.　Pendulum.　Time.

One Day.　Lever.　Time.

Gilt or Nickel.

The *same eight-day* movement is used as in the Chalet.

### 1155

### 1156　　1157

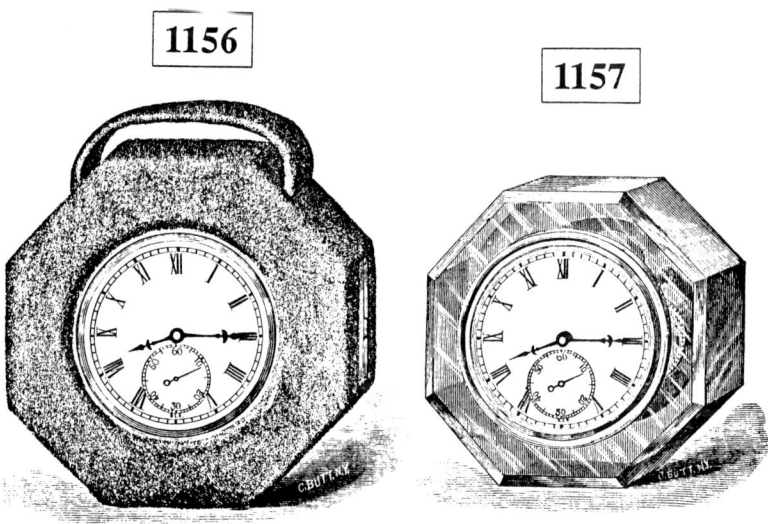

1880

### LA BANNIERE.

Height, 9 inches　　3 inch Dial.

Gilt or Nickel.

One Day.　Lever.　Time.

### THE JEWEL.

1889

We ask special attention to this Clock, being an entire new departure from anything ever produced in either America or Europe. We recommend it to the trade with the assurance that it will compare favorably in QUALITY, and in BEAUTY OF DESIGN with the choicest French novelties.

Height, 3½ inches.　　Porcelain Dial, 2½ inches.

1 Day.　Lever.　Time.　　1 Day.　Lever.　Strike Cathedral Bell.

This Clock has a fine Solid Steel Pinion Movement. The Case is Solid Cut Glass, and can be furnished in Crystal (White), Amber, Emerald and Sapphire. With or without Silk Plush or Leather Traveling Cases. Any of these Clocks may be used in any of the Plush or Leather Cases.

# WELCH

**1158**

## REGULATOR, A.
POLISHED BLACK WALNUT. MAHOGANY. ANTIQUE OAK. GLASS SIDES.
8 Day. Time. Extra Fine Movement. Runs with one Weight. Dead Beat Pin Escapement. Jeweled.
Three Cell Mercurial Compensating Pendulum. 12 inch Porcelain Dial, with Sweep Second. Depth, 12½ inches.
Movement Boxed. Height, 108 inches. Width at Base, 32½ inches.

1885

## REGULATOR, B.
STANDING.
Case, Movement, Dial and Dimensions precisely the same as described in A.
SWEEP SECOND.
With Metal Compensating Pendulum

# INDEX

## ANSONIA

| | |
|---|---:|
| Aida | 32 |
| Aladdin Night Light | 65 |
| Ansonia | 45 |
| Ansonia Banjo No. 1 | 16 |
| Ansonia Banjo No. 2 | 16 |
| Ansonia Banjo No. 3 | 16 |
| Antique Hall Clock | 68 |
| Antique Hanging Clock | 40 |
| Antique Standing No. 1 | 68 |
| Antique Standing No. 2 | 67 |
| Antique Standing No. 3 | 67 |
| Antique Standing No. 4 | 67 |
| Apex | 24 |
| Apollo | 59 |
| Arcadia Ball Swing | 70 |
| Army | 65 |
| Baby | 63 |
| Bagdad | 41 |
| Bahia | 47 |
| Bard | 36 |
| Barge | 63 |
| Bee Ink | 49 |
| Belgrade | 61 |
| Bon-Ton | 19 |
| Bonanza | 19 |
| Bouquet Lever | 46 |
| Brilliant | 19 |
| Brooklyn | 46 |
| Bicycle | 64 |
| Butterfly Ink | 49 |
| Cabinet Antique | 18 |
| Cabinet Antique No. 1 | 18 |
| Cabinet D | 18 |
| Cabinet E | 18 |
| Cabinet F | 18 |
| Calendar No. 3056 | 64 |
| Campania | 50 |
| Candelabra No. 1066 | 53 |
| Candelabra No. 1156 | 52 |
| Candelabra No. 1158 | 51 |
| Candelabra No. 1163 | 50 |
| Candelabra No. 1179 | 54 |
| Capitol | 39 |
| Carlos | 61 |
| Cigar lighter | 62 |

## ANSONIA

| | |
|---|---:|
| Colonel | 43 |
| Comet | 19 |
| Commerce | 48 |
| Comp. Merc. Pendulum | 66 |
| Cortez | 29 |
| Countess | 31 |
| Crystal Palace No 1 Extra | 20 |
| Crystal Palace No. 2 Extra | 20 |
| Crystal Regulator No. 1 | 21 |
| Crystal Regulator No. 2 | 21 |
| Crystal Regulator No. 3 | 21 |
| Crystal Regulator No. 4 | 21 |
| Crystal Regulator No. 5 | 21 |
| Crystal Regulator No. 6 | 21 |
| Crystal Regulator No. 7 | 22 |
| Crystal Regulator No. 8 | 22 |
| Cupid | 34 |
| Diana Ball Swing | 71 |
| Diana Swing | 71 |
| Dinorah | 23 |
| Dispatch | 43 |
| Don Cæsar | 28 |
| Don Cæsar & Don Juan | 28 |
| Don Juan | 28 |
| Double Figure Swing | 69 |
| Drop Extra | 42 |
| Drop Extra Calendar | 42 |
| Eagle | 62 |
| Earl | 22 |
| Echo | 63 |
| Elite | 52 |
| Elliptical Carriage | 19 |
| Emperor | 22 |
| Emporia | 51 |
| Envoy | 23 |
| Eros | 32 |
| Etruria and No. 1184 | 57 |
| Etruscan | 59 |
| Eureka w/ Figure 1085 | 60 |
| Eureka w/ Group 1099 | 59 |
| Fairies | 65 |
| Fantasy | 33 |
| Faust | 30 |
| Fisher | 26 |
| Fisher & Falconer Swing | 69 |

## ANSONIA

| | |
|---|---:|
| Fisher and Hunter | 26 |
| Fisher or Hunter Swing | 72 |
| Fisherman Swing | 70 |
| Floral | 23 |
| Florida & Group No. 1089 | 56 |
| Florida w/ Figure No 1078 | 56 |
| Florida w/ Group No. 1090 | 56 |
| Forrest | 48 |
| Fortuna Ball Swing | 70 |
| Foyer No. 1 | 39 |
| Foyer No. 2 | 38 |
| Franconia | 54 |
| Gallery 18" dial | 39 |
| Gallery 24" dial | 42 |
| General | 40 |
| Germanic | 50 |
| Gloria Ball Swing | 71 |
| Gonfalon | 22 |
| Good Luck | 63 |
| Good Luck Inkstand | 49 |
| Grandfather | 65 |
| Habana | 42 |
| Hanging Monogram | 45 |
| Harp | 62 |
| Hebe and Pizarro | 58 |
| Hector | 59 |
| Helmsmen | 20 |
| Henry IV | 55 |
| Hunter | 26 |
| Huntress Swing | 72 |
| Idyl | 32 |
| Jumper No. 1 | 17 |
| Jumper No. 2 | 17 |
| Jumper No. 3 | 17 |
| Juno Ball Swing | 72 |
| Juno Swing | 72 |
| Jupiter | 23 |
| Kobi Cal. 10" | 46 |
| Latonia w/ Figure No. 1091 | 52 |
| Latonia w/ Figure No. 1012 | 60 |
| Lily Ink | 49 |
| Lucia | 25 |
| Lydia | 53 |
| Mahogany Swing No. 1 | 73 |
| Mahogany Swing No. 2 | 73 |

313

## ANSONIA

| | |
|---|---|
| Mahogany Swing No. 3 | 73 |
| Majestic | 52 |
| Major | 37 |
| Martha | 23 |
| Mecca | 46 |
| Melody | 27 |
| Melody and Motion | 27 |
| Mercury | 33 |
| Mexico | 44 |
| Miranda | 53 |
| Mirror Swing | 69 |
| Mississippi | 48 |
| Monarch | 61 |
| Motion | 27 |
| Mozart | 31 |
| Musician | 34 |
| Navy | 65 |
| New York | 45 |
| Newton | 31 |
| Night Light | 65 |
| Niobe | 40 |
| No. 1053 | 59 |
| Norma | 25 |
| Novelty No. 27 | 64 |
| Novelty No. 44 | 64 |
| Office Regulator | 44 |
| Olympia and Chloris | 58 |
| Opera | 30 |
| Opera Fan | 64 |
| Paper Weight | 62 |
| Para | 47 |
| Peoria | 51 |
| Philosopher | 32 |
| Pizarro | 29 |
| Pizarro and Cortez | 29 |
| Poet | 34 |
| Prompt | 41 |
| Prosperity | 36 |
| Psyche | 62 |
| Queen Anne | 48 |
| Queen Charlotte | 47 |
| Queen Elizabeth | 48 |
| Queen Isabella | 46 |
| Queen Jane | 47 |
| Queen Mary | 47 |
| Quenn Mab | 47 |
| Reflector | 37 |
| Regal | 25 |
| Regent | 53 |
| Regulator A | 42 |
| Regulator No. 4 | 38 |
| Regulator No. 8 | 68 |
| Regulatro No. 9 | 41 |
| Regulatro No. 11 | 66 |
| Regulator No. 14 | 39 |

## ANSONIA

| | |
|---|---|
| Regulator No. 15 | 66 |
| Regulator No. 16 | 38 |
| Regulator No. 17 | 38 |
| Regulator No. 18 | 37 |
| Renaissance | 54 |
| Reubens | 30 |
| Roman Gothic | 44 |
| Royal | 55 |
| Santa Fe | 37 |
| Sappho | 57 |
| Satellite | 19 |
| Senator | 18 |
| Shakespeare | 34 |
| Sibyl and Gloria | 58 |
| Sibyl and Industry | 55 |
| Sibyl and Winter | 57 |
| Sirius | 24 |
| Solo | 36 |
| Sonnet | 62 |
| Standard Regulator | 45 |
| Suitor | 30 |
| Summer | 35 |
| Summer and Winter | 35 |
| Superba and No. 1180 | 54 |
| Superba and No. 1181 | 60 |
| Swing No. 1 | 17 |
| Swing No. 2 | 17 |
| Symbol | 24 |
| Tampico | 44 |
| Tauric | 50 |
| Tenor | 36 |
| Terpsichore | 33 |
| Teutonic | 51 |
| Triumph | 61 |
| Trotter | 63 |
| Turret | 61 |
| Ulster | 44 |
| Undine and Circe | 55 |
| Undine and Flora | 55 |
| Upton | 44 |
| Utopia | 22, 45 |
| Victory | 35 |
| Vocalists | 33 |
| Waltz | 31 |
| What-Not | 37 |
| Windsor | 61 |
| Winter | 35 |
| York | 43 |

## BOSTON

| | |
|---|---|
| Alhambra | 76 |
| Athens | 75 |
| Candelabra | 74 |

## BOSTON

| | |
|---|---|
| Crystal | 75 |
| Cyprus | 75 |
| Delos | 76 |
| Delphus | 75 |
| Edon | 74 |
| Juno | 76 |
| Locomotive | 76 |
| Queen Anne | 76 |
| Sparta | 75 |

## CHELSEA

| | |
|---|---|
| Barometer on Base | 80 |
| Base and Ball Feet | 82 |
| Carved Base-Style No. 1 | 81 |
| Carved Base-Style No. 2 | 81 |
| Carved Yacht Wheel Clock | 79 |
| Chelsea | 79 |
| Clock & Barometer Desk Sets | 80 |
| Clock on Base | 80 |
| Doric | 84 |
| Empire Mahogany | 83 |
| Gothic | 84 |
| Lever Wall Clock | 84 |
| Mahogany Dome Clock | 83 |
| No. 1 Pendulum | 78 |
| Pedestal Mahogany | 82 |
| Regulator No. 3 | 78 |
| Regulator No. 5 | 77 |
| Special Grand | 81 |
| Special Auto Clock | 78 |
| Tambour Clock Style No. 1 | 82 |
| Tambour Clock Style No. 2 | 82 |
| Wardroom Clock | 80 |
| Windsor Clock | 83 |
| Yacht Wheel | 79 |

## GILBERT

| | |
|---|---|
| Abingdon | 86 |
| Admiral | 95 |
| Admiral Calendar | 94 |
| Algonquin | 86 |
| America | 97 |
| Anchor Lever | 102 |
| Annesley | 91 |
| Arden | 90 |
| Armadale | 89 |
| Asbury | 104 |
| Avalon | 91 |
| Bancroft | 106 |
| Barksdale | 96 |

## GILBERT

| | |
|---|---|
| Berkshire | 93 |
| Bordeaux | 89 |
| Brass Lever | 98 |
| Brighton | 104 |
| Cambridge | 98 |
| Chantelle | 90 |
| Coaching | 113 |
| Consort | 95 |
| Constitution | 103 |
| Curfew | 110 |
| Curfew-Visible Escapement | 110 |
| Defender | 103 |
| Drexel | 99 |
| Drop Octagon Calendar | 94 |
| Dundee | 99 |
| Durham | 99 |
| Elberon | 92 |
| Falmouth | 88 |
| Floretta | 113 |
| Gallery 12 inch | 103 |
| Gallery 18 inch | 107 |
| Girard | 98 |
| Glenwood | 106 |
| Golf | 112 |
| Isabella | 88 |
| Itasca | 97 |
| Janeiro | 96 |
| Lenox | 93 |
| Linden | 108 |
| Lobby | 109 |
| Magdeleine | 87 |
| Maine | 93 |
| Melody | 110 |
| Mosonic Lever | 102 |
| Newport | 104 |
| Observatory | 96 |
| Octagon Drop | 97, 100 |
| Octagon Drop-8 inch | 95 |
| Octagon Drop-10 inch | 95 |
| Octagon Drop-12 inch | 95 |
| Office Drop Calendar | 93 |
| Oriental | 92 |
| Orleans | 87 |
| Ossa | 94 |
| Oxford | 98 |
| Parachute | 102 |
| Pierre | 90 |
| Porcelain No. 244 | 111 |
| Porcelain No. 414 | 111 |
| Porcelain No. 416 | 111 |
| Porcelain No. 419 | 111 |
| Porcelain No. 422 | 112 |
| Porcelain No. 423 | 112 |
| Porcelain No. 434 | 113 |
| Porcelain No. 437 | 113 |

## GILBERT

| | |
|---|---|
| Regulator B | 100 |
| Regulator B Calendar | 94 |
| Regulator G | 95 |
| Regulator No. 1 | 102 |
| Regulator No. 2 | 100 |
| Regulator No. 3 | 102 |
| Regulator No. 4 | 101 |
| Regulator No. 5 | 101 |
| Regulator No. 6 | 101 |
| Regulator No. 7 | 114 |
| Regulator No. 8 Hanging | 107 |
| Regulator No. 8 Standing | 115 |
| Regulator No. 9 | 114 |
| Regulator No. 10 Hanging | 109 |
| Regulator No. 12 Hanging | 109 |
| Regulator No. 12 Standing | 115 |
| Regulator No. 14 | 103 |
| Regulator No. 16 Hanging | 107 |
| Regulator No. 16 Standing | 115 |
| Regulator No. 18 | 114 |
| Regulator No. 20 Hanging | 105 |
| Regulator No. 21 | 108 |
| Regulator No. 22 Hanging | 105 |
| Regulator No. 64 | 108 |
| Regulator No. 65 | 109 |
| Regulator No. 66 | 107 |
| Resolute | 97 |
| Round Corner Lever | 99 |
| Ruskin | 91 |
| Salvia | 112 |
| Saratoga | 104 |
| Sharon | 92 |
| Shield | 97 |
| Star Drop | 97 |
| Stockwell | 106 |
| Thalia | 86 |
| Thespian | 108 |
| Torreon | 88 |
| Touraine | 89 |
| Una Drop | 106 |
| Venice | 87 |
| Violette | 85 |
| Walnut Hanging Weight | 100 |
| Walnut Pendant | 100 |
| Washington | 96 |

## HOWARD

| | |
|---|---|
| Astronomical No. 74 | 127 |
| Bayonne | 131 |
| Marble dial No. 20 | 132 |
| Marble dial No. 21 | 132 |
| Marble dial No. 29 | 132 |
| Marble dial No. 33 | 131 |

## HOWARD

| | |
|---|---|
| Marble dial No. 34 | 131 |
| Marble dial No. 35 | 131 |
| Marble dial No. 63 | 131 |
| Marble dial No. 138 | 132 |
| Marble dial No. 140 | 132 |
| No. 1 | 117 |
| No. 2 | 117 |
| No. 3 | 117 |
| No. 4 | 117 |
| No. 5 | 117 |
| No. 6 | 118 |
| No. 7 | 118 |
| No. 8 | 118 |
| No. 9 | 118 |
| No. 10 | 118 |
| No. 11 | 119 |
| No. 12 | 119 |
| No. 13 | 119 |
| No. 14 | 119 |
| No. 15 | 130 |
| No. 16 | 130 |
| No. 17 | 130 |
| No. 18 | 130 |
| No. 19 | 130 |
| No. 22 | 139 |
| No. 23 | 133 |
| No. 24 | 139 |
| No. 25 | 133 |
| No. 27 | 130 |
| No. 28 | 130 |
| No. 36 | 128 |
| No. 38 | 120 |
| No. 39 | 120 |
| No. 40 | 120 |
| No. 41 | 120 |
| No. 42 | 120 |
| No. 43 | 133 |
| No. 44 | 126 |
| No. 45 | 135 |
| No. 46 | 135 |
| No. 47 | 134 |
| No. 48 | 134 |
| No. 49 | 124 |
| No. 57 | 125 |
| No. 58 | 123 |
| No. 59 | 128 |
| No. 61 | 135 |
| No. 67 | 125, 126 |
| No. 68 | 134 |
| No. 69 | 122 |
| No. 70 | 123 |
| No. 70 Program Contact | 129 |
| No. 71 | 128 |
| No. 72 | 122 |
| No. 75 | 121 |

## HOWARD

| | |
|---|---|
| No. 77 | 136 |
| No. 79 | 136 |
| No. 80 | 136 |
| No. 81 | 137 |
| No. 82-13 in. | 137 |
| No. 82-15 in. | 137 |
| No. 83 | 138 |
| No. 84 | 138 |
| No. 85 | 121 |
| No. 86 | 123 |
| No. 86A | 123 |
| No. 87 | 138 |
| No. 88 | 139 |
| No. 89 | 127 |
| No. 95 | 124 |
| No. 99 | 122 |
| No. 100 | 122 |
| No. 101 | 129 |
| No. 123A | 122 |
| Watch Clock | 129 |

## INGRAHAM

| | |
|---|---|
| Dew Drop | 142 |
| Herald | 142 |
| Highland | 142 |
| Index Calendar | 143 |
| Ingraham Calendar | 143 |
| Ionic Calendar | 143 |
| Iota | 144 |
| Lustre | 144 |
| Lyric | 141 |
| Ormond | 141 |
| Reflector | 144 |
| Reflector 24-hour dial | 144 |
| Reliance | 141 |

## ITHACA

| | |
|---|---|
| Melrose | 147 |
| No. 0 Bank | 146 |
| No. 1 Regulator | 145 |
| No. 2 Bank | 146 |
| No. 3 1/2 Parlor | 147 |
| No. 3 Vienna | 146 |
| No. 5 1/2 Hanging Belgrade | 147 |
| No. 7 Hanging Cottage | 148 |
| No. 9 Shelf Cottage | 148 |
| No. 12 Hanging Kildare | 147 |
| No. 14 Granger | 148 |

## KROEBER

| | |
|---|---|
| Alabama | 155 |
| Arctic | 152 |
| Arizona | 155 |
| Ballet Dancer | 154 |
| Bermuda | 158 |
| Blanche | 163 |
| Brass Lever | 151 |
| Bristol | 150 |
| Bronze No. 280 | 156 |
| Cabinet | 158 |
| Calcutta | 158 |
| California | 155 |
| Candelabra No. 4 | 163 |
| China No. 11 | 160 |
| China No. 12 | 160 |
| China No. 13 | 161 |
| China No. 14 | 160 |
| China No. 15 | 162 |
| China No. 16 | 163 |
| China No. 17 | 162 |
| China No. 18 | 162 |
| China No. 19 | 162 |
| China No. 20 | 161 |
| China No. 21 | 162 |
| China No. 22 | 160 |
| China No. 23 | 161 |
| China Set No. 31 | 161 |
| China Set No. 32 | 163 |
| Club Night | 154 |
| Condé | 159 |
| Conquest | 158 |
| Drop Octagon | 151 |
| Easel No. 1 | 156 |
| Hartford | 151 |
| Indian | 156 |
| Laura | 162 |
| Lillian | 163 |
| Milford | 150 |
| Montana | 156 |
| Mountain Boy | 157 |
| Noiseless Rotary No. 1 | 157 |
| Noiseless Rotary No. 3 | 157 |
| Pacific | 152 |
| Parthia | 153 |
| Polaris | 158 |
| Pompadour | 159 |
| Round Corner | 151 |
| Ruth | 163 |
| Santa Barbara | 156 |
| Scythia | 153 |
| Solid Oak Lever | 152 |
| Tete A Tete | 154 |
| Texas | 158 |
| Trump | 157 |
| Vases No. 3 | 161 |

## KROEBER

| | |
|---|---|
| Ventura | 155 |
| Versailles | 159 |
| Vienna No. 1 | 151 |
| Vienna No. 2 | 151 |
| Vienna No. 26 | 150 |
| Vienna No. 27 | 150 |
| Voltaire | 159 |
| Watermill | 154 |
| Windmill | 154 |

## NEW HAVEN

| | |
|---|---|
| Admiral | 173 |
| Apollo | 183 |
| Austrian | 176 |
| Bank Regulator | 179 |
| Barometer Regulator | 166 |
| Braddock | 179 |
| Blake | 179 |
| Cambria (Spring) | 165 |
| Canton | 166 |
| Champion | 185 |
| Chime No. 1 | 186 |
| Chime No. 3 | 186 |
| Chime No. 5 | 186 |
| Chippendale | 187 |
| Clyde | 185 |
| Columbia | 175 |
| Corsair | 183 |
| Countess | 185 |
| Danube | 184 |
| Drop Octagon | 180, 181 |
| Drop Octagon No. 2 | 178 |
| Drop Octagon, R.C. | 182 |
| Dulce | 177 |
| Eagle | 166 |
| Elbe | 184 |
| Elfrida | 179 |
| Enquirer | 176 |
| Flying Pendulum Clock | 164 |
| Gambia | 172 |
| Giant | 188 |
| Grecian | 174 |
| Hall Clock No. 1 | 187 |
| Harvard | 178 |
| Mosaic Drop | 180 |
| Maywood | 175 |
| New York | 174 |
| No. 25 | 170 |
| No. 25, Gilt | 170 |
| No. 26 | 171 |
| No. 26, Gilt | 171 |
| No. 503 | 184 |
| Oakdale | 183 |

## NEW HAVEN

| | |
|---|---|
| Obi | 172 |
| Occidental | 183 |
| Office No. 1 | 171 |
| Office No. 2 | 170 |
| Parisian | 184 |
| Regulator | 178 |
| Regulator No. 0 | 168 |
| Regulator No. 00 | 168 |
| Regulator No. 2 | 168 |
| Regulator No. 3 | 167 |
| Regulator No. 8 | 187 |
| Regulator No. 9 | 169 |
| Regulator No. 10 Hanging | 172 |
| Regulator No. 10 Standing | 188 |
| Regulator, B. B. (Spring) | 165 |
| Regulator, C. C. | 177 |
| Regulator, D. R. | 165 |
| Round Corner Octagon | 178 |
| Salem | 173 |
| Small Drop Octagon | 181, 182 |
| Small Drop Octagon Gilt | 182 |
| Small Drop Octagon, R.C. | 182 |
| Small Regulator | 178 |
| Standard TIme No. 1 | 167 |
| Standard Time No. 2 | 167 |
| Standish | 173 |
| Stanton-Oak | 176 |
| Tampa-Oak | 177 |
| Thornton-Hanging | 173 |
| Thornton-Standing | 188 |
| Trojan | 175 |
| Tyne | 185 |
| Vamoose | 174 |
| Winnipeg Regulator | 169 |
| Wood Lever | 177 |
| Yatch Lever | 176 |

## NEW YORK STANDARD

| | |
|---|---|
| No. 1 | 189 |
| No. 10 | 189 |
| No. 16 | 189 |
| No. 20 | 189 |

## PARKER & WHIPPLE

| | |
|---|---|
| Double Bell No. 98 | 192 |
| No. 1 | 193 |
| No. 2 | 190 |
| No. 5 | 190 |

## PARKER & WHIPPLE

| | |
|---|---|
| No. 11 | 190 |
| No. 300 | 191 |
| No. 601 | 191 |
| No. 604 | 191 |
| No. 40 Time | 193 |
| Parker Princess Alarm | 193 |
| Parker Rotary No. 109 | 193 |
| Rotary Hammer No. 61 | 191 |
| Rotary Hammer No. 63 | 191 |
| Rotary Hammer No. 100 | 192 |
| Rotary Hammer No. 102 | 192 |
| Rotary Hammer No. 103 | 193 |
| Rotary Hammer No. 104 | 192 |
| Rotary Hammer No. 106 | 191 |
| Tornado Intermittent | 193 |

## RUSSELL & JONES

| | |
|---|---|
| Berkshire Drop | 196 |
| Drop Octagon | 195 |
| Drop Octagon, R.C. | 195 |
| Drop Regulator | 195 |
| Easel | 197 |
| Eiffel Tower | 197 |
| Garland | 195 |
| Hand | 197 |
| Highland Drop | 196 |
| Lowell Drop | 196 |
| Plantation | 197 |
| Regulator A | 194 |
| Regulator B | 194 |
| Yesso | 194 |

## SETH THOMAS

| | |
|---|---|
| Arcade | 226 |
| Atlas | 232 |
| Bee | 233 |
| Bona | 236 |
| Boston | 234 |
| Brighton | 223 |
| Buffalo | 234 |
| Colin | 236 |
| Column Weight | 231 |
| Continental | 231 |
| Cupid Empire | 206 |
| Dallas | 233 |
| Doric No. 0 | 228 |
| Doric No. 00 | 228 |

## SETH THOMAS

| | |
|---|---|
| Doric No. 1 | 228 |
| Doric No. 2 | 228 |
| Doric No. 3 | 228 |
| Double Dial Clock | 222 |
| Drop Octagon Calendar | 204 |
| Drop Octagon | 224, 225 |
| Drop Octagon, Gilt | 225 |
| Duchess | 229 |
| Duchess with Lion | 229 |
| Duke | 229 |
| Emperor Set | 205 |
| Empire and Lion set | 209 |
| Empire No. 7 with Bust | 209 |
| Empire No. 7 with Urn | 209 |
| Empire No. 8 | 206 |
| Empire No. 9 | 206 |
| Empire No. 10 | 210 |
| Empire No. 11 | 210 |
| Empire No. 12 | 210 |
| Empire No. 14 | 206 |
| Empire No. 15 | 210 |
| Empire No. 16 | 209 |
| Empire No. 20 | 210 |
| Empire No. 22 | 208 |
| Empire No. 23 | 208 |
| Empire No. 25 | 208 |
| Empire No. 26 | 208 |
| Empire No. 29 | 208 |
| Empire No. 31 | 205 |
| Empire No. 32 | 205 |
| Empire No. 41 | 206 |
| Empire No. 47 | 209 |
| Empire No. 48 | 207 |
| Empire No. 49 | 207 |
| Empire No. 50 | 198 |
| Empire No. 60 | 207 |
| Empire No. 65 | 207 |
| Empire No. 148 | 207 |
| Empire No. 160 | 207 |
| Erie | 234 |
| Flake | 204 |
| Flora | 216 |
| Gallery | 221 |
| Garfield | 232 |
| Globe | 224 |
| Goethe | 235 |
| Gothic No. 0 | 227 |
| Gothic No. 00 | 227 |
| Gothic No. 1 | 227 |
| Gothic No. 2 | 227 |
| Gothic No. 3 | 227 |
| Hall Clock No. 28 | 238 |
| Hecla | 232 |
| Hudson | 226 |
| Joe | 236 |

## SETH THOMAS

| | |
|---|---|
| Jupiter | 215 |
| La Fleur | 230 |
| La Norma | 230 |
| La Reine | 229 |
| Lily | 230 |
| Lincoln | 232 |
| Litchfield | 221 |
| Lobby 14" | 220 |
| Lobby 18" | 226 |
| Lunar | 215 |
| Marcy | 212 |
| Meditation | 230 |
| Minster | 233 |
| Mode Alarm | 236 |
| Mozart | 235 |
| Night Clock "A" | 235 |
| Night Clock "B" | 235 |
| Normandy | 233 |
| Office Calendar No. 1 | 199 |
| Office Calendar No. 2 | 199 |
| Office Calendar No. 3 | 199 |
| Office Calendar No. 4 | 199 |
| Office Calendar No. 5 | 201 |
| Office Calendar No. 6 | 201 |
| Office Calendar No. 7 | 201 |
| Office Calendar No. 8 | 202 |
| Office Calendar No. 9 | 202 |
| Office Calendar No. 10 | 203 |
| Office Calendar No. 11 | 203 |
| Office No. 1 | 222 |
| Office No. 2 | 225 |
| Office No. 3 | 225 |
| Office No. 5 | 221 |
| Office No. 6 | 220 |
| Omaha | 234 |
| Orchid No. 6 | 208 |
| Oregon | 234 |
| Paddock | 236 |
| Parlor Calendar No. 1 | 231 |
| Parlor Calendar No. 5 | 200 |
| Parlor Calendar No. 6 | 200 |
| Parlor Calendar No. 7 | 200 |
| Parlor Calendar No. 8 | 200 |
| Parlor Calendar No. 11 | 204 |
| Petite | 236 |
| Pittsburgh | 233 |
| Precision Clock | 237 |
| Queen Anne | 213 |
| Queen Anne No. 1 | 213 |
| Regulator No. 1 | 211 |
| Regulator No. 1 Extra | 218 |
| Regulator No. 2 | 211, 214 |
| Regulator No. 3 | 218 |
| Regulator No. 4 | 212 |
| Regulator No. 5 | 212 |

## SETH THOMAS

| | |
|---|---|
| Regulator No. 6 | 215 |
| Regulator No. 7 | 219 |
| Regulator No. 8 | 219 |
| Regulator No. 10 | 214 |
| Regulator No. 11, Hanging | 219 |
| Regulator No. 11, Standing | 238 |
| Regulator No. 12 | 214 |
| Regulator No. 13 | 238 |
| Regulator No. 14 | 237 |
| Regulator No. 15 | 237 |
| Regulator No. 16 | 217 |
| Regulator No. 17 | 213 |
| Regulator No. 18 | 224 |
| Regulator No. 19 | 217 |
| Regulator No. 20 | 220 |
| Regulator No. 25 | 221 |
| Regulator No. 30 | 222 |
| Regulator No. 31 | 220 |
| Regulator No. 32 | 216 |
| Regulator No. 60 | 223 |
| Regulator No. 62 | 223 |
| Regulator No. 63 | 218 |
| Rex | 229 |
| Rio | 223 |
| Schiller | 235 |
| Shakespeare | 235 |
| Signet | 214 |
| Sterling | 233 |
| Suez | 216 |
| Thirty day office | 226 |
| Vista | 230 |
| Wagner | 235 |
| World | 224 |
| Yorktown | 231 |

## TERRY

| | |
|---|---|
| Eight Day Chapel Spring | 240 |
| Eight Day Mantel Spring | 241 |
| Eight Day Rose Column | 241 |
| Eight Day Spring | 241 |
| Gothic | 241 |
| Octagon Top Drop | 240 |
| One Day Chapel Spring | 240 |
| One Day Mantel Spring | 240 |
| Round Top Drop | 240 |

## TIFFANY NEVER WIND

| | |
|---|---|
| Never Wind No. 1000 | 242 |
| Never Wind No. 1500 | 242 |
| Never Wind No. 2000 | 242 |

## WALTHAM

| | |
|---|---|
| Chime Dial No. 1 | 255 |
| Chime Dial No. 2 | 256 |
| Five Tube Movement | 255 |
| Half-hour Strike Dial | 256 |
| Harvard | 246 |
| Lyre | 246 |
| Nine Tube Movement | 255 |
| No. 11 | 249 |
| No. 12 | 249 |
| No. 13 | 244 |
| No. 14 | 244 |
| No. 15 | 244 |
| No. 16 | 244 |
| No. 21 | 257 |
| No. 22 | 257 |
| No. 23 | 257 |
| No. 24 | 257 |
| No. 25 | 257 |
| No. 31 | 246 |
| No. 33 | 245 |
| No. 34 | 245 |
| No. 1470 | 246 |
| No. 1500 | 247 |
| No. 1525 | 247 |
| No. 1543 | 247 |
| No. 1546 | 247 |
| No. 1553 | 248 |
| No. 1556 | 248 |
| No. 1570 | 248 |
| No. 8514 | 247 |
| Pattern No. 2 | 250 |
| Pattern No. 6 | 254 |
| Pattern No. 7 | 254 |
| Pattern No. 14 | 252 |
| Pattern No. 35 | 253 |
| Pattern No. 40 | 253 |
| Pattern No. 60 | 253 |
| Pattern No. 62 | 251 |
| Pattern No. 63 | 251 |
| Pattern No. 64 | 251 |
| Pattern No. 65 | 250 |
| Pattern No. 67 | 250 |
| Pattern No. 68 | 254 |
| Pattern No. 69 | 252 |
| Pattern No. 71 | 252 |
| Pattern No. 81 | 249 |
| Pattern No. 82 | 243 |
| Solid Bronze | 247 |
| Waltham Abbot Lyre | 246 |
| Waltham Chronometer | 245, 248 |
| Waltham Curtis Girandole | 245 |

# WARREN

Figure 6 Table Clock ......... 258
Type "A" Master Clock ....... 258
Type "B" Master Clock ....... 258

# WATERBURY

Admiral ............................ 268
Admiral Calendar ............... 282
Alburtis ............................ 270
Antwerp ........................... 271
Arch Top .......................... 290
Atlanta ............................. 285
Aude ................................ 272
Augusta ............................ 285
Automatic ......................... 292
Avignon ............................ 270
Banjo ............................... 292
Bath ................................. 291
Beetle .............................. 293
Birmingham ...................... 291
Bordeaux .......................... 270
Buffalo ............................. 266
Bulb-A Ball Watch ............. 294
Calais ............................... 269
Calendar No. 22 ................. 263
Calendar No. 26 ................. 263
Calendar No. 30 ................. 265
Calendar No. 32 ................. 265
Calendar No. 34 ................. 263
Calendar No. 36 ................. 264
Calendar No. 38 ................. 264
Calendar No. 40 ................. 264
Calendar No. 42 ................. 264
Calendar No. 43 ................. 267
Calendar No. 44 ................. 263
Candelabra No. 11 .............. 288
Cantal .............................. 272
Catawba ........................... 288
Cavalier ............................ 289
Chalons ............................ 270
Charente ........................... 272
Cheshire ........................... 287
Chesterton ........................ 291
Combine, Calendar ............. 266
Consort ............................ 286
Consort, Calendar .............. 267
Continental ....................... 262
Daisy ............................... 294
Dauphin ........................... 289
Digby ............................... 286
Drop Octagon, R. C. ........... 265
Druggist ........................... 292
Durbin ............................. 289
Dwarf .............................. 293

# WATERBURY

Elmont ............................. 291
English Drop No. 1 ............. 268
English Drop No. 2 ............. 268
English Drop No. 3 ............. 268
Extra Carved Walnut ........... 290
Falmouth .......................... 269
Fang ................................ 293
Flanders ........................... 269
Freeport ........................... 276
Gallery ............................. 287
Gibson ............................. 265
Glenwood ......................... 266
Gothic ............................. 290
Gothic Pillar ..................... 290
Hall Clock No. 19 ............... 302
Hall Clock No. 55 ............... 302
Hall Clock No. 58 ............... 296
Hall Clock No. 68 ............... 296
Hall Clock No. 72 ............... 296
Hall Clock No. 73 ............... 301
Hall Clock No. 74 ............... 299
Hall Clock No. 76 ............... 300
Hall Clock No. 77 ............... 300
Hall Clock No. 78 ............... 299
Hall Clock No. 79 ............... 302
Hornet ............................. 293
Laundry ........................... 292
Lily ................................. 294
Lobby .............................. 287
London ............................ 291
Lyceum ............................ 287
Mikado ............................ 292
Mogul ............................. 272
Montgomery .................... 291
Nevers ............................. 271
Niagara ............................ 267
Organ Grinder ................... 262
Orleans ............................ 272
Paris ................................ 269
Pearl ................................ 294
Peking ............................. 286
Pendant ............................ 285
Rajah ............................... 289
Regent ............................. 286
Regent, Calendar ............... 266
Regulator No. 3 ................. 277
Regulator No. 4 ................. 283
Regulator No. 5 ................. 277
Regulator No. 6 ................. 298
Regulator No. 7 ................. 278
Regulator No. 8 ................. 298
Regulator No. 9 ................. 276
Regulator No. 10 ............... 298
Regulator No. 11 ............... 277
Regulator No. 12 ............... 278

# WATERBURY

Regulator No. 13 ............... 299
Regulator No. 16 ............... 297
Regulator No. 17 ............... 279
Regulator No. 20 ............... 281
Regulator No. 21 ............... 279
Regulator "24 inch" ........... 282
Regulator No. 51 ............... 278
Regulator No. 52 ............... 300
Regulator No. 54 ............... 279
Regulator No. 57 ............... 280
Regulator No. 60 ............... 280
Regulator No. 61 ............... 301
Regulator No. 65 ............... 280
Regulator No. 66 ............... 281
Regulator No. 67 ............... 285
Regulator No. 69 ............... 301
Regulator No. 70 ............... 283
Regulator No. 71 ............... 297
Regulator No. 80 ............... 281
Regulator No. 81 ............... 282
Rennes ............................ 272
Romp .............................. 293
Round Gothic ................... 290
Round Top ....................... 290
Sambo ............................. 262
Sartoris ........................... 289
Savoy ............................. 271
Ship's Bell No. 16 .............. 295
Ship's Bell No. 17 .............. 295
Ship's Bell No. 18 .............. 295
Siam ............................... 286
Southampton ................... 291
Spider ............................. 293
Springfield ...................... 267
Stand For Bulb ................. 294
Steamer .......................... 292
Study No. 2 ..................... 284
Study No. 3 Spring ............ 284
Study No. 3 Weight ........... 284
Study No. 4 ..................... 284
Study No. 5 ..................... 284
Topsey ........................... 262
Valdora .......................... 273
Vale ............................... 273
Valencia ......................... 273
Valiant ........................... 288
Vancouver ...................... 288
Vandyke ......................... 274
Vassar ............................ 275
Venango ......................... 275
Verndale ......................... 274
Verona ........................... 273
Versailles ........................ 271
Vesper ........................... 275
Vicksburg ....................... 288

## WATERBURY

| | |
|---|---|
| Victor | 275 |
| Virgins | 274 |
| Votaw | 274 |
| Waltham | 287 |
| Walton | 267 |
| Willard No. 1 | 260 |
| Willard No. 2 | 260 |
| Willard No. 3 | 260 |
| Willard No. 4 | 261 |
| Willard No. 5 | 260 |
| Willard No. 6 | 261 |
| Willard No. 7 | 261 |
| Willard No. 9 | 261 |
| Willard No. 10 | 261 |
| Willard No. 11 | 260 |
| Yarmount | 282 |

## WELCH

| | |
|---|---|
| Armour | 311 |
| Auber, B. W. | 304 |
| Bicycle | 310 |
| Cary, V. P. | 309 |
| Eveline | 309 |
| Fire Bug | 310 |
| Gerster, V. P. | 309 |
| Good Luck | 310 |
| Judic | 309 |
| Mikado | 310 |
| La Banniere | 311 |
| L'Imperial | 311 |
| Nilsson | 308 |
| Papera, V. P. | 309 |
| Patti No. 1 | 308 |
| Patti No. 2 | 308 |
| Regulator A | 312 |
| Regulator B | 312 |
| Regulator Calendar No. 2 | 304 |
| Regulator No. 7 | 306 |
| Regulator No. 8 | 306 |
| Regulator, (I) Eye | 305 |
| Regulator, (J) Jay | 307 |
| Regulator, E | 307 |
| Regulator, F | 307 |
| Regulator, G | 305 |
| Regulator, H | 305 |
| Scalchi | 208 |
| Sinico | 306 |
| The Jewel | 311 |
| Wagner, B. W. | 304 |

# IMPORTANT TIPS, from page 9

to pay.

When considering condition and value of an antique clock, I feel there are certain criteria which should be used when examining the piece. By asking a few specific questions pertaining to specific areas of the clock and depending upon the answers, you can determine the clock's value. Your conclusion will be based on sound judgement and will be the correct one.

Simply, a clock can be broken down into five areas for easy examining. These areas are:
- **Case**
- **Dial & Hands**
- **Movement**
- **Maker's Label**
- **Tablet or Decorative Glass (If any)**

Even though the clock is technically the movement, it is usually the **CASE** that catches the eye first. Whether the clock is used as a timepiece, a decorator's accessory or is a model added to a series of other clocks to finish a collection, the condition of the case is important. Like other antiques, clock cases are viewed differently by different people. To the purist, the beauty and value of the piece is the patina the surface has acquired over the years as a result of much polishing and cleaning. The nicks, scratches, and color shading only add to the warmth and richness of the piece. Others look at the same piece and only see a surface with an old finish that needs striping and "redoing." For them it needs to be made "like new." Restoration? Preservation? Everyone has his own idea on how far to go with "fixing" the case. Usually the less you do the better. A case can be refinished to such an extent as to lessen its value significantly - yet a slight restoration done by a professional can enhance its value while preserving its integrity.

Depending upon the style of the clock, manufacturers used a variety of materials to construct clock cases. <u>Woods</u> - both solids and veneers, <u>Metals</u> including iron, tin, brass, bronze, nickel and copper and gold gilt, <u>Marbles</u>, and <u>Ceramics</u> including china, porcelain and terra cotta were all used in clock manufacturing.

There are important points to consider when examining each kind of case. When considering a clock with a **wooden case**, consider these points:
- Does the door fit the case properly and is it original?
- Is the backboard intact and original?
- Is there significant structural damage (cracking, splitting, missing, etc.)?
- Is there any missing or replaced ornamentation (hardware, finials, etc.)?
- Has the case been refinished?

The first four points are easy to assess. The last point may require more thorough examination. The importance of the originality and condition of the case has been made earlier. Should it be made to look "like new"? This is a subjective matter. Consider the following. Often antique wood cases are dirty from oils and heating and cooking gases which give them a dull appearance to say the least. On items such as a British Longcase clock from the 18th Century a simple waxing with a quality paste wax will remove the surface dirt without removing the patina - the wax and oils that give true antiques a subtle shading and age. Nothing more needs to be done to this case. On more modern clocks with the same surface dirt, a slightly different approach might be taken. A gentle application of a product like Murphy Oil Soap or other mild soap cleaner might be tried. Be sure to check that the finish is not shellac as shellac often whitens when a water based cleaner is used. More resistant dirt may be attacked with a product like Cotton Cleanser or mechanic's hand cleanser applied with a towel then buffed. The clock should be allowed to dry for a few weeks and then waxed.

Commonly a shellac finish is encountered that has alligatored or checkered, or a clock is found that has repeatedly been French polished with shellac over and over again. Either condition might be addressed by carefully rubbing with medium steel wool saturated with alcohol, a solvent for shellac. When the layers of loose shellac have been loosened and put into solution immediately wipe off while wet with a fresh paper towel. Do not try to remove all the finish, but simply all the bad shellac on the top. In this instance, immediately wax with a quality paste wax. A beautiful shellac finish will result. Other solvents such as acetone or lacquer thinner work the same way for a lacquer finish, but lacquer is not as fast drying as shellac and must be allowed to cure before waxing.

Finally, if the finish is absolutely destroyed,

painted over, or already removed, it probably should be refinished; but don't give up too quickly on an old finish as the procedures mentioned above are usually adequate to renew any old finish and the results are usually better than refinishing. However if your decision is that the finish must be "redone", that is stripped, always try to use the least abusive stripper to remove as little patina as possible. This also will remove as little filler from the grain of the wood as possible. At this point, missing veneer can be replaced, missing trim pieces applied and any nail holes that may appear in moldings can be filled.

Putting the finish back on a clock takes time and is dirty work. Open grained woods such as walnut, red oak and mahogany must be filled. The correct stain must be chosen. Stains that are too light won't look right as most antiques were on the dark side by current standards. Finally, multiple coats of finish must be applied and scrubbed out with fine grade steel wool. This is where most refinishing jobs fail. The first coat of finish looks good so that is all that is done. Wrong! Use many coats for a deep, rich finish. Finally, after the last coat of finish, rub with the finest steel wool and wax.

One feature often found on wooden clock cases is stenciling. Around 1825, American clockmakers realized that the American buying public was becoming increasingly interested in purchasing furniture styles that leaned heavily toward the Empire look. This style incorporated many distinctive features, one of which was tastefully done stenciled designs, usually in gold, applied to prominent surfaces of the piece. Examples are extant of stenciling on tables, trunks, pianos, wardrobes, chairs, and many other pieces from this period. Many American clocks made between 1825 and 1845 were also stenciled. Today these clocks are found with original stenciling in good, average or poor condition, or they have been restenciled. Buyers pay a premium for stenciled clocks with good to excellent original stenciling. When the stenciling is poor, it detracts greatly from the value of a stenciled clock. If the stenciling is in such poor condition that it needs to be restored, the following information on stenciling may be helpful.

Clockmakers, furniture makers, and clock case makers used stenciled designs on clock columns and splats (the decorative piece on top of the case which usually sits between the corner blocks), and to a lesser extent, on the case itself. Stenciling provided a fairly quick way to apply, what appeared to be, a fairly complex design in a relatively short time, yet looked stylish and provided the piece with a distinctive appearance. Stenciled motifs on American clocks show an astonishing variety of patterns, subjects, and treatments, but the usual themes are patriotic, taken from nature, or varying curves and designs formed to fit a difficult space - such as the columns of a clock.

Stenciling, as done on clocks in America from around 1825 to 1845, on both cases and clock glasses (tablets), was done by carefully cutting a design in a sturdy piece of paper so that the major portion of the subject would be readily apparent to the eye, though often supporting pieces, called bridges, were needed to connect areas so that the design did not appear to just be a large opening in the stencil paper. Often, designs were created by using several cut stencils and overlaying them one at a time. For example, a leaf stencil could be cut with the shape of the leaf making the opening in the paper, then a second stencil was cut for the veins, and a third stencil was cut for the supporting stem, and so on, thus creating an attractive and somewhat realistic design. Some stencils, for clock tablets, were cut from a single sheet of paper, as was done by the prolific stenciller William Fenn, whose original stencils survive today in the American Clock and Watch Museum in Bristol, Connecticut.

The actual design was executed by coating the area to be stenciled with a thin, uniform coat of varnish. This was allowed to dry to a tack such that when the stencil was placed on it, the stencil would be held in place by the tacky surface, yet would allow for the stencil to be lifted without pulling off the varnish underneath. This meant the varnish had to be almost dry to the touch, but retain just enough tackiness to hold the stencil and accept the bronzing powder into the varnish. Bronzing powder, still available today, is extremely fine metallic powder that can be picked up in small quantities by wrapping velvet around the index finger and dipping it into a shallow container of powder, then carefully rubbing it through the stencil design.

Good stenciling is a delicate art and requires care to avoid the many pitfalls, yet it is simple enough that the untrained can master the fundamentals in a few hours of practice, which made it especially attractive, no doubt, to the clockmakers of the 1820's-1840's who needed inexpensive decoration for their clock cases. In addition to being easily done and mass produced (once cut, a stencil can be used hundreds of times), stenciling provided a decoration that apparently appealed to the ladies' tastes, and was a decorative element that was not easily disfigured in the peddler's wagon as the earlier cast plaster splats were.

A well executed stencilled design imparts depth as well as beauty, as the application of the metallic powders in various shades and thicknesses allows the design to have a dimensional effect similar to modern airbrushing. And since it was almost always done on a dark background (preferably black), the effect of depth is heightened and the play of light and shadow is emphasized.

There are many stencilled pieces extant from these early decades which are still vibrant and attractive over 150 years later, so we may add longevity to our list of advantages of stencilled designs. And remember, clocks with original stenciling in good to excellent condition are sought after. Consider wisely any attempt at restoration.

Concluding, to refinish a wooden case or not has been debated for years. Should the finish be left as found, discounting any damage and dirt as badges of antiquity or should the case be put back the way it looked when new? This dilemma of restoration will continue. Remember though the highest prices paid are for those clocks found untampered with and in ***original mint condition.***

Metal Clocks do not usually have the minor nicks and dings associated with many wooden clock cases. They tend either to survive in relatively good condition or suffer major damage. Incompetent repairers often glue broken parts together, painting over the damage to disguise their faulty work. Therefore, never paint a metal case as future buyers will suspect you are hiding something. I strongly recommend you do nothing to cover the original surface, no matter what the condition. It has taken many years for the patina to develop, just as on a wooden case; and that originality is important to serious collectors and investment buyers.

Most **white metal** and some **brass** ornamental clocks' cases are silver or gold plated. Great care must be taken when cleaning these surfaces. The thin layer of silver or gold can be easily removed by cleaning solutions and strong rubbing. If you believe cleaning is necessary, hire the services of a professional and avoid the expense of re-gilding or re-plating.

In the case of **Marble Clocks** and **Ceramic Clocks** including **Porcelain** and **China**, condition is the most important factor. If such a clock has been chipped or broken and glued back together it is severely reduced in value. Restoration to damaged porcelain clocks can be performed by professional restorers. Before taking your clock to be restored, check the restorer's references, ask dealers or collectors that you know who have had work done by the restorer. Once you have the necessary information, contact the restorer and ask for an appointment to see a sample of their work. If you are given excuses which prevent you from seeing their work, then find a different restorer. You should be allowed to see samples of their work before you have your clock repaired by them.

Restorations today are made with materials that are air dried, not fired in a kiln. The materials that are used vary from acrylic lacquers to epoxies. A professional restoration should be invisible to the naked eye. Back in the late seventies and early eighties many dealers and collectors were successfully using black lights (ultraviolet) to detect repairs. What would be visible under the black light was the lead of the paints and glazes. By the late eighties most paints and glazes were lead free which has made black light detection some-what obsolete. Black lights are no longer foolproof.

One sure way to detect a restoration is to take a straight pin or safety pin and push the tip of the pin into the area that is in question. If the area has been repaired, the pin will leave a slight pin mark. This will not hurt the restoration. If the suspect area is truly china or porcelain, the pin will not leave any mark.

Today's rate for restoration by an expert can vary from $30.00 to $60.00 an INCH. This cost varies depending on the type of damage and the type of design that must be reproduced.

It is important to understand that care must be taken with restored items. Since all materials are air dried, the colors and glazes that are used will never be as hard as the original kiln fired finish. The restored area can be scratched, cut into, nicked, or partially removed by any sharp or semi-sharp contact. Cleaning a restored article should be done by washing gently, do not soak the article.

Color is another important factor in estimating the value of china and porcelain clocks. Bright reds, blues, and frosted pinks are usually considered to be the most desirable. In order to receive top prices, clocks must be in *original mint condition* and in good running order.

After looking over the case, the **DIAL** should be carefully studied. Look for these important signals:
- Is the dial indigenous to the style and period of the clock?
- Does the dial fit the case properly without signs of extra holes or tampering?
- Are the dial holes in perfect alignment with the winding arbor(s) and center arbor?
- Are there any signs of restoration, repainting, or replacement?
- Are there any markings on the dial? (signature, maker's name, logos, etc.) Are they original?
- Are the **hands** original? Are they a matched set and the correct length for the numerals and chapter ring on the dial?

Clock dials are made of many materials, each with different characteristics as to their deterioration, upkeep and restoration. In general, the greatest danger posed to clock dials is rapid changes in temperature or humidity. Let me explain by discussing several types of dials as found on American and European clocks.

Wooden dials were used for Black Forest clocks and early American shelf and tall case clocks. They are especially susceptible to sudden changes in temperature or humidity. Wood can quickly absorb moisture and expand - the paint is not so quick to respond and as a result we see horizontal strips of paint missing on some of the old wooden dials. The paint has lost its elasticity and cracks and breaks off rather than flex with the expanding dial. Likewise, if the dial is taken from a warm house and put in a cold garage, the same thing will result. The American wooden dials often had raised gesso decorations in the corners or spandrels. These were originally covered in gold leaf and made for a very handsome pattern. Since no protective coating was put over the gold leaf, most of it has been worn off by now, leaving only the raised areas with some small amounts of gold leaf at the base of the gesso. If you look closely at some of these spandrels, you can see what incredibly fine work was done and how beautiful it was originally. The Black Forest dials often were prepared with coats of gesso before applying the paint. This gave a very smooth surface and prevented the grain of the wood from showing through. That is why when you see paint missing, it appears white underneath.

Paper dials were used more recently and there are many good replacements available. It is difficult to "restore" a paper dial, because the ink used to replace or go over the numbers tends to "bleed" into the surrounding paper. Sometimes, however, it can be successfully accomplished.

Brass and silvered dials were the stock in trade for tall case clocks before the painted dials appeared in the 1770's. With the advent of lacquer, it is possible to prevent them from tarnishing and they do not need to be cleaned so often. The dials must be completely disassembled, most of the brass parts polished and lacquered, and the silvered chapter ring and any other parts to be silvered must be cleaned down to the brass in order for a silvering solution to be applied and adhere. The engraved numerals and time track are filled in with a hot wax and levelled off. Then the components of the dial are lacquered and reassembled. For smaller dials such as bracket clock dials, ship's bell, banjo, etc. a similar process is used. If the numbers are not engraved, they must be applied by hand and then the entire dial lacquered. These dials have held up very well over the years, tarnish being the biggest problem. It is important to scrupulously wash out all of the brass cleaner, as residues will corrode the brass. A greenish powder is an indication that residue has been left in the crevices. All things considered, these dials are the best survivors of all.

Zinc dials were used for thousands of American shelf and wall clocks. However, zinc oxidizes producing a fine powder, which continues to deteriorate the paint once the integrity of the surface is damaged. The layer of paint is often very thin, not like the painted tall case clock dials of the early 1800's. Once it starts to deteriorate, it generally continues. Many of these dials need to be repainted, but some can be restored if the remaining paint is sound and adheres well to the dial. Best to have it evaluated by a professional dial painter.

Porcelain dials are a particular problem. The porcelain was originally kiln fired onto copper, and when pieces of the porcelain chip and fall out, it is not possible to re-fire the dial, consequently, today's restorers use a method called "cold work." Restoration is not easy and while it can be done, the result is not always perfect. The difficulty arises when trying to match the color and gloss to the rest of the dial. Renumbering of the dial can also be difficult, especially if colors must be matched. The restored area will never be as hard as the original fired porcelain dial. However given to a professional restorer, using the correct products and methods, repairs can be made with pleasing results. If you can't live with the damage on the dial, you have two options. One, restore the dial; or two, contact a company that makes or sells replacement dials and replace the dial on the clock. If you do replace it, be sure to save the old dial. The same rules that apply to the detection of repairs to porcelain clock cases apply to porcelain dials.

Iron dials are the basis for most of the painted English and American tall case clock dials, some banjo dials, and some bracket clock dials. The English coated some with tin to prevent rusting, or they used a red lead-based paint as a first coat. Many of the American tall case clock dials from the 1820's had no-primer coat on them at all, and as a result are deteriorating faster than the dials manufactured in the late 1770's. Usually the first sign of damage is the place where the dial posts attach to the back of the dial. They are made of brass, the dial of iron or steel, and the two metals contract and expand at different rates. Also the dial posts sometimes loosen up and this contributes to loose paint as well.

Over the years, many well-intentioned people have tried to improve their clock dials by repainting the fading numbers and filling in missing paint, or they have tried to improve the value of the clock by adding an important maker's name. Regrettably, they have actually devalued their clock. If you see a dial that is dark brown in color, chances are it has been heavily coated with shellac to obscure poor repair work. Many of the moon dials were covered with shellac when new to preserve them, and it has. The shellac can be removed with solvent alcohol and it will not damage the underlying paint. This refers only to painted moon dials without gold leaf. The moon dials made around the turn of the century actually had decals on them - these should not be cleaned with alcohol. Sometimes you will see a light brown haze around the painted decoration in the spandrel - this is also shellac.

To restore or not to restore a dial - a question often asked and one that cannot easily be answered. All of us want to find clocks in mint condition. After two hundred years almost nothing is left in mint condition, least of all something that has changed hands many times, travelled hundreds of miles, and suffered untold indignities. If your purpose is to buy and sell clocks and make the greatest profit, don't buy one with flaking paint and worn off numbers, because in order to sell it' you will have to pay someone to restore the dial. There is a point where flaking paint is unattractive and a hindrance to selling a clock. To my mind, a little missing paint, some dirt and numbers that are faded is not a detriment to a dial. If the dial is in imminent danger of losing some important information, like the maker's name, by all means do something to stop the deterioration. A good restorer will document a dial's restoration so there is no doubt that the name was there before restoration. Check with a clock museum or with a friend - that has had restoration done to a dial before you send it off to be restored. Make sure the dial painter understands exactly what you want done. (Obviously, there are times when a dial must be repainted, such as one that is blistered from being in a fire.) Retain as much of the original as you can - this maintains its antique value, even though there has been some restoration, providing it is well done. Never wax your dial - the polymers and silicones will hinder any future restoration. Restoration is sometimes necessary, but generally, I tell people to leave their dials as is.

Clocks that are in *original mint condition* always have a dial that is indigenous to the style and period of the piece. The dial is properly fastened to the case in perfect alignment with the movement; and the dial retains its original finish with small signs of wear, very minor hairline cracks, and very slightly faded numerals.

The mechanism or **MOVEMENT** in a clock is the next logical place to proceed with the investigation. Technically it is the most complicated part of a clock, yet it too can be methodically checked. The key points to cover when examining a clock movement

are:
- Does the movement fit the case (back board) without signs of extra holes or tampering?
- Are there any missing pieces? Broken teeth? Signs of major repairs?
- Is the movement signed or stamped with a maker's name? If so, is the name consistent with the name on the dial and/or clock label? Remember there are exceptions. Elias Ingraham was a case maker and his firms did not commence manufacture of their own clock movements until 1865 - five years after the firm of E. Ingraham & Company was formed. Earlier clocks, such as those made by the firms of E. & A. Ingrahams and Elias Ingraham & Company, had movements purchased from other manufacturers including Jerome, Pomeroy, and Waterbury. Early purchased movements usually have no maker's name. Later Ingraham-made movements are usually stamped with the firm's name, but are occasionally unmarked.

Movements are generally made of metal, brass and steel. However, two types of wood movements exist, those with wooden plates, pinions and wheels and those with wooden plates and metal pinions and wheels. Clocks with wooden movements are best inspected and passed by a recognized expert. However, in general the "tips" on metal movements that follow can generally be applied to wood movements.

A clock with the incorrect movement is next to impossible to conceal from the trained examiner. When additional holes are discovered near the movement in the backboard, which is generally constructed of soft wood, it means the movement has been either moved or replaced. If moved, this generally occurs as a result of frequent repairs and the holes in the backboard have enlarged - so that the movement can no longer be secured in its original position. The movement legs may have had to be repositioned to resecure the movement. In this instance, the old, large unused holes should match the movement holes perfectly when the movement is repositioned. If this does not occur, the movement has been replaced. Remember, the value of a clock is based on original condition. A clock with an incorrect movement has little value except as a decorator piece, unless it is purchased by someone who has the correct movement. This type of restoration - a correct movement to a case, is an honest restoration, but the best value in the long run is a clock which has little need for repair now or in the past.

Clock movements are machines which require maintenance and repair. How well the maintenance has been carried on and how well necessary repairs have been performed is of real importance. We have all examined movements which have suffered amateur repairs. Most of these poor repair efforts can be reversed by a good bench mechanic, but this kind of work ultimately adds to the cost of the purchase. A movement can be either spring driven or weight driven. **Spring driven** clocks seem to suffer more abuse than their weight driven counterparts, especially when a strong spring has broken and damaged neighboring parts. A good rule of thumb is to run the clock through its strike operation to decide if you can live with the rate of strike and the related noise in the movement. Very few clocks are naturally noisy; a clock that is noisy may have underlying problems. **Weight driven** clocks require special attention be paid to the seatboard to ensure the movement is not a replacement. Often the fit between the dial and the bezel or surround and the location of the dial in the case door are good indications of originality. A broken weight cable allows the weight to fall; this fall usually causes damage to the base of the clock. The potential buyer of a weight clock ought to examine the bottom of the case for any evidence of repair.

If you find a clock that interests you there are a few more questions concerning the movement you should consider:
- Do the winding arbors line up with the holes in the dial?
- Does the pendulum hang in the appropriate place?
- Are there indications that weight(s) or pendulum have damaged the case?
- Will the seller guarantee that the clock will run for the full length of

time that the clock was intended to run?
- Are the weights correct for the clock?
- Are the hands a matched pair and of the correct length for the numerals and chapter ring on the dial?

The clock's **LABEL** is the next area to inspect as it may provide the only means of determining the origin of the clock. It often contains the directions for its care and use. In some instances there are dates corresponding to a maker's patent or improvement in the mechanism. The questions with regard to the label are:
- Is the label intact and legible with only minor signs of shrinkage, wear, fading, etc.?
- Are there signs of replacement with a reproduction label?

The last area to consider is the **TABLET** or **DECORATIVE GLASS** panel that was incorporated into the designs of many shelf and wall clocks. And since the tablet can represent a sizable percentage of the viewing area of the clock, it behooves the prospective buyer to inspect the tablet carefully to be certain of its originality and/or its condition and to see if deterioration has begun.

Tablets are often called "Reverse Glass Paintings." In this type of decoration paint is applied to the rear side of the glass. In addition, the sequence of steps is the reverse of the procedure normally used on opaque surfaces such as wood, metal or paper. Hence the name "reverse glass painting." There are three points to note in examining a glass tablet:
- Does the tablet appear to be original?
- Are there any signs of restoration?
- Are there any cracks or flaking?

Obviously, the ideal circumstance would be to find a clock with the tablet in 100% original condition with no signs of flaking or lifting already underway. Unfortunately, however, the reality is that seldom is an antique clock purchased today with a perfect tablet. The buyer therefore must have some knowledge about reverse glass painting and have the ability to determine on the spot the condition of the tablet.

- Is the tablet original with no flaking or lifting?
- Is there a small amount of flaking or lifting already in progress?
- Is it a tablet with only minor flaking and missing areas but restorable by a professional... or
- Is it a tablet so far deteriorated as to be indiscernible as to its original design and not possible to be accurately restored?
- Is the tablet so poorly repainted as to be immediately obvious to even the novice collector?
- Is the tablet missing and/or replaced with plain glass of inferior quality?

If the tablet appears to be in good condition and original (look at the putty or retaining wood strips to help determine this) then, of course, nothing need be done. For your own records, it may be wise to photograph the tablet while in this good condition, in the event that some future damage occurs and restoration is necessary. A tablet painter will then have an excellent copy to reproduce.

Keep in mind, some tablets are more restorable than others. The size and area of the design which is flaking and the underlying primary material i.e. gold leaf, or bronze are some of the factors determining whether restoration is possible. If some flaking has occurred, it would again be wise to photograph the tablet, showing what is left of the design. At this time a decision must be made whether you can live with the tablet in it's present condition or whether you want the tablet restored or repainted. If you feel the tablet must be attended to first consider this option. After photographing the tablet, carefully remove it and safely store it in a sturdy box. Have a professional restorer paint a copy tablet of the original tablet from the photographs and place that new tablet on the clock. This way you have both preserved the originality of the clock as well as its value and you will have a "perfect" tablet on your clock. In the opinion of one expert reverse glass restorer, this is the preferable approach. An exception to this might be when the original glass is puttied so securely into the door that attempting to remove it could possibly result in breaking it. Then remove the complete door and have the restorer repaint the glass without removing it from the door.

If the tablet is a poorly painted replacement or is missing altogether, then of course, this too should be considered before purchasing the clock. Finding a suitable replacement could be a problem. Some catalog parts companies sell new tablets of limited designs. For accuracy, it might be necessary to rely on a professional tablet painter who knows clocks well and has a library of designs to draw from. The National Association of Watch and Clock Collectors in Columbia, Pennsylvania, The American Clock & Watch Museum in Bristol, Connecticut; and the two volumes of "*Clock Decorating Stencils of Mid-19th Century*" by William B. Fenn offer many excellent opportunities to study and become familiar with clocks and original tablets.

By familiarizing yourself with the criteria mentioned for the **case**, the **dial and hands**, the **movement**, the **maker's label** and finally the **tablet** or **decorative glass;** you will make a well informed decision based on sound judgement as to the clock's condition and therefore it's value. Know your subject, familiarize yourself with this book and remember my tips.

The clock that meets most or all of these standards and passes most or all of the questions posed can be deemed in *original mint condition*. By verifying the answers to the important questions, your decision is a simple one. If the answers to the questions are suspect, remember that you will be asked the same questions when you decide to sell. An expert in clock inspection gains his expertise through research, by reading all the information he can find about clocks and through experience gained through personally inspecting and studying clocks of all types and manufactures in all conditions. The more you know, the happier you will be with your purchases.

But what about those clocks that didn't stand the test of time, poorer condition and lower quality clocks? It is hard to say what these clocks are really worth. There is a market for lower grade clocks, but any controversy surrounding a clock decreases its desirability. Some books written on this subject suggest that clocks are in very short supply, that all clocks are highly collectable and that clocks in poorer condition should be considered - simply adjust the market price slightly lower. I do not share this view, nor do I recommend that you use this guide in that way. There are many more clocks offered for sale each year than there are buyers for them. Most of these are the lower grade variety and are rarely sold to knowledgeable buyers.

I urge you to stand at the loading area at the end of an NAWCC Regional or National MART and witness the large percentage of unsold clocks returning to their homes. Most of these are lower quality clocks, overpriced. Of course, some mint condition clocks also leave these MARTS unsold simply because the owners would rather take them home than accept less than their asking price.

Concern yourself only with clocks in *original mint condition.* Don't settle for anything less than the best. If you have patience to wait for that perfect piece, and have the resources to acquire it, you will be rewarded in the satisfaction of knowing that you own the best.

In essence a clock is an item of furniture or decoration and is a machine designed to convey a measurement of time. Anything beyond this description hints at the underlying reasons why we are enthusiastic collectors. Factors like rarity and desirability peak our interest in a fine piece while the originality and condition of the case, dial, movement and other embellishments influence our decision to own it.

## GRANDFATHER CLOCKS

Many families will invest in one **Grandfather Clock** for their home. Often a new clock is ideal, but sometimes the family decides to buy an older or antique Grandfather Clock. Buying a Grandfather Clocks requires both knowledge and capital. These clocks have been at the top of the horological ladder for many years and therefore fakes and marriages abound in the marketplace. Caution should be exercised, but if a few simple rules are adhered to, you can be sure of the successful purchase of a Grandfather Clock.

First, separate Grandfather Clocks into two categories: hand made clocks and machine made or manufactured clocks. Machine made or manufactured clocks started being produced in the second half of the 19th century. Most strike on rods, tubular bells, or gongs. There were many exact copies of each model produced and they were made by most clock manufacturers. My book, **Longcase Clocks and Standing Regulators, Part 1 - Machine Made Clocks**, is an excellent source of information on these clocks and their value. The rest of the comments in this section apply to hand made Grandfather Clocks.

Hand made Grandfather Clocks have had many names over the centuries with "tall clock" being the most common for American hand made Grandfather Clocks and "longcase clock" the term used to describe the British. These are the earlier clocks made before 1850 for the most part. Most have solid wood trunk doors and strike on a bell.

When examining a hand made Grandfather Clock the things to determine are: Did the case, movement, and dial start life together? How long does the clock run? Is the clock a high or low quality item? What country did it come from? Are there unusual features or complications? Who was the maker? How much restoration is present? All of these things must be taken into consideration to determine a Grandfather Clock's value.

**Originality.** How do you tell if the case, movement, and dial start life together? If the movement and dial have been adapted, there will usually be evidence of the dial feet being moved or extra holes in the front plate of the movement. If the movement has been married to the case, there will often be a new seatboard or alterations to an old seatboard. Additionally, there may be alterations to the case rails (cheeks) that hold the seatboard such as added shims or recent cuts to reduce the rail height. Finally, you may discover extra holes in the seatboard that do not align with the holes in the rails of the case. Stay away from clocks that are not original unless they are <u>very</u> cheap.

**Duration.** Most hand made grandfather clocks run for one day or eight days. One day clocks with brass movements usually have one weight (8 lbs.) which pulls up with a rope or chain. One day clocks with American wood or Black Forest wood plate movements have two light (3 lb.) weights that pull up by a cord or chain. Eight day clocks usually wind through the dial and raise two heavier (12 lb.) weights. One day clocks are usually worth about 60% of the value of an eight day clock.

**Quality.** Grandfather Clocks were made for wealthy to average income families. Quality clocks usually have quality movements, dials, and cases. Stand back and ask the question, does this clock reflect quality or was it made for a lower income individual? High quality clocks are a better investment than low quality clocks with one exception, primitive clocks. Primitive clocks, especially those with original decorative paint can be at the top of today's price scale.

**Country of origin.** This can be most difficult for a beginner to determine with most mistakes being made between American and British clocks. American clock cases are usually much heavier than British cases. American clocks use American species of woods, mahogany being an exception. American clocks often use poplar as a secondary wood while British clocks do not use poplar. Pine is found in both American and British clocks as a secondary wood. British clocks are often of oak, but hand made American clocks are not of oak. Movements can be very similar and while there are tendencies such as cut out plates, no one movement feature is conclusive. Dials were often made in England and sold to American makers to be used on American clocks, so do not conclude that "Birmingham" on the dial or falseplate means the clock is always English. Clock dials that were made in America such as those made by Nolen or Curtis were not exported to Britain, so clocks with those dials will be American. Except for the top London clocks, most American clocks will be worth more than an equivalent British clock. This is because of demand not quality.

**Complications.** Features such as moon dials, rocking ships, center seconds, long duration, chimes, music, etc., all add to the value of a Grandfather Clock.

**Maker and age.** Many hand made Grandfather Clocks are signed by the maker. Be sure the signature on the dial is original, if it is over-painted, don't trust it. Significant makers' clocks are very valuable if the clock is right. From the maker's name research can be done to determine the clock's age. Strangely, unless a clock is very early, pre-1710 if English or pre-1780 if American, the age of such a clock is not terribly significant. Other factors prevail.

**Restoration.** Mint original hand made Grandfather Clocks do <u>not</u> exist. Attempts to restore these clocks to "as new" condition usually reduce the value rather than increase the value. When buying a hand made Grandfather Clock, buy as original a clock as you can afford. Almost all these clocks have had some restoration. Movement repair, if well done, is acceptable, but movement modifications are not. Dials often have been restored by repainting numerals which is acceptable if well done. Totally repainted dials should be avoided. Case restoration is acceptable if well done, but the less the better. If a case has been stripped and refinished, stay away from it.

It takes time and experience to master all these steps in selecting a hand made Grandfather Clock. Attempt each step to the best of your ability. If you are unsure, ask the seller outright to give his or her opinion on each factor that you are unsure. Be cautious if you hear answers such as, "It's a good old clock." Remember you are about to spend a lot of money and deserve to receive a clock of commensurate value.

## BUYING, SELLING, INVESTING, COLLECTING, SPECULATING

There are three basic personalities associated with every clock enthusiast: **Collector, Speculator, Investor.** Whether you consider yourself to be a hobbyist or a professional, the antique clock market of the 2000's must be approached in these terms.

**COLLECTOR** - The collector is an individual who has adopted antique clocks as a hobby. He acquires clocks for pleasure and personal enjoyment. Factors such as aesthetic appeal, historic, or technical aspects of antique clocks are very important considerations to him. The ownership of an antique clock often stimulates peripheral interests while providing a decorative example of a period in history. In his approach to the antique clock market, the collector expects to pay a fair price for a fine example.

**SPECULATOR** - The speculator purchases a clock for the purpose of resale. The motives behind speculation are clearly profit oriented. Speculators sometimes achieve folk hero status in "rags to riches" stories of buying and selling antique clocks. There is always a chance that you will uncover a "great find" at a garage sale and turn a handsome profit upon selling the clock, but overall, speculating in antique clocks is risky for the average person.

**INVESTOR** - The investor approaches the antique clock market in a businesslike manner. His important considerations are capital appreciation, safety, and liquidity. When investing for capital appreciation, ask:

- What is your cost basis?
- What would you do with the money as an alternative investment?

If you purchase a clock at or above the market price, you can not expect to achieve capital appreciation in the short term. The investment must be made at a price that allows room for growth. Compare and contrast alternative investment opportunities at the time of purchase. Factors such as money market rates, inflation, stock and bond prices, etc. should be considered prior to investing in clocks.

The investment "safety" of clocks is another important consideration. The market for antique clocks should continue to grow at a slow steady rate, but liquidity depends solely upon market demands. As in any other market, the selling price of your clock will depend on how long you are willing to wait for the profit margin you are willing to accept. Also, for investment sake, it may be wise to avoid "one of a kind" custom-made clocks. Because there will be nothing to compare it to and no catalog listing for it, determining its true value may prove to be difficult.

## PHOTOGRAPHING CLOCKS

I have already hinted at the wisdom of photographing your clock to preserve evidence of the original state of your clock's glass tablet. But whether you are a collector, investor, or buyer of a single clock, it is smart to have a photographic record of your purchase. Not only will a good photograph help with any restoration projects, but a good quality photograph will serve you well as "proof of purchase" for insurance purposes should any damage or loss occur. Perhaps the easiest way to accomplish this is with a video camera, but a printed photographic record of your collection is the most versatile. A set of photographs, showing the details and special features of your clock, along with a complete written description including dimensions, will enable you to deal with a repair or replacement situation as well as provide prospective buyers and other collectors visual examples of your collection. This information, along with my books and the accompanying price guides, will provide you with all of the pertinent information you will need regarding the value of your clock. Also note that providing photographs of your stolen clocks to the police will aid them in there recovery and allow you to claim them. You will have visual proof of your ownership of these timepieces.

On a personal note, I often receive photographs of clocks from my readers for possible inclusion in my books. All too often, however, these photographs are not of the quality necessary for publication and I must refuse them. By noting the following hints and suggestions, high quality photographs can be achieved which will better serve us all.

In all cases, you will want photographs that show your clock to its best advantage, with details clearly visible. However, you needn't take each clock you own to a professional photographer to obtain "professional" quality photographs. The following tips will help you obtain the best results using your own camera and setting.

<u>Camera</u>. The choice of a camera is purely a personal one. You have used your camera and know its abilities. By following the instructions provided by the camera maker, you will know best how to use that brand and model. I have found that a good quality 35 mm camera is most efficient and a good quality ASA 200 film gives the best overall results. A lower ASA film produces sharper detail but requires more light. A higher ASA film works in lower light, but the results can be grainy. The use of a tripod for your camera can also be advantageous.

<u>Background</u>. A slightly off-white background photographs well for most wooden clocks. The reason for the off-white is that many cameras with automatic exposure control over compensate for the pure white background making the clock appear dark in the photograph. The off-white, in fact, develops as white and the clock appears in good, clear detail. For brass novelty clocks or brass movements, a black background offers a good contrast to the brass item. If using fabric as your background material, be sure any wrinkles are removed before arranging the clock on the fabric. If photographing a wall clock, do not hang it on a decorated surface. That is, avoid a wall-papered, brick, or cement block wall. Again, a white or off-white background works best.

<u>Lighting</u>. The best situation is natural, bright light. Photographing outdoors, on a bright but overcast day, will enable you to take photographs with clear, distinct results without bothersome shadows. If you must take indoor photographs, use high intensity indirect lighting or a flash attachment. However, the high intensity indirect lighting, such as that obtained from 300 watt halogen lights or 500 watt incandescent bulbs and color compensating filter, is preferable as any "hot spots" caused by the lights are actually visible through the camera view finder and can be eliminated before taking the photograph. You must take into consideration that most film chemistry is formulated for daylight and/or camera flash units. To achieve acceptable results when using artificial light (anything other than sunlight or a flash unit) a color correction filter must be used. For example, photos taken using fluorescent lighting will have a noticeably green tint and photos using incandescent lights will be tinted yellow or orange. The filter, which simply screws onto the front of your lens, compensates for the light variation and gives you accurate color as if you were using daylight. If you are serious about using artificial lighting consult your local photo store for the correct filter for the lightbulbs and film you are using. When using a flash, avoid standing directly in front of the clock as light from the flash will bounce back to the camera resulting in "hot spots" or white areas on the photograph. Take several flash photographs from different angles to ensure getting one good quality picture.

<u>Arrangement</u>. Be sure you take photographs that will show the clock to its best advantage. Some advanced collectors who routinely take many photographs for publication, actually construct a permanent "set" complete with high intensity lighting on stands, and wooden fabric covered stages and backdrops. However, this is not necessary. A good quality photograph can be obtained by creating a simple setup for your subject. The accompanying diagram shows one arrangement that works well for small shelf clocks. Obviously, clocks too large for a chair surface can be set on a table or counter-top which has been similarly prepared. Remember, if you stand too far from the subject, the details will be lost. A photograph showing the whole clock is certainly necessary, but so are close-up photographs showing details such as the dial, movement, tablet, label, and any special features, such as carvings or inlays. Also be aware of any possible reflections of your surroundings showing on the clock's glass tablet. This often doesn't appear until the pictures are developed. Again, remember to take several exposures from different angles.

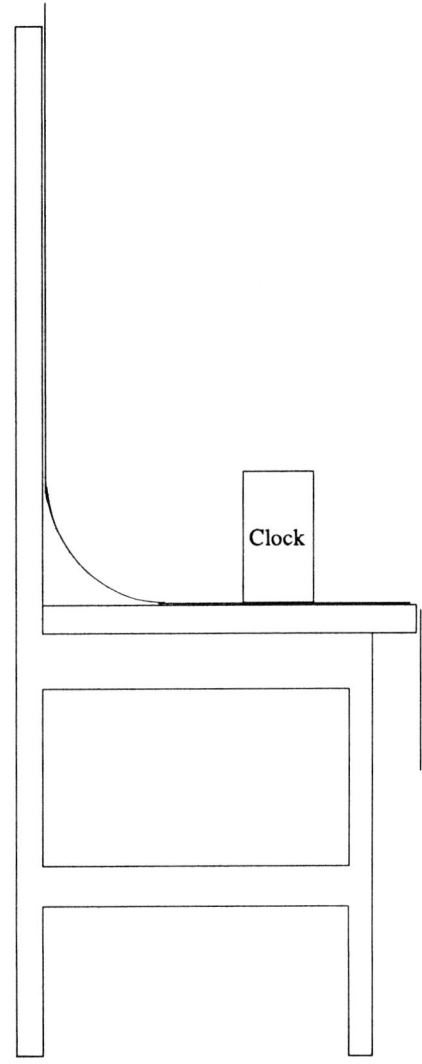

**For small clocks or movements, you can use a flat sturdy chair as shown above. Using a continuous sheet of paper fastened to the top of the chair and arranged as shown for the background, you can create a backdrop behind and under your clock. If using wrinkle-free fabric, arrange it so that it drapes smoothly behind and under the clock smoothing away any wrinkles as you place the clock in position. Determine the color background, by the type of item you are photographing. Black or a dark color for brass items, and off-white for wood cases. Avoid putting the clock too close to the background. Larger clocks can be placed on a table or workbench using the same techniques as detailed here.**

## AUCTIONS, MARTS, ANTIQUE SHOPS, SHOWS and BACKYARDS.

Clocks are everywhere. Almost every household auction flyer includes some clocks. NAWCC MARTS are unquestionably the best source for clocks, while booths in antique shops and at antique shows often have antique clocks featured. Sometimes there are dealers who specialize in clocks and will have an entire booth filled with all styles of timepieces. But your neighborhood is also filled with clocks of all types, age, condition and value. Don't overlook the garage sale down the block or the yard sale up the street. You never know what's for sale.

Perhaps of all these buying opportunities, the **auction** may be the most daunting. Understanding auctions coupled with the following tips will reduce your level of anxiety. Whether it is a local country auction or a big city auction, auctions are important vehicles in the antique clock market and must not be overlooked. There are numerous public and private auctions held across the country each year with nothing but clocks and watches on the inventory. They attract a great deal of publicity and warrant special consideration. Auctions are "happenings" that are attended by all types of collectors, investors, and speculators. Find a reputable auction gallery and attend frequently. They are often an open forum for information exchange and price guidelines.

But beware. Exercise great care in bidding on antique clocks at an auction, and remember it is very important to **always** examine the clock before the bidding starts. Condition and originality are very important if you plan to buy the clock. When you are looking at a clock for the first time there are several procedures you should follow in **checking out a clock.**

> First, take an inventory -- is everything there? Look for:
> - Pendulum
> - Hands
> - Condition of dial -- complete, original or replacement
> - Condition of the movement -- remove the dial to see if the movement is complete, undamaged and is original to the clock; if the auctioneer will not let you examine the movement, either do not bid on the clock or ask the auctioneer to guarantee that the movement is complete, undamaged and original to the clock.
> - Is the case complete -- finials, moldings, veneer, etc.

- Glass
- Key
- Weights (if weight driven)

The second step is to see if the mainsprings are broken. Take the key and wind each of the arbors. If there is no tension felt when winding three or four times, the mainspring is probably broken.

The next testing procedure is to put the pendulum in place and give it a swing. If it is a weight clock, be sure to check to see if the weights are attached. If it does tick, is it a strong sounding tick or is it a weak sound? If the tick is strong, the clock is probably in good running order. If the tick is weak, the movement is probably dirty and/or it could have a powertrain problem -- such as the verge being out of adjustment, bent or broken teeth, or badly worn bushings. If the clock does not tick, there is reason for concern. The clock should not be purchased without further examining the movement to see what is wrong and if the movement could be easily fixed.

The last step in checking out a clock is to activate the strike mechanism. Do this by moving the minute hand slowly to the hour. Move the hand at least one complete revolution because the strike train could be out of adjustment or the strike on the half hour may not be functioning. Listen closely for the sounds of the strike mechanism falling into place during the warning cycle.

From these sounds and the results of the previously mentioned tests, you should be able to form an opinion as to the condition of the clock before the clock is put up for auction.

An important point to keep in mind is that there will be only one winning (highest) bid. If and when your bid is accepted, you must acknowledge the fact that you are the high bidder. This means that none of the other participants felt that the piece was worth as much or more than you did! I remind you of the phrase "Caveat Emptor": buyer beware. The necessary ingredients for success at an auction are **knowledge**, **experience**, and **money**. A deficiency in any of these areas will produce unsatisfactory results.

Armed with the knowledge you have obtained from my "IMPORTANT TIPS" you can approach any or all of these situations and after examining the clock feel confident in your decision to either purchase the clock or let it go to someone else who may not know why it wasn't a good buy!

### BOOKS = KNOWLEDGE

**BOOKS** are the cornerstone of knowledge.

**Books** can greatly enhance the enjoyment that can be derived from collecting watches and clocks. There are many books on all aspects of horology, these include books on the history and development of watches and clocks; directories of makers; the "how-tos" of collecting; the repair and restoration of pieces; and much more. The more extensive your library is on the subject, the greater your reference capacity becomes and the better prepared you'll be in your quest for antique clocks and watches. Time spent in reading about clocks or watches of interest to you will enable you to be a more astute collector. The interested collector will greatly benefit, without great expense, from the development of his own person reference library of horological books that can be quickly turned to when needed. In addition, expanding your horological library is always a good investment. Quite often, the books themselves become rare treasures when copies are not available because they are out of print. For some people the collecting of horological books becomes their area of concentration!

Through my experience of collecting clocks over the years, I have learned to check for certain important key factors each time I consider making a purchase. I have learned that these few simple guidelines, if put to proper use, will save me time, money, and possible disappointment. My "IMPORTANT TIPS" will help you too. This book is designed to be your partner -- a point of reference for all your collecting needs. It is my wish that it will help make your antique clock collecting experience enjoyable and profitable.

Be the Early Bird ...

... Time waits for no one

**Tran Duy Ly**

# LONGCASE CLOCKS AND STANDING REGULATORS
# PART 1 - MACHINE MADE CLOCKS
## by Tran Duy Ly

Published 1994 by Arlington Book Company, Inc.
2706 Elsmore Street, Fairfax, VA. 22031-1409
Phone 703-280-2005 - Fax: 703-280-5300
8½ x 11 - 504 pages - 1150 quality illustrations and photographs.
Smyth sewn hardbound on high gloss enamel text. $69.50

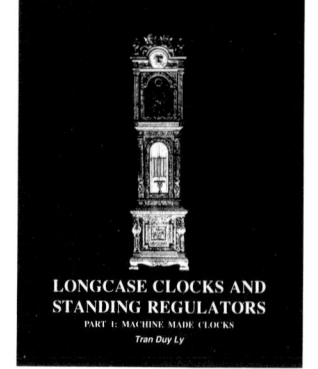

To own a fine standing hall clock or other machine age regulator is the dream of most clock collectors as well as most owners of homes throughout the world. The book, **Long Case Clocks and Standing Regulators, Part 1** will surely be the most important reference available anywhere.

**Long Case Clocks and Standing Regulators** takes the serious collector and investor through hundreds of clocks made by no less than fifty precision companies from several different countries. There is no other work available that provides nearly the coverage of this excellent reference.

The volume begins with a well written and complete list of important buying tips by Thomas Spittler, a recognized authority on European and American long case clocks. He covers important questions and considerations that one should familiarize themselves with before purchasing one of these clocks.

This book not only covers the general description of the various pieces but also provides the technical information necessary to allow the collector to be thoroughly informed prior to making a purchase. More importantly, the subtle differences that make some clocks more desirable and thus more valuable are well presented.

It is quite unlikely that any of us has heard of Haas & Shone, Harris & Harrington, Hirst Brothers & Company or many of the other companies listed in this very large and complete volume. On the other hand, clocks by the well known companies such as E. Howard, Seth Thomas and Colonial are covered in complete detail. Many of their models are in no other reference.

In reading through this book, one realizes that Mr. Ly has composed the most complete and comprehensive reference on these clocks. It is evident that he has spent hundreds of hours researching collections and horological literature to obtain the massive amount of information necessary for this very complete work.

While going through the 504 pages one soon realizes that the machine made, factory produced long case clocks and regulators were produced in vast numbers in both America and Europe, thus, one may find them now in the most unlikely places. **Long Case Clocks and Standing Regulators** serves as the clock collector's best new, indispensable companion in finding and understanding these splendid clocks. Serious collectors, dealers and restorers are fortunate that Tran Duy Ly has taken the lead in horological reference books.

**DAVID MORGAN**, NAWCC President

Available from Horological Booksellers, Part Supplyers or from:

**Arlington Book Company, Inc.** 2706 elsmore street - fairfax, virginia 22031-1409 u.s.a.
PHONE: 703-280-2005 FAX: 703-280-5300 EMAIL: info@arlingtonbooks.com

# NEW HAVEN CLOCKS & WATCHES

With a Special Section on New Haven Movements

by **Tran Duy Ly**

**Published 1997 by Arlington Book Company, Inc.**
2706 Elsmore Street, Fairfax, VA. 22031-1409
Phone 703-280-2005 - Fax: 703-280-5300
8 ½ x 11 - 520 pages - Over 2,200 quality illustrations and photographs.
Smyth sewn hardbound on white gloss enamel text **$55.00**.

I began to seriously collect clocks while living in England in 1974 and most of the clocks I found were American clocks by the New Haven Clock Company and its associated British sales firm Jerome & Co. Many of these clocks had an interesting label showing a factory identified as "The New Haven Clock Company formerly the Jerome Manufacturing Company." I wish I had had a book such as *New Haven Clocks & Watches* by Tran Duy Ly to help me better understand the history of the New Haven Clock Company and identify the many clocks and watches they produced.

Tran Duy Ly has produced *New Haven Clocks & Watches* to complement his other books on Seth Thomas clocks, Waterbury clocks, Ansonia clocks and Welch clocks. In my opinion, these five firms combined to produce over 75% of the clocks manufactured in America. This exciting new volume on New Haven clocks is produced to the usual high standards of the Arlington Book Company. It contains an amazing 38 chapters and over 500 pages on different types of New Haven clocks and a supplementary price guide. The many views provided will help identify correct dials, hands, pendulums, weights, gongs, movements and trim.

What intrigues me most about this book is the 11 page history and chronology written by the noted horologist Chris Bailey. Chris clearly develops the story of the giant of American clockmaking, Chauncey Jerome, and his relationship with Hiram Camp, the first president of the New Haven Clock Company. I have always felt that if Camp had named the firm the Chauncey Jerome Clock Company he would have captured its true identity in the way the Seth Thomas Clock Company did. But this did not happen. This caused problems for the fledgling New Haven Clock Company when it tried to sell its products in England. The British identified the name "Jerome" with quality American clocks and were willing to accept no substitutes. To overcome the situation New Haven used the name of their independent British sales firm, Jerome & Co., and New Haven clocks with the Jerome & Co. label were successfully sold in England from 1856 to 1904. Even today the name "Jerome" in England is bigger than Seth Thomas and E. N. Welch combined.

So it is fitting that the Arlington Book Company has filled a void by producing this long awaited book on the New Haven Clock Company. Tran Duy Ly, more than any other author of horological books in America, has repeatedly produced books that are right on target for clock collectors around the world. Each new book builds on the solid foundation he has established in the past. *New Haven Clocks & Watches*, like his other books, is well written and easy to use. It contains over 2,200 quality illustrations and photographs. The section on kitchen clocks contains the most models ever seen in print. In addition to hundreds of pages of clocks there are even 24 pages of New Haven watches.

I feel the 50 pages of movements and parts showing extremely clear and detailed illustrations of 61 New Haven movements are outstanding. Most of these pages show a front and rear view of the movements and for the first time I can recall, each movement part is clearly identified. For the repairman, these 50 pages alone are worth the price of the book.

In summary, I am truly excited about Tran Duy Ly's new book *New Haven Clocks & Watches*. If you own a New Haven clock and wish to find out more about it, or if you deal in clocks and need to know up-to-date prices, or if you repair clocks and could use wonderful illustrations of New Haven movements, this is the book for you.

**THOMAS J. SPITTLER**
New Carlisle, OH

Available from Horological Booksellers, Part Supplyers or from:

**Arlington Book Company, Inc.** 2706 elsmore street - fairfax, virginia 22031-1409 u.s.a.
PHONE: **703-280-2005** FAX: **703-280-5300** EMAIL: **info@arlingtonbooks.com**

# ANSONIA CLOCKS & WATCHES
## 2nd Edition by Tran Duy Ly

Published 1998 by Arlington Book Company, Inc.
2706 Elsmore Street, Fairfax, VA. 22031-1409
8½ x 11 - 752 pages and 3061 quality illustrations and photographs.
Smyth sewn hardbound on white gloss enamel text $79.50

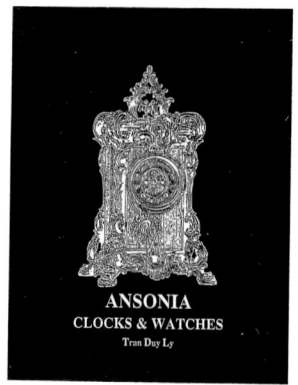

Soon after being asked to write this introduction to the second edition of *Ansonia Clocks* my memory served up a comment by Dr. Steve Petrucelli of the Adams Brown Company to the effect that Tran Duy Ly has singlehandedly revolutionized the clock book business. The import of this statement cannot be under-emphasized and it is a great accomplishment that this revolution has been an ongoing and progressing refinement with the passage of time.

Tran Duy Ly has greatly developed and advanced the principles of organization and pricing first set forth by the pioneers of clock reference books: Roy Ehrhardt with his three *Clock Identifcation and Price-Guide* books, and Andrew and Dalia Miller's *Survey of American Calendar Clocks*. First of Mr. Ly's books to make its appearance in 1984 was the softcover *Clocks - A Guide to Identification and Prices*. This inclusive and well-organized guide is still one of my favorite auction companions even though it is long out of print. In one book the collector could find most any mass-produced clock with detailed illustrations for determination of originality plus a trustworthy pricing reference. My third copy is in poor repair but remains handy for quick reference.

Soon to follow were more ambitious projects involving specific manufacturers as well as an updated two volume general catalog of American clocks. Witness the profusion of publications from Mr. Ly and his Arlington Book Company for collectors and dealers alike: *Seth Thomas Clocks and Movements* (1985), *Ansonia Clocks* (1989), *American Clocks Volume 1* (1989), *Waterbury Clocks* (1989), *American Clocks Volume 2* (1991), *Welch Clocks* (1992), *Calendar Clocks* (1993), *Longcase Clocks and Standing Regulators - Part 1: Machine-Made Clocks* (1994), *Seth Thomas Clocks and Movements - 2nd Edition* (1995), *Calendar Clocks 1997 Update-Featuring Previously Unpublished Photographs and Information* (1997), *Gustav Becker Clocks* (1997), *New Haven Clocks & Watches - With a Special Section on New Haven Movements* (1997), *Gilbert Clocks* (1998), and *Ingraham Clocks & Watches* (1998).

And hence we have come full circle with the second edition of *Ansonia Clocks* found here. Tran Duy Ly, being the experienced and well-respected clock collector and editor-publisher that he is, has kept the best aspects of his previous works while managing to push the horological book industry forward with additional features in his latest publications. Thus, in this new *Ansonia Clocks*, timepieces from the clock company which have never been seen before in the earlier edition are described and depicted. Real photographs of clocks and their respective parts are abundant, clear, and detailed and have been added to the already massive number of original factory catolog illustrations. A large number of photos and specifications for movements are also included and referenced. Of much importance to many collectors will be the new, comprehensive, 49-page section "Side Pieces, Top Ornaments & Candlesticks" which will allow them among other things to name those clocks whose ornamentation has been exchanged.

A national spectrum of the best-known collectors has provided input to this book and has shared their most treasured examples from the Ansonia Clock Company. Much effort has been devoted to realistic pricing suggestions by consulting both auction hammer prices as well as private transactions between the country's top experts; the fruit of such labor may be found in the supplementary price guide to this book. The price guides for this and other books by Tran Duy Ly are updated on a regular basis in an endeavor to keep current with price trends.

As collecting has expanded and matured in the last two decades so have the efforts and interests of Tran Duy Ly. Accordingly, I am quite confident that you will enjoy this new addition to your library and share in the pride of Mr. Ly and his many qualified contributors. Be assured that when buyers and sellers discuss the exchange of an Ansonia clock, the text of reference will already be in your library.

**JOHN TANNER**
Upland, California

Available from Horological Booksellers, Part Supplyers or from:

**Arlington Book Company, Inc.** 2706 elsmore street - fairfax, virginia 22031-1409 u.s.a.
PHONE: **703-280-2005** FAX: **703-280-5300** EMAIL: info@arlingtonbooks.com